Routledge Handbook of Sustainable Real Estate

With the built environment contributing almost half of global greenhouse emissions, there is a pressing need for the property and real estate discipline to thoroughly investigate sustainability concerns. The *Routledge Handbook of Sustainable Real Estate* brings together the latest research of leading academics globally, demonstrating the nature and extent of the impact as well as suggesting means of mitigating humankind's impact and building resilience. Four sections examine the different aspects of sustainable real estate:

- governance and policy
- valuation, investment and finance
- management
- redevelopment and adaptation.

Covering all land uses from residential to commercial, retail and industrial, the *Routledge Handbook of Sustainable Real Estate* is an exciting mixture of received wisdom and emerging ideas and approaches from both the developed and developing world. Academics, upper-level students and researchers will find this book an essential guide to the very best of sustainable real estate research.

Sara Wilkinson is an Associate Professor in the School of Built Environment at the University of Technology Sydney (UTS), Australia, and a fellow of RICS. She is on the editorial boards of five international refereed journals and is the lead author of *Developing Property Sustainably* (Routledge, 2015).

Tim Dixon is the Professorial Chair in Sustainable Futures in the Built Environment at the University of Reading, UK. He is on the editorial boards of four leading international real estate journals and is the co-editor of *Urban Retrofitting for Sustainability: Mapping the Transition to 2050* (Routledge, 2014).

Norm Miller is a Professor and the Ernest W. Hahn Chair Professor of Real Estate Finance at the University of San Diego, USA. He is the founder and editor of *The Journal of Sustainable Real Estate* and board member of several other renowned international real estate journals.

Sarah Sayce is a part-time Professor in Sustainable Real Estate at the University of Reading, UK, Visiting Professor at the Royal Agricultural University, UK, and Emeritus Professor at Kingston University, UK. She is also a visitor at Cass Business School, a Fellow of RICS and on the editorial board of several leading international property journals. She was co-author with Sara Wilkinson of *Developing Property Sustainably.*

"A critical book for all corporate property and mass market housing developers. The chapters show that there are real pockets of deeper innovation and embedded sustainability practice in the market. Niches being developed that will help these companies to have a market advantage as the implications of future climate change and other ecological challenges reach consumer awareness."

—**Dr Dominique Hes**, Director of the Thrive Research Hub, University of Melbourne

"For those involved in real estate either as a user, provider or consultant, there can be no doubt that sustainability is a key challenge that simply cannot be ignored. The *Routledge Handbook of Sustainable Real Estate* brings together the work of highly experienced practitioners who, through their dedicated research and passion for the subject, provide a valuable and informed commentary upon many of the difficult issues that face us in our day-to-day activities. With sections on governance, valuation and investment, management and redevelopment, the book provides a well-balanced font of knowledge that will help students [and] decision- and policy-makers alike. I commend the handbook to those who embrace sustainability in all its forms and also to those who are yet to be persuaded."

—**Trevor Rushton**, FRICS FCABE ACI Arb,
Technical Director, Watts Group Limited

"The *Routledge Handbook of Sustainable Real Estate* provides fascinating insights into the fundamental changes needed to ensure that property makes a substantive contribution to a sustainable future. The book employs a mix of detailed analyses and synoptic reflection to demonstrate the interlinked nature of the economic, social, cultural, organisational, technical and governance influences upon the emergence of a genuinely sustainable built environment. Drawing on international experience and expertise, it highlights the challenges of establishing agreed definitions and measures of sustainable buildings and their features, of incorporating these into real estate policy and practice, and of ensuring that their application produces robust evidence of sustainable buildings' performance. Ways must be found for all real estate actors – developers, investors, occupiers and professional intermediaries such as appraisers – to engage with this process, even though [they] may have ostensibly differing interests. The book shows that a promising start has been made to this process, but that much more needs to be done."

—**John Henneberry**, Professor of Property Development, University of Sheffield

"Working in sustainable real estate, one quickly comes to recognise that it is about a dynamic and active system where the sum – 'place making' – is far bigger than the parts – 'buildings.' As said Sir Patrick Geddes said, 'A city is more than a place in space; it is a drama in time.' Sustainable place making is both integral to and drives how we live and work and enjoy the locations we inhabit. It also offers solutions for many societal and systemic problems faced by our societies today, from climate to health and well-being. The handbook is a comprehensive guide that challenges our understanding of why, how and what sustainable place making and green buildings can and should contribute to the drama of our life."

—**Tatiana Bosteels**, Head of Responsible Property Investment,
Hermes Investment Management

"This must be the most comprehensive collection of research and informed opinion to date on sustainable real estate. Reflecting the international dimension and reinforcing the argument that there are no borders regarding climate change, the editors of the handbook have brought together chapter contributions from leading researchers in sustainability in the built environment.

Developed around key themes of governance and policy; valuation, investment and finance; management; and redevelopment and adaptation issues, the handbook stresses current deficiencies and deficits in issues central to the real estate sector, stressing that, while technology and innovation have created new business landscapes, the professional real estate sector needs to upskill and provide leadership in creating a sustainable built environment. One of the most significant challenges identified in many of the individual chapters is the need to retrofit the existing building stock of both commercial and residential property. A key message from the handbook is the need for sustained activity from public-sector policy makers, building owners, professional surveyors and the financial sector in developing products and delivery vehicles to achieve energy efficiency targets and make the sector investable."

—**Professor Stanley McGreal**, Director of the Built Environment Research Institute, Ulster University

Routledge Handbook of Sustainable Real Estate

Edited by Sara Wilkinson, Tim Dixon, Norm Miller and Sarah Sayce

LONDON AND NEW YORK

First published 2018 by Routledge
2 Park Square, Milton Park, Abingdon, Oxon OX14 4RN

and by Routledge
605 Third Avenue, New York, NY 10017

First issued in paperback 2021

Routledge is an imprint of the Taylor & Francis Group, an informa business

Publisher's Note
The publisher has gone to great lengths to ensure the quality of this reprint but points out that some imperfections in the original copies may be apparent.

British Library Cataloguing-in-Publication Data
A catalogue record for this book is available from the British Library

Library of Congress Cataloging-in-Publication Data
A catalog record for this book has been requested

ISBN 13: 978-1-03-209571-4 (pbk)
ISBN 13: 978-1-138-65509-6 (hbk)

Typeset in Bembo
by Keystroke, Neville Lodge, Tettenhall, Wolverhampton

Contents

Editor biographies

Sara Wilkinson is a Chartered Building Surveyor, a Fellow of the Royal Institution of Chartered Surveyors (RICS) and a member of the Australian Property Institute (API). She has worked in UK and Australian universities for over 26 years. Currently, she is an Associate Professor in the School of Built Environment at the University of Technology Sydney (UTS), Australia. Her PhD examined building adaptation, whilst her MPhil explored the conceptual understanding of green buildings. Her research focus is on sustainability, adaptation of the built environment, retrofit of green roofs and conceptual understanding of sustainability. In 2015, she led City of Sydney-funded project on the "Feasibility of Algae Building Technology in NSW," and she continues to work on a cross-disciplinary project with a prototype panel being tested in 2017. With the Health Faculty, she researches the impacts on health and wellbeing of horticultural therapy on retrofitted green roofs. She is part of a cross-disciplinary team of researchers from four NSW universities investigating urban ecology renewal in NSW for the Environmental Trust NSW. Another project explores whether a mandatory approach towards green roof and walls would work for Australia. She sits on professional committees for the RICS to inform her research and to ensure direct benefit to industry. She sits on the editorial boards of five leading international journals and is the regional editor for the *International Journal of Building Pathology and Adaptation in Australasia*. She has written over 250 publications and books, and is the lead author of *Developing Property Sustainably* (Routledge, 2015). Her research is published in academic, professional journals and, recently, an RICS "Best Practice Guidance Note on Green Roofs and Walls" for RICS practitioners.

Tim Dixon is the professional Chair in Sustainable Futures in the Built Environment at the University of Reading, UK. With more than 30 years' experience in education, training and research in the built environment, he leads the Sustainability in the Built Environment network at the University of Reading and is co-director of the TSBE (Technologies for Sustainable Built Environments) doctoral training centre. He has co-led major UK research council research projects on brownfield land and urban retrofit, and is currently working with local and regional partners to develop a "Reading 2050" smart and sustainable city vision, which is also connected with the UK BIS Future Cities Foresight Programme. Recently, he has worked on funded research projects on smart cities and big data, smart and sustainable districts, and social sustainability for housebuilders. He is a member of the Climate Change Berkshire Group and a member of the All Party Parliamentary Group on Smart Cities. He is also a member of the editorial boards of four leading international real estate journals; a member of the Advisory Board for Local Economy; a member of the review panel for Commonwealth Scholarship Commission; a mentor for the Villiers Park Educational Trust; and a member of the review panel of the RICS Research Paper Series. He was also a member of the international scientific committee for the national "Visions and Pathways

2040 Australia" project on cities. He has written more than 100 papers and books in the field, and is co-editor of *Urban Retrofitting for Sustainability: Mapping the Transition to 2050* (Routledge, 2014).

Norm Miller is a Professor and the Ernest W. Hahn Chair of Real Estate Finance at the University of San Diego, USA, where he has been since the fall of 2007. He has several dozen highly cited papers on sustainability, workplace trends, housing, valuation and forecasting. For much of his academic career, he was at the University of Cincinnati as Academic and Real Estate Center Director with one-year visits at DePaul University and the University of Hawaii and three years at the University of Georgia, where he started his career. He received his PhD in finance from the Ohio State University. He is active on the editorial board of several national/international journals and is a Past President of the American Real Estate Society. Known for his pioneering work on the economics of green and sustainable real estate he, is the founding editor of *The Journal of Sustainable Real Estate* (www.josre.org). He worked as VP of Analytics at CoStar Group, and is a research principal with Collateral Analytics (www.collateralanalytics.com). He is also on the advisory boards of Pathfinders, Measurabl, Surefield and Verdani. His book with David Geltner of MIT, *Commercial Real Estate Analysis and Investment*, is in its third edition and is the leading graduate real estate textbook in the world. He is currently a Homer Hoyt Land Use Institute Faculty and Board member, where he is involved with some premier thought leaders among academics and industry professionals in a think tank setting (www.hoyt.org). He can be reached at nmiller@sandiego.edu, and his personal website is available at www.normmiller.net, where you can download papers and other files.

Sarah Sayce is a part-time Professor in Sustainable Real Estate at the Royal Agricultural University and the University of Reading, UK and Emeritus Professor at Kingston University, UK, where for many years she was Head of the School of Surveying and Planning. She is also a visiting academic to City, University of London. She holds an initial degree and PhD from the University of Reading and is a Fellow of the RICS. She is an active, widely published researcher and public speaker across many aspects of sustainability in the built environment/sustainable property and higher education, and she has extensive research experience. Among the organisations for which she has undertaken research are the UK government, the RICS and the Green Construction Board. She has published several books; the latest, published in 2015, was *Developing Property Sustainably*, co-authored with Sara Wilkinson and Pernille Christensen. She sits on the editorial board of several leading international property journals, and she is also Joint Executive Officer of the Council of Heads of the Built Environment, which is the representative body for the Built Environment Heads of Department of UK Universities. In addition to her academic work, she is also very active in the professional body, currently being an elected member of the RICS and a nominated member of the global and UK valuation boards. Additionally, she has been a property advisor to the Property Working Group of the United Nations Environment Programme Finance Initiative, and she also currently advises the Ethical Property Foundation in relation to their Fairplace Award, which is aimed at driving "triple bottom line" sustainability into workplace property management.

Contributor biographies

Abdallah Al-khawaja was born in Amman, Jordan in 1983. He completed his PhD from Bond University, Australia in 2016 under the supervision of Professor Craig Langston. His thesis title was "Environmental Auditing: Modelling Office Workplace Ecology". He completed his first academic degree in Accounting from Al-Zaytoonah University of Jordan in 2006 before undertaking his Master of Business Administration (Accounting) from Malaysia and ultimately moving to Australia. In 2006, he worked as internal auditor in a tourist company in Jordan, and in 2009 he lectured at the Department of Accounting at Al-Balqa` Applied University, Jordan.

Abdullah Alhamoudi is a Doctoral Researcher in the School of Geography, Earth and Environmental Sciences at the University of Birmingham, UK. He is undertaking study into city development management within Dammam in Saudi Arabia as a case study.

Paul Appleby, BSc (Hons), CEng, FCIBSE, FRSA, is an expert on the integrated sustainable design of buildings and communities. He has worked in the construction industry as a consultant, lecturer and researcher for nearly 50 years, working on award-winning projects with some of the world's leading architects and developers. As well as writing some 70 publications, his book *Integrated Sustainable Design of Buildings* was on the University of Cambridge list of "Top 40 Sustainability Books of 2010". His follow up, *Sustainable Retrofit and Facilities Management,* was published in January 2013. He is a Cabe Built Environment Expert; sits on design panels for the High Speed 2, Design South East, Oxford and Brighton; and is actively involved with the UK and World Green Building Councils.

Susan Bright is a Professor in Land Law at New College, Oxford, UK, and qualified as a commercial property solicitor before becoming an academic lawyer. As an expert in landlord and tenant law, she has developed international expertise on how property governance impacts the ability to make "energy upgrades" to buildings, particularly leasehold properties. Her publications include numerous articles on "green leases" and commercial property, and on how environmental decision making in multi-owned property is structured and mediated by the ownership and management arrangements.

Ben Dalton is a Private Equity Real Estate Analyst and holds an MPhil in Real Estate Finance from the University of Cambridge, UK. He was a student at Girton College in 2015–16. His research interests are mainly in the financial performance of sustainable real estate, which was also the topic of his MPhil dissertation. Prior to his studies at Cambridge, he was a student at the University of Reading's School of Construction Management and Engineering, where he was involved in research on building information modelling and virtual reality.

Ben Elder, BA, BSc, FRICS, ACIArb, is RICS Global Director of Valuation. He is responsible for delivery of the RICS Global Valuation Strategy, which has a key role in securing global financial stability. He is well qualified for the role as an economist and a chartered surveyor with an interest in the interface of economies and property markets. He has been a valuer and respected academic, holding senior positions at Nottingham Trent University and the College of Estate Management. In 2016, he was selected to chair IVSC's Tangible Assets Board and became a member of IVSC's Standards Board. This appointment followed influential periods on the Global Advisory Forum for The Appraisal Foundation and the Advisory Forum Executive to the IVSC.

Jessica Ferm, BA (Hons), MA, PhD, MRTPI, is a Lecturer in Planning and Urban Management at the Bartlett School of Planning, University College London (UCL), UK. Her main research interests are in the overlap between spatial planning and economic development, and the delivery of sustainable outcomes through property development. She is currently working on research projects about "Revealing Local Economies in Suburban London" and RICS-funded research on office-to-residential conversion in the UK. Prior to entering academia, she worked for ten years in planning practice in private consultancy and the public sector. She continues to be actively involved in London planning.

Franz Fuerst is a Reader (Associate Professor) in Real Estate and Housing Finance at the University of Cambridge, UK. He is also a Fellow of CULS and Trinity Hall, Cambridge, where he is also the Director of Studies in Land Economy. At Cambridge, he directs the MPhil in Real Estate Finance and supervises a team of doctoral and postdoctoral researchers on a number of UK and international research projects. His expertise and research interests are principally in the area of green real estate economics, financial analysis of sustainable investments, and portfolio and risk management.

Jeroen van der Heijden is an Associate Professor of Comparative Urban Climate Governance, and holds a joint position at the Australian National University (ANU) and the University of Amsterdam, the Netherlands. He works at the intersections of regulation and governance, policy change, and urban development and transformation. His research aims to improve local, national and international outcomes of urban governance on some of the most pressing challenges of our time: climate change, energy and water use, and a growing and increasingly urbanising world population.

Erwin Heurkens is Assistant Professor in Urban Development Management at the Faculty of Architecture and the Built Environment in the Department Management in the Built Environment at Delft University of Technology, the Netherlands. His focus is on (the management of) public–private partnerships and state–market relations and sustainability in urban development projects. Since his PhD, "Private Sector-led Urban Development Projects" (2012), the comparative international perspective has been key in his research. Learning lessons is critical to his research projects in Delft and Rotterdam, as well as in many publications of which he is the (co-)editor or (co-)author.

Christopher Heywood is an Associate Professor in Property and Management in the Faculty of Architecture, Building and Planning at the University of Melbourne, Australia. From a background in architecture, he now brings a trans-disciplinary perspective to understanding the built environment, its creation, its purposes and its management to achieve those purposes.

His research interests are based in corporate real estate management and affiliated property management disciplines. Among other projects, he is currently working with international colleagues investigating the theory and practice of how organisations align their property and business strategies. He is the co-editor of the *Journal of Corporate Real Estate*.

Neville Hurst, B.Eng, GradCert (HEd), GradDip Property, AAPI, CEA REIV, is a Senior Lecturer in Property Studies at RMIT University, Victoria, Australia and Deputy Head of Learning and Teaching. Originally trained as an electrical engineer, he has worked extensively in the industry and has worked in the property sector as a practitioner and academic for the last 33 years. His research interests focus on real estate industry practices, valuation practices, housing markets and energy-efficient housing in particular. Currently undertaking his PhD, he is investigating real estate agent attitudes and engagement with energy-efficient housing and is working with peak industry bodies to create greater awareness of the importance of sustainable housing to future cities.

Kathryn B. Janda is an interdisciplinary, problem-based Scholar at University College London's Energy Institute and the University of Oxford Environmental Change Institute, UK. She studies the intersection of social and technical systems and is interested in why different organisations and professional groups promote or reject environmental technologies and practices. She has worked in the Energy Analysis Program at Lawrence Berkeley National Laboratory, USA; served as an American Association for the Advancement of Science Environmental Policy Fellow at the US Environmental Protection Agency; and taught Environmental Studies at Oberlin College, USA.

Peter de Jong is a Lecturer in Design and Construction Management at the Faculty of Architecture and the Built Environment in the Department Management in the Built Environment at Delft University of Technology, the Netherlands. His main focus is aimed at the financial viability of buildings, and recently within corporate portfolios. He teaches a wide range of subjects, from cost awareness integrated in the design project (public building, renovation and adaptive reuse) and a management game on urban redevelopment in the bachelor programme, to clients, market and location in a sustainable context, using life cycle analysis in the MSc.

Nils Kok is an Associate Professor in Finance and Real Estate at Maastricht University, the Netherlands. His work on pension funds, commercial real estate, and energy efficiency and sustainability has received numerous government grants and awards. It has appeared in leading academic journals, such as the *American Economic Review*, the *RAND Journal of Economics* and the *Review of Economics and Statistics*, as well as in the global financial and economics media, such as the *Huffington Post*, the *Guardian*, the *Washington Post*, the *Australian Financial Review*, *Das Handelsblatt*, *Le Monde* and *Het Financieele Dagblad*, and in industry publications from such organisations as Bloomberg and Pensions & Investments. He is also the CEO and co-founder of GRESB, a leading global benchmark to assess the sustainability performance of real assets, including real estate portfolios and infrastructure assets. GRESB has rated more than 1,200 REITs and funds on behalf of more than 60 institutional investors who on aggregate represent some $6 trillion.

Craig Langston is a Professor of Construction and Facilities Management at Bond University, Australia. He has a combination of industry and academic experience spanning over 40 years. His research interests include the measurement of sustainable development, adaptive reuse, life cycle costing and productivity. He has held four Australian Research Council Linkage Project grants, amounting to nearly AU$1 million in external competitive funding. He was also the recipient of the Vice-Chancellor's Quality Award (Research Excellence) at Bond University in 2010. He is an international author and has won a number of awards for his research including the Queensland, Australia and Asia Pacific Research Award in the project management discipline in 2016.

Peter Lee is the Senior Postgraduate Tutor in the School of Geography, Earth and Environmental Sciences at the University of Birmingham, UK. He is Director of the Urban Regeneration and Renewal MSc programme. His recent collaborative research with Waseda University focused on resilience in the context of post-Fukushima planning responses.

Nicola Livingstone, BSc, PgCert, PhD, FHEA, is a Lecturer in Real Estate at The Bartlett School of Planning, University College London (UCL), UK. Her background is in real estate; however, her research interests are multi-disciplinary. Her recent work in real estate focuses on property market liquidity and brokerage processes, the emergence of online retailing, sustainability and interpreting the social form of the built environment. In addition to the property market, she also researches the third sector, the political economy of charity and the evolution of food insecurity. She is currently working on RICS-funded research on both rural investment trends and office-to-residential property conversions.

David Parker is an internationally recognised property industry expert on real estate investment trusts (REITs) and a highly-regarded property academic, being a director and adviser to property investment groups including real estate investment trusts, unlisted funds and private property businesses (www.davidparker.com.au). He is currently the inaugural Professor of Property at the University of South Australia; a Visiting Professor at the University of Reading, UK, and at Universiti Tun Hussein Onn Malaysia; a Visiting Fellow at Ulster University, UK; and an Acting Valuation Commissioner of the Land and Environment Court of New South Wales. Author of the authoritative *Global Real Estate Investment Trusts: People, Process and Management* and *International Valuation Standards: A Guide to the Valuation of Real Property Assets*, he may be contacted at davidparker@davidparker.com.au.

Julia Patrick is a Research Associate at the University of Oxford Environmental Change Institute (ECI). As part of the recent WICKED project, she investigated energy management in the retail sector, in particular the use of organisational practices and green leases. Her research and wider interests include energy management generally, the green economy and community energy. She coordinated the Oxfordshire Low Carbon Economy project and has contributed to projects for Oxford City Council, Oxfordshire County Council and ClientEarth. She helped set up Oxford-based Low Carbon Hub and sits on the board of Oxford North Community Renewables (ONCORE). Recent publications are listed at www.eci.ox.ac.uk/people/jpatrick.html.

Dave Pogue is CBRE's Global Director of Corporate Responsibility and oversees CBRE's development, implementation and reporting for all aspects of corporate social responsibility, including environmental stewardship, community engagement and corporate giving. Prior to his current role, he led sustainability programmes for CBRE's property and facilities management portfolio around the globe, managing the development, introduction and implementation of a wide-ranging platform of sustainable practices and policies. His leadership in this area produced an award-winning sustainability platform leveraging thought leadership, service delivery and industry associations to raise worldwide green building standards. Programme achievements included development of CBRE's $1 million Real Green Research Challenge; aggressive endorsement of the US EPA Energy Star programme; the introduction of the Green Knights programme; delivery of co-branded BOMA BEEP training to more than 20,000 attendees; and recognition as the first manager of commercial property to certify more than 500 buildings in the LEED for Existing Buildings rating system.

Hilde Remøy is an Associate Professor in Real Estate Management at the Faculty of Architecture and the Built Environment in the Department Management in the Built Environment at Delft University of Technology, the Netherlands. After an international educational and professional path in Norway, Italy, the Netherlands, and a PhD ("Out of Office", 2010) in Real Estate Management on the theme of office market vacancy and residential transformation, she is now teaching and writing (over 100 publications) on the office market, office space vacancy, office building adaptation and building transformation in the Dutch context, but still with a broader scope.

Spenser J. Robinson, DBA, is the Director of Real Estate and Assistant Professor of Entrepreneurship at Central Michigan University, USA. He publishes regularly in a number of leading journals and is an invited author for the Counselors of Real Estate Financial Council. He is a co-editor of the *Journal of Sustainable Real Estate* and an associate editor for the *American Journal of Business*. He chairs the Research Committee of the American Real Estate Society (ARES) and is an elected member of their board of directors. He is the recipient of multiple research awards, teaching awards and private-sector sustainability research grants.

Vivek Sah is an Associate Professor of Real Estate in the School of Business Administration and the Burnham-Moores Center for Real Estate at the University of San Diego, USA, where he teaches undergraduate and graduate courses in real estate finance, commercial real estate investments, valuation and market analysis. His primary research interests are in the areas of REITs, real estate mutual funds and housing markets. He is well published on REITs and other real estate topics, and his research has been published in leading journals such as *Real Estate Economics*, *The Journal of Real Estate Finance and Economics* and the *Journal of Property Investment & Finance*. He has won awards for his research work. In 2009, his dissertation won the Emerald/EFMD Outstanding Doctoral Research in the property category. In 2014, he won the Best Paper award for his work on "Green REITs" in the *Journal of Real Estate Portfolio Management*.

Robert A. Simons is a Professor at the Maxine Goodman Levin College of Urban Affairs at Cleveland State University, USA. He was a Fulbright scholar in South Africa and received his PhD from the University of North Carolina at Chapel Hill, USA, in City and Regional Planning. A former President of the American Real Estate Society, he teaches real estate development, market analysis and finance. He has published four books and over 60 articles and book chapters on various real estate topics. He is lead editor for *The Journal of Sustainable Real Estate* and has been an expert witness in 85 cases related to real estate and environmental contamination.

Connie Susilawati is Course Coordinator and Senior Lecturer in Property Economics at Queensland University of Technology (QUT), Australia. Her passion for providing real world learning and international opportunities has shown in her teaching, research, leadership and engagement. Her work focuses on sustainable human settlement for both developing and developed countries. She led the economic impact on sustainable housing section of an Australian Research

Council Linkage Project. She offers multiple international education opportunities for both Australian and Indonesian students/leaders in infrastructure, asset management and property development. She was working for the largest development company (Ciputra Group) and led the property programme at Universitas Surabaya, Indonesia.

Clive Warren is an Associate Professor and Head of Property Programs at the University of Queensland Business School, Australia. His research interests include sustainable property, property management, and workplace performance and real property valuation. He has been an external examiner for a number of property programmes throughout Asia and is a frequent presenter at international property conferences. He is the editor-in-chief of *Property Management*, the leading scholarly journal for research in commercial and residential asset management. He is a Green Star Certified Assessor, advises on sustainable property performance and also undertakes assessment of building on behalf of the Green Building Council of Australia (GBCA).

Georgia Warren-Myers, PhD, BPC (Hons), BPD, AAPI (CPV), is a Lecturer in Property in the Faculty of Architecture, Building and Planning at the University of Melbourne, Australia. She is a Certified Practising Valuer, has over ten years' experience in the industry and has worked in the New Zealand and Australian property markets. Her research interests are focused on value, valuation and decision making in the context of sustainability and resilience in the property sector. She continues to be actively involved in the property industry through the Australian Property Institute on their National Property Standards Board and the Victorian Property and Valuation Standards committee. She is presently involved in the project "Advancing New Home Sustainability through Demand-side Empowerment", and is collaborating with Franz Fuerst on value connections in another major project, "Green Lifestyles, Green Buildings, Green Cities".

Agnieszka Zalejska-Jonsson, PhD, is a Researcher in the Division of Building and Real Estate Economics at KTH Royal Institute of Technology, Sweden, and a Director of the Centre for Construction Efficiency. She investigated the comprehensive value of green residential buildings as seen from two perspectives: that of the developer and that of the occupant (the customer). At present, she is involved in projects studying the application of sustainability in organisations and by individuals, property management and the process of effectively commissioning buildings, with a focus on assuring energy efficiency and environmental values during use and operation.

Foreword

The word "handbook" generally implies a type of reference work, or other collection of instructions, that is intended to provide ready reference. However, reading through the extraordinary list of contributors to the *Routledge Handbook of Sustainable Real Estate*, it is clear that the contents will go well beyond that modest characterisation of the term in both context and concepts presented.

Delving further into the ideas, concepts and actual case studies – in particular, how the contributors put forward innovative ideas and practical suggestions for application and further investigation – moves this publication well beyond a "collection of instructions" to an informational and academically rigorous resource reflecting some of the most innovative thought leadership in all aspects of sustainability and the built environment that I have had the pleasure to read.

Who *should* read it and why? Who? Investors, developers, owners, architects, pension fund managers, underwriters, valuers/appraisers, students – the list could go on too long, so I will stop here. It basically includes anyone whose life intersects with real estate; who invests in it; who lives in it; who works in it; who plays in it.

Why? Because the contributors to this book understand that "real estate" is more than just a unique investment opportunity.

Starting from a comprehensive chapter written by the editors, the book is organised into coherent sections ranging from policy, to investment and value, and then to management, and covers various perspectives from many countries, as the line-up of contributors indicates.

Collectively, the authors recognise – and document – the impact of real estate on environmental factors and the need for transformation in the industry; the necessity for international buildings standards whereby property can be assessed more consistently on a global basis; the imperative to retrofit existing buildings, which comprise the vast majority of the building stock now and for at least the next half century, both at the level of the individual building; and, critically, how to scale up to precincts, suburbs and cities.

Further, they offer insights into and perspectives on the difficulty in appropriately valuing high-performing, more sustainable real estate; research the role of green REITS in the investment market; provide strategic insights into green premiums and their impact on real estate; and argue for the need for greater understanding of sustainable "value" by real estate agents and corporate real estate managers. Finally, it is especially useful that the authors recognise the different issues, from energy to finance to value, facing those dealing with commercial markets versus those dealing with residential markets, to which the final section is devoted.

I have had the pleasure of knowing and working directly with several of the contributing authors, but this book has given me the opportunity to enjoy and appreciate the works of others. These are the warriors who continue to explore the issues around sustainability and the built environment with intelligence, passion and courage, despite continued resistance from those of lesser vision and determination. As is cited in the beginning pages of the book, "Here is the problem we face therefore: it takes more time for the majority to believe, accept and act on the above [various tenets of sustainability] than it does for those who have studied sustainability, discovered new truths and recognise the need for action."

I personally recognised the need for action more than a decade ago while working as a valuation consultant, when I created the business case to justify taking a general services administration building located in Portland, Oregon to LEED Silver certification. At that time, I had valued millions of square feet of real estate in the US, Europe, Canada and Australasia, but had not been introduced to the concepts of sustainability. In working on that project, it became acutely clear that architects, engineers and other designers did not "speak the language" of investors, underwriters or, perhaps most of all, the appraisal community.

As a result of this "professional epiphany," I got my LEED AP accreditation and went on to develop the first course on valuing sustainable green buildings for the Appraisal Institute in the US. And while the benefits of a more sustainable built environment seemed quite reasonably to outweigh the costs of such initiatives over time, many colleagues and other market players continue to be unable to move beyond initial costs or to even identify, much less quantify, the multiple and diverse benefits of more sustainable real estate practices. It is this communication gap that still, in my opinion, is one of the greatest challenges we have yet to overcome in moving toward greater sustainability.

Nonetheless, the *Routledge Handbook of Sustainable Real Estate* goes a very long way toward bridging that gap.

The content in this book is provided by some of the most gifted thought leaders today, from multiple disciplines and varying perspectives. They have addressed the most salient issues around sustainability and the many reasons for its adoption in the built environment on a global scale, as well as the diverse and broad challenges this adoption faces from various special-interest groups around the world.

Have they provided all the answers as to how these challenges can be overcome? No. But they have asked all the right questions and will doubtless continue to pursue those answers. And they have provided us – those who have a broader understanding and respect for the plethora of values that sustainable development addresses – with the data, perspectives and suggestions to create a path toward greater sustainability in all aspects of the built environment.

Not only that, but on a more personal level, their innovative approaches, thoughtful analyses and prescient perspectives have re-invigorated my own need and desire to assist in the goal of achieving greater sustainability on a global scale. I hope they will do the same for you.

To that end, I close with the words of Aaron Hurst in his book *The Purpose Economy*: "Everything is in transition and far less permanent than we imagine … Things are done a certain way, until they aren't. You can be the one who makes the change."

Theddi Wright Chappell, CRE, MAI, FRICS, AAPI, LEED AP
CEO, Sustainable Values, Inc.

Acknowledgements

Sara Wilkinson

Sara would like to thank all her family, her co-editors and all the contributing authors to the handbook, as well as all the participants in the various research projects featured. She would also like to extend a heartfelt thank you to Pratichi Chatterjee, who collated and undertook the final formatting of chapters and managed the submission of the manuscript.

Tim Dixon

Tim would like to thank all his family and Cookie the cat for their patience and love during the writing of this handbook.

Norm Miller

This book is partly a result of the data and support of a few primary supporters of sustainable real estate. The CoStar Group and CEO Andrew Florance supported the initial founding of *The Journal of Sustainable Real Estate* and supported the journal for several years. More important to researchers, the firm added data on LEED certification and Energy Star ratings to the enormous commercial real estate database, which has since resulted in quite a bit of research, not otherwise possible without this support. A second organisation that has tremendously helped the academics to advance our knowledge about sustainability has been CBRE under the leadership of the Global Director of Corporate Social Responsibility Dave Pogue. CBRE has provided funding of over $1 million in grants. Some of the recipients are authors of chapters in this book.

Sarah Sayce

Sarah would like to thank her family for their patience and tolerance as, yet again, writing has taken precedence when it should not have done! And especially David for being ever long-suffering.

Part 1
Governance and policy

1

Sustainable real estate

A snapshot of where we are

Sara Wilkinson, Tim Dixon, Sarah Sayce and Norm Miller

1.0 Introduction

Even cynics about climate change should care about sustainable real estate because the market and governments care, not just about climate change, but also about air and water pollution and toxic substances that make us sick. We have knowledge about ecology and the importance of our environments sufficient to prompt real action, but it is the tendency of humans to maintain the status quo and ignore the long term. Rachel Carson wrote her book *Silent Spring,* published in 1962, explaining the problems with farmers using DDT, a strong pesticide that caused many health problems and had numerous unintended side consequences. However, the chemical companies tried to silence her and attacked her truths in the same way tobacco companies attacked the first critics of smoking who had claimed smoking was harmful to health. Eventually, both were proved right, but there remain many truths that are not yet widely accepted, even though strong evidence exists to support them. Among these are some of the following:

- We can use offices and other buildings more efficiently and use energy more effectively by turning things off when not in use.
- We can provide more productive working environments.
- We can do both of the above profitably.
- We can save energy in industrial buildings, apartment buildings and retail buildings while providing more pleasant environments in which to work, live, shop and play, and do so efficiently.
- We can self-produce energy in homes and commercial real estate, and do so with solid returns.
- We can store energy in various ways, and we are on the cusp of even more efficient storage.
- We can design buildings that are passively more efficient and aligned with local sun, climate and weather patterns.
- We can use materials that require less energy input and that are not toxic to humans.
- We can invest in 'greener' buildings and reduce investment risks while gaining rewards.
- We can profitably move up the waste hierarchy to recycling, reusing and resource reducing, and create a culture of doing less harm to our community and planet.
- We can produce better grids and better community systems to share energy, recycling and more.

- We can harness the use of smart and digital technologies to enhance resource efficiency.
- We can produce standards and measurements that, with more disclosure, will let the market work more efficiently in recognising the benefits, both financial and social, that arise from sustainable business strategies.
- We can recognise that all building uses are, fundamentally, connected with human behaviours.
- We can educate ourselves about sustainability, and that is the purpose of this text.

Here is the problem we face therefore: it takes more time for the majority to believe, accept and act on the above than it does for those who have studied sustainability, discovered new truths and recognise the need for action. There is an urge for those converts to shout at the entire world, 'Don't you see that with technology and systems and a caring culture, we can do better and live healthier and be more productive!' However, reality then sets in and the engineer, who knows how to create a better operating system for a building, does not necessarily know how best to communicate to the facilities manager or the vice president of finance, who do not want to hear about kilowatts produced or saved, but would rather hear about the return on investment or the costs of a less productive environment. Alternatively, we face the problem of not knowing where to start. Do I need to learn about the state of the art of toilets, roofs and walls and windows, you might ask? And the answer is probably yes. Do I need to know what passive design is or what tenants really want in their space? And again, the answer is probably yes. Or do I need to know what is driving business decisions or motivating individuals? Emphatically, yes.

Some organisations also have directors of sustainability whose jobs are partially to educate everyone, provide resources on best practices, then measure and monitor performance over time. Yet many do not have such a resource, and it falls upon others in human resources, finance or asset and property management to learn enough to know that improvement is not hard, and that it can be continual. Best of all, for those willing to learn, we discover that we can make the world a better place while saving money *and* becoming more productive at the same time.

The issues at hand have never been so important, and this text comes at a time when the new US President (Donald Trump), in 2017, seems to care less about the long run, the place our children and grandchildren inherit, and more about the short run, the next four years or next year or next quarter. Scott Pruitt, who has historically denied climate change and previously sued the US Environmental Protection Agency (EPA), has been appointed as the EPA's new administrator. Oil pipelines in the US are now being constructed under water reservoirs on open land, and the return of coal jobs has been promised. And in other parts of the world, we have seen some 'rowing back' on international endeavours. The UN Climate Change Conference (COP21) in Paris in 2015 signalled a hope for the future that was followed by far less optimism only a year later at COP22 in Marrakech.

We should sympathise with those whose jobs have been lost to cheaper natural gas or solar energy, but we cannot reverse history, and we should not revert back to what we used to do when fossil fuels were the only practical alternative. Today, we have so much 'low-hanging fruit' in terms of new designs, new technologies and new materials and nascent breakthroughs that can make life better for everyone, now and in the future, but we must invest in learning how to capitalise upon these trends.

That is the overarching aim of this book: to help move us forward, and to help educate and convert those with open minds. Those of you who 'get it' will become ambassadors to the stubborn and less knowledgeable. Those of you who 'get it' will become the enlightened diaspora of sustainability. Thank you for starting and continuing your journey with us. We hope it helps in enabling a more sustainable real estate sector now and in the future.

1.1 Rationale and aims of the handbook

1.1.1 Real estate and climate change

Property (or real estate) is a major global fixed asset. Recent research has estimated the total value of all developed global real estate (comprising retail property, offices, industrials, hotels, residential, other commercial uses and agricultural land) as $217 trillion in 2015 (Figure 1.1). This represents a value of nearly three times the annual global income, 2.7 times the global GDP, and 60 per cent of all main global assets. With a value of more than a third of all global equities and securitised debt, real estate plays a major role in many economies, with much of this stock classified as 'investable', and the majority of the stock being people's homes ($162 trillion by value for residential real estate) (Savills, 2016).

Given the importance of real estate in the global economy, it is not surprising that buildings have come under close scrutiny in the battle to combat climate change impacts at a global scale. The real estate sector as a whole, for example, uses more energy than any other sector and is a growing contributor to greenhouse gas (GHG) emissions (WEF, 2016). Although the basis of calculation differs, Change (2014) suggested that 'buildings' accounted for 32 per cent of total final energy use and 19 per cent of energy-related GHG (primarily CO_2) emissions (including electricity-related) globally in 2010. Recent projections suggest that building-related CO_2 emissions are set to increase by 56 per cent to 2030 (WEF, 2016) and that the majority of this growth will be in developing countries (UNEP, 2009). Buildings also use a significant proportion of raw materials globally (40 per cent) and generate significant amounts of waste (25–40 per cent)

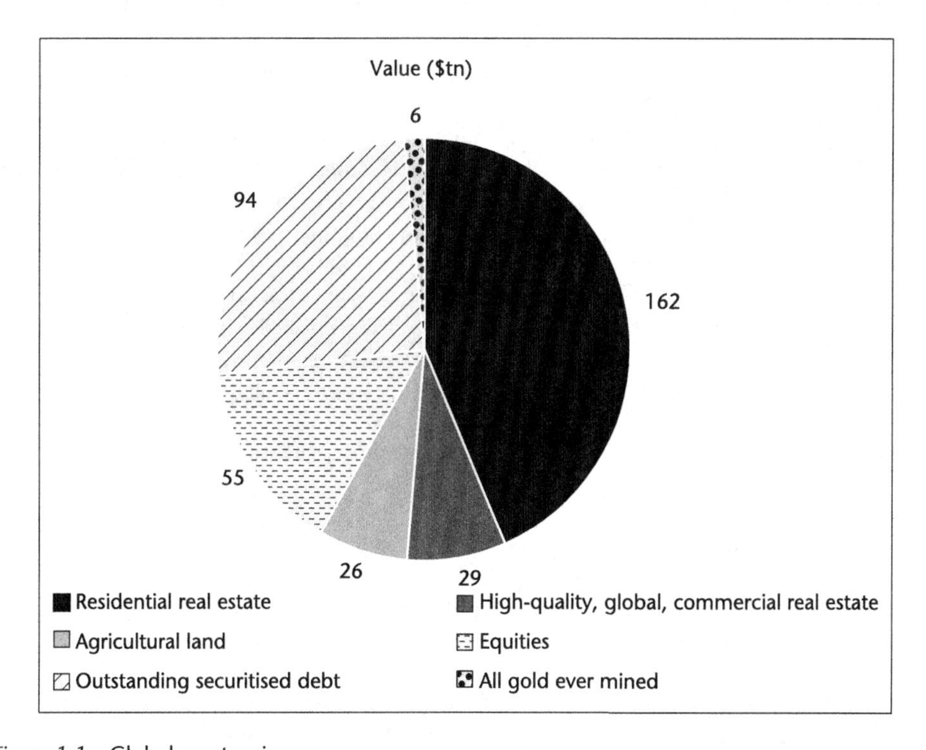

Figure 1.1 Global asset universe

Source: Savills (2016)

(WEF, 2016). If this is not reversed, some of the worst fears regarding climate change are likely to be realised with devastating human consequences, as deltas flood and deserts spread.

A growing global population, which is set to be eight billion by 2030, with the majority of people living in cities (60 per cent), will also exert further pressures of growth on the construction and real estate sectors. The World Economic Forum (WEF, 2016), for example, estimate that the largest 750 cities will require by 2030:

- 260 million new homes;
- 540 million m^2 of new office space; and
- 60 million new jobs in industry.

This will have important implications for climate change, given that COP21 in Paris in 2015 agreed to limit global warming below 2° C and ideally 1.5° C. Substantial efforts will be required from the real estate and construction sectors, therefore, and the World Bank (quoted in WEF, 2016) estimates that to meet the 2° C threshold, a 36 per cent reduction in total carbon emissions will be required by 2030.

1.1.2 Sustainable real estate: A growing trend?

The response of the real estate sector to increasing climate change pressures has been to focus even more strongly on environmental, social and governance (ESG) issues, and essentially to try and protect and enhance the value of real estate assets from risks associated with these issues. There has therefore been an increasing focus on 'sustainable real estate'. But what does this term mean?

As Sayce, Ellison and Parnell (2007) and Dixon (2010) point out, there is a lack of agreed definition internationally. Indeed, the terms 'green' and 'sustainable' may also mean different things to different people in different national contexts. This could, for example, relate simply to achieving a minimum environmental performance standard. One useful definition that perhaps best captures the perspective in this book, however, is by the UK Green Building Council (UKGBC, 2008), who suggest that 'sustainable buildings' are those that

> (1) are resource efficient (physical resources, energy, water, etc); (2) have zero or very low emissions (CO_2, other greenhouse gases, etc); (3) contribute positively to societal development and well-being; and (4) contribute positively to the economic performance of their owners/ beneficiaries and to national economic development more generally.

In a purer academic sense, and drawing on recent literature, Berardi (2013: 76) also offers a helpful holistic definition:

> A sustainable building can be defined as a healthy facility designed and built in a cradle-to-grave resource-efficient manner, using ecological principles, social equity, and life-cycle quality value, and which promotes a sense of sustainable community. According to this, a sustainable building should increase:
>
> - demand for safe building, flexibility, market and economic value;
> - neutralisation of environmental impacts by including its context and its regeneration;
> - human well-being, occupants' satisfaction and stakeholders' rights;
> - social equity, aesthetics improvements, and preservation of cultural values.

There is also a clear difference in the evolution of thinking on green and sustainable buildings, partly created by cultural differences between North America/Australasia, where the term 'green' is commonplace, and the UK and Europe, where the term 'sustainable' often tends to be used. This, however, is not an exclusive distinction, and the terms have frequently been used interchangeably. However, the main differences between the terms can also partly be related to whether the focus is on new build ('green') or new build plus existing ('sustainable') (Dixon, 2010).

In using the term 'sustainable property' in this book, it is important to note therefore that there are various shades of 'green' or 'sustainability' in a real estate, or property, context. In a sense, some definitions (for example, UKGBC, 2008) might be considered to be aspirational, because ultimately the reality of real estate investing means that greater importance is more frequently attached to environmental impact than, for example, to the social dimension (Dixon, 2010).

When we look at the global picture, current data suggests that there is a growth in sustainable buildings primarily driven by client and market demand and with lower operating costs, corporate responsibility and branding/PR also being key drivers (Dodge Data & Analytics, 2016). For example, currently 40–48 per cent of new commercial buildings are 'green' compared with only 2 per cent in 2005, and this is expected to rise to about 55 per cent in 2020 as new regulations come on stream (WEF, 2016). But this, in turn, raises the question: if regulations push new commercial (and residential) buildings towards high 'green' or 'sustainable' standards, does their commercial advantage (if indeed it exists, as discussed at length in the book) disappear, because green becomes the new 'norm'?

Generally, global green building rates are doubling every three years. Much of the demand for sustainable buildings also comes from the developed world, but increasingly there is growing demand from such countries as Mexico, Brazil, China and India (Dodge Data & Analytics, 2016). Recently, the Global Sustainable Buildings Index (Baker & McKenzie, 2016) found that the Netherlands, France, Germany and the UK lead the way towards sustainability through certifications, CO_2 reduction targets and market-led initiatives, with other countries, including Australia, Canada and Australia, following.

Clearly, much of the focus around sustainable buildings is on new build. But the typical lifespan of a building is 40–120 years (IEA, 2011) and is likely to be some eight times longer than a typical appliance, for example (Jennings, Hirst and Gambhir, 2011). In the UK, for example, less than 1–2 per cent of total building stock each year is new build (Dixon, 2009; Stafford, Gorse and Shao, 2011), and in some countries up to 70 per cent of total 2010 building stock will still be in use in 2050 (Better Buildings Partnership, 2010); renovation and refurbishment rates are 2.9–5 per cent of existing stock for domestic buildings and 2–8 per cent for commercial stock, depending on the sector (Hartless, 2004; Stafford et al., 2011). Jennings et al. (2011) estimate that, in England, assuming a retrofit rate of 1.5 per cent of total building stock, it will take until 2062 to completely refurbish 22.7 million buildings. At this rate, hitting current international targets on climate change caps is simply not possible.

Retrofitting (or 'the process of making planned interventions in a building to install or replace elements or systems which are designed to improve energy and/or water and waste performance' (Dixon, 2017: 42) will be an important part of the sustainable real estate sector's response to environmental challenges. Indeed, the WEF (2016) suggest that 13 per cent of total global carbon emissions will come from retrofitted buildings. So a major challenge will be to find the ways, the means, the money and the motivations at government and individual level to effect this transformation of existing stock.

1.1.3 Rationale and aims of the book

The development and operation of sustainable buildings is an important part of the real estate and construction sectors' response to growing threats from climate change and environmental risk.

It is also about the social role that buildings play in health, well-being and cultural terms and in establishing places for people. Given the important role of real estate assets globally, the potential impact that climate change and environmental risk will have on those assets and on the people, organisations and businesses who occupy those assets over the next 30–40 years is immense.

After all, it was Sir Winston Churchill (1874–1965) who spoke about the bomb-damaged House of Commons in London in 1943: 'We shape our buildings, and afterwards they shape us.'[1] This quotation captures the essence of how we need to understand buildings and real estate as fundamental parts of our everyday lives, and so finding the best way to understand our complex relationship with the environment comes through understanding the buildings in which we live, work and play and our relationship with those same buildings.

The overall aim of this international handbook is therefore to bring together the expressed views of academics and other experts from around the world and from different disciplinary and professional backgrounds and show *why* we need to tackle climate change and other environmental and related socio-economic risks in the sector, *how* these risks are impacting buildings and the sector as a whole and *what* can be done to help the sector move towards a sustainable future. In doing this, the book aims to be truly international in scope, drawing on research and solutions from a range of countries in the developed and developing world, which include England, Australia, the United States, Europe, the Netherlands and Sweden.

The handbook focuses not only on property and land markets (both commercial and residential) and associated value and valuation concepts, but also on the role of, and impact on, corporate real estate, property management and property investment processes, as well as key issues relating to building occupants, which are all set in the context of sustainable real estate. Although it is not possible to cover all the aspects of sustainable real estate, the handbook focuses on the following key themes:

- **Governance and policy:** How can we best create and develop regulations and standards to encourage sustainable real estate?
- **Valuation, investment and finance:** What is the relationship between sustainable buildings and value? How can we best value and price sustainable real estate for a variety of purposes? What global valuation standards are required for a changing world? What is the role of responsible property investing (RPI) in sustainable real estate? How can financial vehicles support a sustainable real estate market?
- **Management:** How can we best mainstream sustainable thinking in corporate real estate and facilities management? What is the role of the workplace in thinking about sustainable real estate?
- **Redevelopment and adaptation:** How and why should we retrofit our existing buildings for the future?

1.2 Content of the handbook

Part 1 Governance and policy

Chapter 2 Sustainable real estate and corporate responsibility. Norm Miller and Dave Pogue.

This chapter sets the scene on an optimistic note, concluding that career opportunities in corporate social responsibility and sustainability overall are growing fast. Commercial real estate sits at the nexus of many of the most complex issues we face. In a resource-challenged world, the development and occupancy of real estate has captured the attention of environmentalists, regulators, product designers, architects, investors and occupants alike. By many measures, commercial real estate is one of the most important and visible components in the war on waste and carbon dioxide emissions. The US Green Building Council concludes that commercial real estate accounts for 12 per cent of US water use and 71 per cent of US electricity use, and produces 65 per cent of US waste and 39 per cent of US CO_2 emissions. Corporations, faced with a growing demand to understand, mitigate and report on the environmental impact of their activities, and these statistics command attention and create challenges and opportunities. As a result, corporate social responsibility (CSR) has developed into a field with substantial and diverse career opportunities. Directors of sustainability focus on real estate and internal operations and much of their work involves measuring, monitoring and reporting. While the early professionals wrote their own job descriptions, universities and professional organisations are sharing resources and starting specialised educational programmes. In this way, the future for those who want to do something beneficial to society and for themselves is bright.

Chapter 3 Energy upgrades in commercial property: Minimum energy efficiency standards, compliance pathways, and leases in the UK. Julia Patrick, Susan Bright and Kathryn B. Janda.

Improving the environmental performance of the commercial built environment is a 'wicked' problem, lacking a simplistic or straightforward response, particularly where space is rented. Traditional leases may inhibit rather than promote joint landlord–tenant action on energy management, including building improvements, reinforcing the so-called 'split incentive'. The introduction of statutory minimum energy efficiency standards (MEES) for commercial buildings in the UK is designed to address some of these challenges, by requiring building improvements for rented buildings below a minimum standard. Against this background, this chapter explores how MEES highlight the diverse – and potentially contradictory – roles of 'green lease' clauses in shaping the relationship between landlord and tenant, and in encouraging better energy management. Drawing on interviews and document analysis from the UK-based WICKED project, it examines the role of green lease clauses in managing compliance with MEES, and the extent to which these developments may promote or hinder increased energy efficiency measures and cooperation between landlords and tenants.

Chapter 4 Voluntary programs for low-carbon building development and transformation: Lessons from the United States. Jeroen van der Heijden.

Traditional regulatory interventions, such as building codes and planning legislation, often fall short with regard to accelerating the transition towards sustainable built environments. Seeking to overcome their problems, governments around the globe are increasingly relying on governance innovations, as are firms and civil society organisations. These innovative governance instruments often seek compliance through positive incentives rather than deterrence and voluntary participation rather than mandatory subjection. The opportunities and constraints of these voluntary programmes for urban sustainability have been more theorised than studied empirically. Seeking to increase knowledge on their performance, this chapter draws lessons from an empirical study of eight such programmes for commercial and residential buildings in the United States.

Chapter 5 Sustainable office retrofit in Melbourne. Sara Wilkinson.

In 2008, the City of Melbourne initiated an office retrofit programme, known as the 1200 Buildings Program, to reduce total building-related greenhouse gas emissions in the central business district (CBD). Given that 87 per cent of buildings we will have in 2050 are already built, retrofit is vital to mitigate the impacts of climate change and global warming. City authorities have to lead on promoting and incentivising sustainable building adaptation. Following the 2013 and 2011 City of Melbourne 1200 Buildings Melbourne Retrofit Surveys, this chapter analyses data to ascertain the patterns of retrofit as the city transitions to low carbon. Based on earlier studies of 7,393 Melbourne office building retrofits from 1998 to 2008 and 1,453 retrofits from 2009 to 2011, this chapter investigates building attributes previously found to be important and provides further insight into retrofit trends in the Melbourne office market, the lessons of which may be transferable.

Chapter 6 International standards: Key to unlocking the value of green buildings? Ben Elder.

In trying to understand sustainable real estate and the challenges facing the property industry in delivering a sustainable real estate product, it is critical to understand the context and delivery mechanisms within which the real estate product is placed. By doing this, efforts can focus on the key change elements more effectively. This chapter seeks to understand the place of real estate in the economic systems that allocate scarce resources between competing needs and the steps that have been taken by interested parties, including professional bodies and notably the international community, to improve knowledge within the economic systems that allocate scarce resources to ensure that sustainable features of real estate are recognised and priced appropriately into the market. It is essential to recognise that the delivery of sustainable real estate does not take place in a vacuum and that technology and innovation are happening all around our focus on sustainable real estate and the pace of change is increasing.

Part 2 Valuation, investment and finance

Chapter 7 Valuing sustainability in commercial property in Australia. Georgia Warren-Myers.

The ever-increasing need for sustainability adoption is imperative to mitigate and minimise future impacts from climate change and finite resources. However, sustainability implementation is inhibited due to challenges in valuing property's sustainability characteristics. Understanding, assessing and taking account of the value contribution is crucial to gaining widespread adoption and implementation of sustainability in the built environment. To date in property valuation, the lack of key financial correlations between sustainability and economic return have inhibited valuers from being able to accurately reflect the impact of sustainability on value. This largely results from valuers' inability to assess, compare and value sustainability due to a lack of transparent data and their inadequate knowledge of sustainability, sustainability rating tools and analysis techniques. This chapter examines the challenges facing Australian valuers in understanding the context of value in relation to sustainability; sustainability measurement, comparison and quantification in valuation; value considerations in valuation; and valuers' knowledge of sustainability and the need for further professional development.

Chapter 8 Existing building retrofits: Economic payoff. Norm Miller and Nils Kok.

Retrofitting existing buildings for energy efficiency and improved work environments often has a greater payoff than building new. We will never be able to achieve significant improvements in the overall efficiency and productivity of the building stock unless we retrofit existing buildings. Existing buildings that work well as retrofits allow for maximum natural light use and so some older deep-set-window stone or brick buildings are not as adaptable. The ideal timing for retrofitting is when a major tenant has moved out and the building is substantially empty. The scale of the payoffs depends on the climate of the market where the building is located, local utility rates and the local demand for better buildings. Here we review a major US-based study of office buildings in major markets.

Chapter 9 Building sustainability into valuation and worth. Sarah Sayce.

The process of developing awareness of sustainability as a social, economic and environmental concern has been gradual but escalating. From a situation some 30 years ago, at the time of the Brundtland Report (Brundtland, 1987) when the concept of sustainability and sustainable development had barely entered the professional vocabulary, despite being critical to any form of value judgement, it is now firmly embedded in the educational process of those seeking to qualify as valuers via competency testing. However, the testing of a baseline general knowledge of sustainability is not the same as appreciating its actual or potential role in valuation; even this may not reflect what is apparent from the marketplace. The progress towards deeper knowledge and integration of that knowledge into valuation practice has been challenging. In this chapter, the progress is charted within a predominantly UK and European perspective, and the role of the valuer, conventionally viewed as market 'taker' not market maker, is challenged. The chapter concludes that the point has been reached at which a simple awareness of sustainability issues is insufficient; it is now time for valuers to take a more proactive stance from their initial interface with their clients to their reporting process.

Chapter 10 Green REITs. David Parker.

Sustainability has become an increasingly important part of the management of real estate investment trusts (REITs) around the world over the last decade, manifest in various aspects including property selection, property management, asset management, fund management, performance measurement and REIT marketing. Despite the increased importance of sustainability to the REIT industry worldwide, surprisingly little academic research has been undertaken into REITs and sustainability. This chapter provides a literature review of published research undertaken to date concerning REITs and sustainability, identifies the themes evident through such published research, draws together the conclusions of such published research and suggests areas for further research.

Chapter 11 US green REITs. Vivek Sah and Norm Miller.

Green REITs remain a somewhat arbitrarily defined group that could include a variety of certifications or features. Green REITs may not always exhibit abnormal returns or higher market capitalisations relative to net asset value, but they do tend to exhibit less volatility in returns. There is some evidence that the market values these lower-risk returns and/or the socially responsible aspect of green efforts. REITs must decide whether to go for building-level certifications like Energy Star, LEED or BREEAM, as well as the newer WELL designation. There are also various carbon-based initiatives from GreenPrint and others, and portfolio-level designations like GRESB. The proliferation of so many overlapping standards as these and others might be an obstacle to broader market transparency on the sustainability of an REIT portfolio comprised of several different certification proportions. Eventually, the market will become more standardised in how green is measured, and more REITs will go green.

Chapter 12 The 'green value' proposition in real estate: A meta-analysis. Ben Dalton and Franz Fuerst.

In recent years, mandatory government-led environmental rating systems have gained traction in several countries. At the same time, there has been a proliferation of voluntary eco-labels, such as BREEAM in the UK, Green Star in Australia and LEED in the US. The very existence of the voluntary labels is indicative of a market-led environmental agenda. Any voluntary initiatives that exceed regulatory requirements and national building codes could potentially create 'green value', which should, at least hypothetically, be capitalised into prices and rents. The existence of a green premium would also reflect consumer willingness to pay, which studies have found to be primarily related to increased energy efficiency; therefore, a premium may also indicate the ability to successfully and credibly convey a property's energy efficiency. Amongst the issues that may hinder energy-efficient investment include those that stem from principal–agent problems and a 'vicious circle of blame'. Finding evidence for a green premium, and analysing its dynamics within a transaction setting, may provide a clearer understanding of the incentives available to stakeholders. Evidence-based policy depends on reliable and robust analytical results, particularly in innovative areas such as green real estate finance. However, the growing body of literature on the green premium is disjointed and at least partly inconclusive. The general incentives and disincentives of energy efficiency and broader sustainability are now widely researched, but the empirical studies are often limited in terms of geography and time periods analysed. Hardly any studies have tried to consolidate the burgeoning green premium literature and place the individual studies in a larger context. This chapter attempts to achieve this objective via a meta-analysis of green premium studies in a real estate context and illustrates the implications arising from the green premium consensus on property investment using a simple discounted cash flow (DCF) model.

Chapter 13 Sustainability and housing value in Victoria, Australia. Neville Hurst and Sara Wilkinson.

Shelter for humans, in the form of housing, is essential for survival. In developed countries, it is often seen as a place of habitation and of wealth creation (Logan and Molotch, 2007). However, in the use, construction and adaptation of housing, there is significant environmental impact (Bardhan, Jaffee, Kroll & Wallace, 2014). Australian housing contributes around 20 per cent of the country's greenhouse gas emissions (Your Home, 2013). This chapter considers environmental issues from the perspective of real estate markets and, as market facilitators, real estate agents. It does so from an Australian perspective from one of its most populated states: Victoria. Given the access to the market that agents have, they should be aware of the importance of sustainability measures in the housing market. In addition, given their position of influence, it is conceivable that they can influence the market in a positive way. This chapter explores this relationship and whether sustainability is important in the Victorian residential real estate markets.

Part 3 Management

Chapter 14 Corporate real estate management: The missing link in sustainable real estate. Christopher Heywood.

Using three ideas, this chapter reviews the state of the art for corporate real estate management (CREM) and sustainability:

1. Cadman's (2000) 'vicious circle of blame' or its virtuous antidote locates real estate occupiers (CREM) as important actors in sustainable real estate.
2. A consumption-based perspective locating CREM as the ultimate demand point in the commercial real estate system makes it critical in the operation of that system.
3. CREM as a field operates at the intersection of business and real estate systems. As such, CREM leverages both business and real estate intentions with regard to sustainability.

The chapter's considerations include occupiers' roles in sustainable real estate, occupiers' propensity to pay for sustainability and the impact of changing workplaces practices. Several provocations are also considered, including occupiers' churn (their perpetual search for new premises), being complicit in wasteful real estate practices and choosing CRE locations on eco-property considerations like carbon in business supply chains.

Chapter 15 The burgeoning influence of sustainability in managing UK retail property. Jessica Ferm and Nicola Livingstone.

This chapter assesses the strategic management considerations, drivers and challenges experienced when developing and integrating an approach to sustainable real estate in the UK retail market. Approaches to sustainable property are changing globally, and there is a growing body of research in relation to responses within the office sector, but little from the perspective of retailers. This chapter begins to address this lacuna by first reviewing the existing academic and professional literature on the drivers and barriers to progress in promoting sustainable real estate in retailing. The research then considers the perspectives of relevant actors in the real estate market, including prime retail occupiers, commercial agents and landlords, drawing on evidence collated from an initial scoping study, based on short, structured questionnaires. The research reflects on the extent to which sustainability is currently influencing decision making for retailers and the anticipated effects for real estate management moving into the future, across a spectrum of issues, from consideration of the general effects of legislation to sector-specific changes, such as changing consumer awareness and demand and online retailing.

Chapter 16 Sustainable facilities management. Paul Appleby.

This chapter sets the scene by outlining the key areas of policy and legislation that have driven the sustainability agenda in the management of non-residential and multi-residential property, with particular attention to UN and EU initiatives and their impact on countries such as the UK. This is followed by an examination of the main areas of sustainable facilities management, including modern practices in the integrated management of energy, water, waste, air quality, indoor environment, hygiene and ecology, including how these relate to facilities management and purchasing strategies.

Chapter 17 Building Energy Efficiency Certificates and commercial property: The Australian experience. Clive Warren.

This chapter explores the introduction of Building Energy Efficiency Certificates (BEEC) in Australia and their impact on energy efficiency in commercial buildings. In 2010, the Australian government evaluated a number of options in order to encourage greater energy efficiency in office properties. While some owners were using voluntary energy rating schemes, in order for Australia to meet its international greenhouse gas emission targets, greater energy savings were needed. Legislation requiring public reporting of energy efficiency has been used as a tool to encourage energy conservation in a number of countries. BEECs provide a comparison tool when buildings or substantial parts of buildings are offered for sale or lease, thus enabling potential purchasers to compare one building with another. After five years of operation, it is possible now to evaluate what potential energy savings have been achieved as a result of this scheme.

Chapter 18 Workplace ecology. Craig Langston and Abdallah Al-khawaja.

Climate change is one of the greatest environmental challenges facing the world today. In the last four decades we have seen, and will continue to see, a growing interest in ecology and increasing attention in the protection of the environment as one form of climate change mitigation. This interest can be translated into the built environment and the ecology of the workplace. This chapter makes the case for workplace 'ecology' and builds a unique framework based on the principles of environmental auditing in the context of organisation, space and technology provision. This framework highlights the relationship between organisation, space and technology domains via the quantification of workforce satisfaction, comfort and productivity, respectively, and will assist in measuring the success of business enterprise to create 'healthy' work environments for their people. The assessment of workplace ecology can be achieved by using a structured survey of participants in the workplace, at all levels of responsibility, to determine an overall consensus of satisfaction, comfort and productivity specific to an individual in the context of their job responsibility and its inherent complexity. This research makes a significant contribution to the current body of literature by exploring the nature and strength of the above relationships, as well as providing a means of assessing overall workplace performance as an arithmetic mean of individual perception. The connection between satisfaction and comfort can be alternatively defined as 'happiness', the connection between satisfaction and productivity as 'empowerment' and the connection between comfort and productivity as 'efficiency'.

Workplace ecology, just like environmental ecology, is a balance of factors that contribute to the health of an 'ecosystem' that is fundamental to corporate success and continuous improvement. This chapter develops a five-star rating system for office workplaces based on an environmental audit procedure integrating organisational, spatial and technological attributes into a novel workplace ecology model.

Chapter 19 Creating a green index based on tenant demand for sustainable office buildings and features. Spenser J. Robinson and Robert A. Simons.

This chapter reports on the creation and implementation of a new green office building rating tool. It is intended to provide a market-driven green scoring index for US office buildings that includes buildings below the level of LEED. The index draws from a series of studies summarised in this chapter, including a focus group study to define green office building features, a survey of 708 office tenant surveys for demand and pricing of individual green office building features and analysis of over 2,200 tenant leases through hedonic analysis of rent rolls. The index also includes qualitative input from institutional industry leaders on how the index could be useful to them in practice. The chapter details the creation and first steps towards implementation of the scoring system. A variety of proposed models are analysed, discussed and optimised.

Part 4 Redevelopment and adaptation

Chapter 20 Multi-stakeholder partnership promotes sustainable housing supply chain. Connie Susilawati.

All new houses in Australia have to include energy-efficient housing features to meet the National Construction Code minimum requirements, which are recorded in the local government database. The dominant household energy uses are heating and cooling, and although building designs can increase the energy efficiency and performance of a house, it is neither measured nor communicated to stakeholders. Some researchers have endorsed a building passport as a user manual to improve occupants' understanding of their house's energy efficiency and performance; however, a building passport is not currently available in the housing transaction database. Homebuyers have limited access to sustainable housing feature information and therefore may not use this information as a price determinant in their purchasing decision. It may need the active involvement of the government, financiers and property professionals to inform and promote the sustainable features. Well-informed buyers can increase the demand for sustainable housing and so impact the value of housing if they include energy-efficient and sustainable features.

Chapter 21 Scaling up commercial property energy retrofitting: What needs to be done? Tim Dixon.

Progress in retrofitting the UK's commercial properties continues to be slow and fragmented. Recent research from the UK and the USA suggests that radical changes are needed to drive large-scale retrofitting and that new and innovative models of financing can create new opportunities. This chapter draws on research (including EPSRC Retrofit 2050: 2010–2014) to offer insights into the terminology of retrofit and the changes in UK policy and practice that are needed to 'scale up' activity in the sector as a whole. The chapter also touches on some of the issues associated with scaling up commercial property action at city level. The chapter, which focuses primarily on energy retrofitting, reviews and synthesises key published research into commercial property retrofitting in the UK and the USA and also draws on policy and practice from the EU and Australia. The chapter examines key conceptualisations and characteristics of commercial property retrofitting, drawing on themes from the multi-level perspective (MLP) to identify key actors and relationships. The chapter summarises key findings from recent research and suggests that there are a number of policy and practice measures that need to be implemented in the UK (and elsewhere) for commercial property retrofitting to succeed at scale. These include improved funding vehicles for retrofit; better transparency in actual energy performance; and consistency in measurement, verification and assessment standards.

Chapter 22 Sustainable building conversion and issues relating to durability. Hilde Remøy and Peter de Jong.

Buildings account for 40 per cent of the total energy consumption in the European Union, 50 per cent of all materials extracted and 40 per cent of waste generation. In the EU, more than 1,000 km^2 of land is subject to 'land take' every year for housing, industry, roads or recreational purposes. Eighty-seven per cent of the real estate needed until 2050 is already built, and industrial and cultural heritage is threatened by obsolescence, dilapidation and demolition. It follows that much of the demand for new space has to be accommodated in the current stock. Adaptive reuse of the existing building stock and new adaptive buildings could be the key. Within office real estate, some markets have high levels of vacancy and obsolescence. New buildings drive out underperforming buildings. Ways for dealing with obsolete office stock are consolidation (doing nothing, waiting for better times or hoping others will solve the problem), demolition, within-use adaptation or adaptive reuse. Potentially, half of the existing stock can be converted to new uses; however, adaptive reuse is not taking place on a large scale. Financial gains are the most important criteria for building owners and developers considering adaptive reuse, though environmental and societal issues are becoming increasingly important. This chapter aims to develop a decision-making process for long-term real estate decisions.

Chapter 23 Sustainable urban redevelopment in the Netherlands. Erwin Heurkens.

Consensus on effective strategies and partnerships for delivering sustainable urban redevelopment projects in the Netherlands has yet to be reached. Although there is growing expectance of developing real estate in compliance with BREEAM certifications, it seems that scaling up such projects to an urban area level, taking into account far more complex social, environmental and economic issues, is a bridge too far. However, climate-adaptive and circular urban development projects are some examples of how sustainability is taking foothold in Dutch practice. This chapter explores two broad development approaches and corresponding development strategies for sustainable urban redevelopment. It compares promising and contrasting Dutch case studies in Rotterdam and Amsterdam, which serve as examples to understand how sustainable urban areas can be possibly delivered by developing formal and informal public–private relationships.

Chapter 24 Smart growth and real estate development in Saudi Arabia. Abdullah Alhamoudi and Peter Lee.

Many urban areas are facing significant changes, particularly from rapid urbanisation and economic growth. In this context, there is an increasing concern for sustainability in real estate development worldwide. For instance, there is a desire to solve real estate development issues by amending current planning policies and regulations, especially those that encourage smart growth principles. This chapter firstly reviews literature related to the smart growth approach and its ten underlying principles. Several examples from Saudi Arabia are then put forward that engage with these principles to differing extents. This is followed by a focus on several barriers to real estate development using smart growth principles. The final section looks at the direction of travel for smart growth in real estate development in order to achieve more sustainable real estate development outcomes.

Chapter 25 Applying sustainability in practice: An example of a new urban development in Sweden. Agnieszka Zalejska-Jonsson.

The concept of sustainability and the challenges of applying its principles have been the subject of a long debate. The definition of sustainability, the understanding of consequences and the values driving the decisions vary between individuals and organisations. In order to be able to implement sustainability successfully in the built environment, all stakeholders must understand the implications and share the same priorities. However, consensus in these matters is a challenge. This chapter discusses the conflicts that arise between environmental, social and economic sustainability goals from the perspective of the various stakeholders. Drawing on the experience and lessons learned from a case study of a sustainable urban development, the challenges related to the definition of sustainability and the application of the term in practice are illustrated. In the case study, attention is drawn to the risks of vagueness in the definition of sustainability in projects, as well as the consequences of disharmony in perceptions of sustainability by different stakeholders.

Chapter 26 Sustainable real estate: Where to next? Sara Wilkinson, Tim Dixon, Sarah Sayce and Norm Miller.

In this final chapter, the editors review the contents of the handbook and consider what lies ahead. They ask whether the real estate sector really can make a difference and if it can help reduce carbon emissions globally. The stakes are high: although the Paris Agreement on climate change is in place, there is now substantial uncertainty over the continued role of the USA following the election of President Trump. Moreover, research (PwC, 2016) concludes that, to prevent warming in excess of 2° C, the global economy needs to cuts its carbon intensity (tCO_2/\$m GDP) by 6.3 per cent a year, every year from now to 2100. Simply fulfilling the Paris COP21 Agreement also requires a decarbonisation rate (reduction in carbon intensity) of 3 per cent per year, which is more than double the business-as-usual rate of 1.3 per cent (2000–2014) (PwC, 2016). Responding to these immense challenges will require agility, responsiveness and ability to deal with 'wild-card' or 'black-swan' events (Dixon et al., forthcoming). The chapter highlights key findings from the four parts of this handbook and cross-cutting trends that have emerged in this book, and the short term and medium/longer term trends that will be important in shaping the sector over the next 10–20 years.

1.3 Conclusions

In detailing the contents of this handbook, it is apparent that the scale of issues that the real estate industry faces is large. However, although only recently recognised, the issues are not new. Successful public health programmes, the industrial and technological revolutions of the nineteenth century and rising real wealth have all combined to place pressures on the planet, but it was not until the seminal works of authors such as Carson, referred to above, and, critically, the publication of the Brundtland Report in 1987 (Brundtland, 1987) that governments and corporations began to register the urgency of the agenda. However, recognising an agenda and finding solutions are two very different things.

From a real estate perspective, governments, professional bodies and organisations around the world have taken different approaches towards addressing environmental issues. While most governments have taken a strong lead in respect of new build standards, the recognition of the role of retrofitting of existing buildings is only now beginning to take centre stage. In some cases, such as the UK and Australia, a start is being made on imposing minimum energy standards for existing stock (Warren, Chapter 17) but primarily the ambition is that improving our stock should be market-led and many aspects of how this is now happening are explored throughout the book. So, Chapter 4 (Van der Heijden) explores the role of voluntary certification as a powerful tool, while both Warren-Myers (Chapter 7) and Sayce (Chapter 9) recognise the role of valuers in supporting the green value case (see also Dalton and Fuerst, Chapter 12). Parker (Chapter 10) and Sah and Miller (Chapter 11) explore the aspect of funding, which is crucial to investment in retrofit, many aspects of which are explored, notably in relation to successful initiatives, such as those detailed in Chapter 5 (Wilkinson), Chapter 23 (Heurkens) and Chapter 25 (Zalejska-Jonsson).

It is perhaps invidious to single out the actions of individual stakeholder groups or initiatives in driving the required transformation. We have attempted to draw through perspectives that call on different scales and different stages of the building life cycle. Collectively, we hope that the handbook will provide a lens on how a shared understanding built on better knowledge can be influential in addressing some of the challenges set out at the beginning of this chapter. It does not present complete solutions – far from it; but we hope it represents perspectives to educate, inform and provide a platform for future debate.

Note

1 HC Deb 28 October 1943 vol 393 cc403–73.

References

Baker & McKenzie (2016) *Global Sustainable Buildings Index 2016*. Baker & McKenzie, London. (Accessed December 2016: http://globalrepublications.bakermckenzie.com/sustainabilityindex/).

Bardhan, A., Jaffee, D., Kroll, C. and Wallace, N. (2014) 'Energy efficiency retrofits for US housing: Removing the bottlenecks', *Regional Science and Urban Economics*, 47, 45–60.

Berardi, U. (2013) 'Clarifying the new interpretations of the concept of sustainable building', *Sustainable Cities and Society*, 8, October, 72–78.

Better Buildings Partnership (2010) *Low Carbon Retrofit Toolkit: A Roadmap to Success*. BBP, London. (Accessed May 2012: www.betterbuildingspartnership.co.uk/download/bbp_low_carbon_retrofit_toolkit.pdf).

Brundtland, G. H. (1987) *Report of the World Commission on Environment and Development: Our Common Future*. United Nations, New York.

Carson, R. (1962) *Silent Spring*. Houghton Mifflin, Boston, MA.

Change, I. C. (2014) Impacts, adaptation, and vulnerability. Part B: Regional aspects. *Contribution of Working Group II to the Fifth Assessment Report of the Intergovernmental Panel on Climate Change*. Cambridge University Press, Cambridge and New York.

Dixon, T. (2009) 'Urban land and property ownership patterns in the UK: Trends and forces for change', *Land Use Policy*, 26, Supplement 1, 43–53.

Dixon, T. (2010) 'Sustainable commercial property', in Calvello, A. A. (ed.) *Environmental Alpha: Institutional Investors and Climate Change*. Wiley, London, pp. 265–296.

Dixon, T. (2017) '"City-wide or city-blind?" An analysis of retrofit practices in the UK commercial property sector', in Eames, M., Dixon, T., Lannon, S. and Hunt, M. (eds) *Retrofitting Cities for Tomorrow's World*. Wiley, London.

Dixon, T., Green, S. and Connaughton, J. (forthcoming) *Sustainable Futures in the Built Environment to 2050: A Foresight Approach to Construction and Development*. Wiley, Oxford.

Dodge Data & Analytics (2016) *World Green Building Trends 2016*. Dodge Data & Analytics, Bedford, MA. (Accessed December 2016: http://fidic.org/sites/default/files/World%20Green%20Building%20Trends%202016%20SmartMarket%20Report%20FINAL.pdf).

Hartless, R. (2004) *ENPER-TEBUC Project Final Report of Task B4 Energy Performance of Buildings: Application of Energy Performance Regulations to Existing Buildings*. Building Research Establishment, Waterford.

IEA (2011) 'Analysis of buildings energy policies in the IEA countries'. Grantham Institute/International Energy Agency Workshop: The Reduction of Global Carbon Emissions in the Building Sector to 2050. IEA, London.

Jennings, M., Hirst, N. and Gambhir, A. (2011) *Reduction of Carbon Dioxide Emissions in the Global Building Sector to 2050*. Imperial College, London Grantham Institute for Climate Change Report GR3.

Logan, J. R. and Molotch, H. (2007) *Urban Fortunes: The Political Economy of Place*. University of California Press, Oakland, CA.

PwC (2016) *Low Carbon Economy Index*, October 2016. (Accessed November 2016: www.pwc.com/hu/hu/kiadvanyok/assets/pdf/low_carbon_economy_index.pdf).

Savills (2016) *Around the World in Dollars and Cents*. Savills, London. (Accessed December 2016: www.savills.co.uk/research_articles/188297/198667-0).
Sayce, S., Ellison, L. and Parnell, P. (2007) 'Understanding investment drivers for UK sustainable property', *Building Research & Information*, 35(6), 629–643.
Stafford, A., Gorse, C. and Shao, L. (2011) *The Retrofit Challenge: Delivering Low Carbon Buildings*. Centre for Low Carbon Futures, Leeds.
UKGBC (2008) *UK Green Building Council Consultation: Code for Sustainable Buildings Task Group*. UKGBC, London.
UNEP (2009) *Buildings and Climate Change: Summary for Decision Makers*. UNEP, Nairobi.
WEF (2016) *Environmental Sustainability Principles for the Real Estate Industry*. World Economic Forum, Geneva. (Accessed December 2016: www3.weforum.org/docs/GAC16/CRE_Sustainability.pdf).
Your Home (2013) 'Energy'. (Accessed February 2016: www.yourhome.gov.au/energy).

2

Sustainable real estate and corporate responsibility

Norm Miller and Dave Pogue

2.0 Introduction

Commercial real estate sits at the nexus of many of the most complex issues facing the world today. In a resource-challenged world, the development and occupancy of real estate has captured the attention of environmentalists, regulators, product designers, architects, investors and occupants alike. By many measures, commercial real estate is one of the most important and visible components in the war on waste and carbon dioxide emissions. Data provided by the United States Green Building Council (USGBC) identifies commercial real estate as responsible for a substantial proportion of the nation's water use, waste, CO_2 emissions and electricity use (see Figure 2.1).

As corporations are faced with a growing demand to understand, mitigate and report on the environmental impact of their activities, statistics like these command attention, creating both challenges and opportunities. How corporations specifically address these through their real estate choices will be the focus of this chapter.

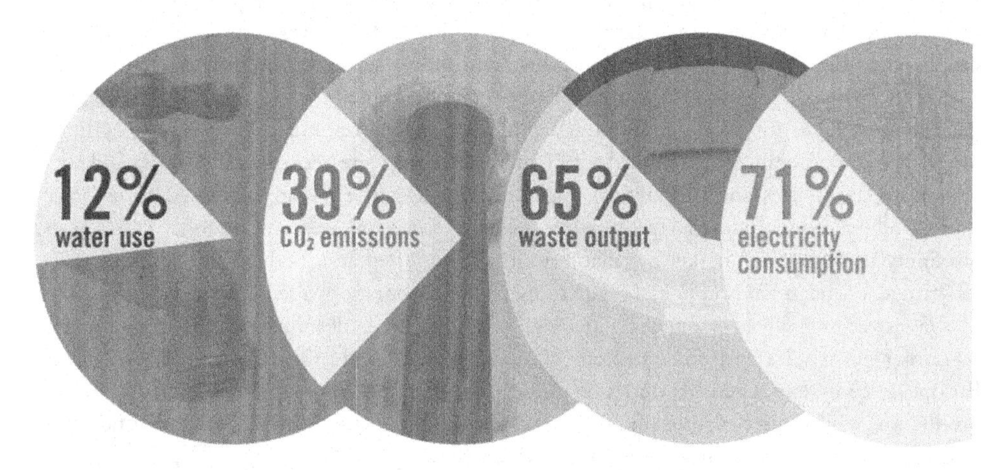

Figure 2.1 USGBC approximate range of building effects, which vary by use, 2016

Source: USGBC (2016)

Milton Friedman, Nobel Laureate and noted free market economist, famously declared that "there is one and only one social responsibility of business—to use its resources and engage in activities designed to increase its profits" in his 1962 book *Capitalism and Freedom* (Friedman, 1962, 55), later made available to a broader audience in a *New York Times* article published in 1970 (Friedman, 1970). Much has occurred since that time, and most commercial real estate players now believe that their responsibilities are significantly broader and more complex in a rapidly changing and more transparent world economy. In a context of resource depletion, environmental degradation, 24/7 news cycles and vigilant social media, corporations today realize that their social responsibilities run much deeper and encompass multiple aspects beyond shareholder value and profits (Deloitte Center for Financial Services, 2017). The topic is made even more complex by the rapid globalization of the world economy and the growth of very powerful multi-national corporations operating around the world (Weinreb, 2017). In this chapter, we will examine many of these issues and discover that many of the central elements of sustainability actually occur within the confines and context of the commercial real estate owned and occupied by these corporations.

2.1 Corporate social responsibility evolved

As noted above, the concept of corporations assuming responsibility for anything beyond their own profitability is relatively new. Though far better than the robber baron days of an earlier century, corporations of the mid-twentieth century were still primarily focused on the bottom line. At the same time that Friedman was espousing profit-first corporate thinking, a rising environmental consciousness was beginning to emerge (Beeman, 1995). In 1970, the same year as Friedman's article, the United States federal government established the Environmental Protection Agency (EPA). Corporations and others could no longer indiscriminately discharge pollutants into the air. This was followed by the Clean Water Act in 1972, extending similar restrictions to the nation's waterways. 1970 saw the introduction of Earth Day, which led to even more people beginning to embrace environmentalism. Further, the introduction of the Endangered Species Act of 1973 added an additional important piece to the environmental regulatory structure, and, taken together, these acts and actions helped make the decade of the 1970s one of the most significant in history for environmental action (Encyclopedia.com, 2017a). Importantly, corporations now had to consider significantly greater responsibilities beyond simple profitability. Protecting the environment was now a legal responsibility as well.

As this was occurring in the US, another important phenomenon was beginning to accelerate: the globalization of the world economy (World Trade Organization, 2008). At the beginning of the 1980s, corporations based in the US and elsewhere were seeking ways to expand their operations, both sales and production, into all corners of the globe. US-based firms began aggressively moving industrial production overseas for many reasons, such as labor costs, proximity to expanding markets, availability of raw resources and, perhaps not coincidentally, less stringent foreign environmental regulations. Other corporations also began to take advantage of the expanding world economy. Japanese firms began to dominate world markets with consumer electronics and efficient auto choices (World Trade Organization, 2008). Many European firms expanded around the globe, emphasizing innovative design and quality production while entire regions and nations exploited inexpensive labor for manual tasks and dominated world markets in such items as textiles, clothing and assembly.

This transition to a world economy created significant benefits to consumers and corporations alike. Access to cheaper products allowed more people to acquire more material goods and improve their living standards (Griswold, 2000). The demand for these goods also increased demand for labor in many emerging economies, providing jobs and income to millions of people, propelling many of them into the growing "middle class" who were in turn able to acquire more material goods and improve their lives. And the global firms participating in this new economy were also thriving, bringing expanded profits and shareholder wealth.

However, the expansion in the world economy did not come without price or consequence. In many ways, the decisions to manufacture goods offshore were decisions to also offshore significant environmental risks (Short, Rice and Lindeberg, 2001). No single episode more starkly revealed this than the 1984 Union Carbide disaster, still considered the world's largest industrial accident, in Bhopal, India (Hart and Milstein, 2003, 25). During the nights of December 2–3, deadly toxins were released that exposed roughly 500,000 people to methyl isocyanate. The death toll remains unknown, but estimates were as high as 20,000 (Encyclopedia Britannica, 2016). Questions still linger regarding work conditions, safety measures and Union Carbide's lack of response to earlier accidents and warnings (Ice, 1991). Catastrophic environmental failures, even a world away, resonated with an American audience. A second seminal incident occurred on March 24, 1989 as a heavily laden oil tanker, the *Exxon Valdez*, ran aground at Prince William Sound on the Alaskan coast, spilling as much as 11 million gallons of crude oil and spoiling nearly 285 miles of coastline, killing hundreds of thousands of sea animals and birds while severely impacting the local economy for years (Short et al., 2001). Each of these examples demonstrates how a significant environmental accident can seriously damage the responsible corporation's reputation along with shareholder value.

People began examining the question of a corporation's broader responsibilities to more carefully conduct their business and were looking for a way to understand and judge a corporation's actions in a number of important areas, such as environmental impact, corporate governance, social welfare, diversity and corporate giving. As more and more consumers became personally concerned about these issues and came to realize that their own purchasing decisions had consequences, they began demanding the companies they supported through those purchases share their same values and behave in a more broadly ethical way (Encyclopedia.com, 2017b).

The 1980s also saw new regional economic alliances formed to provide additional support to their industries and economies. Evolving from its formation in 1957, the European Economic Community (EEC), commonly called the European Common Market, had grown from its original six member states to a dozen by 1987 (European Commission, 2016a). The year prior, the Single European Act was signed, ultimately leading to the establishment of the European Union through the Maastricht Treaty, signed in 1992 and effective November 1993. The European Union created a complex relationship between originally 12 European nations, including a common currency and myriad trade, production and economic standards (Euro, 2016).

While the European nations were creating the EU, the North American nations of Canada, Mexico and the United States created the North American Free Trade Agreement (NAFTA) to similarly eliminate trade barriers and make commercial globalization easier. The agreement was politically sensitive in each nation, particularly with labor and environmental groups, but was ultimately approved and went into effect January 1, 1994, only two months after the EU was formally in place (US Customs and Border Protection, 2016a). Two key elements of the NAFTA agreement were the North American Agreement on Labor Cooperation (NAALC) and the North American Agreement on Environmental Cooperation (NAAEC) to protect both workers and the environment (US Customs and Border Protection, 2016b).

The 1990s also first saw the expanded concerns around global climate change and the role of carbon emissions in accelerating the warming of the planet. While world scientists had been examining this phenomenon since the early 1980s, the formation of the Intergovernmental Panel on Climate Change (IPCC) in 1988 by the World Meteorological Organization (WMO) and the United Nations Environment Programme (UNEP) gave a coordinated mission to the efforts. The initial task for the group was to "prepare a comprehensive review and recommendations with respect to the state of knowledge of the science of climate change; the social and economic impact of climate change; and possible response strategies and elements for inclusion in a possible future international convention on climate" (IPCC, 2016). The first IPCC assessment report of 1990, published in 1991, established the fact that carbon emissions from human activity were accelerating the concentration of atmospheric carbon dioxide (CO_2) and that both global temperatures and sea levels would rise significantly over the next century unless steps were taken to address the issue (IPCC, 1991). Since this initial report, the IPCC has issued four additional assessments, each with more detailed concerns and greater certainty of both human involvement and the importance of proactive measures to mitigate (Union of Concerned Scientists, 2017).

Corporations, particularly those engaged in any form of manufacturing, were again challenged to address their operating practices as now it had been determined that most acts of production, in as much as they include the use of fossil fuels, are in themselves a major factor in global climate change (Semans, Juliani and de Fontaine, 2007). The globalization of business has created significant opportunities for growth and success, but with those opportunities have also come significant new operating realities that call for new ways of thinking, new standards of operating and new ways of communicating with stakeholders.

Firms began developing strategies around these concerns and began to issue public documents describing their efforts along with their financial reporting. For example, Forest Trends Ecosystem Marketplace announced that familiar names such as L'Oréal, General Motors and Delta Airlines are beginning to "offset" as a part of a carbon reduction strategy (GreenBiz, 2016). Others such as Barclays, Disney, Microsoft and Swiss Re are charging their business divisions a fee based on the emissions they emit, which "incentivizes them to reduce their carbon footprint while also raising money that then can be reinvested in energy efficiency." Although some firms had been including modest environmental sections in their annual reports since the 1970s, these efforts were generally viewed as self-promotional and were not specifically linked to corporate performance (World Bank, 2014). Ben & Jerry's commissioned a "social auditor" in 1989 to work with their staff to create what was called a "Stakeholders Report" on their 1988 activities. The report divided the stakeholders into five separate groups, including communities, customers, employees, suppliers and investors and looked at such diverse issues as community outreach, philanthropic giving, and environmental awareness and global awareness, many of which are still commonly used as foundational reporting protocols today. The B&J effort is generally believed to be the first such report ever published in the United States.

Other firms began publishing statements and reports during the 1990s. Many energy and chemical firms were eager to demonstrate their environmental concerns and practices, and one of the first firms credited with reporting their efforts was Shell Canada in 1991 (Maharaj and Herremans, 2008). Although more and more firms reported their efforts, there still was no official format and the reporting lacked any third-party verification or certification. By the end of the 1990s, new groups formed and formats developed to provide such verification and certification, including Social Accounting International (SAI) and the Global Reporting Initiative (GRI), both founded in 1997; the Forest Stewardship Council; the International Federation of Organic Agriculture; and the Dutch Max Havelaar Foundation, among others. GRI has become one of the more prominent sustainability reporting protocols among larger corporations. Their first reporting year was 2000, and during 2001, 14 firms reported on their efforts. By 2013, that number had grown to more than 4,100 worldwide (KPMG, 2013).

In addition to responding to these reporting protocols today, thousands of companies across all sectors undertake the production and publication of a comprehensive report detailing their efforts in the areas of corporate responsibility (GRI, 2017). These reports generally focus on those same issues chosen by Ben & Jerry's in their original report in 1989, including community outreach, philanthropic giving and environmental stewardship, as well as corporate governance and diversity.

2.2 Modern corporate social responsibility: Measuring and reporting to multiple stakeholders

As corporate social responsibility (CSR) reporting has become more and more sophisticated and the scrutiny on corporate behavior even more pronounced, a more targeted and nuanced approach to corporate behavior and strategy has begun to gain traction among some of the most proactive corporate entities. In an attempt to not only measure and mitigate social and environmental costs inherently caused by their corporate actions, in other words to "do less bad", a number of firms have begun to actively seek ways to "do more good" through a more thoughtful approach to their business (Moore, 2014). Dubbed "shared values", this approach attempts to not only improve upon the way a business process has been done, but to completely re-engineer the thinking within the company to move from a process that is harmful to the local environment and population to one that actually benefits all stakeholders (Porter and Kramer, 2011).

One example is Coca-Cola, who have pledged to become water neutral by 2020 (Coca-Cola Company, 2012). Since Coca-Cola are essentially a reseller of water in various forms, the concerns of impending water shortages and spreading drought, exacerbated by global climate change, have serious consequences for their future business. In addition to significantly reducing water use during processing, they are working in local communities to improve the water source and supply through various programs of sanitation, recycling, watershed management and water use efficiency (Coca-Cola Company, 2017).

A second example is Starbucks. One of the leading causes of carbon dioxide emissions is deforestation, and a great deal of previously forested land has been put into coffee production to satisfy the significant growth of this firm. Recognizing the impact their production has had in many tropical regions, Starbucks began seeking better ways to grow and produce their coffee. They first developed and implemented a climate change mitigation strategy in 2004. This strategy involves a variety of actions, including the introduction of more shade-grown varieties, the restoration and protection of natural habitat and the process of capturing the carbon value of those protected assets (Starbucks Coffee Company, 2016).

As the importance of carbon emissions and greenhouse gases (GHG) in global climate change also became clearer, a heightened focus was put upon reporting a firm's carbon output, or "carbon footprint", as well. In 2012, in the wake of the British Petroleum Gulf of Mexico oil spill, the United States Environmental Protection Agency started the Greenhouse Gas Reporting Program in which 85 percent of the nation's leading emitters are required to report their GHG emissions (EPA, 2016). This number will rise over time to include all of the largest emitters. Also, as of 2012, the 1,100 firms publicly traded on the London Stock Exchange were required to annually report their GHG emissions. And throughout the EU, there is a requirement for all industrial installations and aircraft operators to monitor and report their annual emissions (European Commission, 2016b). Taken together, the need for greater accountability, mitigation and transparency around carbon and GHG emissions is becoming a more important aspect of corporate compliance strategy.

As with CSR reporting, new formats and protocols have emerged to guide and oversee the reporting of corporate carbon activities as well. One of the most prominent is the Carbon Disclosure Project (CDP). Founded in London in 2000 in response to the Kyoto Accord, CDP has emerged as the leading protocol for reporting carbon activities. In 2015, more than 3,000 firms worldwide reported their activity through CDP, including 822 institutional investors holding $95 trillion in assets (CDP, 2016). In 2012, all but one firm in the US S&P 500 reported their sustainability and carbon activities through either GRI or CDP, and many reported through both (CDP, 2016).

Adding further urgency to the reporting requirements for listed US corporations was the issuance of guidance from the Securities and Exchange Commission (SEC) on February 10, 2010, regarding reporting requirements related to global climate change (Haque, 2015). Citing various international, federal and state legislation and regulations as well as actions by the insurance industry to mitigate their potential financial losses from potential climate-related catastrophes, the SEC was now requiring that publicly traded firms in the US include a discussion about the potential impacts, both positive and negative, of global change upon their business as part of their public financial reports. As a guideline, the SEC suggested reporting upon the impact of legislation and regulations, international accords, consequences of business trends and the physical impacts of climate change. They recognized that some of these consequences could actually be beneficial to certain firms, and the guidance specified that such positive opportunities be enumerated as well.

As companies respond to the increasing and shifting challenges of measuring, mitigating and reporting on these various social and environmental responsibilities, they are faced with the challenge of how best to accomplish this. In most cases, the traditional corporate structure did not address the issue, and often there was no single person within the firm properly trained or experienced to create, monitor and report on a program sufficient to meet these new demands. Many firms have started by engaging an outside consultant to aid in the process. A number of the larger consulting firms, including most of the larger accounting firms, such as KPMG's Sustainability Services Group and PwC's Sustainable Business Solutions Group, have developed sophisticated specialty units just for this purpose.

Among the larger corporations, however, most have chosen to direct their own efforts, although often with the assistance of consultants as advisors. Over time, most have now created the position of director of CSR or chief sustainability officer. Each firm must decide where this position will sit within the organization and to whom it will report. More and more often, as sustainability and CSR gain in importance, the person in this position is now either one of the corporation's most senior executives or reports directly to them. Although there are no fixed rules, in most instances the structure, reporting lines and authority are directly related to the nature of the company, the specific industry involved and the relative importance each factor places on the topic. Oil and other extraction-related firms, large multi-national consumer products companies and manufacturing companies with long and complicated supply chains have the most risk and perhaps the greatest reason to actively embrace this topic.

Although the position of director of CSR is evolving as the general topic is evolving, this position is primarily responsible for leading the firm's efforts in establishing a roadmap to achieve the desired outcomes. Although each firm must decide what they want to achieve and how to accomplish that, most firms will follow a similar path to establish their road map. The specific steps are as follows:

- *Values*. Most firms already have a clear understanding of their corporate values and the development of a sustainability program or larger all-encompassing CSR program must emanate from this core corporate ethos.

- *Principles.* From the broad statement of values come more focused principles that guide the daily activity of the firm. The alignment of the details of a sustainability or CSR program must always consider the existing corporate principles.
- *Goals.* The setting of short- and long-term goals begins to give structure to the process. Establishing clear outcomes with milestone events helps keep the process on track.
- *Strategies.* Once goals have been agreed to, the firm can develop a specific strategy on how to achieve them.
- *Tactics.* This is the practical application of the strategy: who, what, when and how. Campaigns, programs, funding and communication plans are all included here.
- *Metrics.* No program can be successful without measurement and verification. Generally over time, the quality and quantity of things measured becomes more sophisticated and meaningful.

As the position of director of CSR has evolved, the career path of people engaged has also evolved. Many currently in this position started somewhere else within the firm and transitioned, oftentimes due to either personal interest or a specific early success. Curiosity and commitment to sustainability often led to suggestions inside a firm that led to new positions and strategies. In the future, it will be more common to see professionals being directly hired from universities with specialized sustainable business programs, as has already started to take place in the US.

The process of developing and implementing a specific corporate sustainability strategy is often complex and dynamic. As a firm moves through this process, they often find that their attitudes and approaches change and events unfold. There are several examples of how firms typically transition along the path of sustainability through various identified stages.

The authors conclude that "there remains disagreement among managers regarding the specific meaning of and motivation of enterprise level sustainability. For some managers it is a moral mandate, for others a legal requirement. For still others sustainability is perceived as a cost of doing business—a necessary evil to maintain legitimacy and the right to operate. A few firms have begun to frame sustainability as a business opportunity offering avenues for lowering cost and risk or even growing revenues and market share through innovations. All of these aspects are balanced in their model with the nexus being sustainable value" (Hart and Milstein, 2003).

Another model is provided by McGraw-Hill (now Dodge Data & Analytics) in their collaborative work with Siemens in *2012 Greening of Corporate America: The Influence of Sustainability on Transforming Business Strategy* (McGraw-Hill and Siemens 2012).Their model shows a five-stage progression, from an initial stage at which the corporation perceives sustainability as a cost with no part in the organizational mission, through a middle stage at which sustainability is perceived as consistent with the profit mission of the firm, to the final stage at which sustainability is an integral part of the organization at every level. McGraw-Hill and Siemens conducted a six-year longitudinal survey of US corporate attitudes surrounding sustainability and asked their respondents over time to rate themselves along this line of progression. The survey included 203 responses, all from firms with more than $250 million in revenue (McGraw-Hill and Siemens, 2012). They describe the stages of sustainability in Figure 2.2.

Several interesting facts and trends emerged from the 2012 survey. The first is the status and title of the actual respondent. The first survey in 2006 did not even include a response category for chief sustainability officer (CSO). By 2009, 22 percent of the respondents described themselves as a CSO. By the 2012 survey, that number had increased to 32 percent. This would seem to mark a significant shift in the importance corporate America now places in this role. Second, far fewer firms, only 2 percent, now consider their efforts at sustainability at the lowest stage of commitment (Stage 1). And while the number of firms considering themselves at the middle

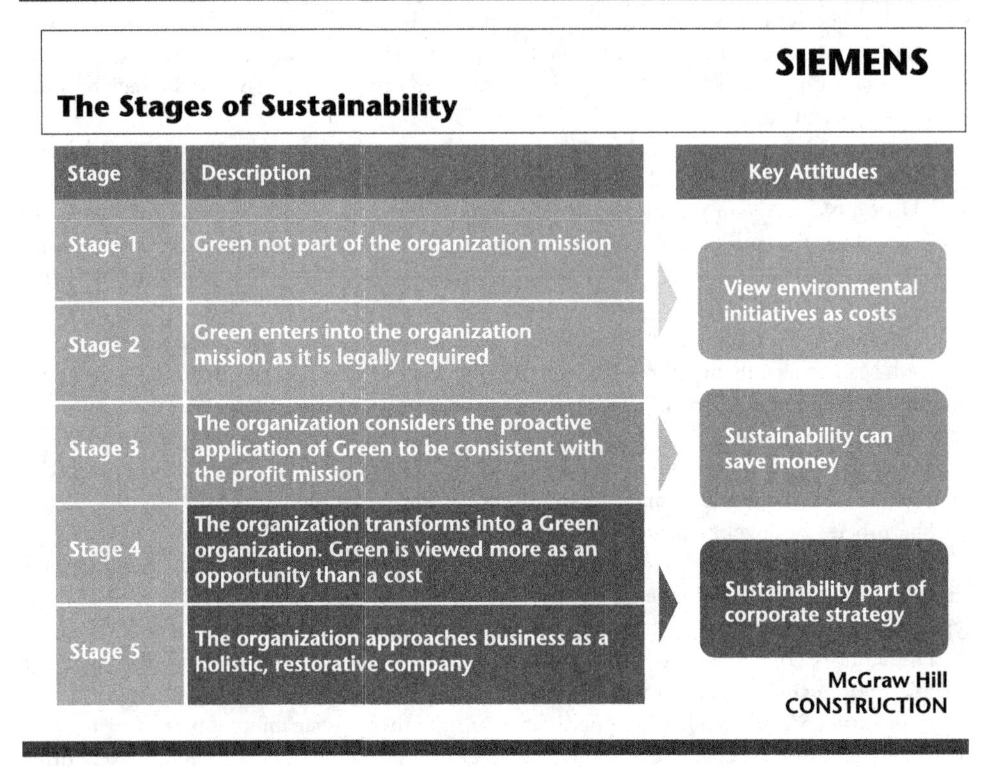

Figure 2.2 The stages of sustainability

Source: McGraw-Hill and Siemens (2012)

stage (Stage 3) has not changed at all, the numbers regarding themselves as being at Stages 4 and 5 have each significantly increased, as now more than a third of firms consider sustainability as part of their business model while another 9 per cent declare that sustainability is a fully integrated part of their operations (McGraw-Hill and Siemens, 2012).

Another survey conducted and released at the end of 2011 by McKinsey & Company (2011) examined similar issues and questions around corporate sustainability practices in a global context. This survey asked more than 3,200 global firms a variety of questions about the role sustainability plays in their organizations, including motivations, actions, commitments and resources provided. This was also a longitudinal study and followed a similar research project finished in 2010. As did McGraw-Hill and Siemens, McKinsey found a generally growing commitment around sustainable practices and a belief that good sustainable practices make a positive contribution to a firm's long- and short-term value. This study looked at the most important drivers of sustainable actions and found that the number one reason was economic, but was followed closely by reputational benefits. Alignment with corporate goals and new growth opportunities rounded out the top four.

As sustainability and sound environmental practices have increased over time, there has been an expansion in the benefits sought and claimed for these practices. As noted in the McKinsey

% of respondents[1]

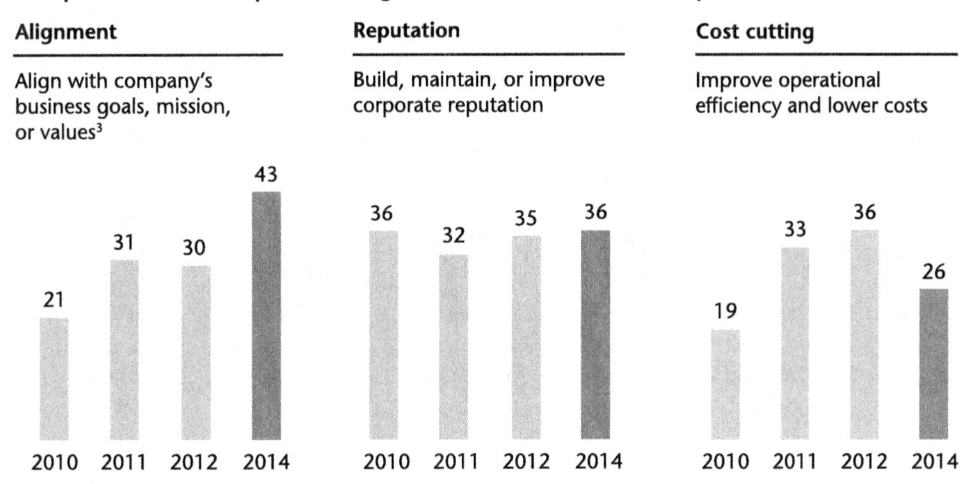

[1] In 2010, n = 1,749; in 2011, n = 2,956; in 2012, n = 3,847; and in 2014, n = 2,904. The survey was not run in 2013.
[2] Out of 12 reasons that were presented as answer choices in the question.
[3] From 2010 to 2012, the answer choice was "Align with company's business goals."

Figure 2.3 Top three reasons that respondents' organizations address sustainability

Source: McKinsey & Company (2011)

study, economics are always a top consideration, as is compliance with regulatory requirements. More frequently now, however, firms will also cite a number of other expected benefits, including reputation, business alignment, recruitment and retention of employees and improved working conditions for employees as additional benefits gained. The McKinsey study also revealed that more than two-thirds of corporate CEOs now include sustainability as a top agenda item, with one-quarter describing it as a top three agenda priority. See Figure 2.3 below and also Figure 2.4, which lists out the types of efforts by category reported by McKinsey in 2011.

One of the more interesting aspects of the McKinsey study was their efforts to determine which firms, by industry, were actually most engaged in sustainability, a group they called "sustainability leaders". To qualify for this designation, the firm needed to successfully meet three criteria. The first was that the CEO declared that sustainability was in their top three agenda items, the second was that they had a formal program to address this issue and the third was that they believed that they were effective at accomplishing their goals.

Perhaps surprisingly, the industries most represented were extraction, transportation and manufacturing. In other words, the type of firms some might think most responsible for environmental damages are in fact the ones most actively addressing this issue. Questions remain unanswered as to why this is so. Are these firms focused on their reputation and perhaps regulatory pressures, or are they genuinely dedicated to developing better processes and practices to proactively address the negative impacts of their businesses? Is this an example of a firm's attempt to "do less harm", or can we see examples of shared value where firms are proactively seeking better ways to produce and deliver their products and services?

A 2015 survey by Dodge Data & Analytics found a deceleration in the commitment to CSR, with only 32 percent of firms exploring involvement in sustainability strategies and only

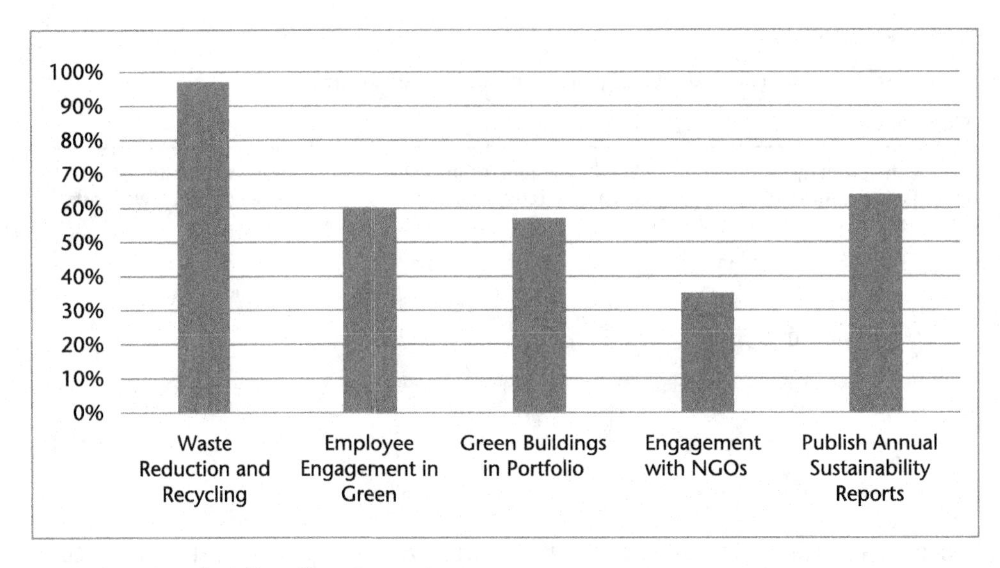

Figure 2.4 Sustainability efforts by category.

Source: McKinsey & Company (2011)

23 percent who were largely or fully committed. In 2017, with President Trump appointing an EPA director (Scott Pruitt) who has largely opposed the EPA, we will see little pressure from the federal government for greater environmental responsibility in the near term (McKinsey & Company, 2012).

As earlier noted, commercial real estate has been cited as an important focus of many of the issues related to and raised around sustainable activities. Many firms have begun to identify their occupied space as not only a place where they can make a meaningful difference through more sustainable choices, but also as a way to publicly demonstrate their concerns and actions, such as Toyota, who recently opened their new North American headquarters in Plano, Texas that was built to be powered almost entirely by renewable sources (Ecotech Institute, 2016). As the identification and rating of "green" buildings has become more prevalent through such means as EPA Energy Star ratings/labels and USGCB or other green building protocols, some companies have become more and more interested in seeking those type buildings. This was demonstrated in the findings of both the McGraw-Hill (now Dodge Data & Analytics) and CBRE/USD/CoreNet surveys of corporate executives referenced in two studies (McKinsey & Company, 2011). The Dodge study found an increasing commitment to seeking green building choices through the 2014 survey, but a decline as of 2015. The CBRE/USD/CoreNet study found an even higher response, with nearly 70 percent of respondents planning to pursue green building certifications for all or some of the buildings they own. But this was based on a survey that is now dated, and perhaps, as in the more recent Dodge survey, commitment has not increased as expected.

Focusing on green building choices allows a firm to make a very public statement about their commitment to sustainability. While often firms may not choose to pursue certified space for all their offices, they often will visibly choose such a building as their corporate headquarters site. This then often becomes part of their ethical CSR reporting package and can also play a role in recruiting and supply chain functions as these divisions often occupy space in the headquarters facilities.

Unless tethered to a coordinated and well-communicated corporate strategy of sustainable activities, this decision in isolation, as well as other disjointed efforts, can result in charges of "greenwashing", the deceptive practice of promoting the perception that a firm's actions or products are environmentally friendly when in fact they are not. The US, through the Federal Trade Commission, currently provides only voluntary guidelines for environmental marketing claims, often leaving the public confused and ultimately creating skepticism among consumers. Greenwashing takes a variety of forms. Underwriters Laboratories (UL) has identified the "seven sins of greenwashing", which comprise the sin of the hidden tradeoff, the sin of no proof, the sin of vagueness, the sin of worshiping false labels, the sin of irrelevance, the sin of the lesser of two evils and the sin of fibbing (UL, 2010).

Some examples of greenwashing include the following:

- Clairol's Herbal Essences, which claim a "truly organic experience" but contain sulfate, propylene, glycol and D&C red number 33, which are not really that organic;
- food products such as cereals whose boxes claim they have vitamins, antioxidants and fiber but contain pesticide residue in their vitamin C-packed berries;
- Dean Foods' switch from "organic" to "natural", an unregulated and relatively meaningless term, and Dean's decision not to inform major retailers of the switch;
- the "green" hotel industry, which attracts eco-tourists willing to pay for environmentally responsible accommodations—a recent independent study by TerraChoice Environmental Marketing found that 99 percent of all products and hotel services that call themselves "green" do not live up to their claims (Knufken, 2010).

Real estate decisions, like all others, should be part of a comprehensive, coordinated corporate strategy aimed at identifying, mitigating, reducing and reporting on a corporation's environmental impact. Transparency and a genuine effort to make measurable progress over time is the ultimate goal.

2.3 Conclusions

Careers in CSR and sustainability in general are exploding. According to the Center for Career Education at Columbia University, CSR is the new big idea in the business world that has developed into a field with "substantial and diverse career opportunities" (Columbia SIPA Office of Career Services, 2014). Directors of sustainability must focus on real estate and internal operations to do their jobs well, and a lot of the work involves measuring, monitoring and reporting. The early professionals had to write their own job descriptions, but universities and professional organizations are sharing resources and starting specialized educational programs. The future for those who want to do something beneficial to society and for themselves is bright.

Appendix

There are several organizations that each provide tools and scorecards for reporting. Some of these are listed in Table 2.1 below, which describes those known and established within the United States and Canada.

Table 2.1 A sample of available report cards for CSR, GHG and sustainable objectives

Organization	*Measuring System*	*Primary Focus*	*Website*
An offshoot of the United Nations with base in Amsterdam, this Global Reporting Initiative group was founded in Boston in 1997.	GRI: Global Reporting Initiative	Governance, ethics and integrity, anti-corruption, labor practices, GHG emissions, environmental impacts.	www.globalreporting.org
CDP: Carbon Disclosure Project	Reports by topic are aggregated by the CDP to country and other groupings.	Climate change, water, deforestation and GHG emissions.	www.cdp.net/en-US
Clinton Climate Initiative	Not a formal system, but with awards for outstanding work.	Climate Change, Cities and C40 Project.	www.clintonfoundation.org
Cradle to Cradle Innovation Institute	Five categories: material health, material reutilization, renewable energy and carbon management, water stewardship and social fairness.	Products as designed and manufactured.	www.c2ccertified.org/ and www.mbdc.com/cradle-to-cradle/cradle-to-cradle-certified-program/certification-overview
Green Globes (USA and Canada) grew out of BREEAM and CSA	Flexible system for new or existing structures.	Design, operations and management, building intelligence.	www.greenglobes.com
Living Building Institute	Net zero energy, waste and water over a 12-month period.	All buildings: the most rigorous of all standards.	http://living-future.org/lbc
ULI Greenprint Center for Building Performance	Greenprint Performance Report	Energy use intensity, GHG emissions and water use.	See Greenprint under www.uli.org
US Environmental Protection Agency	Energy Star: relative scoring compared to historical benchmarks resulting in a 0 to 100 score.	Energy consumption of buildings and products in operation. Need a 75 or higher score to achieve label status.	www.energystar.gov
USGBC	LEED with numerous programs for new construction, existing buildings, core and shell, commercial interiors, neighborhoods, schools and more.	A 100 plus point system that results in a possible score from certified, silver or gold to platinum based upon sustainable sites, water efficiency, energy and atmosphere, materials and resources, and indoor environmental quality.	www.usgbc.org

Note: The genesis of the system was the UK's Building Research Establishment's Environmental Assessment Method (BREEAM), which launched in 1990. In 1996, the Canadian Standards Association (CSA) published BREEAM Canada for Existing Buildings.

Interviews

Interview 2.1: Brendan Owens, Chief of Engineering for the USGBC

1. *Why has LEED been so successful?*

   I don't think there is a simple answer to this question. From my perspective, LEED's success is a combination of a commonly shared and well-articulated mission, a pragmatic and integrated approach to solving a problem, perfect timing and luck. Breaking that down a bit: buildings are part of the problem, so they're part of the solution. LEED was introduced into the market at a time when practitioners were ready to understand and internalize concerns over sustainability and energy and engage in a more integrated, interesting and fun conversation about their work. LEED presented a framework that was both simple and actionable but also, and in my opinion more importantly, provided building practitioners with a reminder of the greater good our collective work can serve. We (USGBC) also enjoyed a bit of luck along the way. Anybody who thinks it was their good decisions and only their good decisions that make anything like LEED happen is deluding themselves.

2. *Do you think that the Government Services Administration (GSA) now approving Green Globes as an acceptable standard will impact the use of LEED at all?*

   There are a number of different layers that could be read into this question, it seems. From a purely numbers perspective, it seems unlikely. Nothing, that I'm aware of, was stopping government agencies from using Green Globes in the past. In fact, if I'm remembering correctly, the Department of Veteran Affairs has one or two Green Globes buildings. So, the GSA assessment didn't "enable" the use of Green Globes. The GSA, EPA and DOD and Interior and Agriculture have been using LEED for the past 14 years and will continue because they have become very good at using LEED.

3. *It seems that for new construction in most private office markets, LEED is a must. Do you see the same trend in the retail, industrial and residential markets? Why have these been so much slower to adopt more energy efficiency and more sustainable strategies?*

   We've already seen strong portfolio commitments from many retail groups [www.usgbc.org/leed/certification/programs/volume]. Additionally, the relatively new LEED Manufacturing User Group is really gaining traction [www.usgbc.org/articles/greener-options-manufacturing-sector]. I think that each of these sectors had challenges that were unique to them that we've only recently created solutions for. Principally, the volume program for retail (where brands tend to have standard prototypes and where the economies of scale make it possible for LEED certification to be much more streamlined) and MUG for industrial (where we've convened a group of folks who are working to solve the technical issues encountered when applying a tool that has its roots in the commercial world). Residential is much more individual consumer driven, and it's proven to be a harder nut to crack. Time will tell.

Interview 2.2: Matt Ellis, CEO of Measurabl

Matt Ellis is the founder and CEO of Measurabl, a software company providing sustainability reporting solutions (www.measurabl.com). While working on CBRE's sustainability efforts, he heard the complaints from directors of sustainability that reporting was taking up too much time and that various reports required the same type of information but in different formats. This generated an idea, and the brief interview below explains what one vendor is trying to do to make reporting easier.

1. *What is Measurabl? Why did you think the market needed a tool like this?*
   Measurabl is B2B reporting software for non-financial sustainability data. We use a TurboTax-style wizard to guide users through the completion of complex qualitative information and automated populating insertions that allow users to automatically sync their quantitative data from any source e.g., utility companies, bill pay, travel and waste providers. Sustainability reporting has been growing at a blistering pace, around 40 percent to 200 percent year over year, depending on which standard of reporting you adopt. It is now de rigueur for public and private organizations of all types around the world: 90 percent of Global 500s, 70 percent of Global 2,500s and every company in the supply chain of companies like Walmart, Johnson & Johnson and Apple provide reports. This reporting is complex, onerous and expensive. Measurabl has created a solution that makes it easy. Our industry-validated software is modeled directly off services like TurboTax with the same impressive value proposition: complex forms and reports generated in hours instead of months, dramatic cost savings and improved accuracy. By lowering the cost and complexity of aggregating, managing and formatting sustainability information, organizations of any type can now disclose sustainability performance regardless of their size, expertise or resources.
2. *Which scoring systems do you plan to serve in the short run to the long run?*
   Measurabl's public launch will support the Global Real Estate Sustainability Benchmark (GRESB), which is the premier global standard for real estate asset owners, managers and investors. Measurabl will simultaneously pilot wizards for the Carbon Disclosure Project (CDP) and the Dow Jones Sustainability Indices (DJSI), both of which count thousands of organizations among their reporters. Measurabl will scale its platform to support numerous other surveys and reports such as GRI, SASB and PGH, as well as build out a big data platform to process all the information captured through its reporting service.
3. *Who else, if anyone, does something similar to Measurabl?*
   A few companies (One Report, Ecovadis) compete in niche verticals. A handful of enterprise software companies (SAP, Oracle, SoFi, Enablon) offer expensive and invasive software. Measurabl is the first and only holistic (qualitative and quantitative) approach to sustainability reporting. We are the only solution that has no barrier to entry (on-demand), whereas all

competitors require lengthy multi-month installs and are priced above the buying power of most target decision makers.

4 *Do you see the market becoming more transparent as a result of such reporting?*
Absolutely! The number one thing stopping organizations from reporting is they don't know how or where to start. By giving them a tool that guides them through every step of the way, we can overcome that barrier. The number two barrier is fear and confusion—fear that investors, consumers or regulators will punish them for not performing better, and confusion about where to get the data, how to format it and how to respond to questions. Measurabl supports the automatic import of data from a huge variety of sources, and by doing so, allows companies to benchmark and assess performance before they file a report.

5 *Anything else you wish to add?*
Our mission is to make sustainability reporting something any organization can do, regardless of size, location, expertise or resources. To do that, we must make it easy, affordable and accurate. If we can pull it off, stakeholders and organizations all over the world will be able to improve on sustainability performance.

Interview 2.3: Brenna S. Walraven, Managing Director with USAA Real Estate Company

1 *Why has LEED been so successful compared to other efforts around the world?*
LEED's success has been driven by successful leadership [and] sound strategy, starting with a conference and certification, adding business cases and advocacy outreach and dedicated volunteers, who believe in the mission. These, coupled with good marketing, have made it the most successful.

2 *Why have you decided to certify so many buildings at USAA?*
We've certified as a way to ensure performance (i.e., third-party verification) and credibility. Second, we used LEED as a platform for a national sustainability execution effort—a tool for the process—much like we've used Energy Star. Lastly, because larger tenants require or prefer LEED-certified space and increasingly investors prefer sustainability that is certified.

3 *Do you think it is possible to game the system?*
Of course, gaming is possible in just about everything, including LEED, but only at the margins. In the end, you can't "do nothing" and achieve certification. Energy performance and being proximate to transit is much more sustainable and can't be gamed. This being said, the real game changer is LEED Existing Building Operations and Management (EBOM)—building LEED is more sustainable than non-LEED, but what really makes a building sustainable is how it's operated.

4 *Do you think that the GSA now approving Green Globes as an acceptable standard will impact the use of LEED at all?*
GSA's approval of Green Globes as another "arrow in the quiver" or tool on the tool belt is positive. By definition, not every building can be LEED (only the top 35 percent or so), and Green Globes allows all buildings to improve over status quo. And no, I don't think it will affect LEED at all.

5 *It seems that for new construction in the private market, LEED is fairly common for office products. Do you see the same trend eventually in the retail, industrial and residential markets? Why have these been so much slower to adopt more energy efficiency and more sustainable strategies?*
I dismiss the premise that product types other than office are not shifting towards increased efficiency and sustainability. LEED will most certainly add industrial to its dominance along with office—it will just take time. Many build to LEED standards today, but don't pay the fees to get the actual certification. I think the least likely to adopt LEED and yet still be more sustainable are multifamily and retail. For multifamily, there are other alternatives to LEED that are very viable and well used: for example, NMHB's green certification or city and regional green programs. The Equity Residential REIT is doing incredible things and never seeks LEED. Retailers, too, seemed to be more focused on energy and Energy Star, and overall don't seem to particularly need LEED—look at Walmart as a great example. They are using daylighting (to lower energy consumption and lower costs and tracking increased sales in "naturally lit areas" of their stores), energy co-gen (including solar) and also pushing suppliers to dramatically reduce packaging waste. Now they are also pushing agricultural suppliers to be environmentally responsible. They sell the most organic cotton clothing and sustainably caught fish in the world. So they don't need LEED certification as a tool for sustainability.

References

Beeman, R 1995, "Friends of the land and the rise of environmentalism, 1940–1954". *Journal of Agricultural and Environmental Ethics*, 8(1), pp. 1–16.

CDP 2016, "About us". Available at: www.cdp.net/en/info/about-us (viewed June 23, 2016).

Coca-Cola Company 2012, "Coca-Cola releases 2011–2012 Global Sustainability Report". Available at: www.coca-colacompany.com/press-center/press-releases/coca-cola-releases-2011-2012-global-sustainability-report (viewed June 22, 2016).

Coca-Cola Company 2017, "Collaborating to replenish the water we use". Available at: www.coca-colacompany.com/stories/collaborating-to-replenish-the-water-we-use (viewed February 8, 2017).

Columbia SIPA Office of Career Services 2014, *Career Opportunities in Corporate Social Responsibility.* Available at: https://sipa.columbia.edu/sites/default/files/Corporate_Social_Responsibility.pdf (viewed November 12, 2017).

Deloitte Center for Financial Services 2017, *Breakthrough for Sustainability in Commercial Real Estate.* Available at: http://deloitte.wsj.com/cfo/files/2014/11/sustainable_real_estate.pdf (viewed February 5, 2017).

Ecotech Institute 2016, "4 companies making major strides in sustainability". Available at: www.ecotechinstitute.com/ecotech-news/4-companies-making-major-strides-in-sustainability (viewed August 18, 2016).

Encyclopedia.com 2017a, "Environmental movement". Available at: www.encyclopedia.com/earth-and-environment/ecology-and-environmentalism/environmental-studies/environmental-movement (viewed February 5, 2017).

Encyclopedia.com 2017b, "Corporate social responsibility". Available at: www.encyclopedia.com/social-sciences-and-law/economics-business-and-labor/businesses-and-occupations/corporate-social (viewed February 5, 2017).

Encyclopedia Britannica 2016, "Categorical imperative".

EPA 2016, "Learn about the Greenhouse Gas Reporting Program (GHGRP)". Available at: www.epa.gov/ghgreporting/learn-about-greenhouse-gas-reporting-program-ghgrp (viewed June 22, 2016).

Euro 2016, "Maastricht Treaty". Available at: www.euro-dollar-currency.com/maastricht-treaty.htm (viewed April 16, 2016).

European Commission 2016a, "The history of the European Union". Available at: http://europa.eu/about-eu/eu-history/index_en.htm (viewed April 16, 2016).

European Commission 2016b, "Progress made in cutting emissions". Available at: https://ec.europa.eu/clima/policies/strategies/progress_en (viewed June 23, 2016).

Friedman, F 1970, "A Friedman doctrine". *New York Times*. Available at: www.nytimes.com/1970/09/13/archives/a-friedman-doctrine-the-social-responsibility-of-business-is-to.html (viewed April 9, 2016).

Friedman, M 1962, *Capitalism and Freedom*, University of Chicago Press, Chicago.

GreenBiz 2016, "Companies are tackling climate emissions in creative ways". Available at: www.greenbiz.com/article/companies-are-tackling-climate-emissions-creative-ways (viewed August 1, 2016).

Griswold, D 2000, *The Blessings and Challenges of Globalization*, Cato Institute. Available at: www.cato.org/publications/commentary/blessings-challenges-globalization (viewed February 5, 2017).

Haque, S 2015, "Corporate climate change-related auditing and disclosure practices: are companies doing enough?" In Rahim, M M & Idowu, S O (eds), *Social Audit Regulation: Development, Challenges and Opportunities*, Springer, London, pp. 169–185.

Hart, S & Milstein, M 2003, "Creating sustainable value". *The Academy of Management Executive*, 17(2), pp. 56–67.

Ice, R 1991, "Corporate publics and rhetorical strategies: the case of Union Carbide's Bhopal crisis". *Management Communication Quarterly*, 4(3). Available at: http://journals.sagepub.com/doi/abs/10.1177/0893318991004003004 (viewed February 5, 2017).

IPCC 1991, *Climate Change: The IPCC Scientific Assessment*. Cambridge University Press, Cambridge. Available at: www.ipcc.ch/publications_and_data/publications_ipcc_first_assessment_1990_wg1.shtml (viewed April 16, 2016).

IPCC 2016, "Organization". Available at: www.ipcc.ch/organization/organization.shtml (viewed April 16, 2016).

Knufken, D 2010, "The top 25 greenwashed products in America". *Business Pundit*. Available at: www.businesspundit.com/the-top-25-greenwashed-products-in-america (viewed February 10, 2017).

KPMG 2013, *The KPMG Survey of Corporate Responsibility Reporting 2013*. Available at: https://assets.kpmg.com/content/dam/kpmg/pdf/2015/08/kpmg-survey-of-corporate-responsibility-reporting-2013 (viewed February 5, 2017).

Maharaj, R & Herremans, I 2008, "Shell Canada: over a decade of sustainable development reporting experience". *Corporate Governance: The International Journal of Business in Society*, 8(3), pp. 235–247.

McGraw-Hill and Siemens 2012, *2012 Greening of Corporate America: The Influence of Sustainability on Transforming Business Strategy*. McGraw-Hill. Available at: https://w3.usa.siemens.com/buildingtechnologies/us/en/events/industry-reports/reports/Documents/ Greening_Corporate_America_2012_(Siemens-MHC)-Industry-Report.pdf (viewed November 12, 2017).

McKinsey & Company 2011, "Sustainability's strategic worth: McKinsey global survey results". Available at: www.mckinsey.com/business-functions/sustainability-and-resource-productivity/our-insights/sustainabilitys-strategic-worth-mckinsey-global-survey-results (viewed October 9, 2017).

Moore, C 2014, *Corporate Social Responsibility and Creating Shared Value: What's the Difference?* Heifer International. Available at: https://sharedvalue.org/sites/default/files/resource-files/ CFR-047%20Corporate%20Social%20Responsibility%20White%20Paper_FINAL.pdf (viewed May 14, 2014).

Porter, M & Kramer, M 2011, "Creating shared value". *Harvard Business Review*. Available at: https://hbr.org/2011/01/the-big-idea-creating-shared-value (viewed April 19, 2016).

Semans, T, Juliani, T & de Fontaine, A 2007, "A climate of change: manufacturing must rise to the risks and opportunities of climate change". *Center for Climate and Energy Solutions*. Available at: www.c2es.org/newsroom/articles/climate-change-manufacturing-must-rise-risks-and-opportunities-climate-change (viewed February 6, 2017).

Short, J, Rice, S & Lindeberg, M 2001, "The *Exxon Valdez* oil spill: how much oil remains?" *Alaska Fisheries Science Center*. Available at: www.afsc.noaa.gov/Quarterly/jas2001/feature_jas01.htm (viewed April 16, 2016).

Starbucks Coffee Company 2016, "Tackling climate change". Available at: www.starbucks.com/responsibility/environment/climate-change (viewed June 22, 2016).

UL 2010, *The Sins of Greenwashing: Home and Family Edition*. Available at: http://sinsofgreenwashing.com/index.html (viewed June 23, 2016).

Union of Concerned Scientists 2017, "The IPCC: who are they and why do their climate reports matter?" Available at: www.ucsusa.org/global_warming/science_and_impacts/science/ipcc-backgrounder.html (viewed February 6, 2017).

US Customs and Border Protection 2016a, "NAFTA". Available at: www.cbp.gov/trade/nafta (viewed April 16, 2016).

US Customs and Border Protection 2016b, "Trilateral/NAFTA side agreements". Available at: www.cbp.gov/trade/nafta/guide-customs-procedures/contacts-additional-assistance/trilateral-nafta-side-agreements (viewed April 16, 2016).

US Green Building Council 2016, *LEED v4 for Building Operations and Maintenance*. Available at: www. usgbc.org/resources/leed-v4-building-operations-and-maintenance-current-version (viewed November 12, 2017).

Weinreb, E 2017, "The expanding role of sustainability leadership". *GreenBiz*. Available at: www.greenbiz.com/article/expanding-role-sustainability-leadership (viewed February 5, 2017).

World Bank 2014, "Corporate social responsibility or corporate self-promotion?" Available at: http://blogs.worldbank.org/futuredevelopment/corporate-social-responsibility-or-corporate-self-promotion (viewed November 26, 2014).

World Trade Organization 2008, *World Trade Report 2008*. Available at: www.wto.org/english/res_e/booksp_e/anrep_e/world_trade_report08_e.pdf (viewed February 5, 2017).

3

Energy upgrades in commercial property

Minimum energy efficiency standards, compliance pathways, and leases in the UK

Julia Patrick, Susan Bright and Kathryn B. Janda

3.0 Introduction

Meeting the UK's carbon reduction targets will require significant improvements to the energy performance of the commercial built environment (DECC 2011). For rented properties, the "split incentive" has been highlighted as a particular challenge, as the misalignment of investment costs and financial savings can mean neither landlords (building owners) nor tenants (building occupiers and energy bill payers) have sufficient incentives to invest in energy efficiency improvements (DECC 2012). UK leases tend to reinforce the split incentive, in particular by not allowing landlords to recover the costs of energy efficiency improvements from occupiers, and by giving landlords only limited rights of access to tenant premises to carry out improvements (Bright 2008).

The UK government has passed regulations introducing statutory minimum energy efficiency standards (MEES) for rented commercial buildings to overcome some of these challenges, with the aim of securing improvements to the least efficient buildings (Energy Efficiency (Private Rented Property) (England and Wales) Regulations 2015: hereafter the MEES Regulations).

This chapter explores how industry actors in the UK – including lawyers, landlords, tenants, and property managers – perceive that the MEES Regulations interact with lease clauses and energy efficiency upgrades. Drawing on interviews and document analysis from the UK-based WICKED project,[1] it examines the potential contribution of lease clauses to managing compliance with the MEES Regulations in the commercial sector, and the extent to which the interaction between the MEES Regulations and leases may support or undermine the objectives of MEES to secure energy upgrades to the UK's least efficient properties.

The chapter starts by providing an overview of the relevant literature on energy efficiency upgrades, MEES, and leases. It goes on to describe the research approach briefly. Key findings from industry participants are then presented and discussed, including three "compliance pathways" to MEES: active, protective, and avoidant. These varying approaches have different impacts on how MEES may manifest in practice, including some unintended consequences. The final section concludes by suggesting implications for industry participants and policymakers, both within the UK and beyond.

3.1 Background: Energy efficiency and commercial buildings

This section begins with a brief discussion of the literature on energy efficiency renovations in the commercial sector. Next it describes MEES, before going on to consider the interaction of MEES and leases.

3.1.1 Energy efficiency opportunities and challenges in commercial buildings

Improving the energy efficiency of existing rented commercial buildings is an important element of – and opportunity for – meeting the UK's carbon reduction targets (DECC 2014a). About 12 per cent of the UK's emissions are attributed to energy used for non-domestic buildings (DECC 2011) and estimates suggest that around 60 per cent of today's non-domestic buildings will exist in 2050 (Carbon Trust 2009). Moreover, in the retail sector (the focus of the WICKED project), whilst energy management is considered a particularly complex challenge, it is also seen as a significant opportunity (BRC 2014; BCSC and CBRE 2015).

The general need for energy efficiency investment in the building stock is thus recognised at a national level, both by the UK government and by industry bodies. However, views on building- and portfolio-level opportunities and drivers for energy efficiency investments differ. A recent industry report by the British Council of Shopping Centres (BCSC) and CBRE (an international commercial property consulting and management firm) suggests that there is a compelling business case for energy efficiency upgrades in shopping centres, whilst recognising a number of barriers including availability of capital, limited awareness of costs and benefits, and the role of fixed service charges (BCSC and CBRE 2015). Conversely, Elliot, Bull and Mallaburn (2015: 667) suggest that the *lack* of a compelling business case is the "primary barrier" to energy efficiency investments. At the same time, some have argued that risk avoidance, in particular avoiding the future risk of obsolescence, can or should drive investment decisions (JLL 2013; Sayce, Ellison and Parnell 2007). BCSC and CBRE (2015: 3) suggest that "the extent to which a centre does or does not possess optimal energy using equipment is a useful gauge of its susceptibility to value erosion or price-chipping by prospective acquirers". They argue that "ultimately it is prospective enhancement in asset value that is most likely to drive change" (BCSC and CBRE 2015: 4). Elliot et al.'s 2015 study of investment decisions in this context, however, did not find any evidence that these considerations were in practice driving more sustainable commercial property investments. Others have suggested that such investment decisions go beyond mere financial considerations and are driven instead by a range of strategic considerations, of which risk is one (Cooremans 2011).

The complexity of energy efficiency renovation is magnified in rented buildings, which make up over half of the UK's commercial buildings sector (PIA 2015). Here, landlord and tenant interests and split incentives create an additional layer of problems. The problem of the split incentive is well known (DECC 2012; EC 2011). It is highlighted as a particular barrier in the government's consultation on the MEES Regulations, where it is defined as a situation where "the costs of energy efficiency improvements are borne by landlords, while the benefits (lower energy bills) accrue to current or future tenants" (DECC 2014b: 18). The government's impact assessment in relation to the MEES Regulations explains: "In principle, in a well-functioning market, rent levels should fully reflect differences in a property's energy efficiency thus overcoming this split incentive issue. However, [in] the presence of other market failures, such as imperfect information on the costs and benefits associated with energy efficiency measures, rents may not fully reflect differences in energy efficiency. This leaves landlords with little incentive to make energy efficiency improvements" (DECC 2015a: 15).

3.1.2 Policy context: MEES

Against this background, MEES are being introduced "to tackle the very least energy efficient properties" (DECC 2014b: 10) and "drive improvements in the energy efficiency of buildings in the non-domestic sector" (DECC 2014b: 11). The focus of MEES is only on rental property, and the MEES Regulations provide a regulatory impetus for landlords to implement energy efficiency upgrades. Operating alongside a range of other government measures (DECC 2014a), MEES is the first government initiative to explicitly address issues associated with rented commercial properties, including split incentives between landlord and tenants.

The MEES Regulations will make it unlawful, from April 2018, to let non-domestic properties (including lease renewals or extensions) that fall below the minimum standard of an E Energy Performance Certificate (EPC) rating, unless all relevant energy efficiency improvements (broadly, those recommended by a surveyor with a payback period of seven years or less) have been carried out (Energy Efficiency [Private Rented Property] [England and Wales] Regulations 2015). EPCs rate properties across seven categories, from A (best performing) to G (worst performing), based on age, size, and fabric of the building, and are required when buildings are constructed, let, or sold (DECC 2014a). Landlords are exempted from the prohibition on letting where tenants or third parties have refused consent to relevant energy efficiency improvements (the consent exemption) or where such an improvement would adversely affect the value of the relevant property by more than 5 per cent (the devaluation exemption). From 2023, the prohibition and related provisions will apply not just to properties that are newly let, but also to those that continue to be let. Importantly, the EPC ratings are an "asset rating" (an estimate of what amount of energy a building *should* use based on a model of its fabric and equipment) rather than an "operational rating" (based on actual metered energy use). This focus provides an impetus for physical renovation rather than encouraging advances in energy management and control strategies. It has been suggested that improving EPC ratings, therefore, is only one part of the picture and does not necessarily lead to reduced energy consumption (JLL and BBP 2012).

Almost a fifth of UK commercial properties are rated F or G (DECC 2014b) and will be affected by the MEES Regulations. For these properties to continue to be let, the regulations result in an indirect requirement to carry out improvements, unless an exemption applies. A further 19 per cent of properties currently carry an E rating (Cushman & Wakefield 2016). As ongoing changes to the EPC methodology could mean that a current E falls to F or G (BPF 2014) and the regulatory standard may be raised going forward, landlords may also need to consider whether improvements are needed for these E-rated properties.

In the original consultation on MEES, the government envisaged the operation of a green financing arrangement as a key mechanism for facilitating improvements (DECC 2014b). This would mean no upfront costs to the landlord ("to ensure that any regulations do not impose disproportionate burdens on business" [DECC 2014b: 12]) and payback through energy savings over time, creating "a win-win opportunity for both landlords and tenants" with energy savings for the bill payer and an improved building for the landlord (DECC 2014b: 4). In practice, this green financing deal has to-date failed to materialise due, it has been suggested, to a lack of financial backing (UKGBC 2016). Instead, therefore, the regulations refer to a seven-year payback period as the alternative "financial arrangement" (Regulation 28; Energy Act 2011: section 49[4][b]). This means, broadly, that landlords are only required to carry out improvements that would pay for themselves through energy cost savings after seven years or less. No alternative mechanism is suggested for ensuring landlords do not have to bear upfront costs.

More broadly, it has been suggested that MEES may help to reinforce links between the sustainability of buildings – as expressed through EPCs – and asset value. In preparation for the start of MEES in April 2018, according to BCSC and CBRE (2015: 20), commercial landlords are reviewing portfolios "to determine their risks and investment required to mitigate them". BCSC and CBRE (2015: 20) go on to say: "As would be expected, the lettability, or otherwise, of a building has a significant impact on its value. We are seeing valuers building in allowances for improvement works in their appraisals and purchasers proposing price chips during acquisition negotiations". It has also been suggested that MEES will add weight to obsolescence risk as a driver (JLL 2013). The BCSC and CBRE (2015: 19) suggest: "Depending on the fabric and systems, and the extent of work required to meet minimum standards, some buildings could be considered obsolete if improvements cannot be made cost effectively". In other words, MEES provides a clear benchmark for which buildings may be considered "obsolete" in the future.

3.1.3 The interaction of MEES and leases

As noted above, the legal arrangements between landlords and tenants, as set out in the lease, commonly reflect – and arguably reinforce – the split incentive problems (DECC 2015a; Dixon, Britnell and Watson 2014; Bright 2008). In response, in the UK, the industry-led Better Buildings Partnership (BBP) and its members have sought to address the role of leases in energy (and broader environmental) performance by promoting the use of "green" clauses and developing a "Green Lease Toolkit". Building on the concept of "green" leases first proposed in Australia in 2006 (Woodford 2007), the Green Lease Toolkit identifies the "split responsibility/incentive … in the procurement, control and use of resources" as a "key barrier to the improved Environmental Performance of commercial buildings" and includes model clauses and guidance to "help overcome this challenge by providing a framework for engagement on environmental issues" (BBP 2013: 2; Janda, Bright, Patrick, Wilkinson and Dixon 2016). More recently, the concept of green clauses has been reflected more widely across the industry with the development of the "Model Commercial Lease" (MCL 2016a).

The introduction of the MEES Regulations is prompting debates about the interaction between MEES and lease clauses, in terms of how existing lease clauses help or hinder MEES compliance and how new lease clauses might do so; for example, whether they allow landlord access to carry out improvements (Farnell 2015; BSDR 2014; Patrick, Bright and Jaksch 2015). Indeed, both the BBP and Model Commercial Lease have considered whether changes should be introduced to their model clauses in light of MEES (Botten 2016; MCL 2015). As a result, the Model Commercial Lease recently announced changes to their standard lease, including greater landlord control over EPC production and, from 2023, a landlord right of entry to carry out works to sub-standard properties subject to tenant consent (the refusal of which would allow the landlord to rely on the relevant MEES exemption) (MCL 2016b).

Williams (2015) has written on the potential implications of MEES for new lease clauses from a practitioner's perspective. Table 3.1 summarises the key points, which provide a useful overview of the potential implications of MEES in relation to seven different types of lease clauses. This overview focuses largely on the landlord's perspective, although it identifies a particular implication for tenants in relation to statutory compliance provisions, suggesting that tenants may seek to exclude any MEES-related obligations or costs from this provision in the future. Moreover, BCSC and CBRE (2015: 30) have reported that retailers have "requested during negotiations that clauses be included that exclude works to improve a centre's EPC rating from service charges". BCSC and CBRE (2015: 30) suggest that these types of leasing responses "may further emphasise the split incentive between owner and retailer, re-enforcing the industry stalemate."

Table 3.1 Potential implications of MEES for new lease clauses

Type of lease clause	*Potential implications of MEES for new clauses*
Service charges Typically, these cover repair and maintenance, but not improvements.	Landlords may wish to include provisions that expressly allow them to recover the costs of energy efficiency improvements.
Yield-up (or reinstatement) Typically, this relates to the tenant's repairing obligations at the end of the lease.	There have been suggestions that provisions could require tenants to ensure the EPC rating at yield-up is the minimum required under MEES, or no lower than at the start of the term. It is suggested that this might in practice require improvements and therefore be unfair to tenants.
Statutory compliance Typically, this requires tenants to comply with statutory obligations in relation to demised premises.	Tenants may wish to seek a specific carve-out from these provisions in relation to MEES.
EPC production Although currently rare, some leases provide for how and when a tenant obtains an EPC.	It is expected that landlords will seek to include provisions that control when and how an EPC is produced in the future, including requiring tenants to use the landlord's choice of assessor or asking the landlord to obtain EPCs on their behalf.
Alterations Typically, this allows tenants to make non-structural alterations with landlord consent.	Landlords may wish to prevent alterations that could adversely affect a building's environmental performance or EPC rating.
Landlord's right of access to carry out works Lease provisions are likely to vary, but improvements (rather than maintenance) are typically excluded.	On the one hand, landlords may wish to include rights of access to carry out improvement works; on the other hand, they may seek to restrict their rights of access in order to be eligible for the consent exemption under MEES.
Rent review Typically, these assume a notional letting on the date of review based on certain assumptions – for example, the property is fit for occupation or may lawfully be used.	It remains unclear whether any additional provisions will be needed to ensure MEES does not adversely affect rent review for landlords.

Source: Adapted from Williams (2015)

3.2 Research approach and methods

MEES arose in the UK with a public consultation in July 2014 and a response to the consultation published in February 2015 (DECC 2014b; DECC 2015b). The emergence of MEES coincided almost exactly with the two-year period of the WICKED project's funding (July 2014–June 2016). One of the WICKED project's research streams addresses organisational energy management practices (Janda 2016), and this stream includes a focus on the role and impact of leases in relation to energy management (Janda et al. 2016; Patrick and Bright 2016). WICKED project researchers in this area carried out semi-structured interviews with industry experts and participants between January and November 2015, which were well timed to capture initial industry thinking about the introduction of MEES with respect to leasing. The sampling of interviewees is best described as a snowball sample reflecting a number of influences: a socio-technical model of participants in the retail sector (Janda, Bottrill and Layberry 2014); the need to synthesise with other WICKED project research themes investigating portfolio-wide energy and building datasets (Janda et al. 2015); and convenience and accessibility. The research team carried out 29 interviews with 38 representatives of 25 different organisations, including retailers (three), property owners (three), letting and property management companies (four), law firms and legal experts (11, including five who typically act for BBP members, four who typically act for retailers and/ or non-BBP property companies, and two independent experts with a broad view of the market), and industry intermediaries and experts (four). Although the inclusion of lawyers as interviewees provided an opportunity, in some cases, to capture experience amongst SMEs as well as larger companies and lettings in the secondary market, overall respondents tended to represent, or have insights into, larger organisations with national or international portfolios of prime properties.

Interviews were recorded and transcribed, and responses were coded (using NVivo software) and analysed thematically. Thematic analysis of interview transcripts was supplemented by document analysis of company strategy reports, green lease clauses in company templates, and model green lease clauses promoted by industry partnerships and a review of policy documents and industry reports. Given the relatively small samples within each category of respondent and the different capacities in which respondents answered questions (at times drawing on their own experience, at other times commenting on wider market practice), the views captured cannot be considered necessarily representative or comprehensive, but they nevertheless reflect a range of different industry views and responses on newly emerging industry and policy issues.

3.3 Industry insights

This section summarises interview responses from the WICKED project relating to MEES, in particular the link between leases and MEES, the role of lease clauses in helping to manage compliance with the MEES Regulations, and the extent to which this interaction may promote or hinder the objectives of MEES to secure energy efficiency improvements to the UK's least efficient properties.

Many interviewees commented on how MEES is now causing landlords, tenants, and their advisers to consider the role of leases in relation to compliance and to environmental improvements more generally. As one in-house lawyer commented, "There's MEES out there, but there's also the lease provisions, and you've got to read the two together" (Interviewee 28).

Overall, interviewees identified broadly with three responses and corresponding potential roles for lease clauses (which are not mutually exclusive):

1. *Active*: From a landlord's point of view, an active approach sets out to identify action necessary to improve F- and G-rated properties (i.e., the worst performing properties, below the minimum standard set by the MEES Regulations). This approach most directly works to meet the government's objectives under MEES, in terms of securing improvements to the least efficient buildings.
2. *Protective*: From a landlord's point of view, a protective approach seeks to preserve the asset rating of properties that already comply with MEES through requiring tenants to maintain (or not worsen) the environmental performance or rating of their premises. The protective approach seeks to ensure that currently compliant buildings do not add to the stock of "least efficient" buildings by falling below the minimum standard.
3. *Avoidant*: For both landlords and tenants, an avoidant approach seeks to avoid or delay action and/or associated exposure to costs in relation to MEES compliance. This approach secures technical compliance with the MEES Regulations at the portfolio level, but risks preventing or delaying improvements to sub-standard properties in the building stock.

Interview findings in relation to these three categories are described further below and are summarised in Table 3.2 on page 47 (the summary not necessarily reflecting the full range of nuances expressed by different interviewees).

3.3.1 Active approach to MEES

Respondents who described an active approach to MEES (typically representing larger property companies operating in the prime market) highlighted ways of using leases and other mechanisms to support landlord action to carry out energy efficiency improvements to comply with MEES and to ensure that F- and G-rated properties are improved to meet the minimum standard.

First, a number of interviewees highlighted the potential role of MEES in reinforcing economic drivers of energy efficiency, in particular the link between MEES compliance and preserving asset value. Amongst these, most seemed uncertain about the link, with some suggesting that value implications would depend on market dynamics, such as the relative bargaining positions of landlords and tenants. Two interviewees suggested that sub-standard properties were likely to attract a "brown discount" under MEES, suggesting that this is one driver of an active approach. This link was explained by one lawyer as follows:

> Some landlords are now saying it's worth investing in stock, not because this necessarily makes it more valuable, but because it is less likely to reduce in value. MEES in particular are driving this as an issue and are "chipping off the price" (Interviewee 23).

An active approach to MEES may, therefore, be important to preserve asset value, depending on market circumstances. Views on the role of the lease in facilitating such an approach are, however, mixed.

One view was that landlords adopting an active approach may seek greater rights of access to enable them to carry out improvements during a lease (thus avoiding the need to plan a void in order to do the works). One lawyer explained: "with MEES coming in ... there might well be a driver now much more for the landlord to make improvements to the building and have access to the tenant areas to do that" (Interviewee 32).

The same lawyer suggested, however, that increased rights of access might not mean very much on their own, without dealing with the issues of cost allocation and potential disruption to the tenant's business:

> As soon as there is a clause that allows landlords to carry out improvements, the tenant will go straight to the service charge schedule and make sure there's a specific exclusion or will definitely come out in the negotiation as to who is paying for this. There are also cases that we can give the landlord right of access but the tenant will say, "when reasonable"... "not affect quiet enjoyment". And if you're talking about major improvement works, the landlord will have to get in and disrupt the tenant. We can explain to the landlord that there can be a right of access but whether you can go in there and carry out these improvements is another matter; because tenants will not want to be disrupted to a major extent ... and if these works are expensive, then a right of access in the lease may not mean much at the end of the day (Interviewee 32).

Related to this, respondents suggested that landlords might increasingly attempt to recover costs for the energy efficiency improvements needed under MEES either through existing lease provisions or through new, explicit lease clauses. One legal expert (with an overview of the wider market) commented that cost recovery is "going to come into sharper focus because of MEES" (Interviewee 34).

In terms of the role of the lease in cost recovery, there were different views about whether improvement costs could be recovered under typical service charge provisions. One professional support lawyer suggested: "Actually, when we look at our leases ... because they're drafted quite widely in what landlords can put through the service charge, you think actually it's wide enough already" (Interviewee 26). By contrast, one retailer in-house lawyer expressed a different view: "Normally you can install new things where it's not economic and viable to repair but generally it shouldn't be improvements" (Interviewee 28). An alternative view was that, in practice, sometimes landlords may include improvement costs in service charges irrespective of what the lease says. One lawyer mentioned: "Managing agents I've spoken to will put stuff through the service charge without bothering to check whether it's chargeable" (Interviewee 34). Going forward, there were some suggestions that landlords are increasingly attempting to insert explicit environmental improvement cost recovery clauses into new leases, although there was some uncertainty over whether tenants will accept these. These views suggest that leases *may* – one way or another – have a role in supporting an active approach.

Overall, the responses suggest that increased landlord attempts to recover costs – possibly facilitated through lease provisions – could enable energy efficiency improvements that might otherwise not happen and thereby help deliver MEES objectives.

There was also a different view that the lease is not a helpful tool in facilitating positive energy efficiency improvements and that these are more effectively dealt with outside the lease, even under MEES. One property company representative explained:

> Even in relation to MEES, opportunities can happen outside the lease. As soon as it's in the lease, lawyers argue over it (Interviewee 16).

One lawyer put it slightly differently:

> With MEES, parties would rather take a light touch approach – you don't want to spend hours negotiating these provisions and then find that it does not work in practice. It is a difficult balance, and until MEES settles down, we'll see very light touch on documents. I think that's probably right because I don't think the lease can cater for all of this. It will probably have to sit outside the lease so it can be a bit more free-flowing and fluid, and you don't then have to spend time varying the lease when things change (Interviewee 32).

Some interviewees presented yet another perspective, suggesting that cost recovery during the term of a lease might be less of an issue and that the likely landlord response to sub-standard properties under MEES would be to carry out improvements following the end of a lease, and possibly recover costs through charging a higher rent to the incoming tenant. Others suggested that certain landlords may want to pay for improvements themselves to show leadership, or because they perceived clear benefits from investment in terms of higher tenant retention, shorter voids, and preserving the value of their portfolio.

3.3.2 Protective approach to MEES

Alongside the "active" approach, a number of respondents (again, typically representing larger property companies) saw MEES acting as a driver to protect existing asset ratings. As one lawyer put it, "With MEES, it's around EPCs ... Changes to leases will focus very much on preserving an asset rating" (Interviewee 32).

Maintaining EPC ratings was generally discussed in relation to tenant action and seen in terms of both preventing actions that would reduce environmental performance and ensuring investment to improve environmental performance where this would be necessary to maintain the minimum rating of E, given the likely shifting thresholds in the calculation of EPC ratings. One property company representative commented: "What we can say around maintaining an EPC rating and what that actually means on a practical level and a legal level as well is becoming increasingly important" (Interviewee 11).

In relation to alterations, respondents (commenting typically from a landlord's perspective) were concerned that the lease should prevent tenants from carrying out alterations that adversely affect either the property's EPC rating, or more generally, environmental performance of the property. One lawyer described a need to "beef up what the lease says about the tenant carrying out alterations, so that the tenant is prevented from doing things that would have a detrimental impact" (Interviewee 23). Other lawyers highlighted a possible view that alterations clauses commonly require landlords' reasonable consent in any event, and there was an argument that withholding consent would not be considered unreasonable where proposed alterations would adversely affect environmental performance.

In relation to yield-up, respondents seemed uncertain about what provisions might be needed to protect against non-compliance with MEES. The representative of a property management company explained:

> One [link between leases and energy management] that's come up recently is this Minimum Energy Performance Standard [now generally referred to as MEES] and the cost of dilapidation when whoever it is vacates. So if, when we get to 2018, it's the law that you can't let an F or a G, who does that cost lie with and what part does the lease play in that? I think that's something that the industry's not really clear on (Interviewee 4).

Another area where landlords were considering the role of lease clauses was in relation to the production and quality of EPCs. It was suggested that landlords were increasingly including clauses to control when and how EPCs are produced, thus preventing tenants from obtaining EPCs unless legally required to do so (for example, on a subletting) and requiring tenants to use the landlord's choice of EPC assessor or to let the landlord carry out the EPC for them. This trend is likely to continue. The position was summed up by one lawyer as follows:

> I think the way it feeds into leases ... there has been a recognition that EPCs are sometimes good and sometimes bad ... If the tenants do something that changes the EPC rating or might do, then they, the landlord, want to control that process.They want to be able to step in and do the EPC even if it's a tenant sale or letting that's triggered the requirement (Interviewee 38).

3.3.3 Avoidant approach to MEES

Respondents also highlighted a more risk-averse approach, seemingly concerned not with improving the environmental performance of buildings – as envisaged under the MEES Regulations – or preserving it, but with ensuring compliance whilst minimising exposure to MEES and associated costs, thus avoiding or delaying investments in building improvements.

As mentioned above, in response to MEES, some landlords may increasingly seek to use lease clauses to recover improvement costs. In itself, this might be considered helpful in enabling improvements and thereby achieving the objectives of MEES. However, interviewees also suggested that MEES would cause *tenants* to consider more explicitly what happens to the costs of energy improvements, resulting in increased tension between landlords and tenants: landlords aiming to recoup costs, and tenants simultaneously aiming to exclude these. It was suggested that this could potentially undermine attempts to create a more collaborative approach between landlords and tenants, and also create additional legal barriers to improvements, making tenants less likely to agree informally to minor improvements. Interviewees gave different examples of improvements that had been agreed between landlords and tenants in this way, including those that could be recovered by the landlord within the annual service charge budget (suggesting relatively low capital outlay), those with a payback of two years or less, and those with a payback of five years or less, where the cost charged to tenants was spread over the payback period.

One lawyer (who does not typically act for BBP members or large property companies) explained the avoidant approach as follows:

> There have been more comments and questions from clients since MEES. They're not asking how they can sort this out. Landlords are saying, "How can I get tenants to pay?" The tenant is saying, "How can I stop landlords charging me?" ... All are jockeying for position to make sure that whatever happens, they're not paying for it (Interviewee 26).

Similarly, the in-house lawyer for a retailer took the following view: "We want to protect against having to update a landlord's property, at our expense" (Interviewee 28).

In this context, interviewees referred both to service charges and to statutory compliance provisions as potential battlegrounds for cost recovery. For example, one property management company representative commented:

> The landlords will argue that most leases now will have a clause that says the landlord is entitled to charge the tenant the cost for any works that are required to meet a statutory requirement. On the flip side, in general, leases tend to say that any improvements to a property that enhance the asset value are the responsibility of the landlord, and the landlord pays (Interviewee 24).

Other lawyers commented:

> You've then got who picks up the cost ... and we are seeing landlords put in they can recover the costs of environmental improvements. Whether a tenant can strike this out or not depends on how strong a tenant they are (Interviewee 31).

> MEES is the driver that would mean that landlords would be able to negotiate that [cost recovery] in. Whether landlords try to get that in through the back door by defining improvements quite widely ... it remains to be seen how easy it is to negotiate that into the lease. ... With MEES, landlords will argue more heavily that they have to carry out these improvements and that the tenants will benefit from the energy efficiency savings as a result of these improvements. ... As soon as there is a clause that allows landlords to carry out improvements, the tenant will go straight to the service charge schedule and make sure there's a specific exclusion or will definitely come out in the negotiation as to who is paying for this (Interviewee 32).

All of these views seem to paint a picture of a mutual desire by landlords and tenants to avoid the costs of any improvements necessary to comply with MEES.

Similarly, it was suggested by lawyers and one landlord that landlords may seek to limit their access rights, making their right of access to carry out improvements subject to tenant consent, in order to be able to rely on the statutory exemption from MEES, under which landlords are not prohibited from renting out sub-standard properties where tenants have refused consent to relevant energy efficiency improvements. One lawyer explained, "[Y]ou see landlords' minds working … 'How can I get myself into an exemption up to until the premises are vacant?'" (Interviewee 32). Another lawyer commented, "We've been considering from a landlord's perspective how this will interface with MEES. Whether if the landlord has that right they'll be able to rely on the consent exemption … One way to deal with this is to temper it so the landlord only has right of access to deal with improvements that have been consented to by the tenant, and the tenant has absolute discretion" (Interviewee 31).

It was suggested that fundamentally, in relation to cost recovery, there would be "two camps":

> There are two views. There are some landlords who will wish to carry out those improvements and will wish to recover the cost from the tenant; or there are some better informed landlords who accept that if they have to make improvements to the building that they should pay for those. And therefore there will be two camps. The landlords looking to use the exemption in the regulation that allow them not to carry out improvements if they can't get consent. And those who accept if they want to carry out improvements they should pay for those (Interviewee 32).

Overall, interviewees highlighted a range of approaches, and one view was that, in practice, MEES compliance might involve using several of these. As one lawyer explained:

> So I suspect we will see provisions requiring tenants not to do anything that might impact EPCs, we might see ability for landlord to go in and carry out improvements; the tenant will counter that by saying, "You can only have access to the premises when I say and in terms of cost I'm not paying for any improvements", and there will be some arguments – the landlord arguing that if there are any benefits, the tenant should pay (Interviewee 32).

Table 3.2 Summary of issues raised by interviewees: Responses to the introduction of MEES and the role of leases

Types of responses	*Summary of interviewee views*
Active approach (improving sub-standard properties)	**Landlord perspective** (typically larger companies in the prime market) MEES may reinforce the economic driver to preserve asset value, depending on market dynamics. Landlords may seek greater rights of access to tenant premises to carry out improvements. Landlords may increasingly seek to recover costs of improvements, which may involve more explicit cost recovery clauses in leases. Alternatively, landlords may deal with improvements and related cost recovery outside the lease, or at the end of a lease term, and/or may accept that they should pay for improvements. The extent to which landlords will be able to recover costs from tenants either through the service charge or rent is uncertain.

(continued)

Table 3.2 Summary of issues raised by interviewees: Responses to the introduction of MEES and the role of leases *(continued)*

Types of responses	*Summary of interviewee views*
Protective approach (preserving minimum standards)	**Landlord perspective** Landlords will seek to prevent tenant alterations that reduce environmental performance and/or EPC ratings through appropriate lease clauses. In relation to yield-up/reinstatement, landlords may similarly wish to secure a certain level of environmental performance, possibly through the lease, but whether and how this would work is unclear. Preserving minimum EPC ratings may involve investment to *improve* environmental performance due to changing ratings thresholds. Landlords are increasingly including lease clauses to control when and how EPCs are produced by tenants.
Avoidant approach (ensuring compliance whilst avoiding improvements and/or associated costs)	**Landlord perspective** (typically the wider market, possibly not including larger landlords) Landlords may increasingly seek to recover the cost of improvements, including through more explicit lease provisions. Landlords may seek to limit their rights of access to tenant premises under the lease, to rely on the MEES exemption where tenants refuse consent to improvements. **Tenant perspective** Tenants may increasingly seek to exclude explicitly the cost of improvements, including from service charge provisions and/or statutory compliance provisions. Whether landlords are able to recover costs will depend on market dynamics. These avoidant approaches may undermine attempts to create a more collaborative approach. Tensions may be magnified from 2023, when MEES will apply to existing lets.
Other issues	Fit-out arrangements (before the start of the lease) may present an important opportunity for engaging with tenants to comply with MEES. There is a risk that sub-standard properties may be offloaded, possibly to those with less capacity to deal with MEES and energy management generally. There are questions about how MEES will be enforced. It is not clear what the consequences will be where landlords are unable to access upfront cash to fund improvements. There are a number of issues related to EPC methodology and logistics, which will make monitoring ongoing compliance with MEES challenging.

The same lawyer also commented that the issues may be magnified from 2023, when MEES apply to existing lets as well as new (and renewal) ones:

> I suspect landlords will just have to take it on the chin and improve the building and pay for it. The problems are going to be when we come to 2023 when it's all leases and you have to start targeting existing leases; that's when we're going to have problems and arguments are going to start (Interviewee 32).

3.3.4 Other MEES issues

A complete description of the interview responses is beyond the scope of this chapter. However, we note that interviewees raised a range of other issues related to MEES including:

- the role of fit-out as a key opportunity for engaging with tenants to respond to MEES;
- the risk that sub-standard properties are offloaded, possibly to those with less capacity to deal with MEES and energy management generally;
- questions about how MEES will be enforced;
- the consequences of landlords' inability to access upfront cash to fund improvements; and
- the challenges of EPC methodology and logistics to monitor ongoing compliance with MEES.

The issues are picked up further, where relevant, in the discussion and conclusions below.

3.4 Discussion

The findings describe a variety of industry views on MEES and the roles of leases in relation to MEES. This section summarises and discusses these roles, as well as their implication for energy efficiency improvements of sub-standard buildings and the prospects of meeting the objectives of the MEES Regulations.

In summary, interview responses highlighted three broad responses to the MEES Regulations and three corresponding – potentially conflicting – roles for lease clauses. MEES are causing some – typically larger – landlords to adopt an active approach in seeking improvements to F- and G-rated buildings. At the same time, MEES has focused the attention of these landlords on the need to protect ratings and the environmental performance of assets, both through preventing detrimental activity and through encouraging positive improvements by tenants. Lease clauses are seen as playing an existing or potential role to enable these two approaches, although other mechanisms, such as fit-out, may also need to play a role. To what extent these approaches will involve cost recovery (or more cost recovery than at present) or whether landlords will have to "take it on the chin" (Interviewee 32) remains to be seen.

On the other hand, there are suggestions that MEES may reinforce tensions between landlords and tenants around cost allocation for improvements, with both seeking to avoid exposure under MEES. In this context, lease provisions may serve to clarify cost allocations more explicitly, with both landlords and tenants seeking to pass on or avoid exposure to costs, and they may restrict landlord access to carry out improvements. These developments may reduce the likelihood of energy efficiency upgrades within the lease, and they may make informal collaboration and arrangements to support improvements outside the lease less likely.

In terms of the role of lease clauses, it was noticeable that when discussing MEES, interviewees tended to focus on the allocation of responsibility and cost between landlords and tenants and on the prevention of certain activity. Other clauses that are typically referred to in the context of green leases – such as cooperation and general sustainability clauses, data sharing, and regular landlord–tenant meetings (Janda et al. 2016) – did not seem to be seen by respondents as relevant to MEES. Failure to associate these broader green clauses with MEES may partly reflect the asset-based (rather than operational) focus of EPCs, which are used as the measure for MEES. However, for those property owners who adopt an active approach, cooperation with their tenants and data sharing may assume new significance, for example, to enable identification of "relevant improvements" under MEES, including those with payback periods of seven years or less. Collaboration may also be needed to deal with the technical and logistical challenges of assessing actual and potential EPC ratings, for example, related to proposed alterations, and in particular in relation to E-rated properties. Moreover, tenants for whom subletting is a possibility may take an active interest in MEES compliance as they themselves would be caught by the MEES prohibition in the event of subletting a sub-standard property.

In terms of the broader landlord–tenant relationship, the issue of cost recovery – who should pay for improvements – featured strongly in the interviews is at the heart of the split incentive problem. It is clear from the early policy documents that the UK government initially intended MEES to tackle this split incentive. In practice, however, policy shifted and the MEES Regulations do not address the underlying misalignment of fiscal incentives. The WICKED interviews revealed the continuing misalignment of incentives in ongoing uncertainty about the link between environmental performance and rental and capital values and the tensions surrounding cost recovery. It may be that, as questions around the link between value and building performance become more settled within the commercial property industry, the underlying split incentive will be reduced and the lease will assume a clearer role in supporting positive compliance, preserving asset value, and avoiding obsolescence (cp. Elliot et al. 2015). In the meantime, however, MEES may inadvertently put extra pressure on the historically adversarial landlord–tenant relationship and undermine moves to encourage more collaborative approaches.

Ultimately, where parties adopt an avoidant approach may of course serve only to delay improvements to sub-standard properties; as existing leases come to an end, landlords are unlikely to be able to rely on the consent exemption to avoid improvements and will have to bear the costs of such improvements before being able to re-let. Whether such costs could be recouped through higher asset or rental values – or whether some landlords may choose to offload sub-standard properties and/or re-let sub-standard properties in contravention of the MEES Regulations – remains to be seen.

To the extent that landlords are simply unable to access the cash needed for relevant improvements and tenants refuse to share the costs, it is unclear what the implications under MEES are – including how the government will be able to enforce compliance in these circumstances. In the absence of a workable finance arrangement (Green Deal or otherwise), the government has arguably failed in its promise to ensure landlords do not face upfront costs (DECC 2014b). Offloading sub-standard properties at a discount to property owners who do have the cash to fund improvements may be one way that improvements are ultimately achieved; however, if offloading results in a secondary, sub-standard market, it may further hamper the successful delivery of MEES objectives.

By mandating minimum standards, the MEES Regulations may well result in active compliance, the harnessing of leases as an active compliance tool, and resulting improvements to much of today's sub-standard commercial building stock. However, MEES only go so far (Elliot et al. 2015): upgrading the UK's commercial building stock to meet the UK's carbon reduction targets will require companies to go beyond compliance and, of course, the application of MEES is confined to commercial properties that are let and does not affect owner-occupied buildings that are sold, for example. In addition, the lease's role in supporting energy efficiency improvements may be limited, even if it is used for active compliance. As many interviewees suggested, positive, proactive, informal, and creative collaboration between landlords and tenants outside the lease will be needed to lift standards across the commercial built environment as a whole. Finally, reductions in energy consumption will require improved operational strategies and management in addition to improvements to the fabric of buildings (JLL and BBP 2012).

3.5 Conclusions

Throughout the WICKED project, the role of leases in relation to energy management and energy efficiency has been questioned by industry participants, including retailers (tenants), property owners (landlords), letting and property management companies, law firms and legal experts, and industry intermediaries and experts. However, MEES are creating an arena that demands consideration of the implications of lease clauses from a regulatory compliance perspective. MEES have highlighted the diverse roles of lease clauses and their potential to support active, protective, and avoidant approaches to compliance. There are many detailed questions that have yet to be explored further in relation to MEES. For example, how will MEES in 2023 affect buildings let on long leases pre-dating MEES? What are the implications of MEES for rent review and statutory lease renewals (cp. CO_2 Estates 2015)? How will MEES interact with "reasonable consent" clauses in leases? The UK Department of Energy and Climate Change will need to consider some of the issues and questions raised – in particular, what MEES will mean for property owners who lack access to upfront cash to carry out improvements.

The implications for individual property owners will vary across different types of company, buildings, and landlord–tenant relationships. However, the findings discussed here suggest that there is a good case for an active approach, which seeks to plan for improvements combined with a protective approach, which prevents the deterioration of assets' environmental performance, provided that investments in energy efficiency are reflected in the avoidance of obsolescence or – perhaps less dramatically – a price differential or "brown discount". Property companies and their occupiers may find this a more compelling proposition if value questions are further clarified across the industry. Professional valuers and industry bodies such as the Royal Institution of Chartered Surveyors (RICS) have a key role here, and further guidance from BBP will also be welcome.

Beyond the UK, barriers to energy efficiency investments, including split incentives and appropriate regulatory and market responses, have been recognised and discussed widely, including across the EU, Australia, and the US (EC 2011; Coalition for Energy Savings 2015; CAEPB 2016; Australian Government 2015; McKinsey & Company 2009). Most policy responses, however, take the form of focusing on minimum standards when building, maintenance, or refurbishment occurs (Coalition for Energy Savings 2015; CAEPB 2016; LEAF 2016) or on mandatory disclosure requirements (for example, the Building Energy Efficiency Certificate [BEEC] in Australia required under the Building Energy Efficiency Disclosure Act 2010). Only the UK and Ireland are – it would appear – introducing minimum standards specifically for *rented* properties (Coalition for Energy Savings 2015). In the UK's case, this means making it unlawful to let sub-standard properties. The wide range of measures and responses across different countries suggests that many of the challenges outlined in this chapter are not UK-specific and that much further work by governments, companies, and industry bodies will be needed to enable improvements to the commercial building stock across the EU and worldwide. The UK's pioneering approach, tying the requirement for minimum standards to the letting of property and thereby creating clear trigger points for improvements, will no doubt be followed with interest by industry and policymakers across the globe.

Note

Since this chapter was written, the Department of Business, Energy and Industrial Strategy (BEIS) has published non-statutory guidance on MEES (BEIS 2017). This does not affect the findings or issues described in this chapter, but provides some further detail on the MEES

Regulations, including, for example, EPC requirements, application of MEES to subletting, and the operation of statutory lease renewals.

Acknowledgements

We gratefully acknowledge the help and assistance of all those who gave up their time for an interview and without whom this chapter would not have been possible. Thank you also to Chris Botten, Programme Manager for the Better Buildings Partnership, who shared his knowledge on and insights into MEES throughout the WICKED project and reviewed an early draft of this chapter. The research for and preparation of this chapter was supported by the UK Engineering and Physical Sciences Research Council under the WICKED research project, grant number EP/L024357/1.

Note

1 The Oxford University-based, RCUK Energy Programme-funded WICKED project (Working with Infrastructure, Creation of Knowledge and Energy Strategy Development) is an interdisciplinary investigation of energy management practices and issues across the UK retail sector. Part of this project set out to investigate the role and impact of leases in relation to energy management. See www.energy.ox.ac/wicked, Janda et al. (2016) and Patrick and Bright (2016).

References

Australian Government. (2015). *2015 commercial building disclosure legal framework*. Canberra: Australian Government. [online] Available at http://cbd.gov.au/overview-of-the-program/legal-framework [accessed 26 May 2016]

BEIS 2017. *The non-domestic Private Rented Property minimum standard: Landlord guidance*. [online] Available at: www.gov.uk/government/publications/the-non-domestic-private-rented-property-minimum-standard-landlord-guidance [accessed 24 August 2016]

Better Buildings Partnership (BBP) (2013). Green Lease Toolkit. [online] Available at: www.betterbuildingspartnership.co.uk/green-lease-toolkit [accessed 15 April 2016]

Botten, C. (2016). Personal communication. 8 March.

Bright, S. (2008). Drafting green leases. *Conveyancer*, 72(6), 498–517

British Council of Shopping Centres (BCSC) and CBRE (2015). Sustainable shopping centres: Energy, performance and value. [online] Available at: www.bcsc.org.uk/news/view?id=67&x%5b0%5d=news/list [accessed 15 April 2016]

British Property Federation (BPF) (2014). *A British Property Federation response to: Private Rented Sector Minimum Energy Efficiency Standard Regulations (Non-Domestic) (England and Wales)*. [online] Available at: www.bpf.org.uk/sites/default/files/resources/BPF-non-domestic-MEES-Consultation-Response.pdf [accessed 20 April 2016]

British Retail Consortium (BRC) (2014). *A better retailing climate: Driving resource efficiency*. [online] Available at: www.brc.org.uk/ePublications/ABRC_Driving_Resource_Efficiency/index.html#/4/ [accessed 15 April 2016]

BSDR (2014). *Green Leases and Minimum Energy Performance Standards*. [online] Available at: www.bsdr.com/green-leases-and-minimum-energy-performance-standards.aspx [accessed 15 April 2016]

CAEPB (Concerted Action Energy Performance of Buildings) (2016). *2016: Implementing the Energy Performance of Buildings Directive (EPBD)*. [online] www.epbd-ca.eu/ca-outcomes/2011-2015 [accessed 6 June 2016]

Carbon Trust (2009). *Building the future, today*. [online] Available at: www.carbontrust.com/media/77252/ctc765_building_the_future__today.pdf [accessed 15 April 2016]

CO_2 Estates (2015). *MEES: The Implications for Rent Reviews, Lease Renewals and Valuation*. [online] Available at: www.co2estates.com/mees-the-implications-for-rent-reviews-lease-renewals-and-valuation/ [accessed 5 May 2016]

Coalition for Energy Savings (2015). *Putting energy efficiency first: Addressing the barriers to energy efficiency. Analysis of the National Energy Efficiency Action Plans in the context of Article 19 of EU Energy Efficiency*

Directive. [online] Available at http://energycoalition.eu/sites/default/files/20150923_Putting%20EE1%20Addressing%20barriers%20EED%20Art%2019%20FINAL.pdf [accessed 25 May 2016]

Cooremans, C. (2011). Make it strategic! Financial investment logic is not enough. *Energy Efficiency*, 4(4), 473–492

Cushman & Wakefield (2016). *Act now to ensure commercial property buildings comply with Energy Act*. Press release, 17 February [online] Available at: www.cushmanwakefield.co.uk/en-gb/news/2016/02/act-now-to-ensure-commercial-property-buildings-comply-with-energy-act/ [accessed 13 April 2016]

Department of Energy and Climate Change (DECC) (2011). *The carbon plan: Delivering our low carbon future*. [online] Available at: www.gov.uk/government/uploads/system/uploads/attachment_data/file/47613/3702-the-carbon-plan-delivering-our-low-carbon-future.pdf [accessed 22 April 2016]

Department of Energy and Climate Change (DECC) (2012). *The Energy Efficiency Strategy: The Energy Efficiency Opportunity in the UK*. [online] Available at: www.gov.uk/government/uploads/system/uploads/attachment_data/file/65602/6927-energy-efficiency-strategy--the-energy-efficiency.pdf [accessed 15 April 2016]

Department of Energy and Climate Change (DECC) (2014a). *The UK's National Energy Efficiency Action Plan and Building Renovation Strategy*. [online] Available at: www.gov.uk/government/publications/the-uks-national-energy-efficiency-action-plan-and-building-renovation-strategy [accessed 22 March 2016]

Department of Energy and Climate Change (DECC) (2014b). *Private Rented Sector Minimum Energy Efficiency Standard Regulations (Non-Domestic): (England and Wales): Consultation on implementation of the Energy Act 2011 provision for energy efficiency regulation of the non-domestic private rented sector*. [online] Available at: www.gov.uk/government/uploads/system/uploads/attachment_data/file/338398/Non-Domestic_PRS_Regulations_Consultation__v1_51__No_Tracks_Final_Version_30_07_14.pdf [accessed 15 April 2016]

Department of Energy and Climate Change (DECC) (2015a). *Final stage impact assessment for the private rented sector regulations*. [online] Available at: www.gov.uk/government/uploads/system/uploads/attachment_data/file/335073/Consultation_Stage_Impact_Assessment_for_the_PRS_Regulations.pdf [accessed 15 April 2016]

Department of Energy and Climate Change (DECC) (2015b). *Private Rented Sector Minimum Energy Efficiency Standard Regulations (Non-Domestic): Government response to 22 July 2014 Consultation on the non-domestic private rented sector energy efficiency regulations (England and Wales)*. [online] Available at: www.gov.uk/government/uploads/system/uploads/attachment_data/file/401378/Non_Dom_PRS_Energy_Efficiency_Regulations_-_Gov_Response__FINAL_1_1__04_02_15_.pdf [accessed 26 April 2016]

Dixon T., Britnell, J., and Watson, G. B. (2014). "City-wide" or "city-blind?" An analysis of emergent retrofit practices in the UK commercial property sector. WP2014/1. Retrofit 2050 Working Paper. [online] Available at: www.retrofit2050.org.uk/sites/default/files/resources/citywidecityblind.pdf [accessed 25 April 2016]

Elliot, B., Bull, R,. and Mallaburn, P. (2015). A new lease of life? Investigating UK property investor attitudes to low carbon investment decisions in commercial buildings. *Energy Efficiency*, 8(4), 667–680

Energy Act (2011). [online] Available at: www.legislation.gov.uk/ukpga/2011/16/contents/enacted [accessed 21 April 2016]

European Commission (EC) (2011). *Energy Efficiency Plan 2011*. COM (2011) 109, European Commission, Brussels. [online] Available at: http://ec.europa.eu/clima/policies/strategies/2050/docs/efficiency_plan_en.pdf [accessed 15 April 2016]

Farnell, B. (2015). MEES: Opening the floodgates for green leases? UK Green Building Council Blogs. [blog] Available at: www.ukgbc.org/resources/blog/mees-opening-floodgates-green-leases [accessed 15 April 2016]

Janda, K. B. (2016). "Who does what with data? A WICKED approach to energy strategies." In *Proceedings of Improving Energy Efficiency in Commercial Buildings Conference (IEECB)*, 16–18 March, 2016 (Frankfurt, Germany). Joint Research Commission

Janda, K. B., Bottrill, C., and Layberry, R. (2014). Learning from the "data poor:" Energy management in understudied organizations. *Journal of Property Investment & Finance*, 32(4), 424–442

Janda, K., Bright, S., Patrick, J., Wilkinson, S. and Dixon, T. (2016). The evolution of green leases: Towards inter-organizational environmental governance. *Building Research & Information*. [online] Available at: http://dx.doi.org/10.1080/09613218.2016.1142811 [accessed 22 April 2016]

Janda, K. B., Patrick, J., Granell, R., Bright, S., Wallom, D., and Layberry, R. (2015). A WICKED approach to retail sector energy management. In *Proceedings of ECEEE Summer Study*, 1–6 June

(Toulon/Hyères, France). Vol. 1: Foundations of Future Energy Policy, pp. 185–195. Stockholm: European Council for an Energy-Efficient Economy.

Jones Lang LaSalle (JLL) (2013). *From obsolescence to resilience: Creating value through strategic refurbishment and asset management.* [online] Available at: www.jll.co.uk/united-kingdom/en-gb/Documents/JLL Obsolescence-Resilience.pdf [accessed 15 April 2016]

Jones Lang LaSalle (JLL) and Better Buildings Partnership (BBP) (2012). *A tale of two buildings: Are EPCs a true indicator of energy efficiency?* [online] Available at: www.jll.co.uk/united-kingdom/en-gb/Research/JLL_BBP_tale_of_two_buildings.pdf [accessed 20 May 2016]

Low Energy Apartment Futures (LEAF) (2016). *National and EU Policy Recommendations: Recommendations for local, national and EU policy on retrofitting multi-occupancy, mixed tenure buildings.* [online] Available at: www.lowenergyapartments.eu/project-findings/policy-recommendations/ [accessed 6 June 2016]

McKinsey & Company (2009). *Unlocking energy efficiency in the US economy.* [online] Available at: www.greenbuildinglawblog.com/uploads/file/mckinseyUS_energy_efficiency_full_report.pdf [accessed 25 May 2016]

Model Commercial Lease (MCL) (2015). *Consultation on proposed amendments to the Model Commercial Lease to take account of the Minimum Energy Efficiency Standard regulations.* 8 October. [online] Available at: http://modelcommerciallease.co.uk/mcl-news/consultation-on-proposed-amendments-to-the-model-commercial-lease-to-take-account-of-the-minimum-energy-efficiency-standard-regulations/ [accessed 20 April 2016]

Model Commercial Lease (MCL) (2016a). [online] Available at: http://modelcommerciallease.co.uk [accessed 15 April 2016]t

Model Commercial Lease (MCL) (2016b). *Changes to version 1.2 of the Model Commercial Lease.* [online] Available at: http://modelcommerciallease.co.uk/july-2016-changes/ [accessed 15 July 2016]

Patrick, J., and Bright, S. (2016). WICKED insights into the role of green leases. *The Conveyancer and Property Lawyer*, 4, 264–285.

Patrick, J., Bright, S., and Jaksch, C. (2015). Still going green? *Estates Gazette,* 29 August, 1534, 50–52

Property Industry Alliance (PIA) (2015). *Property Data Report 2015.* [online] Available at: www.ipf.org.uk/resourceLibrary/pia-property-data-report-2015.html [accessed 15 April 2016]

Sayce, S., Ellison, L., and Parnell, P. (2007). Understanding investment drivers for UK sustainable property. *Building Research & Information*, 35(6), 629–643

UK Green Building Council (UKGBC) (2016). Retrofit: Non-domestic buildings. [online] Available at: www.ukgbc.org/resources/key-topics/new-build-and-retrofit/retrofit-non-domestic-buildings [accessed 20 May 2016]

Williams, P. (2015), The effect on commercial leases of the Minimum Energy Efficiency Standard: Part 2, 1 September [online] Available at: www.falcolegaltraining.co.uk/commentary/effect-commercial-leases-minimum-energy-efficiency-standard-part-2/ [accessed 20 April 2016]

Woodford, L. (2007). The Green Lease Schedule. In *Proceedings of ECEEE Summer Study*, 4–9 June (Colle Sur Loop, France). Panel 3 Local and Regional Activities, pp. 547–556. Stockholm: European Council for an Energy-Efficient Economy

4

Voluntary programs for low-carbon building development and transformation

Lessons from the United States

Jeroen van der Heijden

4.0 Introduction

Governing the transition to a resource-efficient, low-carbon, environmentally sustainable built environment is one of the world's key environmental challenges (IPCC, 2014). To regulate and govern the built environment, governments around the globe have traditionally relied on building codes and planning legislation. Whilst these interventions have contributed to achieving a relatively structurally safe and healthy built environment, they seem less capable of accelerating the transition to resource-efficient and low-carbon buildings and cities. Seeking to overcome problems with traditional governance interventions, governments have begun to trial innovative governance instruments, as have firms and civil society organisations. Often, these innovative instruments rely on voluntary participation and 'beyond-compliance' performance. Examples of such voluntary programs include certification and classification programs, action networks, and innovative forms of financing (Van der Heijden, 2014).

The opportunities and constraints of voluntary programs for urban sustainability have been more theorised than studied empirically (Bulkeley, Castan Broto, & Edwards, 2015; Van der Heijden, 2014). This chapter seeks to increase our knowledge on their performance in the area of urban sustainability. The chapter first briefly reviews the literature on voluntary programs for increased (urban) sustainability, seeking to understand what may be expected of them. The chapter then studies eight voluntary programs for urban sustainability in the United States and assesses their performance. This set includes programs for commercial buildings and programs for residential buildings. The chapter concludes by identifying the main lessons learnt from the empirical study.

4.1 Voluntary programs for urban sustainability: High expectations

Traditional, mandatory governance instruments such as building codes and zoning legislation are often found to fall short as regards in accelerating the transition to resource-efficient and environmentally sustainable buildings and cities. Key problems are the difficulty of ensuring that regulatory interventions keep up with the speed of technological development and insights into the relationship between urbanisation and climate change; the tendency to exempt existing buildings and infrastructure from new and amended regulation; the institutional capital required to develop, implement, monitor, and enforce regulatory interventions; and the political barriers that need to be overcome when implementing or amending mandatory regulation (Van der Heijden, 2014). Understanding these problems, governments, firms, and civil society organisations around the globe have begun trialling voluntary programs for resource-efficient and environmentally sustainable buildings in both the commercial and residential sectors.

A classic example is the voluntary Green Lights program run by the United States' Environmental Protection Agency (EPA), implemented in the early 1990s (EPA, 1994). Through Green Lights, the EPA sought to overcome initial resistance to and unfamiliarity with energy-efficient lighting in commercial buildings. It made visible the ease of reducing energy consumption to building users and supported them in generating knowledge relevant for running their business. Green Lights participants committed to installing energy-efficient lighting in 90 per cent of their facilities—but only where this was profitable to do so. In return, the EPA provided participants with tools (software predominantly) to carry out assessments and keep track of energy savings, helped them connect with lighting retrofitting services, and pointed out potential funding opportunities (Moon & Ko, 2013).

Green Lights fits a larger trend of climate governance innovations (Holley, Gunningham, & Shearing, 2012; Wurzel, Zito, & Jordan, 2013). This trend is characterised by, first, a shift away from government as the sole authority in governing climate change and environment-related problems towards the involvement of public- and private-sector stakeholders for that purpose. Second, there is an interest in governance instruments that encourage self-organisation, market solutions, or both as substitutes for or complements to mandatory command-and-control style instruments. Third, there is a shift towards instruments that reward voluntary compliance as opposed to enforcing mandated behaviour.

The broader trend of the use of voluntary programs for resource-efficient and environmentally sustainable buildings and cities has achieved much acclaim for what it is potentially able to achieve in terms of increased resource efficiency and reduced carbon intensity of buildings and cities (Bingham, 2006; Blanco, 2013; Hohn & Neuer, 2006). A first expectation relates to the often collaborative development processes of these programs. By involving a wide range of stakeholders, such as business and civil society representatives, their tacit knowledge can be used. This is expected to result in programs that are 'smarter' than traditional governance instruments that are developed by somewhat distant bureaucrats (De Búrca & Scott, 2006; Lobel, 2012). Also, by involving a range of stakeholders, programs can be developed through a consensus-building process that allows for a reflection on the advantages and disadvantages for the various parties involved. This is expected to bridge their diverse and sometimes rival views (Bulkeley & Mol, 2003; Holley et al., 2012). It is further expected to increase the acceptance of the programs, and correspondingly, to improve compliance with them (Scott & Trubek, 2002; Walters, 2004).

In terms of the design of voluntary programs, scholars focus particularly on the move away from traditional deterrence-based hard-law instruments that penalise non-compliance, such as building codes, to soft-law instruments that reward compliance and provide positive incentives. Such positive incentives come, for example, in the form of information, the ability to market compliant behaviour, or some form of financial compensation. These positive incentives are, again, expected to increase compliance (Scott & Trubek, 2002). Scholars further expect that compliance with these programs is more likely because individuals and firms commit voluntarily to them rather than being mandated to do so (Borck & Coglianese, 2009). Finally, they highlight a move away from prescriptive rules that specify how compliance should be achieved towards the use of performance-based standards that allow those subject to these instruments (some) flexibility in how to comply. This is expected to make them more willing to move beyond mere bottom-line compliance (Carrigan & Coglianese, 2011; Jänicke & Jörgens, 2006).

That being said, in praise of voluntary programs, to date they have been more theorised than studied empirically (Bulkeley et al., 2015; Van der Heijden, 2014). The limited empirical knowledge base is not univocal about whether or not these innovative governance instruments and voluntary programs for urban sustainability live up to their high normative expectations. Whilst some studies indeed report positive results (Evans, Joas, Sundback, & Thobald, 2005), others have found that they are not always able to attract large numbers of participants, fail to change the behaviour of their participants, or have a negligible overall impact because they set lenient requirements (Gupta, Pfeffer, Verrest, & Ros-Tonen, 2015; Read & Pekkanen, 2009). Also, the broader literature on voluntary programs as a governance answer to climate change and environmental problems points to both successes and failures (Van der Heijden, 2012).

Below, a series of voluntary programs for urban sustainability in the United States are assessed in light of the expectations expressed in the literature with a view to increasing our knowledge of their performance. The empirical research presented builds on a larger research project on voluntary programs for low-carbon buildings and cities (Van der Heijden, 2017).

4.2 Experiences with voluntary programs for low-carbon buildings

The eight programs that are discussed below were chosen from a larger pool of voluntary programs for urban sustainability because they are either a United States-first type of program or a world-first type of program. They can broadly be clustered in three groups: certification and classification programs, action networks, and innovative forms of financing. The set includes programs for commercial buildings and programs for residential buildings. Table 4.1 gives a summary of the eight programs studied.

4.2.1 Certification and classification programs

Certification and classification programs allow the assessment of the performance of buildings or city districts against a set of criteria such as energy efficiency or carbon emissions (Van der Heijden, 2015). If criteria are met, a certificate is issued to indicate compliance. Different forms of certification exist. The first form certifies the expected performance of a building design ('as designed'), the second form certifies the performance of a building built ('as built'), and the final form certifies the achieved performance after a specified period of use ('in operation'). The first two forms are dominant, and the latter was developed in response to critiques of these two. Buildings are often not built in compliance with their certified design; during construction, many flaws might be made that program administrators or their inspectors do not notice, or user behaviour may undo the low-carbon credentials of a building. All of this may result in a certified design or completed building not meeting its expected performance. Such problems are often reported in the literature, as are problems of developers and property owners gaming the programs by seeking easy but not necessarily low-carbon solutions to achieve high classes of certification (Pérez-Lombard, Ortiz, González, & Maestre, 2009; Van der Heijden, 2015).

Popular certification and classification programs in the United States are Leadership in Energy and Environmental Design (LEED) and Energy Star for Buildings for the commercial sector and Energy Star for Homes for the residential sector. LEED was developed and implemented by the United States Green Building Council (USGBC) in 2000. The USGBC is a non-profit organisation, which includes representatives of the construction and property sectors, as well as

Table 4.1 Summary of voluntary programs studied

Name (type and year of introduction)	*Brief description*	*Actual participants as a percentage of targeted pool of participants*	*Average performance by participants*	*Performance in perspective*
Better Buildings Challenge (action network, 2011)	Program for reduced commercial property-related carbon emissions. Brings together the United States federal government and commercial property owners; sole focus on existing commercial property. It aims for a 20 per cent reduction of emissions as of 2010 by 2020.	Less than 1 per cent (representing 4 per cent of commercial building space).	2 per cent energy consumption reduction (with a quarter of participants having achieved at least 10 per cent reductions).	Negligible† compared to the energy consumption of commercial buildings in the United States.
Billion Dollar Green Challenge (innovative financing, 2011)	Program that challenges colleges and universities to set up self-managed revolving loan funds for university building retrofits; sole focus on existing commercial property.	1 per cent.	Over 20 per cent energy consumption reduction.	Negligible compared to the energy consumption of educational facilities in the United States.
Energy Efficient Mortgage (innovative financing, 1995)	Voluntary mortgages for (future) homeowners to make energy efficiency improvements to new or existing residential property.	Less than 1 per cent.	Unknown (no data available).	Negligible compared to residential building energy consumption in the United States.
Energy Star for Buildings (certification and classification, 1999)	Program for energy-efficient commercial property. Applied in new and existing commercial property market.	3 per cent.	Over 20 per cent less energy consumption than conventional commercial buildings (but paper performance*).	Approximately 1 per cent of commercial building energy consumption in the United States.
Energy Star for Homes (certification and classification, 1999)	Program for energy-efficient residential property. Predominantly applied in new residential property market.	7 per cent.	Over 20 per cent less energy consumption than conventional residential buildings (but paper performance*).	Approximately 1.5 per cent of residential building energy consumption in the United States.

LEED (certification and classification, 2000)	Holistic sustainable building program (comparable to BREEAM). Predominantly applied in new commercial property market.	3 per cent (of new office buildings).	Over 20 per cent less energy consumption than conventional commercial buildings (but paper performance*).	Approximately 0.5 per cent of commercial building energy consumption in the United States.
PACE (innovative financing, 2008)	Program that allows municipalities to issue bonds for commercial building retrofits. Applied in 30 states in the United States.	Less than 1 per cent (of commercial property).	62 per cent energy consumption reduction.	Negligible compared to commercial building energy consumption in the United States.
Retrofit Chicago (action network, 2012)	Program for reduced commercial property-related carbon emissions. Brings together the City of Chicago government and the city's major property owners; strong focus on existing commercial property. Spin-off of the Better Buildings Challenge. It aims for a 20 per cent reduction of energy consumption as of 2010 by 2017.	Over 5 per cent (of major office buildings).	7 per cent energy consumption reduction.	Approximately 1.5 per cent of Chicago's office building energy consumption.

† The qualitative descriptor 'negligible' indicates a maximum of 0.5 per cent (for a more elaborate discussion on assessment, see Van der Heijden, 2017).
* The term 'paper performance' indicates that these are expected reductions, not observed reductions; most certificates have been issued certificates for the design of a building and not the performance of a building in operation. This issue is further explained in the text.
Abbreviations: LEED=Leadership in Energy and Environmental Design; PACE=Property Assessed Clean Energy.

government. LEED was launched because various actors in the construction and property sectors in the United States were aware of a potential market for low-carbon buildings for which owners or users are willing to pay a premium. Since its introduction, LEED has become the dominant voluntary program for urban sustainability in the United States (Yudelson & Meyer, 2013). At first glance, LEED has achieved impressive results. By 2015, more than 40,000 projects representing some 350 billion square metres of built-up space had been certified throughout the United States. When looking behind the scenes of LEED, a different picture emerges, however.

Of the LEED certificates issued in the United States, 92 per cent relate to new development projects and 8 per cent to existing buildings. LEED allows for certification of building designs ('as designed'), completed buildings ('as built'), and buildings in use ('in operation'); the latter category makes up about 7 per cent of all certificates issued in the United States. Certificates are issued in four classes based on a summarised weighted outcome of all credits awarded to a building: Platinum, issued to 10 per cent of certified buildings (indicating that at least 75 per cent of the maximum number of credits possible was awarded); Gold, issued to 30 per cent (at least 55 per cent of credits); Silver, issued to 35 per cent (at least 45 per cent of credits); and Certified, issued to 25 per cent (at least 35 per cent of credits). This indicates that those seeking certification prefer low levels of certification over higher ones and prefer new construction projects over retrofitting existing buildings.

LEED sets fairly low criteria. In terms of energy efficiency, for example, an 'as-designed' or 'as-built' building is required to show at least a 14 per cent energy efficiency improvement against a baseline building; or to meet the energy efficiency requirements set in the American Society of Heating, Refrigerating, and Air-Conditioning Engineers (ASHRAE) Standard 90.1 (Roderick, McEwan, Wheatley, & Alonso, 2009). The baseline is based on conventional construction practice in the United States and the ASHRAE standard is widely applied throughout the United States to govern building energy efficiency—sometimes as a voluntary standard, sometimes included in mandatory state and local building regulation as discussed before. These requirements are, generally, not considered to be very challenging; meeting the ASHRAE standard, for example, in many states implies merely complying with mandatory requirements (Roderick et al., 2009; Yudelson & Meyer, 2013).

Buildings certified 'as designed' or 'as built' are repeatedly found to not live up to their certified expected performance; this may be a result of changes in design after certification, construction flaws that go unnoticed by LEED inspectors, or because occupants do not use the building as expected (Newsham, Mancini, & Birt, 2009; Todd, Pyke, & Tufts, 2013). Also problematic is that LEED allows for gaming. The possibility to mix and match criteria under LEED of achieving a specific class of certification was critiqued by interviewees for undermining the possibility of achieving deep decarbonisation of the built environment (Hoffman & Henn, 2009). Whilst this possibility to mix and match allows property developers and property owners to come up with solutions that best suit their buildings, the risk is that they choose low-cost solutions over high-cost solutions—irrespective whether the solution chosen contributes to improved energy efficiency or reduced carbon intensity. 'The road to green certification is paved with low-hanging fruit', the introduction to an article on LEED on a sustainable building website explains. 'This cheat sheet with 22 shortcuts will get you to LEED certification without a lot of trouble' (Seville, 2011). Whilst the author of the article uses the idea of cheating tongue-in-cheek, the 22 shortcuts relate to aspects such as, 'If [the] house [you seek to get certified] is big, make sure you have lots of rooms that can be classified as bedrooms to offset the point penalty for larger homes', and 'Install one efficient showerhead (just one!) per stall, high-efficiency toilets and lavatories (all of them) and get 6 [credits]'. All together, these shortcuts add up to 70 LEED credits, which would get most projects close to—if not over—the 85-credit threshold needed to achieve the highest class of certification: Platinum.

Equally problematic, so argued some interviewees, is that LEED allows for the certification of buildings that contribute to carbon-intensive urban behaviour. Can casinos in the Nevada desert and parking garages in city centres truly be considered low-carbon and environmentally sustainable buildings (Alter, 2008; *USA Today*, 2013)? This is, then, the broader context in which the performance of LEED needs to be considered. When getting behind the descriptive performance statistics, it becomes clear that a building-by-building performance assessment is needed to understand LEED-certified buildings' carbon intensities. But even without a building-by-building performance assessment, a general insight into the performance of LEED can be obtained. More than 800,000 commercial buildings have been constructed in the United States since LEED was launched,[1] of which some 25,000 have been LEED certified (3 per cent). For residential buildings, the numbers are even less promising: Some 21 million homes and housing units have been developed in this period in the United States, of which less than 0.5 per cent have been LEED certified.[2] Furthermore, most of these certificates reflect 'as designed' or 'as built' in the lower levels of classification.

A broadly similar narrative unfolds for Energy Star for Homes and Energy Star for Buildings. These certification and classification programs were developed and implemented by the EPA in 1995 based on its experience with Green Light (discussed above). The agency reports that since their introduction, they have certified 1.6 million homes and 28,000 commercial buildings. Again, this represents only a small fraction of buildings constructed since the programs were introduced: 7 per cent of homes and 3 per cent of commercial buildings.[3] Keep in mind that, as with LEED, this only represents certification as a percentage of buildings built since the program was introduced. Compared to the full building stock in the United States, these numbers reduce even further: at best, 1 per cent of all existing homes and 0.5 per cent of existing commercial buildings are Energy Star certified. Interviewees were again critical that those seeking Energy Star certification might game the system or seek low classes of certification over higher ones, and that due to changes in construction practice or building user behaviour, Energy Star buildings may not live up to their certified performance (Bloom, Nobe, & Nobe, 2011; Shrestha & Kulkarni, 2013).

4.2.2 Action networks

Action networks seek to generate and disseminate knowledge on how to construct and retrofit low-carbon buildings and city districts, and how building user behaviour can be modified to reduce building-related carbon emissions (Van der Heijden, 2016b). Voluntary programs in this category create knowledge 'on the go' by actually developing or retrofitting sustainable buildings or by changing building user behaviour (Evans et al., 2005; Gollagher & Hartz-Karp, 2013). By lifting restrictive building codes to allow knowledge generation in a 'regulation-light' situation, for example, or by pooling resources so that the risks of losing time and money invested are not carried by a single participant, they create secure and supportive environments for participants.

A highly ambitious action network in the United States is the Better Buildings Challenge, which has a focus on the commercial property market. It is part of a larger United States Department of Energy's Better Buildings Initiative (The White House, 2011a). Through the initiative, the federal government seeks a 20 per cent reduction in the United States commercial property sector's energy consumption as of 2010 by 2020. The Better Buildings Challenge is key to achieving that ambition. When launched, it was presented as 'one of the fastest, easiest, and cheapest ways to save money, cut down on harmful pollution, and create good jobs' (The White

House, 2011b). Participants in the challenge commit to reducing their energy consumption by 20 per cent or more over the period of ten years, compared to a base year. They may set this base year up to three years before joining the challenge and agree to conduct an energy efficiency assessment of their building portfolio in order to set their personal benchmark. They agree further to develop and implement an energy plan to achieve their stated ambitions and to report results—data that shows their energy performance, but also insights into how they have achieved improvements. In return, the department supports them with technical and administrative assistance, connects them with firms that can help them save energy, and publicly recognises their participation and performance. The department further transforms their experiences into knowledge into documents and makes this knowledge available through a program website.

Most buildings that the challenge seeks to upgrade were built under lenient mandatory building energy requirements, and participants should find it easy to harvest the 'low-hanging fruit' in terms of energy upgrades (IEA, 2013). Some indeed do: the department highlights that the best performers have achieved energy consumption reductions of up to 40 per cent—and they have achieved this through conventional and low-tech building envelope upgrades and retrofitted duct systems (U.S. Department of Energy, 2014, 2015). In sheer numbers, the program's performance also appears impressive. By 2015, 3.5 billion square feet of commercial space was committed to the challenge, representing 32,000 properties, and in 2014, a total of 58 trillion Btus of energy savings were achieved. On average, participants had reduced their energy intensity by 2 per cent compared to their baseline year, about 14 per cent of participants had already achieved reductions of 20 per cent or more, and another 25 per cent had achieved reductions of at least 10 per cent (U.S. Department of Energy, 2015).

The question, again, is: what do these numbers mean in relative terms? The answer, again, is: not a lot. Current participants in the challenge represent about 4 per cent of commercial building space and 0.6 per cent of the number of commercial buildings—indicating that predominantly large property owners participate. The relative energy savings achieved represents less than 0.3 per cent of the 20 quadrillion Btus of commercial building-related energy consumption in the United States in 2014.[4] What explains how a program that has it all on paper—generating and sharing knowledge, supporting property owners in acquiring funds for their retrofits, and having the full support of the federal government—achieves such negligible results? It appears that the program's dominant focus on rewarding leadership is only of interest to a small group of property owners. 'We want to make the case that not everyone has to start at square one', the director of the challenge explained about the underlying strategy. 'We want other organizations to follow what the leaders have done' (quoted in Zimmerman, 2012). The first participants were selected by the department because of their past performance of being leaders in the area of environmental sustainability—including organisations such as 3M and Lend Lease, who in 2015 were among the absolute top performers in the challenge. 'The hope is that the success that they've had becomes solutions that other organizations can use so that leaders in the Better Buildings Challenge are serving as mentors for other organizations', the director clarified further (quoted in Lack, 2012). This is why the department actively transforms the experiences of participants into knowledge documents that can be used by others. However, these others might very well face different problems than the leaders of the pack. The leaders are major corporations with large property portfolios. They have a corporate image to live up to, they have dedicated property staff, and they can replicate experiences throughout their property portfolio, so explained interviewees. The large majority of commercial property owners only own a single or a few often relatively small buildings and do not have dedicated property staff, and upgrading the energy efficiency of their businesses considerably interrupts their daily business or that of their tenants. The lessons drawn from the Better Building Challenge might very well not resonate with a large majority of property owners.

Comparable experiences were reported for another action program, Retrofit Chicago, which also addresses the commercial property market. It seeks to reduce the energy consumption of office buildings in Chicago by at least 20 per cent over a period of five years. It was launched by the City of Chicago in 2012, and builds on close collaboration between the city council and participating commercial property owners, seeks to generate and share knowledge on office retrofits, and rewards participants with media exposure in local, national, and international forums. By 2015, some 45 property owners—represented by a small number of professional property managers—had committed 50 buildings to the program, and in its first three years of implementation, participants had reduced their building-related energy consumption by, on average, 7 per cent as of the 2010 baseline. Again, this was not exceptional performance, interviewees argued. They were further critical of the fact that participants were allowed to choose or set a baseline well before the action network was introduced. It is likely that reported reductions are a result of building improvements made or planned by participants in the period between the baseline year and the launch of the action network. This misleadingly suggests that the action network has performed rather better than it actually has (Lydersen, 2012; NRDC, 2014).

4.2.3 Innovative forms of financing

Innovative forms of financing are particularly concerned with the difficulties that property developers and property owners face in obtaining funds for the development or retrofitting of low-carbon buildings or city districts (Van der Heijden, 2016a). Although the academic literature reports that low-carbon buildings come with considerable cost savings and possibly even with an increase in market value, banks and other financial institutions are hesitant to provide loans and mortgages for low-carbon buildings. They fear that the low-carbon credentials of these buildings are not reflected in their market value, which implies a risk if building owners cannot pay back the loans and mortgages provided (Junghans & Dorsch, 2015; World Bank, 2011). Banks further fear that anticipated cost savings will not materialise when building owners show 'prebound'-type behaviour—a situation where they feel that they can use more energy now that they are in an energy-efficient building (Sunikka-Blank & Galvin, 2012). Likewise, building owners might be concerned that they will not see a return on their investments in low-carbon buildings because they do not own a property long enough, or they might not value long-term gains over short-term costs, a problem known as hyperbolic discounting (Ameli & Brandt, 2015).

Seeking to overcome these two specific problems, city governments have begun to act as intermediary between finance suppliers and building owners. In doing so, they seek to take away the risk experienced by finance suppliers by generating funds on behalf of building owners, or by directly borrowing funds from finance suppliers, and then making these available to building owners. City governments have, normally, a better risk profile for finance providers than individual building owners. Funds are, normally, recouped through new property taxes on the property. If a property changes ownership, the new owner becomes responsible for paying back funds. This design was first introduced in the United States as a nationwide innovative governance instrument in 2008 as Property Assessed Clean Energy (PACE). PACE helps commercial property owners to access long-term loans for energy retrofits and upgrades.[5] Loans are sought from local governments, and PACE provides a framework and guidelines for government and property owners to come to agreement on how loans are being used.

PACE allows local governments and property owners to enter into tailored agreements that take into account the specific circumstances of participants and their property. Once they have entered into an agreement, the local government issues a bond on behalf of the property owner that can be purchased by a third-party finance provider. After obtaining funds, the local government supplies these to the property owner, who uses them for energy efficiency retrofits and upgrades following the agreement. The local government recoups these funds—with interest—through an additional property tax on the property, and pays back the funds and interest to the third-party finance provider. To be allowed to impose this additional property tax, PACE requires that state and local governments enact specific legislation. By 2015, 30 states had in place such legislation, with 13 states actively using it (PACE Now, 2015). Administrators of local PACE programs in San Francisco and Sacramento considered that one of the main strengths of the program is that it ties funds provided to the building and not to the building owner. The funds provided are recouped through the additional property tax, which implies that the duty to repay funds moves to a new property owner if the building changes ownership. Another main strength is the involvement of the local government as the risk-taker. The repayment obligation is secured through property taxes, which allows the government to advertise the bonds over a 20-year repayment period at competitive rates. By 2015, seven years after its launch, PACE in the United States had been applied to 350 commercial projects and had generated $130 million in private funds.[6] Whilst the energy efficiency improvements of these projects are impressive—62 per cent on average compared to the pre-retrofitting situation—the total number of projects PACE has been applied to is negligible when compared to the United States' full commercial building stock (see above). The administrators of the local PACE programs considered that they faced a demand problem because building owners do not prioritise building retrofits.

Another popular innovative form of financing low-carbon building development and transformation are revolving loan funds. Such funds consist of a sum of money dedicated to achieving a specific result (for example, building retrofits) and are 'revolving' because, once they have been paid back to the central fund, it can issue new loans to other projects (Boyd, 2013; Chou, Hammer, & Levine, 2014). A unique take on revolving loan funds is the Billion Dollar Green Challenge in the United States, which has a focus on, predominantly, the university sector. It was implemented by the Sustainable Endowments Institute, a project within the Rockefeller Foundation, in 2011 and challenges (prospective) participants to invest a total of $1 billion in self-managed revolving loan funds to finance energy efficiency improvements to educational facilities. Targeted participants are educational organisations (universities and colleges) and environmental non-profits. The challenge does not provide funds itself, but seeks to make the concept and advantages of revolving loan funds better known throughout the United States' education sector. It supports participants with guidelines on how to set up self-managed revolving loan funds, and with computer software to track their own performance, compare it with that of other participants, and give case studies from other participants on how to improve the energy efficiency of educational buildings (Bornstein, 2015). The challenge does not require its participants to achieve a specified increase in energy efficiency. By 2015, some 50 universities and colleges participated in the challenge with $110 million committed to self-managed revolving loan funds; they represent a mere 1 per cent of the close to 5,000 universities and colleges in the United States.[7]

A third innovative form of financing are energy efficiency mortgages, which are predominantly applied in the residential property sector. The United States Federal Housing Administration, for example, implemented the Energy Efficient Mortgage Program in 1995. It recognises that homeowners can reduce their utility expenses through energy retrofits or upgrades of their houses—making them more capable of paying back mortgages—and allows those who seek doing so to top up their approved mortgage. It considers energy savings as additional disposable income that improves borrowers' ability to pay and thus reduces the risk for finance providers (U.S. Department of Housing and Urban Development, 1995). Homeowners are allowed additional mortgages that are less than the expected savings from energy upgrades of new or existing homes; estimated energy savings must be determined through an acknowledged system or by an acknowledged energy consultant. Maximum mortgage limits are set to 5 per cent of the property value. Mortgages issued under the Energy Efficient Mortgage Program are insured by the Federal Housing Administration, which secures finance providers against loan default. Typically, these mortgages are marketed as a win–win opportunity: homeowners can save costs by reducing their energy consumption, which at the same time reduces the carbon emissions of their homes (Kats, Menkin, Dommu, & DeBold, 2012).

Yet again, the impact of the Energy Efficiency Mortgage Program is negligible. The Federal Housing Administration issued fewer than 6,000 mortgages under the program in the period of 2006–2014 (Federal Housing Administration, 2011, 2012, 2013, 2014). To put this number in perspective, in its best year thus far, 2011, out of close to 15 million mortgages issued throughout the United States, a mere 1,065 were issued under the program, representing fewer than 1 in 10,000 mortgages issued that year. Reviews highlight a number of problems that explain this low uptake. On the demand side, there is limited interest in energy efficiency retrofits. Homeowners simply do not demand these mortgages because they are generally not interested in retrofitting their homes. Those who are interested in doing so often have funds for retrofits, or consider the paperwork and other related administrative efforts too much hassle for the relatively small (additional) mortgage provided (Kolstad, 2014). On the supply side, finance providers are found to also be hesitant. One of their recurring arguments is that technical energy retrofits and upgrades do not guarantee energy savings by consumers because consumer behaviour plays an important role as well. If consumers use more energy because they feel this is acceptable since they (now) occupy an energy-efficient house, their cost savings reduce, making them less capable of paying back the additional mortgage (Allen, Barth, & Yago, 2012).

4.3 Discussion and conclusions

This chapter has highlighted that various organisations in the United States have introduced voluntary programs for urban sustainability, seeking to overcome the shortfalls of traditional, mandatory governance interventions. However, these voluntary programs have, at best, resulted in some pockets of good practice. Despite having been in place since the mid-1990s, they have not considerably accelerated the transition to urban sustainability in the United States—neither in the commercial building sector, nor in the residential building sector.

The main finding of this study is that the voluntary programs studied fail in attracting participants. This indicates that whether or not these programs require their participants to make considerable urban sustainability improvements or whether or not individual participants showcase exemplary performance, the programs' overall impact on reduced resource consumption and carbon emissions at the building and city level is negligible. This is a worrisome finding. The administrators of these programs do their utmost to make low-carbon development and transformation look possible and financially viable, but have at the very best attracted only the top segment of the leaders in the construction and property sectors. This indicates that not much should be expected of such voluntary programs in terms of attracting the majority of these sectors, let alone their laggards. This problem seems, in part, to be a result of the tendency of program administrators to target the leaders in the industry, assuming that once they are 'in', the rest will follow. Whilst such ripple effects may take place in homogeneous markets where individuals and organisations feel much pressure to act in the same way as their peers, the construction and property sectors are likely too heterogeneous to result in such processes (Rogers, Medina, Rivera, & Wiley, 2005). Administrators may want to be more aware of the heterogeneity of these sectors and develop different programs that target different market segments or promote these programs in different ways to different market segments. Another part of the problem seems to be related to the voluntariness and open-endedness of most of the instruments studied. These combined characteristics may not capture the sense of urgency that is needed in achieving urban sustainability on a large scale and in a timely manner (Hoffmann, 2011).

That being said, the practices studied have resulted in a wealth of examples that indicate that resource-efficient and environmentally sustainable buildings are possible with current technology and can be achieved relatively rapidly—and often at a net-cost benefit. The question, then, is: how to move on? If neither mandatory requirements nor these voluntary programs are likely to accelerate the transition to urban sustainability, then the answer is perhaps somewhere in the middle (Van der Heijden, 2017). Combining aspects of traditional mandatory and voluntary programs may result in hybrids that build on the strengths of both approaches and help overcome their weaknesses. One may think of considering voluntary programs as a test area for mandatory local and state construction regulation and legislation. They could be operational as voluntary instruments for some years, and, after a set period, the best performing could be included in the mandatory framework. Such 'rolling rule regimes' have been referred to earlier as a possible answer to complex environmental risks (Sabel, Fung, Karkkainen, Cohen, & Rogers, 2000). One could also think of moving from the currently voluntary opting-in approach that these instruments rely on to a voluntary opting-out approach. Why not supply all households who seek a mortgage with an energy efficiency mortgage, which they can turn down if they really do not want the energy efficiency of their homes improved? Opting in to these mortgages may be too high a barrier to begin with, and in other areas of behavioural interventions, opting-out systems have shown better results than opting-in systems (Kosters & Van der Heijden, 2015). Yet another alternative would be to actively design synergies between various interventions. A PACE-like instrument that seeks to address commercial property owners could work together with an instrument that seeks to change the behaviour of tenants of such commercial property.[8]

To conclude, this chapter has provided a unique insight into the performance of voluntary programs for urban sustainability in the United States. Contrary to the high expectations that are often expressed about these governance instruments, this chapter found that a series of eight real-world examples of such voluntary programs have had a negligible overall impact in reducing resource consumption and carbon emissions at building and city level. Whilst this study comes, as with any study, with caveats (as discussed before), it challenges us to be more critical about the highly normative expectations expressed about voluntary programs for urban sustainability. Future scholarship may wish to assess whether the trend of negligible uptake and impact found in this study holds more broadly for the United States, and whether it also holds in other countries and regions.

Notes

1 Statistical data on the voluntary programs throughout this section was obtained from the United States' Census Bureau, www.census.gov (31 March 2016), unless otherwise stated.
2 Data from www.leed.usgbc.org and www.census.gov and www.eia.gov (31 March 2016).
3 Data from www.energystar.gov and www.census.gov (31 March 2016).
4 Additional data from www.eia.gov (31 March 2016).

5 PACE initially applied to commercial and residential property, but as a result of the subprime mortgage crisis of 2008, the United States mortgage authorities (Fannie Mae and Freddie Mac) refused to finance mortgages under PACE (Bird & Hernandez, 2012; Kirkpatrick & Bennear, 2014; Sichtermann, 2011).
6 It had generated close to another $1 billion for residential property. Data from: www.pacenow.org (31 March 2016).
7 Data from: www.nces.ed.gov (31 March 2016).
8 The Chicago Green Office Challenge is an example of a tenant-oriented instrument; see www.greenpsf.com/go/community/challenge/chicago (31 March 2016).

References

Allen, F., Barth, J., & Yago, G. (2012). *Fixing the Housing Market*. Philadelphia: Wharton School Publishing.

Alter, J. (2008, 2 January). Slate on "Decidedly Dupable" LEED. Retrieved from www.treehugger.com/sustainable-product-design/slate-on-decidedly-dupable-leed.html.

Ameli, N., & Brandt, N. (2015). *What Impedes Household Investment in Energy Efficiency and Renewable Energy?* Paris: OECD Publishing.

Bingham, L. B. (2006). The New Urban Governance. *Review of Policy Research, 23*(4), 815–826.

Bird, S., & Hernandez, D. (2012). Policy Options for the Split Incentive. *Energy Policy, 48*, 506–514.

Blanco, I. (2013). Analysing Urban Governance Networks. *Environment and Planning C, 31*(2), 276–291.

Bloom, B., Nobe, M., & Nobe, M. (2011). Valuing Green Home Designs. *Journal of Sustainable Real Estate, 3*(1), 109–126.

Borck, J., & Coglianese, C. (2009). Voluntary Environmental Programs. *Annual Review of Environmental Resources, 34*, 305–324.

Bornstein, D. (2015, 6 February). Investing in Energy Efficiency Pays Off. *New York Times*. Retrieved from http://opinionator.blogs.nytimes.com/2015/02/06/investing-in-energy-efficiency-pays-off/.

Boyd, S. (2013). Financing and Managing Energy Projects Through Revolving Loan Funds. *Sustainability, 6*(6), 345–352.

Bulkeley, H., & Mol, A. (2003). Participation and Environmental Governance: Consensus, Ambivalence and Debate. *Environmental Values, 12*(2), 143–154.

Bulkeley, H., Castan Broto, V., & Edwards, G. (2015). *An Urban Politics of Climate Change*. London: Routledge.

Carrigan, C., & Coglianese, C. (2011). The Politics of Regulation. *Annual Review of Political Science, 14*, 107–129.

Chou, B., Hammer, B., & Levine, L. (2014). *Using State Revolving Funds to Build Climate-Resilient Communities*. New York: National Resource Defence Council.

De Búrca, G., & Scott, J. (2006). *New Governance and Constitutionalism in Europe and the U.S.* Oxford: Hart.

EPA. (1994). *EPA's Financial Management Status Report and Five-Year Plan*. Washington: U.S. Environmental Protection Agency.

Evans, B., Joas, M., Sundback, S., & Thobald, K. (2005). *Governing Sustainable Cities*. London: Earthscan.

Federal Housing Administration. (2011). *Annual Management Report: Fiscal Year 2011*. Washington, D.C.: U.S. Department of Housing and Urban Development.

Federal Housing Administration. (2012). *Annual Management Report: Fiscal Year 2012*. Washington, D.C.: U.S. Department of Housing and Urban Development.

Federal Housing Administration. (2013). *Annual Management Report: Fiscal Year 2013*. Washington, D.C.: U.S. Department of Housing and Urban Development.

Federal Housing Administration. (2014). *Annual Management Report: Fiscal Year 2014*. Washington, D.C.: U.S. Department of Housing and Urban Development.

Gollagher, M., & Hartz-Karp, J. (2013). The Role of Deliberative Collaborative Governance in Achieving Sustainable Cities. *Sustainability, 5*(6), 2343–2366.

Gupta, R., Pfeffer, K., Verrest, H., & Ros-Tonen, M. (2015). *Geographies of Urban Governance*. New York: Springer.

Hoffman, A. J., & Henn, R. (2009). Overcoming the Social Barriers to Green Building. *Organization & Environment, 32*(4), 390–419.

Hoffmann, M. J. (2011). *Climate Governance at the Crossroads: Experimenting with a Global Response after Kyoto*. Oxford: Oxford University Press.

Hohn, U., & Neuer, B. (2006). New urban governance. *European Planning Studies, 14*(3), 291–298.

Holley, C., Gunningham, N., and Shearing, C. (2012). *The New Environmental Governance*. London: Routledge.

IEA. (2013). *Modernising Building Energy Codes*. Paris: United Nations Development Programme.

IPCC. (2014). *Climate Change 2014*. Cambridge: Cambridge University Press.

Jänicke, M., & Jörgens, H. (2006). New Approaches to Environmental Governance. In M. Jänicke & K. Jacobs (Eds), *Environmental Governance in Global Perspective* (pp. 167–209). Berlin: Freie Universitat Berlin.

Junghans, L., & Dorsch, L. (2015). *Finding the Finance*. Bonn: Germanwatch.

Kats, G., Menkin, A., Dommu, J., & DeBold, M. (2012). *Energy Efficiency Financing*. Washington, D.C.: Capital E.

Kirkpatrick, J., & Bennear, L. (2014). Promoting Clean Energy Investment. *Journal of Environmental Economics and Management, 68*(2), 357–375.

Kolstad, L. (2014). *Designing a Mortgage Process for Energy Efficiency*. Washington, D.C.: American Council for an Energy-Efficient Economy.

Kosters, M., & Van der Heijden, J. (2015). From Mechanism to Virtue. *Evaluation, 21*(3), 276–291.

Lack, B. (2012, 9 January). The Better Buildings Initiative. Retrieved from www.dailyenergyreport.com/the-better-buildings-initiative-why-companies-are-taking-the-challenge.

Lobel, O. (2012). New Governance as Regulatory Governance. In D. Levi-Faur (Ed.), *The Oxford Handbook of Governance* (pp. 65–82). Oxford: Oxford University Press.

Lydersen, K. (2012, 31 August). Retrofit Chicago. *Midwest Energy News*. Retrieved from www.midwestenergynews.com/2012/08/31/retrofit-chicago-is-energy-efficiency-plan-worth-the-hype/.

Moon, S.-G., & Ko, K. (2013). The Effectiveness of U.S. Voluntary Environmental Programs. *International Review of Public Administration, 18*(3), 163–184.

Newsham, G., Mancini, S., & Birt, B. (2009). Do LEED-Certified Buildings Save Energy? Yes, but... *Energy and Buildings, 41*(8), 897–905.

NRDC. (2014). *Retrofit Chicago: Commercial Buildings Initiative*. Chicago: Natural Resources Defence Council.

PACE Now. (2015). *Annual Report 2014*. Pleasantville: PACE Now.

Pérez-Lombard, L., Ortiz, J., González, R., & Maestre, I. R. (2009). Review of Benchmarking, Rating and Labelling Concepts within the Framework of Building Energy Certification Schemes. *Energy and Buildings, 41*(3), 272–278.

Read, B. L. and Pekkanen, R. (Eds). (2009). *Local Organizations and Urban Governance in East and Southeast Asia: Straddling State and Society*. London: Routledge.

Roderick, Y., McEwan, D., Wheatley, C., and Alonso, C. (2009). *A Comparative Study of Building Energy Performance Assessment Between LEED, BREEAM and Green Star Schemes*. Paper presented at the Building Simulation 2009, Glasgow.

Rogers, E. M., Medina, U., Rivera, M., & Wiley, C. (2005). Complex Adaptive Systems and the Diffusion of Innovations. *Innovation Journal, 10*(3), 1–26.

Sabel, C., Fung, A., Karkkainen, B., Cohen, J., & Rogers, J. (2000). *Beyond Backyard Environmentalism*. Boston: Beacon Press.

Scott, J., & Trubek, D. (2002). Mind the Gap. *European Law Journal, 8*(1), 1–18.

Seville, C. (2011, 24 May). How to Cheat at LEED for Homes. Retrieved from www.greenbuildingadvisor.com/blogs/dept/green-building-curmudgeon/how-cheat-leed-homes.

Shrestha, P., & Kulkarni, P. (2013). Factors Influencing Energy Consumption of Energy Star and Non-Energy Star Homes. *Journal of Management in Engineering, 29*(3), 269–278.

Sichtermann, J. (2011). Slowing the Pace of Recovery. *Valparaiso University Law Review, 46*(1), 263–309.

Sunikka-Blank, M., & Galvin, R. (2012). Introducing the Prebound Effect: The Gap Between Energy Performance and Actual Energy Consumption. *Building Research & Information, 40*(3), 260–273.

The White House. (2011a). *Factsheet: Better Buildings Initiative*. Washington, D.C.: The White House.

The White House. (2011b, 2 December). We Can't Wait. Retrieved from https://obamawhitehouse.archives.gov/the-press-office/2011/12/02/we-cant-wait-president-obama-announces-nearly-4-billion-investment-energ.

Todd, J. A., Pyke, C., & Tufts, R. (2013). Implications of Trends in LEED Usage. *Building Research & Information, 41*(4), 384–400.

U.S. Department of Energy. (2014, 9 May). Better Buildings Challenge to Cut Energy Waste Grows by 1 Billion Square Feet. Retrieved from http://energy.gov/articles/better-buildings-challenge-cut-energy-waste-grows-1-billion-square-feet.

U.S. Department of Energy. (2015). *Progress Report 2015*. Washington, D.C.: U.S. Department of Energy.

U.S. Department of Housing and Urban Development. (1995). *Mortgage Letter 95: 46*. Washington, D.C.: U. S. Department of Housing and Urban Development.

USA Today. (2013, 13 June). In U.S. Building Industry, Is It Too Easy to Be Green? Retrieved from www.usatoday.com/story/news/nation/2012/10/24/green-building-leed-certification/1650517/.

Van der Heijden, J. (2012). Voluntary Environmental Governance Arrangements. *Environmental Policies*, *21*(3), 486–509.

Van der Heijden, J. (2014). *Governance for Urban Sustainability and Resilience*. Cheltenham: Edward Elgar.

Van der Heijden, J. (2015). On the Potential of Voluntary Environmental Programmes for the Built Environment: A Critical Analysis of LEED. *Journal of Housing and the Built Environment*, *30*(4), 553–567.

Van der Heijden, J. (2016a). Eco-Financing for Low-Carbon Buildings and Cities: Value and Limits. *Urban Studies* (online first), 1–16. doi:10.1177/0042098016655056.

Van der Heijden, J. (2016b). Experimental Governance for Low-Carbon Buildings and Cities: Value and Limits of Local Action Networks. *Cities*, *53* (April), 1–7. doi:10.1016/j.cities.2015.12.008.

Van der Heijden, J. (2017). *Voluntary Programs for Sustainable Buildings and Cities: Opportunities and Constraints in Decarbonising the Built Environment*. Cambridge: Cambridge University Press.

Walters, W. (2004). Some Critical Notes on Governance. *Studies in Political Economy*, *73* (Spring/Summer), 27–46.

World Bank. (2011). *Climate Change and the World Bank Group*. Washington, D.C.: The World Bank.

Wurzel, R., Zito, A., & Jordan, A. (2013). *Environmental Governance in Europe*. Cheltenham: Edward Elgar.

Yudelson, J., & Meyer, U. (2013). *The World's Greenest Buildings*. London: Routledge.

Zimmerman, G. (2012). Department of Energy: Goal of Better Buildings Challenge is to Cut Down on $60 Billion in Wasted Energy. Retrieved from www.facilitiesnet.com/energyefficiency/article/Department-of-Energy-Goal-of-Better-Buildings-Challenge-is-to-Cut-Down-on-60-Billion-in-Wasted-Energy-Facilities-Management-Energy-Efficiency-Feature--13269.

5

Sustainable office retrofit in Melbourne

Sara Wilkinson

5.0 Introduction

In 1950, the global urban population stood at 30 per cent of total population; in 2015, this figure stood at 56 per cent (or 4.1 billion); by 2050, it is predicted to reach 66 per cent, or 6.3 billion (UN, 2014). Though the number and size of dense cities is rising, they are crucial to humankind's attempts to mitigate the effects of global warming and calamitous levels of climate change. This is because the environmental footprint of people in dense cities is lower than that of those in rural areas and low-density cities (Brand, 2009). In addition, the majority of building stock will be around for many decades; for example, it is estimated that 87 per cent of the UK's 2050 building stock is already built (Kelly, 2008). However, the design, construction and operation of most existing stock predates considerations of sustainability, and with new building adding between a maximum of 1–2 per cent to the total building stock annually (Knott, 2007), it is in retrofit that sustainability in the built environment may be delivered. Other issues include changing work patterns and smaller household composition, and retrofit of our built environment must take these considerations into account (RICS, 2015). A question arises: are voluntary or mandatory measures more likely to bring about the necessary changes?

Recognising the importance of retrofit to reduce total building-related greenhouse gas emissions (GHG), the City of Melbourne initiated the 1200 Buildings Program in 2008. The program encourages owners and tenants to undertake sustainable office retrofit. Like many global cities, the City of Melbourne is seeking to become a carbon-neutral city. Melbourne wants to do this by 2020 by reducing GHG emissions by 50 per cent (or 1,850,000 tonnes), implementing new and renewable energy use, greening the power supply and sequestering of residual emissions to deliver zero-net emissions or offsetting (City of Melbourne, 2003). Reported emissions in 2002 were 3.75 million tonnes of carbon dioxide equivalent (t CO_2-e) with commercial buildings accounting for 59 per cent (2.21 million t CO_2-e). It is estimated that commercial-sector emissions will be 9.9 t CO_2-e per employee by 2020 and a reduction target to 4.1 t CO_2-e was set. This figure represents an emission reduction target of 1,004 kt CO_2-e on 2020 business-as-usual (BAU) emissions. For existing office buildings, a retrofit program of 66 per cent of total stock, or 1,200 buildings, was established. This chapter reports on a detailed analysis of the buildings within the program and whether there has been a shift in the market towards environmental retrofit by assessing the importance of environmental attributes within office building retrofits. A 2014 update (City of Melbourne, 2014) acknowledged that 'collective progress in certain areas has been slow and that targets were based on assumptions that Australia would put a price on carbon and international policy would be in place to drive significant reductions'. They state further that if the current growth rate continues to 2020, total emissions will rise to 7.7 million tonnes, a 60 per cent increase on 2010 levels (City of Melbourne, 2014).

Reliance on legislation is slow and sustainability is a recent driver, especially in Australia. Only in 2006 was energy efficiency for offices introduced into the Building Code of Australia, 22 years after similar UK legislation (Wilkinson, 2014). It will take decades to transform the stock to comply with current code standards. Another recent change was the Mandatory Disclosure Act 2010, which requires an energy rating of all commercial space exceeding 2000m^2 at the point of sale or lease renewal (Warren and Huston, 2011). The energy rating used here is the National Australian Building Energy Rating System (NABERS). Minimum standards for energy efficiency apply to new-build and retrofit projects over a certain amount of work. It is possible that the mandatory disclosure legislation may have a faster impact on sustainable retrofit because the current typical lease term in Australia is five years, and therefore turnover is short term (Warren and Huston, 2011). Given that the first piece of office building sustainability-related legislation was established in 2006, followed by the 1200 Buildings Program and the advent of the Mandatory Disclosure Act in 2010, a question arises: what are the current patterns of retrofit in commercial offices? The definition for 'retrofit' comes from the 1200 Buildings/Survey definition: 'works that improve the environmental (energy, water, waste, IEQ) performance of a building, i.e., replacement of a chiller or boiler, a lighting upgrade and so on'. This definition enables an analysis of retrofits from minor to major works, across and within use. Across-use retrofits are excluded as they fall outside the 1200 Buildings Program.

5.1 Building retrofit and sustainability attributes

Decision making with regards to building retrofit is multifaceted, and many studies (Bullen, 2007; Bullen and Love, 2011; Remøy and Van der Voordt, 2007; Langston, Wong, Hui and Shen 2008) found precise classification of the attributes influencing retrofit to be subjective, complex and challenging. Attributes affecting retrofit are economic, environmental, social, technological, legal and physical (Wilkinson, 2011a; Wilkinson, Remøy and Langston, 2014).

Physical attributes determine whether retrofit is possible and/or desirable, and all studies identified age as important (Ball, 2002; Nutt, Walker, Holliday and Sears, 1976). Some construction forms and materials make retrofit more expensive and/or difficult because of regulatory compliance (Bullen, 2007). Gann and Barlow (1996) found height, construction type and frame condition to be important, while a 2011 Melbourne study confirmed that height is important in office retrofit (Wilkinson, 2011a). Kincaid's (2002) London study discovered that floor size was significant and buildings with unusual floor plates or sizes were harder to adapt compared to open-plan space. The location of the service core was critical as it affected flexibility and ability to sub-divide space (Wilkinson, 2011a; Arge, 2005; Szarejko and Trocka-Leszczynska, 2007); a central location gave greater scope for sub-dividing the floor plate and minimised loss of net lettable area (NLA). Detached buildings were easier and more desirable for retrofit as contractors work faster with less disruption (Isaacs in Baird, Gray, Isaacs, Kernohan and McIndoe, 1996; Wilkinson 2011a). Access, or the number of entry and exit points to buildings, impacted on ease of retrofit (Remøy and Van der Voordt, 2006). With building width, Arge (2005) and Povall and Eley (in Markus, 1979) identified optimal benchmarks as 15–17m, demonstrating that these buildings suited more space configurations and user needs. The technical grid or distance between structural columns also influenced the ease of retrofit (Arge, 2005). Floor strength was vital in a London study; a 3–5 kN/m^2 floor suited retail, office and hospital uses (Kincaid, 2002). Floor strength determined possible land uses, and it was not viable to adapt to office use when floor strength was 3 kN/m^2, unless strengthening or replacement was undertaken.

Technical grid and services equipment influenced whether buildings could accommodate extra capacity (Arge, 2005). Raised floors permitted IT cabling to be upgraded easily and zone-based internet communication technology provision allowed greater flexibility and adaptability, a sought-after attribute (Arge, 2005). Modularity described modules that could be rearranged, replaced, combined or interchanged (Arge, 2005). Buildings with potential for lateral or vertical extension were more adaptable as size could be increased to suit new uses and occupiers (Arge, 2005), an attribute also found to be important in a Melbourne retrofit study (Wilkinson, 2011a). Elasticity, the potential for sub-division for either letting or sale, allowed owners to keep up with changes in demand (Arge, 2005).

Physical location was found to be important in two Melbourne office retrofit studies in terms of proximity to public transport, which is environmentally positive in reducing private car emissions (Wilkinson, 2011a; Wilkinson, 2014). The amount of on-site parking was significant where little or no public transport was available (Douglas, 2006). Land use attributes were found to be important (Arge, 2005), as existing land use impacted the scope for new or changed uses. Though existing planning zones define legally permissible development, sites can be rezoned and a proactive policy approach in Toronto increased retrofit (Heath, 2001).

Swallow (1997) concluded that retrofit was affected by tenure as it influenced parties' willingness to invest. Owners have an interest that lasts into perpetuity, whereas lessees have short-term interests for lease terms, typically five years in Melbourne. Owners can be institutional or private. Institutional owners invest to maximise return on investment and appoint consultants to evaluate retrofit potential. Private owners may or may not use consultants and may be located offshore with less knowledge of market conditions. Private owners hold property for various reasons, such as future development potential, rental income or capital growth, and may engage in less retrofit. Wilkinson's (2011a) study of 7,393 office retrofits found owner type and tenure to be important, with institutional owners most likely to engage in retrofit activity.

Ability to retrofit is affected by occupation. With single tenants, owners can retrofit when leases expire; however, with multiple tenants, leases are unlikely to terminate simultaneously, and the building may be partly empty and not income earning before owners can retrofit. Owners can terminate leases early with agreement, compensating or decanting tenants temporarily or permanently.

Historic listing protects architecturally or socially important buildings for societal benefit (Ball, 2002). Heritage retrofit includes possible additional costs of using traditional materials, techniques and craftspeople (Bullen, 2007). As regards legal issues, retrofit was found to benefit from proactive policies and legislation (Gann and Barlow, 1996; Ball, 2002). Urban regeneration studies in London and Bristol docklands found that proactive policy and legislation enhanced retention of existing stock (Bromley, Tallon and Thomas, 2005).

Negative factors include proximity to motorway and air traffic noise, which make properties less desirable, as do deleterious materials such as asbestos. Negative factors force up costs (Remøy and Van der Voordt, 2006). Binder's (2003) study found that some building types, such as prisons, carried a social stigma, making change of use retrofit problematic.

Retrofit has to be economically viable, though costs can be traded against environmental and social gains when triple-bottom-line accounting theory is used (Kincaid, 2002). Depending on whether the intention is to occupy or let, different features are more or less important. Arge (2005) found owner-occupied stock had higher rates of retrofit; although those retrofits were more expensive, they provided a greater return on investment over the whole life cycle. There has to be demand for economically viable retrofit. Ball (2002) discovered that positive user demand and active marketing by stakeholders were vital in the reuse of vacant stock in Stoke-on-Trent, where demand was for low-cost accommodation for start-up businesses.

Market research is vital, revealing yields, post-retrofit value and investment. Chau, Leung, Yui and Wong (2003) found a positive relationship between retrofit and the value of residential blocks in Hong Kong. Lower vacancy was a positive economic indicator in retrofitted buildings (Ball, 2002). Depending on condition, retrofit can increase quality and rental and capital values (Highfield, 2000; Hyland, Lyons and Lyons, 2013). Quality in office buildings is measured in terms of amenity features, services, fixtures and fittings. The Property Council of Australia building quality matrix grades office buildings, with Premium being highest, followed by A, B, C and D. Earlier Melbourne office retrofit studies found that higher-graded properties, (Premium, A and B) were retrofitted more frequently than lower-grade stock (Wilkinson, 2011b; Wilkinson, 2014), which shows that owners of top-quality office stock see a worthwhile return on investments to maintain their properties to reduce vacancy and churn to attract higher rental and capital values.

Environmental aspects have increased the scope and extent of retrofit (Kincaid, 2002). There is overlap with social, economic and location factors; for example, proximity to public transport provides environmental, location, economic and social benefits. Thus, some attributes can be interpreted on multiple levels. The most important environmental impact is GHG emissions in operational energy use (Douglas, 2006). Buildings evaluated under assessment methods, such as Green Star in Australia, meet standards in respect of environmental criteria, including energy use. Water economy, an important environmental attribute in Australia, is included in Green Star (GBCA, 2010). Most stock was built with little attention to minimising water use, and retrofit offers the opportunity to reduce, recycle, harvest and re-use water. Wilkinson (2011a) found that no environmental attributes had been important between 1998 and 2008 in Melbourne; however, a follow-up study showed that NABERS ratings, a part of mandatory disclosure legislation, and the voluntary Green Star rating had become important from 2009 to 2011 (Wilkinson, 2014).

The transport modes that occupants use affect building sustainability (Kincaid, 2002). Public transport has lower per capita emissions than individuals driving cars, and proximity to public transport features in environmental assessments (Davis Langdon, 2008). However, car parking is a desirable attribute in the Property Council of Australia building quality matrix and is a contradiction in terms of market perceptions of quality and the perception of environmental features. Finally, energy efficiency was mandated in the Building Code of Australia (BCA) in 2006, and a 2011 study found that environmental attributes were not important in Melbourne retrofits from 1998 to 2008 (Wilkinson, 2011a). Table 5.1 summarises the office building attributes found to be important in the literature and the two Melbourne retrofit studies.

5.2 Research aims and objectives

The research analysed office building retrofits reported in the 1200 Buildings Retrofit Survey 2013. The objective was to evaluate the City of Melbourne retrofit database to understand the attributes important in office retrofit. The analysis considered 18 attributes found to be important in a study of 7,393 Melbourne office retrofits from 1998 to 2008 (Wilkinson, 2011a) and a study of 1,453 retrofits from 2009 to 2011 (Wilkinson, 2014). The 18 attributes considered were:

1 Building age
2 Location

Table 5.1 Building retrofit attributes found to be important in Melbourne office retrofits

Factors	*Attribute*	*Melbourne study 1998–2008*	*Melbourne study 2009–2011*
Economic	PCA grade	X	X
	Cost in use	X	
Physical	Building height/number of storeys	X	X
	Service core location	X	X
	Elasticity (able to extend laterally/vertically)	X	
	Degree of attachment to other buildings	X	
	Construction type	X	X
	Age	X	X
	Street frontage (width)	X	X
	Parking	X	
	Site access	X	
Location	Location	X	X
Legal	Ownership – tenure	X	
	Owner type	X	
Social	Heritage listing	X	X
	Aesthetics	X	X
Environmental	NABERS rating		X
	Green Star rating		X

3. PCA grade
4. Type of retrofit
5. Listed/unlisted status
6. Height
7. Attachment to other buildings
8. Potential to extend laterally/vertically
9. Cost in use
10. Owner type
11. Tenure
12. NABERS rating
13. Green Star rating
14. Vertical services location
15. Street frontage
16. Construction type
17. Parking
18. Aesthetics
19. Site access

5.3 Research design and methodology

A building retrofit database was evaluated to ascertain which attributes were most important (Wilkinson et al., 2014). This quantitative approach is not reliant on the subjective views of individuals, and large numbers of retrofits can be evaluated (Silverman and Marvasti, 2008). Other studies adopting a case study approach use comparatively small samples of buildings and qualitative approaches (Blakstad, 2001; Ball, 2002; Kincaid, 2002; Kucik, 2004; Arge, 2005; Remøy and Van der Voordt, 2007). Using the attributes found to be important in these studies, a building retrofit database was designed and populated. Retrofit attributes formed the fields for the database. The database comprised 589 buildings in the City of Melbourne 1200 Buildings Retrofit Survey 2013, supplied by the City of Melbourne. This data was augmented with variables from the Census of Land Use and Employment database, provided by the City of Melbourne. Additional variables, found to be important in studies above, were:

1 Property Council of Australia Grade
2 Age (construction date)
3 Building gross floor area (GFA)
4 Site area
5 Building footprint – ground floor GFA
6 Number of levels/storeys (excluding underground)
7 Heritage grading

Green Star and NABERS ratings were added, where they existed from the project directory (GBCA; NABERS, 2015). Much research is limited by sample size; however, every building in the 1200 Buildings Retrofit Survey 2013 was analysed. The study examined building retrofits in a mature commercial market in the City of Melbourne, Victoria, a distinct geographical location, continuously occupied since 1834. Melbourne is a global city with an estimated 4.44 million people residing in Greater Melbourne, and this makes the findings potentially relevant to other global cities (ABS, 2015).

The analysis used a Principal Component Analysis (PCA) to come up with profiles/categories of buildings, each represented by a component. After the PCA was undertaken, the component scores of the buildings that had past or current retrofit activity were compared to those of the buildings without any retrofit activity. In this way, it was possible to profile buildings that undergo retrofit. The PCA uses triangulation to determine the reliability and validity of results. The attributes in the PCA were:

- Building age/construction date
- Location
- Property Council of Australia Grade
- Type of retrofit
- Heritage status
- Height/number of storeys
- Cost in use
- Owner type
- NABERS rating
- Green Star rating
- Gross floor area (GFA)
- Site area (M^2)
- Building footprint (Ground floor area M^2)

The final stage comprised further analysis of the NABERS ratings data, reviewing awareness of NABERS ratings using the City of Melbourne 1200 Buildings Retrofit Survey 2013. Second, changing the floor area trigger level for mandatory disclosure was explored.

PCA limitations included that the distribution of building attributes for the entire building stock may not be represented in this sample. Green Star ratings were only available for projects that have provided consent for the rating to be published. There were low matches for NABERS and Green Star ratings in the database, indicating low levels of take-up.

5.4 PCA results

PCA is a reliable technique of weighting dimensions in cross-sectional data (Jolliffe, 2002) with the capacity to reveal, untangle and sum up configurations of connection within a dataset (Heikkila, 1992). From a number of original variables, PCA condenses data into a smaller set of new merged factors (Jackson, 2003). Here, PCA reduced the dimensionality of attribute data relating to retrofit. The aim was to ascertain the highest level of variance explained by an interpretable group of factors, and all attributes were appraised. Initially, all variables were input into the PCA to generate a lesser number of components where factors with Eigenvalues exceeding 1.0 were retained. For an extended description of PCA, see Wilkinson (2011a), Hair, Anderson, Tatham and Black (1995) and Tabachnick and Fidell (2001). The factors were rotated using an oblique 'Oblimin' rotation method, the result being a table of specific factors, which includes the loadings of individual building attributes.

Different types of retrofit were contained in the database and all 589 retrofits were analysed. Retrofits comprised anything from fit-out of one or more floors to whole-building retrofit. To determine meaning, the patterns of the factor loadings are assessed—a subjective process (Hair et al., 1995). Following analysis of the loadings across the factors, the minimum threshold for a robust PCA was 0.6, as recommended by Tabachnick and Fidell (2001). The Kaiser-Meyer-Olkin (KMO) measure of sampling adequacy varies between zero and one, and values closer to one are better; a value of 0.6 is a suggested minimum. The Bartlett's test of sphericity, another measure of the robustness of the PCA, has to be significant. Taken together, these tests provide a minimum standard that should be exceeded before a PCA should be conducted. Table 5.2 shows that these criteria were met. With the list of each factor containing high-loading building attribute variables, a PCA can be undertaken and factor names assigned (Table 5.3). For the retained building attributes, the KMO score and a Bartlett's test of sphericity are shown in Table 5.2.

Table 5.3 shows the total variance explained by the PCA of retrofit events in the 1200 Buildings Retrofit Survey 2013, where five components are shown to account for 67.376 per cent of the original variance. Component 1 accounts for 27.889 per cent; Component 2, 12.720 per cent; Component 3, 9.831 per cent; Component 4, 8.934 per cent; and Component 5, 8.002 per cent.

Within the five components in Table 5.3, 14 attributes were found, analysed and used to produce a five component table (Table 5.4). These attribute variables were: in the suburb of Melbourne; in the suburb of East Melbourne; in the suburb of North Melbourne; building is

Table 5.2 KMO score and Bartlett's test

KMO and Bartlett's test		
KMO score		0.677
Bartlett's test	Approx. Chi-Square	2474.122
	Df	91
	Sig	0

Table 5.3 Total variance explained by PCA of retrofit events in the 1200 Buildings Retrofit Survey 2013

Component	*Initial eigenvalues*			*Extraction sums of squared loadings*			*Rotation sums of squared loadings*		
	Total	*% of variance*	*Cumulative %*	*Total*	*% of variance*	*Cumulative %*	*Total*	*% of variance*	*Cumulative %*
1	3.905	27.889	27.889	3.905	27.889	27.889	2.600	18.571	18.571
2	1.781	12.720	40.609	1.781	12.720	40.609	2.292	16.368	34.939
3	1.376	9.831	50.440	1.376	9.831	50.440	1.783	12.739	47.678
4	1.251	8.934	59.374	1.251	8.934	59.374	1.504	10.744	58.422
5	1.120	8.002	67.376	1.120	8.002	67.376	1.253	8.953	67.376

office; owner is corporate; Property Council of Australia grade is Premium, A or B; heritage grade is A or B; building age in 2015; number of storeys (excluding underground); building gross floor area (GFA); site area; building footprint (ground-floor GFA); NABERS rating 3+; Green Star rating 4+. Five components explain 67.376 per cent of the original variance. Table 5.4 shows the five components and how each variable loads on each of the components. Loadings with a magnitude of 0.4 or greater (highlighted in bold) are of interest because they indicate a greater level of importance.

Five attributes were all highly or very highly loaded on Component 1—owner is corporate, number of storeys, building GFA, NABERS rating and Green Star rating (Table 5.4)—and explain 27.889 per cent of the original variance. The first three attributes relate to building

Table 5.4 Five components for PCA

Rotated component matrix	*Component*				
	1	2	3	4	5
In the suburb of Melbourne	0.236	–0.057	–0.156	**0.828**	–0.171
In the suburb of East Melbourne	–0.080	0.035	0.169	–0.207	**0.772**
In the suburb of North Melbourne	0.057	–0.081	–0.232	**–0.755**	–0.156
Building is office	0.320	–0.236	**0.528**	–0.060	0.161
Owner is corporate	**0.746**	0.100	0.279	0.066	0.008
Property Council of Australia grade is Premium, A or B	0.233	0.179	**0.715**	0.105	0.017
Heritage grade is A or B	0.012	–0.049	–0.266	0.267	**0.727**
Building age in 2015	0.042	–0.347	**–0.743**	–0.024	0.203
Number of storeys (excluding underground)	**0.699**	**0.407**	0.198	0.284	–0.046
Building gross floor area (GFA)	**0.435**	**0.689**	0.286	0.168	–0.015
Site area	0.073	**0.802**	0.004	–0.052	0.015
Building footprint (ground floor GFA)	0.028	**0.879**	0.127	0.006	–0.011
NABERS rating 3+	**0.797**	0.054	0.189	0.046	0.003
Green Star rating 4+	**0.707**	0.017	–0.115	–0.026	–0.071

quality and size, which are physical and economic, whilst the fourth and fifth are environmental factors. This component is labelled 'tall, corporate-owned buildings with good energy and environmental ratings'. This finding is not surprising as it is acknowledged that these buildings are likely to be higher-grade stock whose tenants are looking for space that meets expectations and also CSR goals (Douglas, 2006), and therefore is more likely to be retrofitted.

Component 2 comprises three variables: GFA, site area and building footprint. All are size related, and this component is labelled 'Large buildings'. These variables explain 12.720 per cent of the original variance with three attributes. Generally, larger buildings are institutionally owned, are likely to have higher Property Council of Australia grades and undergo more frequent retrofit. Furthermore, they are more likely to trigger the mandatory disclosure requirement for a NABERS rating; when low ratings are found, this prompts some owners to increase ratings to attract certain tenants, especially government tenants who only occupy stock with a 4.5 rating or higher.

Component 3 has three variables—building is office, Property Council of Australia grade and building age—and is labelled 'Newer, top-grade office buildings'. These three variables explain 9.831 per cent of the original variance. This stock is more likely to have sustainability within the original design, and again, having higher quality will attract tenants wanting to occupy more sustainable property.

Component 4 has two variables—in the suburb of Melbourne and in the suburb of North Melbourne—and is labelled 'Buildings in Melbourne and North Melbourne'. These two variables explain 8.934 per cent of the original variance. This shows that buildings in these areas are more likely to be retrofitted and there may be a knock-on effect where owners see neighbours retrofitting and are encouraged to upgrade their stock to attract similar rents and tenants.

Component 5 has two variables—in the suburb of East Melbourne and heritage grade and is labelled 'East Melbourne heritage-grade buildings'. These variables account for 8.002 per cent of the original variance. With this result, it appears that in this area of the city, there are a predominance of heritage buildings where retrofit is undertaken.

Finally, the average component scores for the 'buildings with' and 'buildings without' a past or current retrofit activity showed differences. Buildings with past or current retrofit activities loaded more positively on the first four components and negatively on the fifth, suggesting a different profile.

5.5 NABERS rating analysis

The 2013 survey asked respondents, who were building owners, if they knew their building's NABERS rating. 89.33 per cent did not know the NABERS rating for their building, and 8.5 per cent of those who thought they knew their rating have no record in the 2015 NABERS registry. It is not known who within the organisation responded to the survey. These results indicate that there is low awareness of actual NABERS ratings in the sample. Table 5.5 shows that 15 owners out of 544 increased their ratings during the period, assuming they knew it correctly at the time of the survey.

Of the 589 buildings in the database, several scenarios were run for the number of buildings that would have required a NABERS rating should the trigger level for the mandatory disclosure legislation be amended. Some have suggested reducing the existing 2000m^2 trigger level to 1000m^2 or to 500m^2. The results showed the number of buildings over the threshold increases by 132 per cent if the threshold is reduced to 1000m^2 and by 170 per cent if it is reduced to 500m^2.

Table 5.5 NABERS energy ratings 2013 perceptions and 2015 rating compared

		No rating found	0	1	1.5	2	2.5	3	3.5	4	4.5	5	6
Survey answer to 'What is the NABERS energy rating?' (2013)	1 star	3	1		1								
	1.5 star	1	1										
	2 star	8	1		1	1	1	3			1		
	2.5 star	5						2					
	3 star	6					2	1		1			
	3.5 star	2							3	2	1		1
	4 star	9	1						3	5	2		
	4.5 star	1									5		
	5 star	7										1	
	6 star	3											
	6 plus	1											
	Don't know	486	1			1	1				1		
	No rating	12											
Total		544	5	0	2	2	4	6	6	8	10	1	1

Key
- Answer is a higher rating than NABERS
- Answer is the same as NABERS
- Answer is a lower rating than NABERS
- Answer was a specific rating, but no rating was found

5.6 Conclusions

Melbourne is proactively encouraging sustainable retrofit through the 1200 Buildings Program. Using a database of 589 buildings whose owners responded to the 2013 Retrofit Survey, a PCA showed that five typologies account for 67.376 per cent of retrofits. These typologies are:

1 tall, corporate-owned buildings with good energy ratings
2 large buildings
3 newer, top-grade office buildings
4 Buildings in Melbourne and North Melbourne
5 East Melbourne heritage-grade buildings.

In many respects, these results confirmed findings from Wilkinson's (2011a, 2014) earlier studies where building quality, height, age, heritage status, location and environmental ratings were found to be important (Table 5.1). It is apparent that the City of Melbourne can further encourage other owners and occupants of these building types to continue engaging in retrofit, and contact owners of similar un-retrofitted stock to advise them of the measures taken by others directly competing in this market. Therefore, the low-grade, smaller non-heritage buildings are not being retrofitted. The buildings that are being retrofitted are larger, better-quality buildings, and the logic is 'Tell owners of other similar stock what their direct competition is doing and they may upgrade their buildings'. Given that owners of these typologies are engaging in retrofit, those owners who are not engaged are non-corporate owners, smaller-building, lower-grade

(Grade B, C and D) and non-heritage stock building owners. These trends are likely to be found in other global cities.

The results indicated low awareness in the sample of actual NABERS ratings, which is a concern. Analysis of the trigger level for a NABERS rating under mandatory disclosure legislation showed that the number of buildings over the threshold increased by 132 per cent if the threshold is reduced to $1000m^2$ and 170 per cent if it is reduced to $500m^2$. On 21 June 2016, the Australian government announced the move to lower the mandatory disclosure threshold on commercial buildings from $2000m^2$ to $1000m^2$ (Australian Government, 2016). If the behaviour pattern is similar, this will lead to more retrofit and possibly encourage more owners to upgrade stock to higher ratings. As a result, the achievement of the 1200 Buildings Program target is more likely; however, further reduction of the trigger level for mandatory disclosure in commercial buildings to $500m^2$ is recommended.

A final thought is that there is an opportunity to use an existing policy and data collection resource: building permit data. The proposal would be to amend the system of recording data to include what is undertaken in respect of sustainability (Part J of the BCA) in new build and retrofits. This data would be provided with occupancy permits. The data could be extracted and analysed by year, location, building type, age, construction type, GFA, NLA and number of storeys. Over time, it is possible to use this data to identify and track trends in retrofit of the building stock in respect of energy.

Acknowledgement

The author wishes to thank the City of Melbourne for making the data from the 2013 Retrofit Survey available for this analysis.

References

Arge, K. (2005). 'Adaptable Office Buildings: Theory and Practice.' *Facilities* 23(3): 119–127.

Australian Bureau of Statistics (ABS) (2015). Retrieved on 7 May 2015 from www.abs.gov.au/ausstats/abs@.nsf/Latestproducts/3218.0Main%20Features252013-14?opendocument&tabname=Summary&prodno=3218.0&issue=2013-14&num=&view=.

Australian Government (2016). 'Changes to the Commercial Building Disclosure Program.' Retrieved on 21 July 2016 from http://cbd.gov.au/changes-to-the-commercial-building-disclosure-program.

Baird, G., Gray, J., Isaacs, N., Kernohan, D. and McIndoe, G. (1996). *Building Evaluation Techniques*. McGraw-Hill: Maidenhead.

Ball, R. M. (2002). 'Re-use potential and vacant industrial premises: Revisiting the regeneration issue in Stoke on Trent.' *Journal of Property Research* 19: 93–110.

Binder, M. (2003). *Adaptive Reuse and Sustainable Design: A Holistic Approach for Abandoned Industrial Buildings*. Master of Architecture thesis. The School of Architecture and Interior Design. Cincinnati, University of Cincinnati.

Blakstad, S. H. (2001). *A Strategic Approach to Adaptability in Office Buildings*. PhD thesis. Faculty of Architecture, Planning and Fine Arts, Norwegian University of Science and Technology.

Brand, S. (2009). *Whole Earth Discipline*. Atlantic Books: London.

Bromley, R. D. F., Tallon, A. R. and Thomas, C. J. (2005). 'City centre regeneration through residential development: Contributing to sustainability.' *Urban Studies* 42(13): 2407–2429.

Bullen, P. A. (2007). 'Adaptive reuse and sustainability of offices.' *Facilities* 25: 20–31.

Bullen, P. A. and Love, P. E. (2011). 'A new future for the past: A model for adaptive reuse decision-making.' *Built Environment Project and Asset Management* 1(1): 32–44.

Chau, K. W., Leung, A. Y. T., Yui, C. Y. and Wong, S. K. (2003). 'Estimating the value enhancement effects of refurbishment.' *Facilities* 21(1): 13–19.

City of Melbourne (2003). *Zero Net Emissions by 2020. A Roadmap to a Carbon Neutral City*. City of Melbourne: Melbourne.

City of Melbourne (2014). *Zero Net Emissions Update 2014.* Retrieved on 17 November 2017 from www.melbourne.vic.gov.au/SiteCollectionDocuments/zero-net-emissions-update-2014.pdf.

Davis Langdon (2008). 'Opportunities for existing buildings. Deep emission cuts.' *innovative thinking* 8.

Douglas, J. (2006). *Building Retrofit.* Routledge: London.

Gann, D. M. and Barlow, J. (1996). 'Flexibility in building use: The technical feasibility of converting redundant offices into flats.' *Construction Management & Economics* 14(1): 55–66.

GBCA (2010). 'Green Building Council Australia'. Retrieved on 13 June 2010 from http://new.gbca.org.au.

GBCA (2015). 'Green Building Rating Register.' Retrieved on 11 April 2015 from www.gbca.org.au/project-directory.asp.

Hair, J. F, Anderson, R. E, Tatham, R. L. and Black, W. C. (1995) *Multivariate Data Analysis,* 4th ed. Prentice Hall: Englewood Cliffs, NJ.

Heath, T. (2001). 'Adaptive re-use of offices for residential use: The experiences of London and Toronto.' *Cities* 18(3): 173–184.

Heikkila, E. J. (1992). 'Describing urban structure.' *Review of Urban and Regional Development Studies* 4: 84–101.

Highfield, D. (2000). *Refurbishment and Upgrading of Buildings.* E & FN Spon: London.

Hyland, M., Lyons, R. C. and Lyons, S. (2013). 'The value of domestic building energy efficiency: evidence from Ireland.' *Energy Economics* 40: 943–952.

Jackson, J. E. (2003). *A User's Guide to Principal Component Analysis.* Wiley: Chichester.

Jolliffe, I. T. (2002). *Principal Component Analysis.* 2nd ed. Springer: New York.

Kelly, M. J. (2008). *Britain's Building Stock: A Carbon Challenge.* Retrieved on 7 April 2010 from www.lcmp.eng.cam.ac.uk/wp-content/uploads/081012_kelly.pdf.

Kincaid, D. (2002). *Adapting Buildings for Changing Uses: Guidelines for Change of Use Refurbishment.* Spon Press: London.

Knott, J. (2007). 'Green refurbishments. Where to next?' *RICS Oceania e-News Sustainable*: 5–7.

Kucik, L. M. (2004). *Restoring Life: The Adaptive Reuse of a Sanatorium.* Master of Architecture thesis. School of Architecture and Interior Design. Cincinnati, University of Cincinnati.

Langston, C., Wong, F. K. W., Hui, E. C. M. and Shen, L. (2008). 'Strategic assessment of building adaptive reuse opportunities in Hong Kong.' *Building & Environment* 43(10): 1709–1718.

Markus, A. M. (ed.) (1979) *Building Conversion and Rehabilitation: Designing for Change in Building Use.* Newnes-Butterworth: London.

NABERS (2015). 'National Australian Built Environment Rating System'. Retrieved on 11 April 2015 from www.nabers.gov.au/public/WebPages/ContentStandard.aspx?module=0&template=3&include=homeIntro.htm.

Nutt, B., Walker, B., Holliday, S. and Sears, D. (1976). *Obsolescence in Housing: Theory and Applications.* Saxon House: Farnborough.

Remøy, H. T. and van der Voordt, T. J. M. (2006). 'A new life: Transformation of vacant office buildings into housing.' CIBW70 Trondheim International Symposium, 12–14 June 2006, Trondheim, Norway. Norwegian University of Science and Technology, NTNU.

Remøy, H. T. and van der Voordt, T. J. M. (2007). 'A new life: Conversion of vacant office buildings into housing.' *Facilities* 25(3/4): 88–103.

RICS (2015). *RICS Futures: Our Changing World.* Retrieved on 5 May 2015 from www.youtube.com/watch?v=t6hPnE3giXw.

Silverman, D. and Marvasti, A. (2008). *Doing Qualitative Research: A Comprehensive Guide.* SAGE Publications.

Swallow, P. (1997). 'Managing unoccupied buildings and sites.' *Structural Survey* 15(2): 74–79.

Szarejko, W. and Trocka-Leszczynska, E. (2007). 'Aspect of functionality in modernization of office buildings.' *Facilities* 25(3): 163–170.

Tabachnick, B. G. and Fidell, L. S. (2001). *Using Multivariate Statistics,* 4th ed. Allyn & Bacon: Needham Heights, MA.

United Nations (UN) (2014). *World Urbanization Prospects: The 2014 Revision, Highlights.* Retrieved on 17 November 2016 from https://esa.un.org/unpd/wup/Publications/Files/WUP2014-Highlights.pdf.

Warren, C. M. J. and Huston, S. (2011). 'Promoting energy efficiency in public sector commercial buildings in Australia.' Proceedings of RICS Construction and Property Conference. *COBRA: Construction and Building Research Conference,* Salford, United Kingdom, 12–13 September 2011: 128–134.

Wilkinson, S. J. (2011a). 'The relationship between building retrofit and property attributes.' PhD thesis. http://hdl.handle.net/10536/DRO/DU:30036710.

Wilkinson S. J. (2011b). *Sustainable Retrofit Potential in Lower Quality Office Stock in the Central Business District: CIB Management and Innovation in the Sustainable Built Environment.* Delft University of Technology: Delft, Amsterdam.

Wilkinson, S. J. (2014). *Building Permits and Sustainability: A Method for Measuring the Uptake of Sustainability in the Built Environment over Time* RICS Research Trust Report. Retrieved on 14 October 2017 from www.rics.org/uk/knowledge/research/research-reports/building-permits-and-sustainability/.

Wilkinson, S. J., Remøy, H. T. and Langston, C. (2014). *Sustainable Building Retrofit: Innovations in Decision-Making*. Wiley-Blackwell: Oxford.

6

International standards

Key to unlocking the value of green buildings?

Ben Elder

(The views expressed in this chapter are the views of the author not the RICS.)

6.0 Introduction

In trying to understand sustainable real estate and the challenges that face the property industry in delivering a sustainable real estate product, it is critical that we understand the context and delivery mechanisms within which the real estate product is placed. By doing this, we can focus efforts on the key change elements more effectively.

This chapter seeks to understand the place of real estate in the economic systems that allocate scarce resources between competing needs and the steps that have been taken by interested parties, including professional bodies and notably the international community, to improve knowledge within the economic systems that allocate scarce resources to ensure that sustainable features of real estate are recognised and priced appropriately into the market.

It is essential to recognise that the delivery of sustainable real estate does not take place in a vacuum and that technology and innovation are happening all around our focus on sustainable real estate and that the pace of change is increasing.

There is no apology for the fact that this chapter draws extensively from neo-classical market economic theory. Economics is defined in the *Oxford English Dictionary* as 'the branch of knowledge concerned with production, consumption and the transfer of wealth' (OED, 2017).

It is recognised that there are different economic systems that operate in different countries, but neo-classical market economics has become the dominant method of allocation of scarce resources in almost all countries since the decline of the command system of resource allocation in the late 1980s and early 1990s.

Neo-classical market economics is defined by the online *Business Dictionary* (2016) as:

> The present day dominant school of economic thought built on the foundation laid by the 18th century (classical) theories of Adam Smith (1723–1790) and David Ricardo (1772–1823), and refined by the 19th and 20th century theories of Alfred Marshall (1842–1924), Vilfredo Pareto (1848–1923), John Clark (1847–1938), and Irving Fisher (1867–1947). It is 'classical' in the sense that it is based on the belief that competition leads to an efficient allocation of resources, and regulates economic activity that establishes equilibrium between demand and supply through the operation of market forces. It is "neo" in the sense that it

departs sharply from the classical viewpoint in its analytic approach that places great emphasis on mathematical techniques.

It is recognised that real estate has many characteristics that influence ownership and occupation, but as the focus of this chapter is on the role of real estate in the market, this chapter will focus on the economic characteristics of real estate in an attempt to understand its role and particularly how the sustainable characteristics of real estate can be priced in the market. The key initial focus is on recognised international standards used by professionals engaged in pricing assets and how they exert influence on market activity. International standards provide a common language for business so that business transactions are understandable across borders. They achieve this by creating a framework for the exchange of information that is consistent and comparable, thereby bringing a common understanding and transparency to market information. Examples include International Financial Reporting Standards (IFRS) produced by the International Accounting Standards Board (IASB) and International Valuation Standards (IVS) produced by the International Valuation Standards Council (IVSC). Both the IASB and the IVSC are not-for-profit organisations that have a public interest mandate.

6.1 Economic efficiency through the price system

The basis of neo-classical economic theory is that behaviours and reactions can be modelled as long as certain assumptions are met. One of the key assumptions is the hypothetical economic state of 'perfect competition'. Perfect competition is seen by Myers as a market structure in which the decisions of buyers and sellers have no effect on price (Myers, 2013). Under this assumption, the market economy is entirely or very largely efficient in its allocation of scarce resources to their highest value use. Perfect market assumptions include equal access to information by all market participants, completely rational economic actors and no transaction costs, such as taxes (Farlex Financial Dictionary, 2012). Economists recognise that the equal access to information assumption rarely, if ever, holds true in the real world, but it does provide an important reference point for analysis as a benchmark against which other market situations can be judged.

Within the discipline of economics, a whole new field has developed, examining transactions in which some parties have access to more information than others. This area of economics is generally referred to as the economics of asymmetric information (Myers, 2013). The perfect market conditions may not exist in the real world, but if we are relying on the market to allocate resources to society's highest and best use, it is necessary to take steps that bring the market closer to those defined in the economic theory. To this end, the introduction of international standards has the potential to improve the knowledge of how markets work through greater transparency, allowing market participants to more accurately reflect where and how scarce resources are allocated. This will include real estate and its sustainability characteristics.

The remainder of this chapter is focused on analysing how the development and introduction of a range of real estate-related international standards in the marketplace can assist in the drive to transparency in information and knowledge, and how it encourages the development of properties that display sustainability characteristics. The international standards we will consider are:

- International Financial Reporting Standards (IFRS)
- International Valuation Standards (IVS)
- International Ethics Standards (IES) for the Built Environment

- International Property Measurement Standards (IPMS)
- International Land Measurement Standards (ILMS)
- International Construction Measurement Standards (ICMS).

It is important to understand the background for international standards against which the detail of each standard will be explored.

6.2 Free market versus regulation

Since the collapse of the controlled economies of Eastern Europe between 1989 and 1992, the market has been the primary vehicle used for the allocation of scarce resources (Harrison, 2001). However, there was a failure to recognise that the perfect market does not exist, and the imperfections in market knowledge and information significantly contributed to the 2008 global financial crisis (Bank of England, 2008). The imperfections initially contributed to the global credit (liquidity) crunch in August 2007, which led to the global financial crisis in 2008 (Avgouleas, 2009). Following the near collapse of the global financial market, almost every government has been trying to establish the correct balance between allowing the free market to allocate scarce resources and imposing regulations to control the excesses of the market, but without becoming too bureaucratic to stifle entrepreneurship (G20, 2008).

A key part of regulating the excesses of the market is improving information and transparency in the marketplace, and this is a key goal for international standards.

6.3 Technology

We need to recognise the way that changes in a range of technologies have influenced the way that people and markets behave, particularly in real estate. One of the economic characteristics of real estate is that it is fixed to the ground and therefore cannot be taken to market (Evans, 2008). In traditional theory, this led to the development of local real estate markets (Harvey and Jowsey, 2004). Land, therefore, does not work as a good for which substitution theory applies easily: its fixed location means that real estate markets tended to develop in isolation. The workings of those local real estate markets, perhaps inevitably, were not consistent, as local practice and implementation of any policies was driven by a few individuals operating in virtual isolation. Remember, it was not until the 1960s that reliable motorcars were available to allow frequent and economically viable travel between markets, and telephones were installed in most businesses. These changes revolutionised the workplace, expanding the geographical coverage possible, and, to accommodate these changes, a more consistent approach to the packaging and transfer of market information on a global scale became common.

6.4 Trade globalisation

The political and technological changes outlined in Sections 6.2 and 6.3 have significantly changed the risk profile of undertaking transactions across markets and particularly in real estate markets because of the expansion and normalisation of global trade. The market is now a global market with trillions of dollars being transacted on a daily basis. The real estate markets have followed, with billions of dollars flowing around the globe in cross-border transactions, yet interestingly some of the processes and procedures adopted within markets (particularly the real estate markets) still operate on local practice or interpretation of policies. These inconsistencies add risk to the transaction, and where risk is involved in a transaction, the market takes a premium; if that premium is too high, the resources will flow to an alternative use, yielding the same returns at less risk. International standards play a critical role in reducing risk by creating a common language that brings a transparency to the marketplace; for example, the IVSC definition of market value provides a commonly understood set of criteria that market participants can easily understand.

6.5 International standards for real estate

One of the key elements in controlling or influencing the amount of government regulation has been the increase in the number of international standards created and self-regulated by coalitions of professionals, particularly in the field of real estate, with efforts focused on increasing the transparency in the real estate market to facilitate more informed investment decisions throughout the market.

The importance of this is illustrated by research undertaken by the International Bank for Reconstruction and Development/World Bank (2006) that estimates that between 50 per cent and 70 per cent of any nation's wealth is held in its land and real estate. This gives the valuer unprecedented responsibility to provide the market with accurate advice by which decision making can be improved. This in turn should support more efficient and effective use of resources while simultaneously reducing risk, through increased transparency, deeper knowledge and information. The result logically is an increase in property values. The transparency may also reduce the murky waters where corruption often hides. By reducing risk, the introduction and adoption of international standards can, in certain circumstances, encourage investors to take a longer-term view of their investments. It achieves this by improving market knowledge and providing confidence to the market to make longer-term decisions. This is a key factor in allowing the environmental characteristics of buildings to be appropriately priced and their potential returns over time to be identified.

The introduction of International Valuation Standards also saves time for global market players increasing their productivity. This is achieved through standardisation of information, a key component of almost all industrial production and one that real estate valuation has been slow to adopt. The process of standardisation also enables better and clearer benchmarking to be achieved. For example, the introduction of IPMS will allow true cross-border comparison of the environmental and economic performance of buildings.

6.6 Economic benefits of standardisation

Most industry players operating in market economies are continuously striving to reduce costs or produce the same output from fewer resources. Standardisation of product is a major contributor to this, as it facilitates economies of scale and critically allows for the division of labour by codifying what each element of a job entails and the skills required to undertake that element of the process; for example, the standardisation of information under IFRS facilitates the use of labour with different skill levels at various points in the production process of financial reporting. This process of increased production through the division of labour can be extrapolated across other areas where market information is delivered to decision makers, such as valuation.

Standardisation limits 'time stealers': people or processes that unnecessarily take our time to undertake actions that are unnecessary or low risk to the overall objective. For example, in the past when travelling on overseas business, it was necessary to arrange for funds to be transferred from your home bank to the overseas bank that could facilitate access to the funds transferred. This could be a time-consuming process, involving visits to your home bank and the overseas bank to sign required documentation, possibly taking up to half a working day. Now we tend to use ATMs for the same function. Using a plastic card and entering in 4–6-digit security codes through thousands of machines in almost every country in the world enables us to access our own money in minutes instead of hours and potentially increases individuals' productivity by half a day – all achieved through standardisation of information and operating systems between banks and card operators.

As De Soto (2000: 57) said:

> If standard descriptions of assets were not readily available, anyone who wanted to buy, rent or give credit against an asset would have to expend enormous resources comparing and evaluating it against other assets, which would also lack standard descriptions.

Leading on from this, if there is more than one standard, there is no standard at all. Yet despite the evidence of economic benefit, the real estate industry has been reluctant to embrace and adopt international standards. This is possibly a legacy of the fragmented nature of the real estate industry created by the economic characteristics described above.

6.7 International standards

It is now time to focus on international standards and the benefits that they could have in securing sustainable real estate. Before looking at some of the key international standards, it is helpful to examine what an International Standard is. Referring back to the natural fragmentation of the real estate industry because of its economic characteristics, it is not surprising that 'standard setters' have appeared in many different markets – usually nationally based as governments adopt standards developed by professional bodies in their respective countries. In the valuation profession, this is often synonymous with a 'licensing' process; for example, in the USA, the legislated valuation standards are the Uniform Standards of Professional Appraisal Practice (USPAP). USPAP are developed by the US Appraisal Foundation and delivered state by state under licensing arrangements. As markets expand across borders, we have the situation where professionals are working to more than one standard and, as argued above, more than one standard means no standard. From this foundation, standard-setting organisations throughout the world are coming together to develop high-level international standards. Collectively, these organisations own the standards that are implemented by each organisation through their own professional guidelines. International standards are high-level principles-based standards that are designed to provide consistency and transparency throughout the profession.

Let us take a closer look at some of the obstacles to the implementation of international standards. Some people see 'international' as 'over there and not including us'. We often hear, when speaking about international standards, that the listener is very excited and supportive, but concludes that 'they won't work here – we are different. We have our own culture, legal systems and regulations that make our market different'. This is a fallacy; international standards are designed to work everywhere. They may require a local implementation manual, but the principles will apply everywhere.

6.7.1 International Financial Reporting Standards (IFRS)

International Financial Reporting Standards (IFRS) are set by the International Accounting Standards Board (IASB), which is a not-for-profit public-interest organisation with oversight by a monitoring board of public authorities (IASB, 2016). The mission of the IASB is to develop

IFRS that bring transparency, accountability and efficiency to the financial markets around the world. The work serves the public interest by fostering trust, growth and long-term financial stability in the global economy (IFRS, 2016). IFRS governance and due process are designed to keep the standards independent from special interests while ensuring accountability to stakeholders around the world (IFRS, 2016). Currently, 147 jurisdictions adopt IFRS.

6.7.2 International Valuation Standards (IVS)

International Valuation Standards (IVS) are set by the International Valuation Standards Council (IVSC) (IVSC, 2016). The IVSC is an independent not-for-profit organisation that produces and sets standards for the valuation of assets around the world in the public interest. The organisation's mission is to establish and maintain effective, high-quality international valuation and professional standards and to contribute to the development of the global valuation profession, thereby serving the public interest (IVSC, 2016). IVS are the recommended standards for valuing assets in the IFRS.

IVS are set by an independent standard-setting board within the IVSC whose members are selected from an open application process. All standards are subject to an open public consultation process before adoption.

6.7.3 International Ethics Standards (IES) for the Built Environment

International Ethics Standards (IES) for the real estate and related professions are set by the International Ethics Standards Coalition (IESC) (IESC, 2016). The IESC is a coalition of not-for-profit professional bodies that have come together to create a universal set of ethical principles for real estate and related professions that will provide assurance, consistency and confidence to all users of professional services (IESC, 2016). The standards are set by an independent standard-setting committee whose members are selected following an open application process (IESC, 2016). IES are subject to public consultation before implementation.

6.7.4 International Property Measurement Standards (IPMS)

International Property Measurement Standards (IPMS) are set by the International Property Measurement Standards Consortium (IPMSC) (IPMSC, 2016). This is a coalition of not-for-profit professional bodies that have come together to agree to a single set of standards for the measurement of real estate by use class. This is the first time that a single set of standards for the measurement of real estate has been introduced on an international scale. Currently, the way that real estate assets, such as homes, offices or shopping centres, are measured varies dramatically. Data from JLL interpreted by the RICS (RICS, 2016a) shows that, depending on which standard of measurement is adopted, a 24 per cent variance in the reported floor area is possible. The IPMSC (2016) is working to develop and embed a single property measurement standard. IPMS will ensure that property assets are measured in a consistent way, creating a more transparent marketplace, greater public trust, stronger investor confidence and increased market stability.

6.7.5 International Land Measurement Standards (ILMS)

International Land Measurement Standards (ILMS) are being developed and set by the International Land Measurement Standards Coalition (ILMSC) (RICS, 2016b). The ILMSC is a coalition of not-for-profit professional bodies that are working together to develop and implement standards that reflect best practice in land tenure, focusing on issues that will bring consistency in approach and transparency to the treatment of unregistered land that currently lies outside of formal markets. It is estimated that up to 70 per cent of land and property in the developing world falls into this unregistered category. Success in developing ILMS will support some of the most vulnerable families and individuals in society by providing clear steps to be applied, improving transparency where the unregistered land and or assets meet the market (RICS, 2016b). The standards will be set by an independent standard-setting committee whose members will be selected from an open application process. The standards are subject to public consultation before adoption.

6.7.6 International Construction Measurement Standards (ICMS)

International Construction Measurement Standards (ICMS) are being developed and set by the International Construction Measurement Standards Coalition (ICMSC). The ICMSC is a coalition of over 50 not-for-profit professional bodies (ICMSC, 2016a) that are working together to develop and implement international standards for benchmarking, measuring and reporting construction project costs. At present, the way that construction projects are reported and costed varies significantly between markets (ICMSC, 2016b). These differences make it difficult to compare projects around the world, increasing investment risk and impeding transparency. The ICMSC seeks to develop and implement a common standard for construction measurement, which will enhance transparency, investor confidence and public trust in the sector. The standards will be set by an independent standard-setting committee whose members will be selected from an open application process. The standards are subject to public consultation before adoption.

6.8 Conclusions

The focus of this chapter has been to demonstrate how the development and implementation of international standards increases the quality and transparency in information provided to decision makers in the market, and by doing so to consider how the improved dataset of information can help create cross-border benchmarks (including on environmental matters) that will facilitate the market in allocating scarce resources to their highest and best use. Through improved information and market efficiency, there is the potential for the market to allocate appropriate values to 'environmental characteristics' to a greater level than current market practices allow.

It is important to envisage the totality of the individual standards and how each of them can contribute to the creation of an improved transparency for the market, which aids market efficiency.

It may be helpful to view the standards as a quality circle where each of the standards supports each other in mutual support of transparency (see Figure 6.1).

The provision of more accurate information to the marketplace through international standards allows for better informed decisions to be made by the market. Over time, this will lead to more transparent reporting of how society and investors view the environmental characteristics of buildings and in turn, this will enable such characteristics to be factored into maket prices.

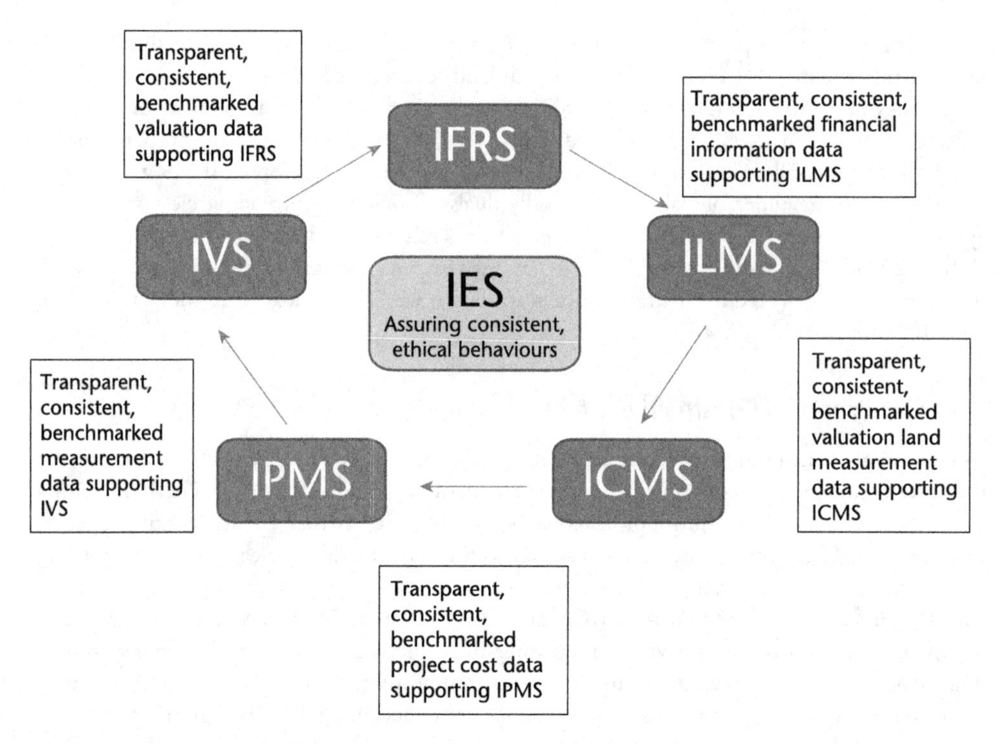

Figure 6.1 International standards quality circle

References

Avgouleas, E., 2009. The global financial crisis, behavioural finance and financial regulation: In search of a new orthodoxy. *Journal of Corporate Law Studies*, 9:1, 23–59

Bank of England, 2008. *Financial stability report*. October 2008, Issue 24. Retrieved on 29 December 2016 from www.bankofengland.co.uk/publications/Documents/fsr/2008/fsrfull0810.pdf.

Business Dictionary, 2016. Definition of neo-classical economics. Retrieved on 29 December 2016 from www.businessdictionary.com/definition/neo-classical-economics.html.

De Soto, H., 2000. *The mystery of capital: Why capitalism succeeds in the West and fails everywhere else*. Black Swan.

Evans, A. W., 2008. *Economics, real estate and the supply of land*. John Wiley & Sons.

Farlex Financial Dictionary, 2012. Retrieved on 29 December 2016 from www.farlex.com.

G20, 1998. *Declaration: Summit on financial markets and the world economy*. 15 November. Retrieved on 14 October 2017 from www.mofa.go.jp/policy/economy/g20_summit/2008/declaration.pdf.

Harrison, M., 2001. *Are command economies unstable? Why did the Soviet economy collapse*? Retrieved on 29 December 2016 from www2.warwick.ac.uk/fac/soc/economics/staff/mharrison/inactive/command.pdf.

Harvey, J. and Jowsey, E., 2004. *Urban land economics*. Palgrave Macmillan.

International Accounting Standards Board (IASB), 2016. Monitoring Board. Retrieved on 29 December 2016 from www.iasplus.com/en/resources/ifrsf/governance/monitoring-board.

International Bank for Reconstruction and Development/World Bank, 2006. *Where is the wealth of nations? Measuring capital for the 21st Century*. Retrieved on 29 December 2016 from http://siteresources.worldbank.org/INTEEI/214578-1110886258964/20748034/All.pdf.

International Construction Measurement Standards Coalition (ICMSC), 2016a. International Construction Measurement Standards. Retrieved on 29 December 2016 from https://icms-coalition.org/coalition-members.

International Construction Measurement Standards Coalition (ICMSC), 2016b. International Construction Measurement Standards. Retrieved on 29 December 2016 from https://icms-coalition.org.

International Ethics Standards Coalition (IESC), 2016. Retrieved on 29 December 2016 from https://ies-coalition.org.

International Financial Reporting Standards (IFRS), 2016. IFRS Foundation. Retrieved on 29 December 2016 from www.ifrs.org/About-us/Pages/IFRS-Foundation-and-IASB.aspx.

International Property Measurement Standards Consortium (IPMSC), 2016. Retrieved on 29 December 2016 from https://ipmsc.org.

International Valuation Standards Council (IVSC), 2016. Retrieved on 29 December 2016 from www.ivsc.org.

Myers, D., 2013. *Construction economics: A new approach*. Routledge.

OED, 2017. Definition of economics. Retrieved on 13 November 2017 from https://en.oxforddictionaries.com/definition/economics.

Royal Institution of Chartered Surveyors (RICS), 2016a. International Standards. Retrieved on 29 December 2016 from www.rics.org/uk/footer/international-standards.

Royal Institution of Chartered Surveyors (RICS), 2016b. International Standards. Retrieved on 29 December 2016 from www.rics.org/uk/footer/international-standards/#Land.

Part 2

Valuation, investment and finance

7

Valuing sustainability in commercial property in Australia

Georgia Warren-Myers

7.0 Introduction

Identifying the value of sustainability and reflecting this in valuation has been considered the 'holy grail' in property valuation and investment. Much research has investigated the value relationships, types of value and implications for practice and the ability of valuers to incorporate sustainability in the valuation process (see Ciora, Maier and Anghel, 2016; Abidoye and Chan, 2016; and Nurick, Le Jeune, Dawber, Flowers and Wilkinson, 2015, and for further discussion on the types of research, see Warren-Myers, 2012 and Sayce, Sundberg and Clements, 2010). To date, no clear answer exists and it is an evolving process of stakeholder knowledge, perception and investment, number of buildings, and valuers' ability to reflect the market's value of sustainability features. Identifying sustainability's value is key for future investment in sustainability. Consequently, it is important to investigate why there is no clear evidence of sustainability being reflected in valuation. Similarly, advice for valuers on how to deal with the complex issues is needed.

First, this chapter examines the valuers' role and their importance in the process and how valuations are conducted. Then it considers sustainability present in valuation, focusing on rating systems, available evidence, integration into valuation and implications for analysis and assessment of value when sustainability is a factor. As the author is Australian and a certified practising valuer in Australia and New Zealand, the context of the chapter is set in Australasia; however, the issues identified and advice provided are generalisable to all countries that have property valuers or real estate appraisers governed by the International Valuation Standards.

7.1 Valuers, value and sustainability

The value of sustainability has, for several years, been a key area of investigation by industry and academia. The relationship between sustainability and value is thought to be pivotal to the mainstreaming of sustainability in property (Lorenz, 2008; Warren-Myers, 2013). Consequently, valuers have an important role in assessing and reflecting the influence that sustainability has on property values. It is important that valuers do not seek to create value associated with sustainability or lead the market in how sustainability should be viewed; the role of a valuer is to reflect the market. Valuers' reflective assessment of value is for a particular asset, at a specific point in time, and takes into account all of the dynamics of the market and all relevant factors that may influence the reactions and actions of the market stakeholders or market conditions (Warren-Myers, 2013; Mallinson and French, 2000; Gallimore, 1996; Adair, Berry and McGreal, 1996). Therefore, the value of sustainability needs to be evident in the market for valuers to reflect it in the assessment of market values for property.

Valuers in their role have certain fundamental requirements, considerations and ethical responsibilities in their assessment of market value (Bellman and Öhman, 2016). Often, valuers' reliance on evidence and considerations of change has resulted in them being accused of being backwards looking and not keeping up with market change (Michl, Sayce and Lorenz, 2016). Sustainability presents a new challenge for valuers and exposes fundamental valuation practices that need to be understood. Valuation is often referred to as an art and a science, with the science referring to the underpinning economic theory, mathematical models and frameworks, which are then standardised and governed by international standards and national and state professional institutes. The 'art' has evolved as a result of the heterogenic characteristics of the property that requires individual investigation, analysis and judgement. Valuers rely on expert intuition, their judgement, in valuation practice, and this is developed over time through strategic knowledge development, experience within the market and the process of valuing properties that leads to intuitive knowledge and the creation of heuristics. Heuristics are known as knowledge shortcuts or 'rules of thumb', which allow the solving of problems and the ability to make judgements quickly and efficiently. However, as a result, these shortcuts can be prone to errors as it relies on the knowledge development and experience of the person (Ashton and Ashton, 1990; Gladwell, 2005; Tversky and Kahnemann, 1974).

Valuation praxis is based on the practical implementation of heuristics, where the experience and knowledge of the market shapes the final result prior to the actual evidence being examined (Tidwell and Gallimore, 2014). Valuers are essentially problem solvers who are required to make decisions and judgements efficiently in a challenging environment where no exact answers can be attained due to the unique characteristics of property (Hogarth, 1981). This reliance on heuristics in valuation practice means that when significant market change occurs, or new technology transformations like sustainability come along, the reliance on extant heuristics fails the valuer in the assessment of market value (Warren-Myers, 2009). Consequently, when this occurs, valuers need to realise and undertake a process to acquire and develop their strategic knowledge base, which initially will be founded on the analysis of trends and macro and micro market analysis. Experience in the market and continual analysis of data, information and results will in time lead to strategic intuition allowing new theories, observations and practices so that heuristics and intuitive knowledge can again be used in the practice of valuation (Warren-Myers, 2009; Gladwell, 2005; Atkinson and Claxton, 2000). This does not happen overnight, and valuers need to be aware of their limitations and engage proactively in the development of knowledge. This is why many valuation- or appraisal-based positions require a period of time – two to three years – working under a senior valuer or appraiser prior to becoming qualified, such as in Australia and New Zealand where a minimum of two years' work experience is required, along with a formal viva voce (Australia) and board assessment and approval (New Zealand). This allows for knowledge development and experience, with senior valuers guiding new valuers in developing expertise. Without this knowledge and understanding of heuristics, valuers will likely ignore or misapply their judgement in regard to new developments, such as those related to sustainability, in the valuation of property assets, leading to implications for the broader market in terms of the reliance on valuers' information, reports and opinions in regard to sustainability.

The value of an asset is the fundamental driver for decision making for property, be it investment in renovation, retrofitting or development. The primary focus is to enhance and increase the value of a property. Consequently, decision making for property asset management, development and investment is anchored to the implications that an activity will have on the value of the asset. The development of and investment in sustainability initiatives for property are perceived to enhance the value (Rydin, 2016; Olubunmi, Xia and Skitmore, 2016; World Green Building Council, 2013; Bowman and Wills, 2008; Madew, 2006; Kats, 2003). However, the quantum of contribution that sustainability makes is still unknown and is not necessarily equivalent to the cost of implementing sustainability initiatives in buildings. This is important in the valuation of a property and reflecting the market's value assessment of sustainability, rather than the cost–benefit or payback calculation.

Many rely on the values reported by valuers. The utilisation of a valuer's knowledge and judgement applied in a single valuation has broad implications for primary, secondary and tertiary stakeholders (Warren-Myers, 2013). For example, a market value valuation may be undertaken for a real estate investment trust (REIT), which then uses this report for internal reporting, external reporting and (potentially) debt financing. Furthermore, the value of the properties is published on websites and in annual reports, then utilised by other parties, such as shareholders or prospective investors, as a decision-making tool on whether to invest (Kieffel, 2012). Consequently, this utilisation of valuers' valuations underlines the importance for valuers to understand and report on the relationship between sustainability and value, reflecting current market attitudes and investment in sustainability. The market has demonstrated a growing interest and investment in sustainability over time; consequently, valuers need to accurately and knowledgeably reflect this change in valuations. If this does not occur, inaccurate valuations, misallocation of funds and mispricing of assets may result, and the valuation profession may be put at risk, with increasingly litigious consequences, further limiting broad-scale investment in sustainability (Warren-Myers, 2013).

The commonly rationalised 'vicious circle of blame' has hindered broad-scale development and investment in sustainability. Key stakeholders – developers, investors, end users (occupiers/owners) and constructors – blame each other as to why they will not or cannot invest in or develop sustainability due to lack of demand (Cadman, 2000). However, as Myers, Reed and Robinson (2007) suggest, valuers have a crucial role in demonstrating the value of sustainability to investors and developers, which might change stakeholders' decision making. Lorenz (2008) identified that for change to occur, clear relationships between sustainability and value were required, and if demonstrated to multiple stakeholders, this would result in a 'virtuous circle' (Hartenberger and Lorenz, 2008) whereby the value of sustainability identified would be a positive driver for sustainability. This would change the extant perception 'Why would you invest in sustainability?' to 'Why wouldn't you invest in sustainability?' because of the positive value effect. However, identification of the quantum of value has been an area of substantial research for academics, industry and government for well over a decade. Furthermore, this has been the nemesis for property valuers and appraisers worldwide.

It is not the role of the valuer to create the value of sustainability or generate the premium that sustainability may add to a property. In essence, in undertaking a valuation, valuers need to consider all aspects, characteristics and factors of the property and its market in the reflection of the market value. Valuers need to be cognisant of changing market perceptions, actions and reactions and be able to reflect this in their market value assessments. In valuation, sustainability comprises one of the many factors considered and the broader market perception and demand for sustainability is a key dynamic for incorporation. As with other factors that are considered in the process of valuation, such as building location, quality, age, rental income, tenant profile and risk factors, sustainability needs to be considered in the comparative analysis of the subject property against other comparable assets. Consequently, this will affect judgements made by the valuer in terms of variables like market rents, capitalisation rate, discount rate, vacancy, rental growth and terminal yields. The ability of the valuer to consider the effect of sustainability upon the market value of a property requires knowledge, analytical comparison skills, experience and judgement for accurate reflection through the valuation (Warren, Bienert and Warren-Myers, 2009; RICS, 2013; Muldavin, 2010; RenoValue, 2016).

For the valuer, sustainability presents a major challenge of up-skilling in a number of knowledge areas. First, it requires understanding the market's perception and treatment of sustainability, for example, what metrics valuers are using to 'quantify' sustainability; where demand comes from; and actions and decision making by investors, owners and tenants. Second, valuers need to understand what sustainability is and how this is then measured, quantified and compared. The prevalent use of rating tools in the property sector has meant that sustainability assessments and comparisons have been simplified by providing a structured approach to sustainability metrics in buildings. However, this adds complications, such as the number of tools in the market, which one to use, the complexities of tools and the use of tools for comparative analysis. This requires valuers to have significant knowledge of the rating systems. Third, valuers need to have enough transparent market evidence to allow comparative analysis with sustainability considerations for valuation. Finally, valuers must compile market knowledge and comparative analysis to develop judgements for decisions in valuation, relying on the valuers' knowledge, experience and heuristics.

7.2 Sustainability assessment for valuation

The assessment of sustainability in property from a valuation perspective presents a range of challenges. Fundamentally, sustainability is no different from other technological advances, such as elevators, heating and air conditioning, made in the design, consumption, occupation and management of property that have shaped and changed the position of the property in the market, and consequently over time the market implications and dynamics become clear. Sustainability is a broader and more encompassing programme, comprising possibilities of multiple factors occurring together, extending the focus of responsibility and accountability and strengthening, enhancing and highlighting existing property objectives. Rather than solutions and initiatives being treated individually, the sustainability agenda has engaged multiple objectives across a triple-bottom-line perspective of environmental, social and economic considerations in the design, construction, occupation and management of property. The uniqueness of property, its design, development and management depends on the surrounding environment, the building, the landlord, the management and the tenants, which are all intertwined in the operation of property. The complexity of sustainability and its understanding, interpretation and perception in regard to property by different stakeholders is immense, consequently making it inherently difficult to equitably compare sustainability levels in different properties. As a result, sustainability assessment and rating tools have dominated the sector in an attempt to classify and identify sustainability and the key initiatives and objectives that can be attained and measured in property. However, the challenge is in identifying which particular attributes, features, initiatives and indicators are the 'right' ones because this is highly subjective (Wilkinson, Sayce and Christensen, 2015a).

There are literally hundreds of rating tools, assessment matrices, performance indicators and weightings for the built environment, many focused on property, developed by various organisations, including governments, not-for-profit groups, academics and industry. However, in the commercial sector, the World Green Building Council has been influential in fostering key tools that have become prevalent and adopted in property markets worldwide. Well-known tools include: Building Research Establishment Environmental Assessment Method (BREEAM), which originated in the UK and forms the fundamental basis for a number of other tools; Leadership in Energy and Environment Design (LEED) and Energy Star, which originated in the USA;

Comprehensive Assessment System for Building Environmental Efficiency (CASBEE), developed in Japan; Green Star and the National Australian Built Environment Rating System (NABERS), created in Australia; and the Deutsche Gesellschaft für Nachhaltiges Bauen (DGNB) tool, developed in Germany. Other prevalent tools in global property markets are discussed by Christensen and Sayce (2015). The development of these assessment tools has been fundamental to engaging the industry and creating market drive to increase sustainability in the built environment. However, the various assessment characteristics, parameters and approaches complicate the comparability of these tools to each other and also from property to property with the same type of rating.

The rating systems are predominantly categorised as either design-based tools or performance-based tools. The design-based tools focus on new buildings and major redevelopment of existing buildings, and assess the potential of the design to perform sustainably in environmental and social categories. Often, these design-based tools are linked with 'as-built' tools that ensure that the design was built, but do not always assess or connect to the actual performance of the building. The performance-based tools, as the name suggests, examine the building's actual performance. The performance-based tools are generally different rating systems, disconnected from the design-based tools. The performance-based tools focus on easily measurable aspects, such as energy and its relationship with greenhouse gas (GHG) emissions; such tools include Energy Star in the US and NABERS (previously known as the Australian Building Greenhouse Rating – ABGR) in Australia. The tools are vastly different in how and what they assess; however, they have provided some common language for property stakeholders to communicate sustainability desires for property investment, ownership and occupation, developing the market for sustainability. Nevertheless, the range of tools and the differences between them have at the same time contributed substantially to market confusion.

In the Australian and New Zealand commercial office markets, there are two separate rating systems, namely Green Star and NABERS. Green Star is a holistic design-based tool, focused predominantly on assessing the design of the building across eight environmental categories, which results in a single star rating that is a one-off award (GBCA, 2016). By contrast, NABERS has separate certifications for the performance of the building for each category: energy efficiency, water efficiency, waste and indoor environment quality (NABERS, 2016). NABERS certifications require 12 months of operational data and are only valid for 12 months. Both tools rate using 'stars', but are administered by different organisations and target different sectors of the market. Green Star aims to rate the top 25 per cent of buildings, focusing on the new-build sector, whereas NABERS can rate any building for which there is 12 months of performance data. Furthermore, NABERS has recently been an integral part of the mandatory Commercial Building Disclosure Program as part of the National Energy Efficiency Program implemented in 2010. As a result, it applies to a much larger proportion of the office building stock in Australia.

Rating tools have provided the industry with a common language to communicate sustainability wants and desires in new and existing buildings. However, complications with the rating star systems have evolved as a result of separation and disconnect between the two tools, confusing industry stakeholders about the complexity, equivalence, role and outcomes of the different tools (Warren-Myers and Reed, 2010; Wilkinson et al., 2015a).

Green Star ratings and NABERS ratings are not equivalent (Table 7.1). A five-star Green Star rating does not provide you with a five-star NABERS rating. The Green Star tool only rates the potential of the building to perform to a certain level. The building's actual performance is highly dependent on the way in which it is subsequently managed, occupied and used. The terminology 'star rating' has confused much of the sector for some time, with common trip-ups, such as agents promoting their building with a '5.5 Green Star rating', when there are no half stars in Green Star and it is an existing building! The disconnect between the two tools has

Table 7.1 Comparison of Green Star and NABERS rating tools

	Green Star/Green Star NZ	*NABERS/NABERSNZ*
Assessment type	Design-based tool, assessed on the potential of the building to perform over a range of categories.	Operation-based tool using 12 months of building performance data to assess efficiency.
Tool type	Initially focused on new builds and major renovated buildings; expanded later to include ratings for as-built, performance, interiors and communities.	Performance based; uses operational data used in the base building, whole building or tenancy.
Rating style	Star based.	Star based.
Rating range	4–6 stars (no half stars).	0–6 stars (half-star increments).
Rating categories	Eight categories (energy, water, management, emissions, indoor environment quality, land use and ecology, materials and transport) plus a bonus innovation category.	Individual ratings are achievable for each category: energy, water, waste and indoor environment quality.
Administrator	Green Building Council of Australia and New Zealand Green Building Council.	Office of Environment and Heritage (NSW government) on behalf of federal, state and territory governments. New Zealand Government – Energy Efficiency and Conservation Authority.
Rating frequency	One-off.	Annual.
Date established	2003 (Australia); 2006 (New Zealand).	1998 (Australia); 2013 (New Zealand).
Building types	Initially offices; expanded to include education, healthcare, industrial, multi-unit residential and public buildings and retail centres.	Initially offices; expanded to include shopping centres, hotels, data centres and homes.
Historical basis	Based on LEED and BREEAM.	Previously known as ABGR (Australian Building Greenhouse Rating).

Sources: GBCA (2016); New Zealand Green Building Council (2016); NABERS (2016); NABERSNZ (2016)

hindered the perception, acceptance and implementation of sustainability ratings initially. However, in recent years, increases in collaborative agreements and Green Leases have attempted to bridge gap between design and performance. The Australian property market in the long term has focused more on the performance-based tool NABERS, primarily because of the tangible measurement and quantification aspects that could be applied across both new and existing office buildings. The mandatory energy efficiency disclosure for commercial property implemented in 2010 through the Building Energy Efficiency Disclosure Act 2010 and Building Energy Efficiency Disclosure Determination 2016 (Australian Government, 2016) has only strengthened and enhanced the use of the NABERS rating tool in the property sector, similar to findings in the UK with their mandatory disclosure programme and the actions of landlords and tenants (Turley and Sayce, 2015). The number of Green Star-rated buildings has steadily increased, as new buildings are added to stock. In major Australian central business districts (CBDs) like Brisbane, Sydney, Melbourne, Perth and Adelaide, a new building would not be built without a Green Star rating, because the use of a rating is now seen as part of market positioning of the building in the marketplace (Wilkinson, van der Heijden and Sayce, 2015b). Thus, both rating systems do have their roles in the Australian office property market.

The Green Star rating tool treats sustainability from a more holistic perspective, using eight environmental categories plus a bonus innovation category by which to assess a building's attributes. This allows for flexibility of design and differentiation between assets, and provides a single holistic star rating. This star rating is one-off and lasts for the life of the building. By contrast, NABERS has separate certifications for each of its four categories, and a user is not required to get certification for all four categories; they only need to get one if they wish. The NABERS certifications are also only valid for a year, so annual re-rating is required. Both tools allow for self-assessment, as you can download the Green Star tool in Excel format and NABERS is an online calculator. Although these ratings do not count when self-assessed, it has caused considerable issues for valuers when presented with information or a manager indicating the property has a star rating, when it is in fact a self-assessed star rating.

Another factor is the complexity of the Green Star tool. The use of the eight environmental categories allows for multiple approaches to achieve a rating, complicating analysis of the building. From the perspective of a valuer, investor or prospective tenant, when investigating two five-star Green Star-rated buildings, do they have the same sustainability attributes? The answer is no; one building may have achieved their five-star rating by focusing on credits available through the energy and water efficiency categories, for instance, whilst the other building might have a more even spread of credits across the indoor environment quality, land use & ecology, management and transport categories. This has major implications for property stakeholders, as the tangible and intangible benefits vary substantially between the two. For example, the first building will be of keen interest to investors and occupants, as the energy and water efficiency will minimise operating expenditure, reduce risk to carbon-related regulation and is tangible. By contrast, the second building will not necessarily have these tangible features to value, and certainly a tenant might value the enhanced indoor environment quality and the transport initiatives in the building, but will they pay a higher rent for the building? As a result, what is the value of the certification and how are the attributes that contribute to that certification considered to affect the value of the property? To further complicate this analysis, access to the information pertaining to how buildings have achieved their rating is not always available, making comparing sustainability characteristics inherently difficult for valuers (and other market stakeholders) when Green Star is utilised. In addition, certification creep is another factor, as Green Star upgrades the version of the tool to ensure that the rating is current and continues to push the boundaries of sustainability in property. This requires understanding of the differences between the versions – for example, a five-star Design Green Star Version 1 rating compared to a five-star Design Green Star Version 3 rating. The multiple categories, criteria and weightings in the rating systems make it inherently difficult to make judgements regarding sustainability in property and its effect; consequently, valuers need to be very knowledgeable about sustainability, understand sustainability rating tools and be able to compare these attributes in the analysis of comparable transactions. Furthermore, they need to be sensitive as to how informed market participants are of the benefits and costs and how willing they are to pay for sustainability through transactional evidence.

The NABERS rating system, due to its use of performance data and the single rating per category, allows for easier comparison between buildings. However, there are still parameters in the NABERS tool that can easily affect the result, such as the assumptions related to building occupation hours, the use of Green Energy (up to a half-star difference) and the GHG co-efficient calculation. The latter, for instance, results in differences between the states in Australia with different GHG co-efficients. What this effectively means is that a higher rating would be achieved by a building in Tasmania (low GHG co-efficient) compared to a building with the same energy use in Melbourne (Victoria) (higher GHG co-efficient as a result of brown coal gas) (Hertzsch and Heywood, 2012; Australian Government, 2014). This does not have a significant implication for valuers, as long as the comparable properties are from within the same state, but it does affect decision making by other stakeholders within the sector, particularly when considering the profile of their portfolio across states and rating levels.

Since the introduction of the sustainability agenda in property and the adoption of sustainability rating tools, there has been a significant question about the value relationship between sustainability and market value. In order to answer this question, valuers need to be able to analyse and compare properties adequately based on their sustainability attributes and ratings in conjunction with a range of other factors that affect the market value of a property. Valuers need to know and understand how the rating systems work, their multi-criteria assessment processes and market perception. It is crucial that valuers understand and are able to reflect sustainability and its effect on market value through critical analysis and comparison. Consequently, valuers need to understand the recognition and value perceived in sustainability by the market and reflect this through their comparison and analysis of transactions in the market.

7.3 Market recognition of sustainability

The research dedicated to identifying the value of sustainability is substantial. There is a plethora of research dedicated to identifying the relationship between sustainability and value comprising cost–benefit analyses; case studies; variations on valuation methodologies taking a normative approach to identifying the value of sustainability; and quantitative studies (see Sayce et al., 2010; Lorenz and Lützkendorf, 2011; and Warren-Myers, 2012 for an overview of the breadth and style of extant research on the relationship between sustainability and value). However, what is required to demonstrate the relationship between sustainability and value and its quantum is evidence. In compiling evidence, there is a critical need to ensure that what is to be valued can also be measured. Consequently, sustainability in the Australian context is measured in commercial real estate through the rating tools Green Star and NABERS. Once a means of quantifying sustainability has been established, there is then a need to rate a number of buildings and subsequent market transactions. Consequently, the popularity of quantitative studies has produced a number of academic studies investigating various markets, mostly commercial office property markets (Miller, Spivey and Florence, 2008, Eichholtz, Kok and Quigley, 2010; Pivo and Fisher, 2010; Fuerst and McAllister, 2011; Reichardt, Fuerst, Rottke and Zeitz, 2012; Dermisi, 2012; Chegut, Eichholtz and Kok, 2014; Surmann, Brunauer and Bienert, 2015; Robinson and Sanderford, 2016), including several recent studies of the Australian market (Newell, MacFarlane and Kok, 2011; Newell, MacFarlane and Walker, 2014; Gabe and Rehm, 2014). However, for such studies to be undertaken, there need to be enough buildings and transactions within the dataset to allow analysis. Certainly, as the number of rated buildings increases, market awareness and stakeholder value-based decision making is easier to ascertain and measure. It is then the role of the valuer to perform the comparative analysis to reflect market sentiment towards sustainability through valuations of market value.

Market recognition of sustainability is reflected through the adoption of and investment in assets identified as sustainable or having sustainability characteristics; in most cases, this is tracked through rating tool certifications. By tracking Green Star and NABERS rating certifications over time, an understanding of market adoption, engagement and investment in sustainability is possible. The initially voluntary nature of the tools in Australia meant a market-led demand for sustainability in the early 2000s, as shown in Figure 7.1, demonstrated through a gradual increase in the number of buildings certified under NABERS and later Green Star. Government

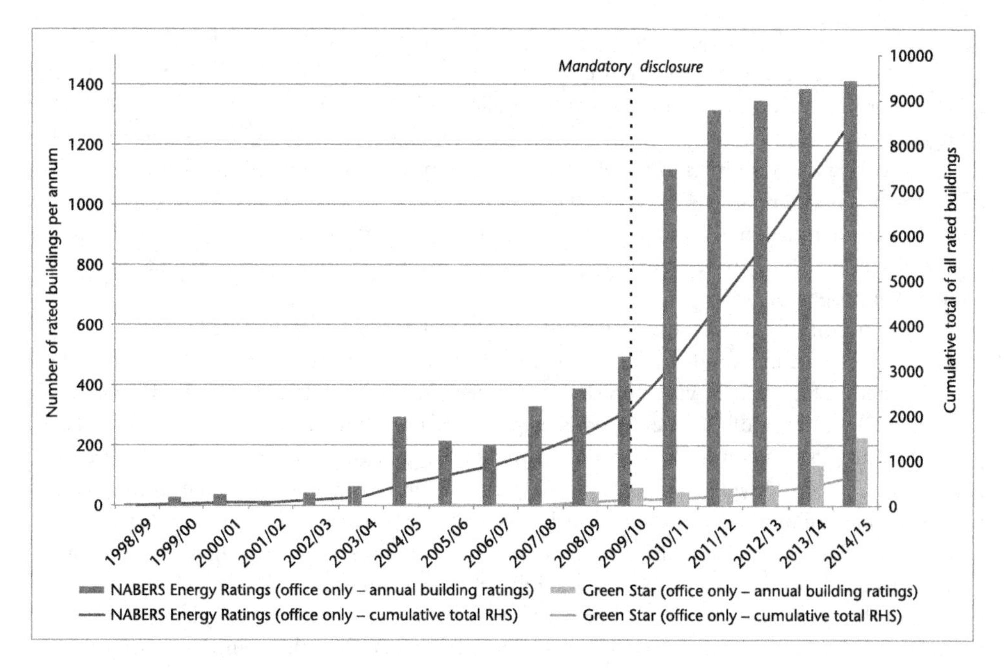

Figure 7.1 The number of buildings in Australia rated with Green Star and NABERS

Sources: NSW Office of Environment and Heritage (2015); NABERS (2016); GBCA (2013); and GBCA (2016)

intervention in the market has resulted in major changes in the profile of certifications, the most notable occurring in 2010 when the mandatory building disclosure programme was launched.

Sustainability assessment in Australia has been in the market for over a decade and in New Zealand just on a decade, with the NABERS performance tool developed in 1998 to measure commercial office buildings' energy efficiency and the Green Star design-based tool developed in 2004. The recent adoption of these tools, shown in Figure 7.1, highlights their evolution and the time they have taken to become established in the market. Between 2004 and 2007, commercial office market interest in sustainability began in Australia. Sustainability was considered a key differentiator and the focus of many owners, investment funds, occupiers and developers as a 'must have' in their commercial office buildings. This predominantly market-led drive for increased sustainability was encouraged by state and federal governments requiring space occupied to be rated, and in some cases local councils indicated office projects would be required to achieve a Green Star rating for development in their municipality. The global financial crisis in 2008 crushed the momentum of sustainability adoption and investment in the Australasian commercial office sector. However, Australia's political climate at the time played an important role with the introduction of the Building Energy Efficiency Disclosure Act in 2010, making it mandatory to disclose office building energy consumption. This applied to all commercial offices for sale or lease greater than 2,000 square metres and required them to attain a Building Energy Efficiency Certificate (BEEC), which comprises primarily a NABERS Energy Rating certification (base building) (Australian Government, 2016). As a result, sustainability maintained its relevance and importance in the sector, and the focus moved away from more holistic approaches to sustainability to focus on energy efficiency, demonstrated by the significant increase in the number of buildings rated with NABERS around 2011 in Figure 7.1. The Australian commercial office market comprises some 4,500 commercial office buildings in eight state capital cities and 17 major towns; however, it does not capture all office buildings of under 1,000 square metres in the CBDs and under 500 square metres in the 17 major towns (PCA, 2016). Approximately 30 per cent of the office buildings in Australia currently have a NABERS rating and 15 per cent have a Green Star rating. The number of unrated buildings still outweighs the number of rated buildings in the Australian property market, although market recognition of sustainability is increasing. Consequently, any quantitative analyses or valuation investigations need to consider the implications of the limited number of rated buildings, their location and attributes and, in particular from a valuation perspective, careful selection and analysis of comparable properties.

The evolution of certifications within the Australian market as shown in Figure 7.1 aligns with the concept of market development and maturity of sustainable buildings within a market demonstrated in Figure 7.2, which utilises the concept from McColl-Kennedy, Kiel, Lusch and Lusch (1992) and McColl-Kennedy and Kiel (2000) of the product evolution within a market place, detailing its initial introduction to the market, its growth phase and then its progression to maturity. Warren-Myers (2009) anticipated that this process would be similar in the market for sustainable buildings. As evidenced in Figure 7.1, the gradual uptake and development of sustainable buildings evolved in the years between 1998 and 2004, with a strong surge and change in the market and the number of rated buildings indicating a move to the growth phase. Given the size of the Australian commercial property market, significant growth is still needed before the market achieves or approaches maturity. The importance for valuers in this relationship is that as the market develops, particularly through the growth phase, the increase in certified buildings in the market will lead to enhanced awareness and knowledge of the certifications. The development of valuers' knowledge and understanding enhances their ability to identify over time value relationships and reporting (Warren-Myers, 2016).

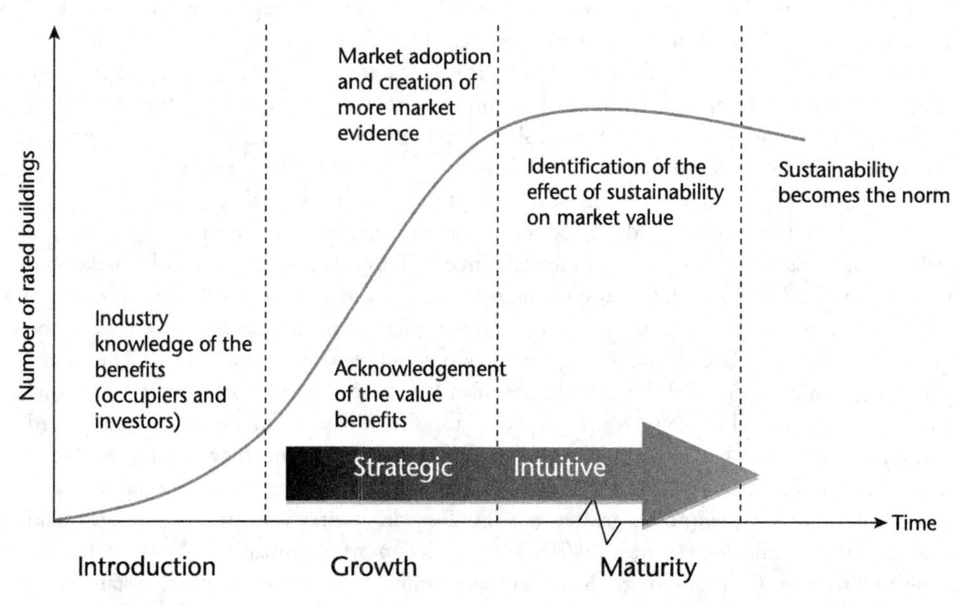

Figure 7.2 Market evolution of sustainability and valuers' knowledge development

Source: Warren-Myers (2009)

7.4 Evidence underpinning the value relationship between sustainability and value

Market evolution for sustainability and increasing numbers of certified buildings enhances the utilisation of quantitative methods to analyse trends and understand market dynamics. Identifying sustainability *premiums* and the *value* of certifications is an area of key interest to industry and academics. This has resulted in quantitative analysis of the property market, with sustainability as a key variable, to investigate whether a relationship between sustainability and value exists and the value premium. However, these quantitative studies require vast amounts of transactional evidence for analysis, and consequently are reliant upon enough rated or certified buildings and transactions to ascertain whether there is a price differential.

Research of this type has typically been conducted in the US, although there are studies from the UK and the Netherlands, as the large databases of building data and information and also a larger market with more rated buildings have allowed thorough analysis to be undertaken. This research has typically found a positive relationship between buildings with a sustainability rating and rental or price premiums (Miller et al., 2008; Eichholtz et al., 2010; Pivo and Fisher, 2010); Fuerst and McAllister, 2011; Bienert and Schiitzenhojlzr, 2011; Reichardt et al., 2012; Dermisi, 2012; Davis, McCord, McCord and Haran, 2015; Robinson and Sanderford, 2016).

Ascertaining the value of these certifications has been the focus of recent investigation in Australia with key studies examining rentals and returns (Newell et al., 2011; Newell et al., 2014, rents (Gabe and Rehm, 2014) and the creation of the MSCI/IPD Green Property Index launched in 2011. The studies by Newell et al. (2011) and Newell et al. (2014) used a sample of 206 NABERS-rated office buildings, 26 Green Star-rated buildings and 160 non-rated buildings from Sydney and Canberra and found positive relationships for buildings with high NABERS and Green Star ratings. The research found that premiums were only noted for higher-rated buildings, whilst lower-rated buildings demonstrated a discount in terms of value and net rent and increases in vacancy, incentives, yields and outgoings. These studies concur with the MSCI/IPD Green Property Index, which has demonstrated premiums associated with Green Star ratings and high-rated NABERS buildings on an annual basis since 2011. Little information can be obtained in regard to how the index was created and the assumptions, controls and judgements made in relation to the creation of the index. All results since the inception of the index have identified a positive result of Green Star-rated buildings, with a 12.9 per cent annualised return, a 6.6 per cent income return and 5.9 per cent capital growth. The results have also demonstrated that the Green Star-rated buildings outperformed the broader CBD office market by 2.7 per cent, whilst office buildings with high NABERS ratings for energy efficiency outperformed other office buildings with lower ratings (MSCI, 2015). The index has shown little change over time and publicly available information confirms the overall performance of the index; however, it does not demonstrate the breakdown between the capital cities or markets.

The premiums found in the studies by Newell et al. (2011) and Newell et al. (2014) and the Green Property Index concur with those found in studies in the US by Miller et al. (2008), Eichholtz et al. (2010), Pivo and Fisher (2010), Wiley, Benefield and Johnson (2010), Fuerst and McAllister (2011) and Reichardt et al. (2012). However, there are substantial limitations to the Australian studies such as the data veracity, sample size, treatment of data and assumptions and judgement made by the researchers. For example, all of the US studies (aforementioned) had substantially larger data sets than the Australian analyses. Furthermore, in the US studies, value premiums ranged from no rental premium for LEED buildings to 17 per cent, depending on the study (indeed, some authors found that further analysis of their original data gave different results) and sale price premiums ranged from no premium to 11.1 per cent among the different studies, whilst Energy Star did demonstrate a more consistent effect on rentals, ranging from 2.1 per cent to 3.3 per cent. Variations in econometric modelling, regression and hedonic pricing models is not unusual and is highly dependent on whether the researcher is investigating a negative (cost-concerned) or positive (value creation) relationship, which subsequently will result in biases in the assumptions and assessment within the modelling process.

The results of the Newell et al. (2011) and the Newell et al. (2014) studies and the Green Porperty Index are at odds with the research undertaken by Gabe and Rehm (2014), who specifically examined the Sydney CBD office market from a tenancy rental perspective, using lease information. This allows a greater understanding of the actual rent paid (net or gross), incentives offered and the like. This study comprised a larger data sample, examining registered leases from January 2009 to July 2011 that provided a data set of 1,526 office lease transactions; after cleansing, a total of 673 observations in NABERS-certified buildings were utilised. The descriptive statistics initially suggested a value premium for higher-rated buildings; however, the authors suggested that this is likely a consequence of the location influencing the results. The research then used a semi-log ordinary least squares hedonic regression model to test whether the rent could be explained as part of a function of unique independent characteristics that is associated with the tenancy space. The regression analysis suggested no significant presence of a value premium for rent in the Sydney CBD analysis. However, low-rated buildings did incur a discount, noting that the confidence interval was 85 per cent. The deviation of the results from those of the Newell et al. (2011) and Newell et al. (2014) studies and the Green Property Index suggests that Australia still requires further investigation to ascertain the quantum of a relationship between sustainability and value. Studies using the increasing amount of data available relating to rated buildings are required to focus on the micro and macro markets to ensure that strong correlations and high confidence can be achieved.

Although the Newell et al. (2011) and Newell et al. (2014) studies and the Green Property Index found premiums associated with sustainability, to valuers this research is complicated due to the mass analysis of attributes and computational calculations. Valuers require more focused studies of markets they work in, and in particular need to be cognisant of the comparative attributes between transactions and able to analyse their implications. Consequently, more focused studies, such as Gabe and Rehm (2014), which examined the Sydney CBD commercial office market and found no clear premium, are seen to provide a clearer direction to valuers. However, such studies still do not replace the need for in-depth comparative analysis of comparable transactions. Dermisi (2012) also undertook a focused study on a specific office market – Chicago – investigating vacancy and rental parameters before and after achieving an LEED existing building certification. The results identified a positive effect on vacancy, yet rental analysis demonstrated varied results dependent on the type of LEED certification. Surmann the et al. (2015) examined commercial property in Germany and found no relationship between energy efficiency and market value. An interesting feature of this research was the examination of the relationship over time, although there was no clear evidence to suggest any such relationship.

The application of this quantitative type of research (involving econometric modelling, regression and hedonic pricing models) in valuation practice is problematic, as the treatment of data, assumptions and judgements made by the researcher, sample size and verifiability are not generally known or accessible to a valuer and can be often biased based on the modelling process used (Warren-Myers, 2012; Muldavin, 2010). These studies, however, do provide a marker to stakeholders in the sector, offering insight and direction as to value relationships. More in-depth market analyses like Gabe and Rehm (2014) and Dermisi (2012) are required to be able to extrapolate out and analyse change, trends, preferences and implications for market sentiment, attitude and value effects that sustainability might have in the commercial office market. However, from a valuation perspective, comprehensive comparison and analysis of buildings with and without sustainability characteristics is still required in order for valuers to be able to reflect any influence that sustainability may have on values.

These studies do provide strategic and informative opinion as to the relationships between sustainability and value, based on empirical analysis. However, there is only a small dataset in the Australian context, which means that there is a need for more research to be undertaken and property stakeholders, owners, investors, occupiers and valuers should not solely base their opinion and judgement on these studies. Furthermore, from a valuation perspective, these studies do not provide the framework within which valuers are able to utilise the information to accurately assess a relationship between sustainability and market value, as the reliability and communication of these studies' specific quantitative results are incomplete and inadequate for use in practice. Valuers should focus on adopting critical analysis techniques in the comparison of local data from their markets and building information that is comparable to the subject property that they are valuing.

7.5 Valuer reflection of value

Sustainability has been a part of the commercial office market for well over a decade in Australia. The number of sustainable rated buildings is significant and increasing. The current number of certified buildings and time in the market suggests that there should now be a variety of market transactions for these buildings. So the valuers' catchcry of 'Not enough evidence' should no longer be justified. However, sustainability reporting in commercial office valuations is sporadic and there is limited understanding of sustainability assessment and its implications for value (Warren-Myers, 2016).

Valuers have a difficult task in ascertaining the value relationship between sustainability and market value; this has to do with a series of factors, some inherent in the nature of practice and others relating to sustainability assessment and interpretation. Challenges for valuers include:

- the complex nature of tangible and intangible benefits of sustainability;
- the market's assessment and value of these tangible and intangible benefits;
- sustainability assessment and the complexities of the current rating systems;
- knowledge development of sustainability and interpretation of market actions; and
- reliance in practice on heuristics.

Lorenz and Lützkendorf (2011) provide guidance for valuers in regard to how sustainability may be integrated, assessed and considered in the valuation process, and the RenoValue toolkit has demonstrated how sustainability can be incorporated and considered in valuation practice (RenoValue, 2016). In addition, the United Nations Environment Program Finance Initiative (UNEP FI, 2014) provides insights into best-practice approaches for investment and portfolio building in relation to sustainability and key sustainability metrics for consideration in the valuation process. However, for valuers to begin to understand the relationship between sustainability and value, they must firstly have enough knowledge about sustainability and its assessment in the market to begin to interpret market behaviour and how this may influence and affect values. Valuers' use of art and science in valuation is underpinned by the development of strategic knowledge, experience and application of heuristics in practice. This has two implications for the valuer: the experience of and exposure to market dynamics, and the development of knowledge. Warren-Myers (2009, 2013, 2016) has explored the relationship between sustainability and value in the valuation profession in Australia (from 2007 to 2015) through a longitudinal study examining valuers' knowledge, perception and understanding of sustainability and the value relationships that they perceive in the market. Valuers' understanding of the relationships between particular valuation variables, sustainability and certain sustainability characteristics has developed over the course of the survey. In particular, saleability, rental growth, risk premium and obsolescence were identified as valuation variables that were being affected by sustainability, whilst sustainability attributes that have an effect on value were clearly identified as energy efficiency and water conservation (Warren-Myers, 2016). The longitudinal research has demonstrated a change from 2007, when valuers relied more on normative theory and research, to 2015, when valuers appear to be better reflecting market sentiments and utilising key metrics from market preferences for sustainability assessment to report on in the valuation process.

The number of rated buildings within the commercial office market has grown considerably over the past decade and a half (Figure 7.1), demonstrating market evolution. Warren-Myers's (2009) hypothesis (Figure 7.2) is that as the number of rated buildings and transactions occurring within a market increase, this would expose valuers to knowledge development opportunities and experience, which would over time lead valuers to create heuristics reflecting the relationship between sustainability and market value. Warren-Myers (2016) determined that, although market evolution had progressed and valuers were demonstrating a strengthening opinion regarding the relationship between sustainability and value, valuers' knowledge was not developing at the same rate as the influx of rated buildings in the market, consequently leading to continuing reservations about the quantum of the relationship between sustainability and value from a valuation perspective. There was certainly change demonstrated over time in valuers' understanding, capacity and relationship perspectives of sustainability within the commercial property market that indicated stronger implied relationships between sustainability and market value. However, the analysis of valuers' knowledge of sustainable rating systems, although it has improved and morphed to reflect market dynamics, shows that the actual knowledge levels of valuers is severely limited (Warren-Myers, 2016). This indicates an urgent need to create further knowledge development programmes and assist valuers to enhance and develop their knowledge of sustainability, whilst also creating tools and approaches to assist in the comparative analysis of properties' sustainability attributes. As the market continues to establish sustainability as a key attribute, valuers need to be able to accurately reflect the market change and sentiment towards sustainability and their current approach and consideration in the valuation process is inhibiting investment in sustainability. Furthermore, the lack of, or misapplication of, knowledge by valuers in the process of determining market values or advising clients in regard to sustainability will likely lead to inaccurate valuations, mispricing of assets, misallocation of capital and increasing litigious implications for the profession. The challenges evident in the sector in ascertaining value connections between sustainability and market value could be exacerbated through unethical practice, which may cause further damage for the mainstreaming of sustainability investment in the property industry.

7.6 Advice for the valuation of property

All buildings, whether they have sustainability features 'classified' by a rating or not, should be considered, compared and assessed, remembering that sustainability is only one of a multitude of characteristics considered in the valuation of property. There have been a number of resources that provide guidance for the assessment and incorporation of sustainability consideration in the valuation process, from academia, industry and the governing bodies of the valuation profession in their jurisdiction. A proportion of this available information and research is not directly applicable to current valuation practice; for example, research suggesting changes to valuation approaches, proposed risk-weighting systems and special value considerations can be utilised for reference, but relying on or utilising this in practice should be substantiated and rationalised with local practice standards and guidelines. Current practice requires a valuer to ascertain the market value; consequently, the valuer has to adhere to the market value definition and to the competence and knowledge standards of the profession.

Goddard (2012), on behalf of RICS Oceania, has developed a guidance document to provide Australian valuers with advice on the key components for investigation in the valuation of a building with sustainable attributes. Similarly specific valuation guidance is also found in RICS' *Sustainability and Commercial Property Valuation* (2013), UNEP FI's *Sustainability Metrics* (2014) and the RenoValue (2016) training programme. In the assessment of sustainability features, attributes and considerations in buildings, the initial step should be to examine how the market is measuring sustainability. This market opinion and sentiment will likely lead towards particular rating tools and systems, although this preference may change over time. So the valuer must be aware of market considerations and objectives relating to sustainability assessment. The valuer should investigate and develop specific knowledge of tools and ratings used in the market, enhancing comparative analysis and rationalisation of decisions in the valuation process to reflect market value sustainability considerations in the valuation. If there are mandatory legislative requirements, valuers should understand these requirements, their implications and how they might affect property assets.

Sustainability and rating information may be provided by the owner or manager; always verify the rating through sighting the actual certificate or checking the online databases. If there are no ratings and/or other initiatives are being undertaken in the building that are not part of a rating system, these should not be discounted but incorporated in the assessment of the property's characteristics. Sustainability is not necessarily focused on resource efficiency and there may be more holistic objectives being integrated into the building that fall outside of the key rating systems. Therefore, there is still a requirement to analyse and consider all aspects in the assessment of a property's characteristics (certifications, sustainability strategies, owner and tenant preferences and other initiatives) and the implications for long-term risk assessments that may affect obsolescence, capital expenditure profiles, rental growth, vacancy and terminal yield assumptions.

Office buildings provide a level of utility; whether sustainable or not, this utility (the provision of office space) remains the same. Consequently, an office property that provides office space for occupation, regardless of whether it is sustainable or not, provides comparable utility, and the market consideration of the asset needs to ensure that market-based assessment is used in the valuation of the asset. This means that the use of special valuation methods or approaches for sustainable buildings is not appropriate and that the traditional valuation approaches – income capitalisation and discounted cash flow, comparison approaches – need to be used. However, the onus is then on the valuer to perform certain tasks such as the comparative analysis components in the valuation process to take into consideration the sustainability characteristics (Lorenz and Lützkendorf, 2011; UNEP FI, 2014; RenoValue, 2016). Furthermore, valuers need to ensure that other traditional factors are not ignored, particularly if sustainability may enhance certain aspects, such as tenant or owner objectives in relation to corporate social responsibility reporting and market sentiment changes.

Any adjustments, judgements or assumptions that are made in the context of sustainability in the valuation process need to be supported by adequate evidence. As with any other determination of rental growth rates, discount rates and capitalisation rates all need to be investigated, researched and rationalised with evidence to support the valuer's decision. As with these traditional considerations, the valuer needs to ensure that the evidence is clear for anything that they believe may be affected or influenced by sustainability, so that their judgement in making assumptions and determinations is sound. Remember that for market value valuation, 'if sustainability features are identified and recognised as having an impact on value, they should be built into the calculations only to the extent that a well-informed buyer would account for them, as evidenced from an analysis of the market' (RenoValue, 2016: 55).

Valuers should be aware of their own limitations and knowledge levels and remember the code of ethics and the standards by which they practice. They should investigate and be guided by standards and practice guidelines in their area/country in regard to the treatment and suggestions concerning sustainability assessment and comparison. In addition, they should develop their own knowledge through taking advantage of continuing professional development opportunities that may be presented by educational institutions, professional bodies, property industry participants or sustainability industry stakeholders.

Governing bodies and those setting the standards are providing valuers with directions as to how to develop knowledge and skills to value buildings with sustainable attributes; however, more is needed. There are a range of professional development options, through continuing professional development events, online modules and literature. The RICS has published several valuation guidance notes or information papers that provide specific approaches as to how to value sustainability and the introduction of the RenoValue guidance documents and training programmes has acknowledged the need for professional development targeted specifically at valuers. The Appraisal Institute in the US and the Australia Property Institute in Australia have also released several papers, books and a programme of professional development sessions and online modules to provide advice on sustainability assessment and valuations. There is also a range of academic papers that examine sustainability assessment, valuation issues, valuation approaches and implications for valuation in a spectrum of journals. Valuers should seek to advance their knowledge through the professional development opportunities offered by professional bodies and through reading guidance and research about sustainability and valuation.

7.7 Conclusions

The consideration of sustainability in property does not require new valuation approaches or specialist considerations through weighting factors or adopting standardised changes to variables in the valuation methods. What *is* required is for the valuer to be able to utilise the fundamentals of practice to make decisions based on market comparison and analysis of the asset's sustainability levels. The complication lies in the ability to assess the tangible and intangible benefits of sustainability. If valuers have limited knowledge of any particular characteristic, be it of the building, the market or sustainability, their ability to *value* that characteristic and accurately reflect its effect on market value is inhibited. There have been suggestions for applying weightings – matrices – modifying the valuation methods to better encapsulate sustainability in the valuation process. However, given that valuers struggle to develop enough strategic knowledge and experience in analysing and making judgements in relation to sustainability, applying new methods or models or weighting might exacerbate the potential for inaccuracy and misjudgement in the application of decisions in the valuation process. Consequently, valuers need to recognise their knowledge shortfalls and endeavour to build and develop their strategic knowledge in the area of sustainability and its relationship with market value, through experience in the market, comprehension of stakeholder decision-making processes and values and market evolution. In time, the relationship between market value and sustainability will become clearer as more detailed analysis, decision-making rationales and attitudes towards sustainability become more transparent. The subsequent creation of heuristics and intuitive knowledge will guide valuers in the assessment and judgement of sustainability in property valuation practice.

References

Abidoye, R. B. and Chan, A. P. (2016) Critical determinants of residential property value: professionals' perspective, *Journal of Facilities Management*, 14(3), pp. 283–300.

Adair, A., Berry, J. L. and McGreal, S. (1996) Valuation of residential property: analysis of participant behaviour, *Journal of Property Valuation and Investment*, 14(1), pp. 20–35.

Ashton, A. H. and Ashton, R. H. (1990) Evidence-responsiveness in professional judgement: effect of positive versus negative evidence and presentation mode, *Organisational Behaviour and Human Decision Processes*, 35, pp. 1–19.

Atkinson, T. and Claxton, G. eds. (2000) *The intuitive practitioner: on the value of not always knowing what one is doing*. Taylor & Francis.

Australian Government (2014) *National Greenhouse Accounts Factors: December 2014*, accessed 2/7/2015 from www.environment.gov.au/climate-change/greenhouse-gas-measurement/publications/national-greenhouse-accounts-factors-dec-2014.

Australian Government (2016) *Commercial Building Disclosure: A National Energy Efficiency Program*, accessed 30/8/2016 from http://cbd.gov.au/overview-of-the-program/legal-framework.

Bellman, L. and Öhman, P. (2016) Authorised property appraisers' perceptions of commercial property valuation, *Journal of Property Investment & Finance*, 34(3), pp. 225–248.

Bienert, S. and Schiitzenhojlzr, C. (2011) Energising property valuation: putting a value on energy-efficient buildings, *The Appraisal Journal*, 79(2), pp. 115–125.

Bowman, R. and Wills, J. (2008) *Valuing green: how green buildings affect property values and getting the valuation method right*. Green Building Council of Australia, Sydney.

Cadman, D. (2000) The vicious circle ofblame, *Upstream*, accessed 17/9/2006 from www.upstreamstrategies.co.uk.

Chegut, A., Eichholtz, P. and Kok, N. (2014) Supply, demand, and the value of green buildings, *Urban Studies*, 51(1), pp. 22–43.

Christensen, P. H. and Sayce, S. L. (2015) Sustainable property reporting and rating tools, in Wilkinson, S. J., Sayce, S. L. and Christensen, P. H., *Developing property sustainably*. Routledge, pp. 203–233.

Ciora, C., Maier, G. and Anghel, I. (2016) Is the higher value of green buildings reflected in current valuation practices? *Journal of Accounting and Management Information Systems*, 15(1), pp. 58–71.

Davis, P. T., McCord, J. A, McCord, M. and Haran, M. (2015) Modelling the effect of energy performance certificate rating on property value in the Belfast housing market, *International Journal of Housing Markets and Analysis*, 8(3), pp. 292–317.

Dermisi, S. (2012) Performance of LEED-Existing Buildings before and after their certification. Paper presented at the *Real Estate Research Institute Conference*. Chicago, IL.

Eichholtz, P., Kok, N. and Quigley, J. (2010) Doing well by doing good: green office buildings, *American Economic Review*, 100(5), pp. 2494–2511.

Fuerst, F. and McAllister, P. (2011) Green noise or green value? Measuring the effects of environmental certification on office values, *Real Estate Economics*, 39(1), pp. 45–69.

Gabe, J. and Rehm, M. (2014). Do tenants pay energy efficiency rent premiums? *Journal of Property Investment & Finance*, 32(4), pp. 333–351.

Gallimore, P. (1996) Confirmation bias in the valuation process: a test for corroborating evidence, *Journal of Property Research*, 13(4), pp. 261–273.

Gladwell, M. (2005) *Blink: the power of thinking without thinking*. Allen Lane.

Goddard, J. (2012) *Sustainability and the valuation of commercial property*. Royal Institution of Chartered Surveyors Oceania.

Green Building Council of Australia (GBCA) (2013) *A Decade of Green Building*. Report for the GreenBuilding Council of Australia, Sydney, accessed 24/4/2016 from www.gbca.org.au/resources/gbcapublications/a-decade-of-green-building.

Green Building Council of Australia (GBCA) (2016) Green Star Project Directory, accessed 24/4/2016 from www.gbca.org.au/project-directory.asp.

Hartenberger, U. and Lorenz, D. (2008) *Breaking the vicious circle of blame: making the business case for sustainable buildings*. RICS Research, Royal Institution of Chartered Surveyors, London.

Hertzsch, E. and Heywood, C. (2012) Climatic influence on life-cycle investing for sustainable refurbishment in Australia. Paper presented at the *18th Annual Pacific Rim Real Estate Society Conference*, Adelaide, Australia, 15–18 January 2012.

Hogarth, R. (1981) Beyond discrete biases: functional and dysfunctional aspects of judgemental heuristics, *Psychological Bulletin*, 90, pp. 197–217.

Lorenz, D. (2008) Breaking the vicious circle of blame. *RICS FiBRE Findings in Built and Rural Environments*, Royal Institution of Chartered Surveyors, London.

Lorenz, D. and Lützkendorf, T. (2011) Sustainability and property valuation: systematisation of existing approaches and recommendations for future action, *Journal of Property Investment & Finance*, 29(6), pp. 644–676.

Kats, G. (2003) The costs and benefits of green buildings, *A Report to California's Sustainable Task Force*, Sustainable Building Taskforce, California.

Kieffel, H. C. (2012) *Sustainability valuation: an oxymoron*? PricewaterhouseCoopers, accessed 23/4/2016 from www.pwc.com/us/en/audit-assurance-services/valuation/publications/assets/pwc-sustainability-valuation.pdf.

Madew, R. (2006) *The dollars and sense of green buildings 2006*. Australian Green Building Council, Sydney.

Mallinson, M. and French, N. (2000) Uncertainty in property valuation: the nature and relevance of uncertainty and how it might be measured and reported, *Journal of Property Investment & Finance*, 18(1), pp. 13–32.

McColl-Kennedy, J. R. and Kiel, G. C. (2000) *Marketing: a strategic approach*. Nelson Thomson Learning.

McColl-Kennedy, J. R., Kiel, G. C., Lusch, R. F. and Lusch, V. N. (1992) *Marketing: concepts and strategies*. Nelson Thomson Learning.

Michl, P., Sayce, S. and Lorenz, D. (2016) Reflecting sustainability in property valuation: a progress report, *Journal of Property Investment & Finance*, 34(6), pp. 552–577.

Miller, N., Spivey, J. and Florence, A. (2008) Does green pay off? *Journal of Real Estate Portfolio Management*, 14(4), pp. 385–400.

MSCI (2015) *The Property Council/IPD Australia Green Property Index*, accessed 20/5/2016 from www.msci.com/real-estate-fact-sheet.

Muldavin, S. R. (2010) *Value beyond cost savings: how to underwrite sustainable properties*. Green Building Finance Consortium.

Myers, G., Reed, R. and Robinson, J. (2007) The relationship between sustainability and the value of office buildings. Paper presented at the *13th Annual PRRES Conference*, January 2007, Fremantle, Western Australia.

NABERS (2016) National Australian Built Environment Rating System, Office of Environment and Heritage, NSW Government, accessed 20/5/2016 from www.nabers.gov.au.

NABERSNZ (2016) About NABERSNZ, Energy Efficiency and Conservation Authority, New Zealand Government, accessed 20/5/2016 from www.nabersnz.govt.nz.

Newell, G., MacFarlane, J. and Kok, N. (2011) *Building better returns: a study of the financial performance of green office buildings in Australia*, Australian Property Institute and Property Funds Association, Sydney.

Newell, G., MacFarlane, J. and Walker, R. (2014) Assessing energy rating premiums in the performance of green office buildings in Australia, *Journal of Property Investment & Finance*, 32(4), pp. 352–370.

New Zealand Green Building Council (2016) *About Green Star*, New Zealand Green Building Council website, accessed 20/5/2016, from www.nzgbc.org.nz.

NSW Office of Environment and Heritage (2015) *NABERS Annual Report 2014/15*. Report to Stakeholders, New South Wales Office of Environment and Heritage, accessed 24/4/2016 from www.nabers.gov.au/AnnualReport/index.html.

Nurick, S., Le Jeune, K., Dawber, E., Flowers, R. and Wilkinson, J. (2015) Incorporating green building features and initiatives into commercial property valuation, *Journal of Sustainable Real Estate*, 7(1), pp. 21–40.

Olubunmi, O. A., Xia, P. B. and Skitmore, M. (2016) Green building incentives: a review, *Renewable and Sustainable Energy Reviews*, 59, pp. 1611–1621.

Pivo, G. and Fisher, J. D. (2010) Income, value and returns in socially responsible office properties, *Journal of Real Estate Research*, 32(3), pp. 243–270.

Property Council of Australia (PCA) (2016) *Office market report*, Property Council of Australia, accessed 20/5/2016 from www.propertycouncil.com.au.

Reichardt, A., Fuerst, F., Rottke, N. B. and Zietz, J. (2012) Sustainable building certification and the rent premium: a panel data approach, *Journal of Real Estate Research*, 34(1), pp. 99–126.

RenoValue (2016) Valuing sustainability, accessed 30/8/2016 from www.scribd.com/doc/312097195/Valuing-Sustainability-English#fullscreen=1.

Robinson, S. J. and Sanderford, A. R. (2016) Green buildings: similar to other premium buildings, *The Journal of Real Estate Finance and Economics*, 52(2), pp. 99–116.

Royal Institution of Chartered Surveyors (RICS) (2013) *Sustainability and commercial property valuation*, (2nd Ed). RICS Global Guidance Note, Royal Institution of Chartered Surveyors, London.

Rydin, Y. (2016) Sustainability and the financialisation of commercial property: making prime and non-prime markets, *Environment and Planning D: Society and Space*, 34(4), pp. 745–762.

Sayce, S., Sundberg, A. and Clements, B. (2010) *Is sustainability reflected in commercial property prices? An analysis of the evidence base*. RICS Research, Research Report January 2010, Royal Institution of Chartered Surveyors, London.

Surmann, M., Brunauer, W. and Bienert, S. (2015) How does energy efficiency influence the market value of office buildings in Germany and does this effect increase over time? *Journal of European Real Estate Research*, (8)3, pp. 243–266.

Tidwell, O. A. and Gallimore, P. (2014) The influence of a decision support tool on real estate valuations, *Journal of Property Research*, 31(1), pp. 45–63.

Turley, M. and Sayce, S. (2015) Energy performance certificates in the context of sustainability and the impact on valuations, *Journal of Property Investment & Finance*, 33(5), pp. 446–455.

Tversky, A. and Kahnemann, D. (1974) Judgement under uncertainty: heuristics and biases, *Science*, 185, pp. 1124–1131.

United Nations Environment Program (UNEP) (2014) *Sustainability metrics: translation and impact on property investment and management*. UNEP FI Working Group Report, May 2014.

Warren, C. M., Bienert, S. and Warren-Myers, G. (2009) Valuation and sustainability are rating tools enough? Paper presented at the *6th Annual European Real Estate Society Conference*, ERES.

Warren-Myers, G. (2009) *Valuation practice issues in commercial property: the relationship between sustainability and market value*. PhD Thesis, University of Melbourne.

Warren-Myers, G. (2012) The value of sustainability in real estate: a review from a valuation perspective, *Journal of Property Investment & Finance*, 30(2), pp. 115–144.

Warren-Myers, G. (2013) Is the valuer the barrier to identifying the value of sustainability?, *Journal of Property Investment & Finance*, 31(4), pp. 345–359.

Warren-Myers, G. (2016) Sustainability evolution in the Australian property market: examining valuers' comprehensive, knowledge and value, *Journal of Property Investment & Finance*, 34(6), pp. 578–601.

Warren-Myers, G. and Reed, R. (2010) Identifying and examining links between sustainability and value: evidence from Australia and New Zealand, *Journal of Sustainable Real Estate*, 2, pp. 201–219.

Wiley, J., Benefield, J. and Johnson, K. (2010) Green design and the market for commercial office space, *Journal of Real Estate Finance and Economics*, 41(2), pp. 228–243.

Wilkinson, S. J., Sayce, S. L. and Christensen, P. H. (2015a) *Developing property sustainably*. Routledge.

Wilkinson, S., van der Heijden, J. J. and Sayce, S. (2015b) Tackling sustainability in the built environment: mandatory or voluntary approaches. The smoking gun? Paper presented at the *RICS COBRA Conference*, UTS Sydney, 8–10 July 2015, accessed 16/10/17 from www.rics.org/au/knowledge/research/conference-papers/hybrid-governance-instruments-for-built-environment-sustainability-and-resilience---a-comparative-perspective.

World Green Building Council (2013) *The business case for green buildings: a review of the costs and benefits for developers, investors and occupants*, World Green Building Council, accessed 30/8/2016 from www.worldgbc.org/files/1513/6608/0674/Business_Case_For_Green_Building_Report_WEB_2013-04-11.pdf.

8

Existing building retrofits

Economic payoff

Norm Miller and Nils Kok

8.0 Introduction

In this chapter, we will be examining the economics of retrofits: their costs, benefits, valuation, and financing. Buildings wear out and become obsolete. The rate at which they wear out is estimated at about 1.5 percent to 2.5 percent per year.[1] Buildings also become obsolete due to technological change and poor design. A building retrofit occurs when modifications are made to the building's systems or structure some time after initial construction and occupation. Retrofits are more than just maintenance. They may involve lighting systems, windows, insulation, HVAC, plumbing, roofing, system controls, walls, and more. Later in the chapter, we will illustrate light, medium, and heavy retrofits and also discuss the causes of obsolescence.

Generally, retrofits, which involve modifications to existing commercial buildings that may improve energy efficiency or decrease energy demand, are completed with the objective of improving financial performance as a result of increased building efficiency.[2] Efficiencies are gained in energy use, water use, and increased productivity of the building for occupants through better layouts, amenities, more natural light, or a variety of other factors. Sometimes, a reduced carbon footprint is an additional objective or simply a prerequisite to achieve green certifications, such as LEED certification, Energy Star ratings, BREEAM, Green Star, CASBEE, DNGB, or other such metrics that are important and required by many tenants today.

If all new construction were to be "green," and if no renovation took place, it would take several decades to improve the energy efficiency and sustainability performance of the existing building stock. Based on CoStar data from 1983 to 2014, approximately 250 million square feet of new industrial space was added each year in the U.S. As of 2016, there was approximately 26 billion square feet of industrial space, so new construction averages only 1 percent of existing industrial space per year. Assuming that the space never wore out or became obsolete, it would take at least 100 years for 50 percent of the stock to have been replaced. If we lose 1 percent per year to obsolescence and deterioration, we will reach the 50 percent level in 50 years, which is still a long time.

In the office market, new construction from 2000 to 2016 added approximately 160 million square feet each year. With a stock of approximately 13 billion square feet of office space, this means that only 1.2 percent of the total stock is added as new office construction per year.

However, a similar amount of office space is lost to conversion, obsolescence, natural disaster, and other occurrences resulting in little or no change in the net total stock.

In the retail market, about 185 million square feet of retail space is added each year. There is a stock of approximately 19 billion square feet of retail space, so there is only 1 percent of new retail construction per year. Similar to office space, some retail space wears out and is converted to other uses.

According to the National Association of Home Builders (NAHB), there are 300,000 to 500,000 multifamily starts per year. With a stock of 25 million units, new construction occurs at a rate of 1.7 percent per year. About 1 to 1.25 percent of multifamily stock is lost to fires, natural disasters, and deterioration; consequently, the net change of multifamily units has been quite small. The point is that only by converting existing buildings to greener buildings will we ever reach a more efficient stock of real estate in a reasonable period of time.

Figure 8.1 depicts new LEED certifications by square feet for newly constructed buildings and for existing buildings in the U.S. In 2009, there was a large increase in the square footage of LEED-certified buildings, both new and existing. Additionally, the percentage of square feet of existing buildings that were LEED certified began to approach and even surpass the percentage of square feet of LEED-certified new buildings. However, there is a lag between LEED registration and approval. According to Chris Pyke, former research director at the USGBC, it generally takes two to three years for newly constructed buildings and one to two years for existing buildings to achieve LEED certification once an application has been submitted.[3] However, the dynamics of the project itself can create longer lags. This lag explains why the U.S. had little new construction and still saw LEED certifications peak well after the height of new construction around 2006 and 2007.

The greening of existing buildings is becoming a more popular and feasible option for building owners. In many cases, retrofitting can be more financially viable than demolishing a building and constructing another in its place. For example, the owners of the "Home on the Range" (HOTR) building, a modest building in Montana that houses the Northern Plains Resource Council and the Western Organization of Resource Councils (WORC), conducted a cost analysis comparing the cost of retrofitting to achieve LEED Platinum status versus demolishing the existing building

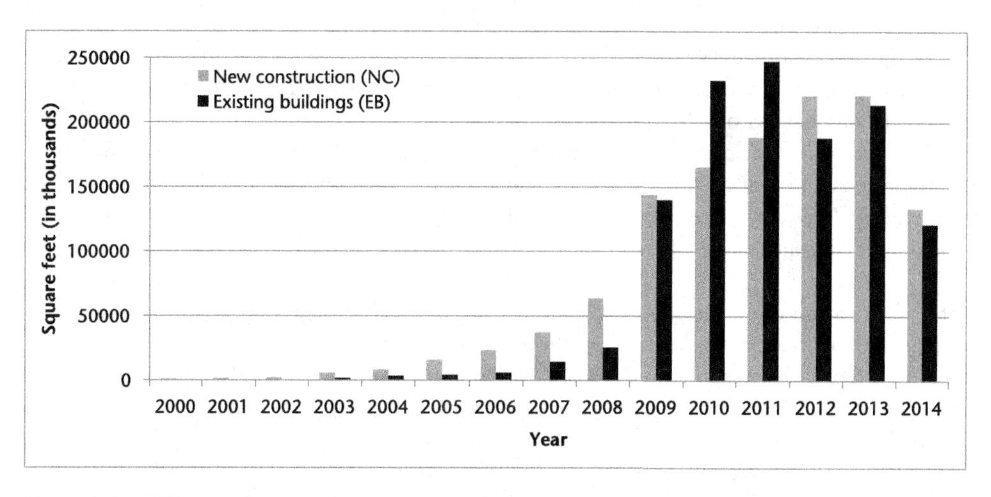

Figure 8.1 LEED certifications by square feet for newly constructed buildings and existing buildings, 2000–2014

Source: LEED certifications by square feet provided direct to author by USGBC. US Green Building Council (USGBC) https://new.usgbc.org/

and constructing a new structure. The owners found that retrofitting the building would cost about $325,000 less than rebuilding (NBI, 2010), well over 10 percent.

8.1 Workplace trends create a need for building retrofits

The shifting nature of the workplace is also contributing to the necessity of retrofits that are both more efficient and more suitable for workers' needs. Organizational theorists describe the pressure to become a "lean enterprise"[4] and to incorporate technological improvements that separate "work ... from time and space" as the primary factors contributing to the evolving patterns of work.[5] The rise of collaborative work and increased pressure for communication and information flow is yielding more meeting spaces, unassigned workspaces, increased use of open floor plans to enhance awareness, and small, individual workspaces. The increased demand for more efficient space use is also resulting in a rise in overall densities within the workplace and increased spatial variety to accommodate a variety of working tasks and styles at the same time. Furthermore, the use of facilities past typical working hours is necessary to accommodate meetings between geographically dispersed participants. The pressure to improve the quality of work life and the goal to attract new workers will result in increased access to daylight and views of the outdoors and "more equitable allocation and workspace features" (Heerwagen, Kelly, and Kampschroer, 2005: 112; see also Miller, 2014).

Because different building features are demanded over time, the changing nature of the workplace contributes to the obsolescence of a building. However, retrofitting is an option that can address these changing demands while extending the life of an existing building. The term "obsolescence" has a variety of definitions within the real estate literature. Below, we will briefly describe five types of obsolescence:

- *Economic or locational obsolescence* results from changes in the "highest and best use" for the land on which the building is located.
- *Technological obsolescence* takes place when improved technological alternatives that have lower operating costs or are more efficient become available. This form of obsolescence also includes the physical deterioration of the building's infrastructure and mechanical or electrical components as a result of age, use and extent of maintenance, weathering, overload, or changes in functions. For example, the invention of the forklift made warehouses with low ceiling heights obsolete.
- *Environmental obsolescence* occurs through the depreciation of the land on which a building is located as a result of high pollution, road congestion, urban decay, or similar factors.
- *Legal obsolescence* is the result of new legislation affecting health, safety, and fire control.
- *Social obsolescence* occurs when social needs or preferences change.

Of the above forms of obsolescence, economic, technological, legal, and social obsolescence can be unpredictable due to a lack of information on future development and the lack of accurate ability to forecast changes in taste. Obsolescence resulting from a building's physical features is more predictable. Note that homes become obsolete at a slower rate than office, retail, or industrial buildings since their function remains relatively constant over time (Thomsen and Van der Flier, 2011). We can, therefore, expect the demand for improved building efficiencies to continue to accelerate over time and to see retrofits as the only viable path for success.

While the empirical research discussed below is based on office property, we should mention that all properties suffer from obsolescence and need occasional retrofitting. We simply do not have much data yet on the costs and benefits of retrofits for other property types except on a case basis. In some instances, the retrofit involves a change of use such as hotels to micro-unit housing or warehouse to loft-style apartments or data storage centers. Here we utilize data from the office market, but the climate data and utility rates apply to all property types, and the costs described below are representative of average commercial property in the US.

8.2 The cost of saving energy

8.2.1 Costs of office retrofits

One of the first steps in assessing whether to implement retrofits is to determine the cost of the features to be installed. Kok, Miller, and Morris (2012) described the cost per square foot for select examples of retrofits (Table 8.1):

- *Plug loads:* The typical office property consumes about 10–20 KBtus per square foot per year for plug load, but that can easily be improved to 4–10 KBtus by replacing outdated appliances and equipment (printers, faxes, computer screens) and adding occupancy sensors that shut off power when there are no occupants (after an appropriate delay). "Vampire kill switches" also shut down the entire suite or floor power when the last person leaves the premises. Importantly, the cost for these strategies is negligible.
- *Lighting:* The typical office property consumes 10–15 KBtus per square foot per year for lighting, with the best practices at 4–7 KBtus. Simply replacing the lights with more modern T5/T8s and motion sensors, adding task-lighting and day-lighting controls, and moving to daytime cleaning will accomplish this energy reduction for a cost of $3–$5 per square foot. Light-emitting diode (LED) lighting is even more efficient, and prices are rapidly dropping. LEDs are twice as efficient as most fluorescent fixtures, so even greater efficiency will soon be possible. Day lighting can be brought in by a variety of new skylights, some with reflectors and sun tracking, as well as light diffusers.
- *Ventilation:* The ideal situation for indoor air quality and energy use reduction is operable windows, but that is considered a deeper retrofit. The typical office property requires 6–10 KBtus per square foot per year and can reduce that to 3–6 KBtus for a cost of $2–$5 per square foot. The work required includes sealing air ducts, optimizing air handlers and terminal units, and better balancing heating and cooling with integration, if possible, with shade controls and windows. In some cases, large fans are brought in and the maximum comfortable temperature can be raised prior to any cooling.
- *Cooling:* Typical office buildings require 15–40 KBtus per square foot per year for cooling, except for those in cooler climate zones. The current best practices are 10–20 KBtus; it costs

Table 8.1 Cost and energy savings of select retrofits

Strategy	*KBtu/SF/Yr (Reduction)*	*Cost/SF*
Plug load	6–15	Negligible
Lighting	6–8	$3–$5
Ventilation	4–5	$2–$5
Cooling	10–15	$3–$7
Heating	3–10	$1–$2
Total	30–50	$10–$20

Source: Kok et al. (2012)

about \$3–\$7 dollars per square foot to reach these with a retrofit. The typical strategies include replacing primary equipment, drying the air prior to cooling, adding large fans, and improving ventilation, so that the equipment capacity can be decreased. Heat gain can be better controlled by shading windows or adding glazing, although this is considered a deeper retrofit.

- *Heating:* The typical office property requires 5–15 KBtus per square foot per year for heating, while the best practices are at 2–8 KBtus. This can be accomplished for just \$1–\$2 per square foot by replacing primary equipment, improving controls, optimizing terminal units, and balancing heating and cooling with more localized controls.
- *Water conservation:* Water flow equipment investments are economically justified when fixtures must be replaced, but there is no reasonable economic payoff at present as water prices are often too low for any kind of significant return on investment or reasonable payback.
- *Deeper retrofits:* For \$10–\$75 per square foot, deeper retrofits can be accomplished, including envelope sealing, improved glazing, additional insulation, and chilled beams or some other form of radiant cooling. Computer-controlled window shades may be considered, along with solar photovoltaic cells or wind turbines. Energy recapture systems can also be employed on elevators. Such strategies typically reduce the energy consumed by 10–25 KBtus and can add energy generation equal to that consumed in some cases.

8.2.2 Climate and energy costs matter for all property types

Energy cost savings are a function of the number of cooling and heating days per year in a region and the costs of energy itself. In states with higher electricity rates, there is a larger benefit to retrofitting; energy savings are larger for each unit of energy saved. Regarding residential properties, the U.S. Department of Energy's (DOE) Weatherization Assistance Program (WAP), a provider of funding for retrofits to low-income homes, reports average savings of 32 percent on homeowners' heating bills (U.S. DOE, 2010). Heating bill savings can be larger in colder climates if improved insulation and reduced infiltration that ease space heating and cooling demands are used (Moomaw and Johnston, 2008). Figure 8.2 illustrates the average retail price of electricity for all sectors across the U.S. as of May 2014. The five states with the lowest average retail price of electricity for all sectors are: Washington State (6.98 cents per kWh), West Virginia

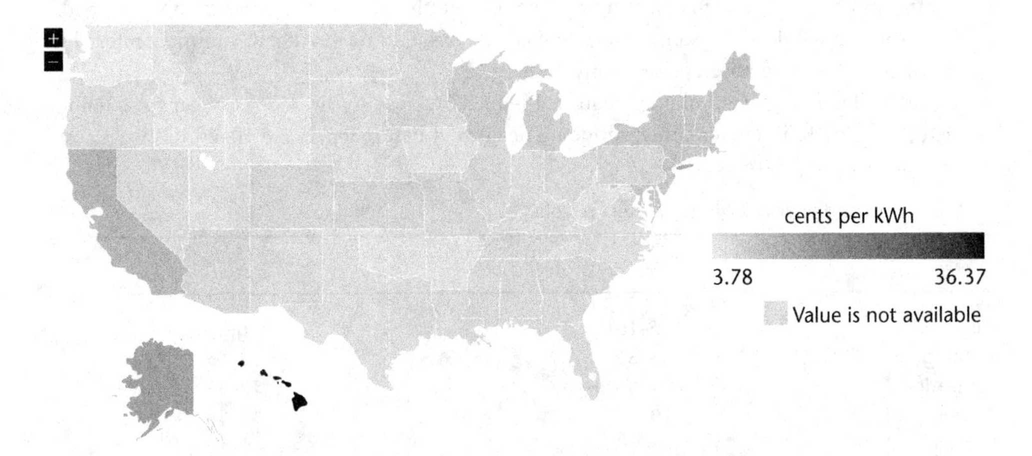

Figure 8.2 U.S. average retail price of electricity, all sectors, May 2014

Source: U.S. Energy Information Administration State Electricity Profiles (2014) www.eia.gov/electricity/state/

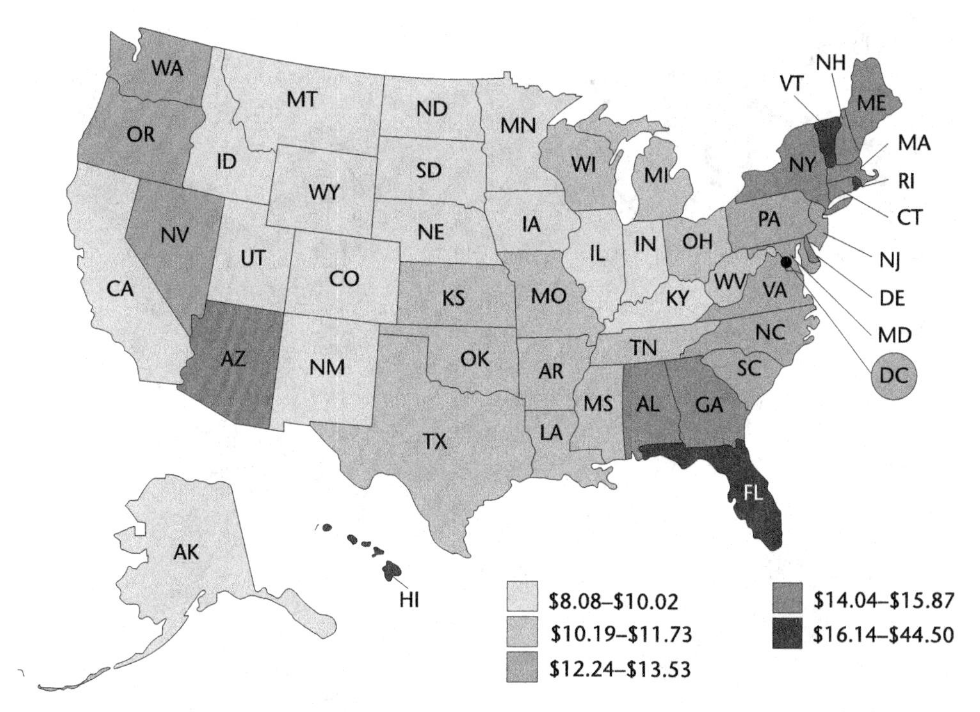

Figure 8.3 U.S. residential natural gas prices (per thousand cubic feet), 2010

Source: U.S. Energy Information Administration, Natural Gas Annual (December 2011)

(7.58 cents per kWh), Idaho (7.63 cents per kWh), Wyoming (7.74 cents per kWh), and Iowa (7.84 cents per kWh). The five states with the highest average retail price of electricity for all sectors are: Rhode Island (15.02 cents per kWh), New York (15.50 cents per kWh), Connecticut (16.38 cents per kWh), Alaska (17.72 cents per kWh), and Hawaii (33.98 cents per kWh).

Figure 8.3 depicts residential natural gas prices.

Figure 8.4 illustrates the U.S. climate zones as of 2003. The National Oceanic and Atmospheric Administration (NOAA) defines Commercial Buildings Energy Consumption Survey (CBECS) climate zones as regions within a state that are as similar as possible in terms of climate. There are five CBECS climate zones based on 30-year averages of heating degree days (HDDs) and cooling degree days (CDDs). An HDD measures how cold a location was over a period of time, relative to a base temperature of 65 degrees Fahrenheit. The measure is calculated by taking the difference between a day's average temperature and the base temperature if the daily average is less than 65. The HDD is equal to zero if the daily average temperature is at least 65. On the other hand, a CDD measures how hot a location was over a period of time, relative to the base temperature of 65 degrees Fahrenheit. It is calculated by taking the difference between the daily average temperature and the base temperature, if the former is greater than the latter. The CDD is equal to zero if the daily average temperature is at most 65. You can expect regions within Zone 1, with fewer than 2,000 CDDs and more than 7,000 HDDs annually, to benefit from more energy-efficient heating systems. On the other hand, regions in Zone 5, with more than 2,000 CDDs and fewer than 4,000 HDDs annually, would benefit from more energy-efficient cooling systems.

Combining information from Figures 8.2 and 8.4, commercial buildings in Florida, which has the highest average retail price of electricity within Zone 5, would accrue the largest benefit

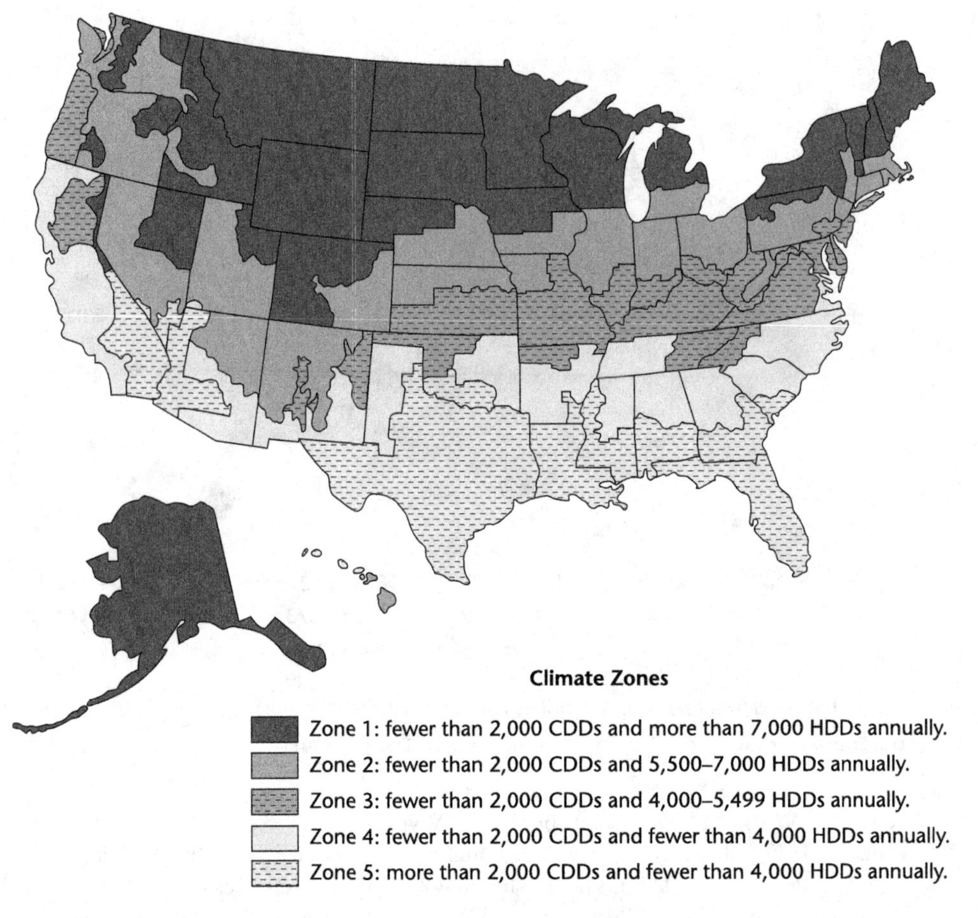

Figure 8.4 U.S. climate zones for the 2003 Commercial Buildings Energy Consumption Survey (CBECS)

Source: U.S. Energy Information Administration (2014) www.eia.gov/consumption/commercial/maps.php

through energy cost savings by installing more energy-efficient cooling systems. Combining information from Figures 8.3 and 8.4, states in the northeast (Maine, New Hampshire, Vermont, New York, and upper Pennsylvania) would benefit the most from improved heating systems, since these states tend to have the highest natural gas prices within Zone 1.

8.2.3 Costs of green certifications for commercial property

If the objective is to enhance a firm's reputation, building business owners must also consider the cost of green certifications. Leonardo Academy Inc. (2008) found that higher certification levels did not result in increased costs, but small sample size errors may have been present. Kats, Alevantis, Berman, Mills, and Perlman (2003), using a data set of 33 LEED-registered projects, found that eight Certified-level buildings, 18 Silver-level buildings, six Gold-level buildings, and one Platinum-level building had an average cost premium of 0.66 percent, 2.11 percent, 1.82 percent, and 6.5 percent, respectively. Note that most of the above buildings had yet to actually receive official certification at the time of the cost analysis; the LEED level was an assessment by

Table 8.2 Estimated cost summary

Building systems	*Market Costs*	*LEED Certified*	*LEED Silver*	*LEED Gold*	*LEED Platinum*	*Living Building*
Site preparation	$49,041	$88,375	$88,375	$88,375	$144,568	$144,568
Substructure	$128,349	$128,349	$147,505	$147,505	$147,505	$147,505
Superstructure	$710,069	$710,069	$839,312	$919,514	$919,514	$919,514
Exterior closure	$574,415	$570,069	$911,213	$927,815	$1,212,736	$1,235,597
Roofing and waterproofing	$115,578	$115,578	$166,981	$287,029	$338,113	$646,980
Interior construction	$890,958	$884,904	$924,290	$897,037	$901,379	$887,586
Conveying systems	$60,381	$60,381	$90,010	$90,010	$90,010	$90,010
Mechanical systems	$667,226	$687,059	$653,710	$636,687	$660,916	$755,639
Electrical systems	$504,928	$517,763	$496,001	$479,756	$463,486	$463,486
Finish work	$127,710	$127,710	$178,794	$178,794	$178,794	$178,794
Subtotals	**$3,828,655**	**$3,890,257**	**$4,496,191**	**$4,652,522**	**$5,057,021**	**$5,469,679**
General conditions 9%	$344,579	$350,137	$404,657	$418,727	$455,132	$492,271
Contractor's fee 4.5%	$187,797	$190,825	$220,538	$228,206	$248,047	$268,288
Design contingency 10%	$436,103	$443,138	$512,139	$529,946	$576,020	$632,024
Subtotal building Construction cost	**$4,797,134**	**$4,874,357**	**$5,633,525**	**$5,829,401**	**$6,336,220**	**$6,862,262**
Subtotal parking garage cost	$4,062,226	$4,062,226	$4,291,335	$4,305,417	$4,283,015	$4,305,417
Construction contingency	$664,452	$670,256	$744,365	$760,111	$796,443	$836,901
Escalation to construction start	$476,190	$480,350	$640,154	$653,696	$684,941	$899,668
Total hard costs	**$10,000,002**	**$10,087,189**	**$11,309,379**	**$11,548,625**	**$12,100,619**	**$12,904,248**

Source: The David and Lucile Packard Foundation (2002)

Table 8.3 Incremental costs from baseline to specific HERS levels

Type of residential building	*Energy Star minimum (HERS 85)*	*HERS 70*	*HERS 65*
Single-family home with gas furnace:	$2,869	$7,136	$9,286
total and per square foot cost	$1.18	$2.94	$3.83
Single-family home with gas boiler:	$2,646	$6,570	$8,160
total and per square foot cost	$1.09	$2.71	$3.36
Single-family home with oil boiler:	$2,371	$6,325	$7,914
total and per square foot cost	$0.98	$2.61	$3.26
Average for all single-family homes:	$2,599	$6,677	$8,453
total and per square foot cost	$1.07	$2.75	$3.49
Multifamily unit with gas furnace:	$1,068	$5,314	NA
total and per square foot cost	$0.71	$3.53	
Multifamily unit with gas boiler:	$1,470	$4,756	NA
total and per square foot cost	$0.98	$3.16	
Multifamily unit with oil boiler:	$1,246	$4,697	NA
total and per square foot cost	$0.83	$3.12	
Average for all multifamily units:	$1,286	$4,922	NA
total and per square foot cost	$0.85	$3.27	

Source: Tolkin et al. (2008)

the architect or client team. Nonetheless, the predictions are assumed to be fairly accurate. These cost premiums, however, do not appear to have accounted for soft costs (Jackson, 2009). Tatari and Kucukvar (2011) found that certification premiums are likely to rise with certification levels. Table 8.2 illustrates the cost of a new building that meets the requirements of various levels of LEED certification; however, this data is some 15 years old now and should be treated with caution. The figures are based on cost estimates adjusted to reflect a $10 million market building located in the Bay Area as a baseline (The David and Lucile Packard Foundation, 2002). The higher the targeted certification level, the higher the hard costs.

8.2.4 Costs of energy rating systems for residential housing

Table 8.3 depicts the incremental costs to obtain a Home Energy Ratings System (HERS) level of 85, 70, and 65 relative to a baseline building, where a lower HERS level indicates a more energy-efficient home. While the incremental costs to reach a HERS level of 70 or 65 from the baseline are relatively high, many builders looking to construct energy-efficient homes are probably building homes that would achieve a HERS level lower than 85. The incremental costs of moving from HERS 85 to HERS 70 for an average single-family home are approximately 1.5 times greater than the incremental costs of moving from a baseline home to HERS 85. The incremental costs of moving from HERS 70 to HERS 65 for an average single-family home are less than half those required to move from HERS 85 to HERS 70 (Tolkin et al., 2008).

8.3 The benefits of retrofitting: Empirical tests in the U.S. office market

The next step in assessing the feasibility of retrofitting is to consider its benefits. Prior published literature on the financial implications of green certification mostly focuses on new construction within the U.S., and results generally indicate a positive relationship between environmental certification and financial outcomes in the marketplace. Binkley (2007), using Boston, MA, as a case study, found that retrofits are profitable. The sum of energy efficiency and greenhouse gas (GHG) emissions savings add between 1 percent and 6.1 percent to the value of the U.S. commercial real estate market. Eichholtz, Kok, and Quigley (2010) documented large and positive effects on market rents and selling prices following environmental certification of office buildings. Relative to a control sample of conventional office buildings, LEED- or Energy Star-rated office buildings' rents per square foot were about 2 percent higher, effective rents were about 6 percent higher, and premiums to selling prices per square foot were as high as 16 percent. Other studies (Miller, Spivey, and Florance, 2008; Fuerst and McAllister, 2009) confirm these findings.

Miller, Pogue, Gough, and Davis (2009) documented that over half of the occupants of environmentally-certified buildings found their employees to be more productive. Interpretation of these results is problematic, though, as these responses cannot control for management style and individual employee characteristics. William Pape, the cofounder of VeriFone, reported that over 18 months following building retrofits that reduced indoor pollutants and improved indoor environmental quality, absenteeism rates declined 40 percent and productivity increased by about 5 percent (Kats et al., 2003). Bendewald, Hutchinson, Muldavin, and Torbert (2014) also cited benefits, including faster absorption of tenants through improved pre-leasing, reduced tenant turnover, competitive lease terms, reduced operating and maintenance costs, increased tenant satisfaction, the receipt of "superior grants" and subsidies, and an increased number of occupants. According to Navigant Consulting and RMI (2014), through a series of case studies it appeared that building owners were motivated to pursue deep retrofits in order to replace inefficient or failing equipment with energy-efficient equipment, maintain building performance, and retain savings to increase net operating income, which can be passed along to tenants. As a result, the building will become more competitive and possibly increase or maintain occupancy. Other motivations included improving market positioning by increasing energy efficiency and improving LEED ratings to attract tenants, assuming that the building is located in a market that has a high demand for green buildings, improving the reputation of the owner, and enhancing workplace wellbeing. Lastly, owners may be motivated to develop a new leasing structure by moving from triple-net to gross leases, which is expected to create more value. This is because the owners will receive directly the benefits of reduced energy costs or reduced waste collection.

Kok et al. (2012) were the first to address the economic implications of LEED certification (following a retrofit), extending the rapidly growing literature on the effects of "green" building in the marketplace. The data in this study were from CoStar and included 374 LEED-certified properties (EBOM) and nearly 600 control properties for comparative purposes and empirical analysis. Many of the buildings in the sample were in the process of being renovated to become more sustainable at the time the EBOM system was published. The authors identified the renovation period as generally running from 2005 to 2009 with certifications received from 2008 to 2011.

The results show that the average rents of the EBOM-certified buildings were below those of the control buildings prior to 2006, but have exceeded the average rents of the control buildings since then. Vacancy rates within EBOM-certified buildings were 7 percent higher than for the control buildings in Orange County, East Bay/Oakland, Denver, Atlanta, Dallas/Fort Worth, and Minneapolis/St. Paul. The control group was matched in terms of the abovementioned filters, but the selection was adjusted such that the ages and sizes of the treated and untreated samples were as similar as possible. The control sample included some 600 properties, after applying the filters on location, age, and size. Comparing the treatment and control groups, green buildings were slightly younger and had a higher renovation propensity, but the differences were limited through the data selection procedure.

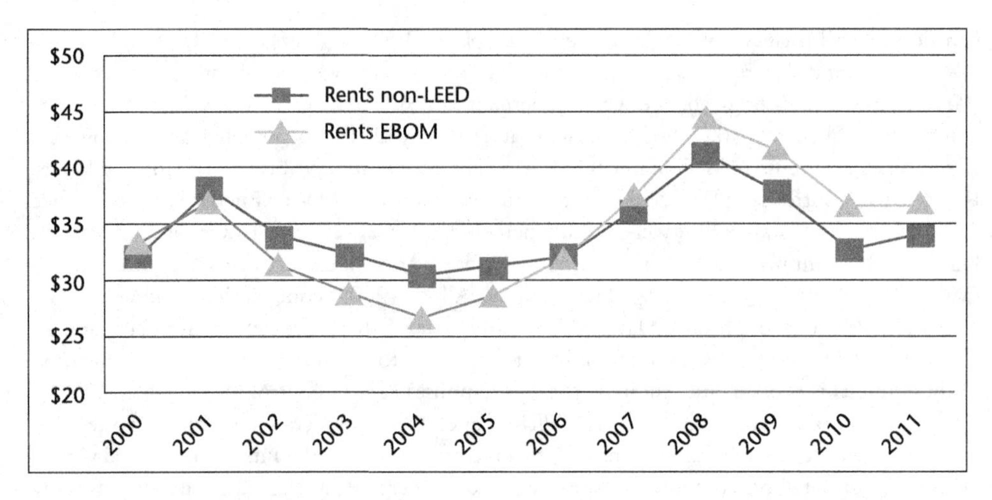

Figure 8.5 Rents prior to and after renovation

Source: Kok et al. (2012)

In Figure 8.5 above, the period prior to renovation is before 2005. Most improvements were completed by 2005 (although some improvements continued after that). Note that the rents of the renovated properties were lower than those of the control sample prior to the renovations. Similarly, the occupancy rates prior to the renovations were lower than those of the control sample. Of significance is the fact that average rents increased faster than for the control group through 2008. While premiums were maintained for the buildings certified by LEED for existing buildings, the rents declined after 2008 at about the same rates as for the control sample. This finding is similar to that of Eichholtz et al. (2010). Figure 8.6 indicates that the occupancy gap narrowed after the improvements but never completely dissipated during the rather soft rental period from 2007 through 2010. Of course, rental and occupancy rates vary by market. Significant rental premiums are observed in the major markets of Washington, D.C., New York City, and Boston. Occupancy rates strongly depend on when the LEED buildings came "on line;" with many of the LEED buildings being renovated during a period of decline, lower occupancy rates continue to be observed for green buildings in quite a few markets.

Relating the logarithm of rent per square foot in commercial office buildings and holding all other hedonic characteristics of the buildings constant, an office building with an LEED EBOM certification rents for a 7 percent premium, on average. Measured attributes of sustainability and energy efficiency are incorporated in property rents, and this seems to have persisted through periods of volatility in the property market.

The statistical results suggest that a green rating is associated with a 9 percent increase in effective rent, an even larger effect.[6] Taken together, the results from both regressions suggest that the occupancy rate of green buildings is about 2 percent higher than in otherwise comparable non-green buildings.

On average, the empirical results suggest a rental premium of $2 per square foot for buildings certified by LEED for existing buildings (i.e., 7 percent of an average rent of some $29 per square foot), which at a capitalization rate of 8 percent (Eichholtz, et al., 2010) results in a value impact of $25 per square foot.

The results are consistent with those findings observed for newly constructed buildings that are LEED certified. Most salient is the fact that the types of office space renovations observed here for improved productivity and energy efficiency apply to a much larger pool of candidate

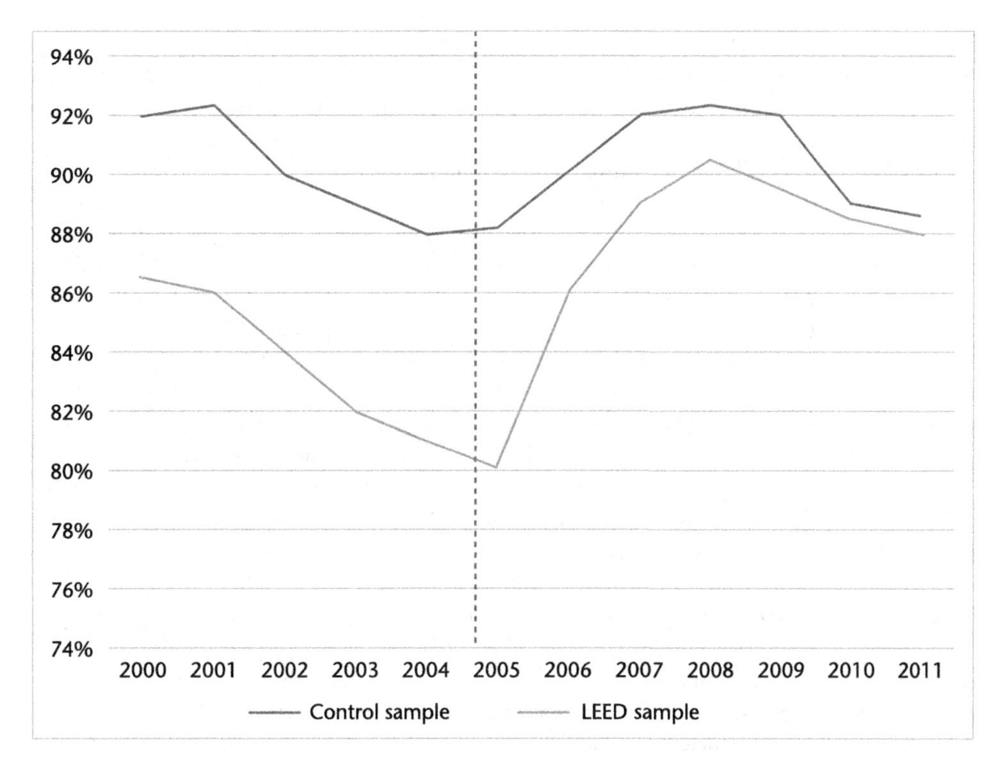

Figure 8.6 Occupancy prior to and after renovation shown approximately at the vertical dashed line

Source: Kok et al. (2012)

properties. These market developments will continue to affect the existing stock of non-certified office buildings, especially as regulatory trends are forcing greater energy consumption transparency upon the commercial real estate market and as tenants report on actions to achieve corporate social responsibility goals via portfolio sustainability reporting tools, such as the Global Reporting Initiative and the Global Real Estate Sustainability Benchmark, and the plethora of building-level benchmarks now available for assessing the sustainability of commercial real estate.

8.4 Making an investment decision

According to the Rocky Mountain Institute (RMI), an energy retrofit for commercial property can be defined as simple/light or deep/heavy. Light energy retrofits typically correspond to improved lighting equipment, better building management systems and modest upgrades, and heating and cooling systems additions. These retrofits should yield a small reduction in energy costs. Deep retrofits typically involve the building envelope: for example, new windows, insulation, and air vent sealing. Deep retrofits require a "whole-building analysis and retrofit construction process" and should result in larger energy savings of at least 50 percent. Yet they have high up-front costs and increased perceived risks. Deep retrofit implementation necessitates a phasing period over several years, compared to simple retrofits. This creates difficulties in attaining the entirety of energy savings (Navigant Consulting, Inc. and RMI, 2014). Energy retrofits can also be classified as medium, referring to the use of three to four measures with payback that is slower than that of the light retrofits, but quicker than that of the deep retrofits.

Examples of medium retrofits include increasing attic insulation and adding new boiler and heating controls.

Firms make capital investment decisions by comparing a potential investment's rate of return and the rate of return on alternative uses of the funds. However, when firms are deciding whether to retrofit a building, the hurdle rate, or minimum rate of return, required by building owners is often higher than the returns from energy cost savings provided by the retrofit, particularly for deep retrofits. The 2009 Corporate Energy Efficiency Survey[7] received responses from 48 large corporations from various industry sectors who have shown commitment to climate and energy concerns. Of the companies that reported their energy savings targets, over half had an energy savings goal of at least 20 percent. Of the 92 percent of firms that reported using specific financial criteria for energy-efficient investments, 52 percent used simple paybacks while 50 percent used internal rate of return (IRR). Fifteen corporations provided information on desired payback periods, a simple measure of financial performance, which is quite common among the engineers and contractors engaged in building renovations. Twelve of these firms sought a payback period of one to three years. Half of the respondents sought an IRR threshold of at least 18 percent. Benson et al. (2011) found that individuals expect returns over a period of three to five years, and that building owners are often hesitant to invest in retrofits and other investments that have payback periods of 15 to 20 years. Kok et al. (2012) surveyed owners of LEED-certified buildings and received 41 responses. Those surveyed reported a typical payback period that is fairly quick, at less than five years. This reflects the preference of commercial building owners for "quick wins" rather than more substantial deep retrofits. About one-third of the respondents expected a payback period of five to ten years. According to the energy efficiency indicators study, a survey of investors and occupants indicated that most retrofits are required to meet an approximate three-and-a-half-year payback period using only energy cost savings (Muldavin, Torbert and Bendewald 2013).

Because there is often a discrepancy between the hurdle rate required by building owners and the simple rate of return on retrofits, a retrofit's valuation should include non-energy benefits, such as the types of products and materials used and any resulting cost premium, as described in the next section. When speaking to developers and investor, one author of this chapter, Norm Miller, frequently asks them what is the payback required or the hurdle rate before they would consider investing in sustainable building features. The payback required by most developers is less than five years and the same for investors. Yet, most investors in commercial real estate have not achieved anything close to such returns on a long-term basis.[8]

8.5 Valuing green retrofit buildings

Sustainability can be considered a performance outcome by doing less harm to the world, saving energy and money, and not just an attribute of properties. Therefore, sustainability should be priced by measuring the dollar value of the associated costs and benefits, not just a payback period calculation using expected energy cost savings. Failing to value non-energy outcomes is often an issue for whole-building retrofits, since these non-energy benefits are highly valued by owners and occupants, and they are often the motivation behind retrofit investments (Loutzenhiser Associates, 2004).

A building's valuation should include the types of products and materials used and any resulting cost premium. Davis Langdon (2007) found no statistically significant difference between the costs per square foot for LEED-certified buildings and non-LEED-certified buildings across five different types of buildings. On the other hand, a 2007 survey by Building Design and Company found that 78 percent of respondents thought that the addition of sustainable features significantly impacted first costs; 32 percent estimated the increased costs to range from 6 to 10 percent; and 41 percent estimated a minimum of an 11 percent increase in costs (Hunt, 2008). Similarly, a 2007 survey of U.S. building professionals by the World Business Council for Sustainable Development (ibid.) estimated costs to rise by an average of 16 percent. When measuring the cost premium for the addition of sustainable features, both the hard costs and the soft costs must be measured. Hard costs are actual construction expenses. Miller et al. (2008) estimated a hard cost of approximately 2.5 percent through a survey of 26 projects. Soft costs include code-compliance delays, design and engineering support costs, modeling, and documentation (Jackson, 2009). Soft-cost estimates are sensitive to the project scale; the relative cost is higher for smaller buildings (Steven Winter Associates, Inc., 2004).

While including energy savings performance remains an important feature of a retrofit valuation, the reliability of the "modeled" outcomes must be considered (Bendewald et al., 2014). Osser, Neuhauser, and Ueno (2012) found that there is a discrepancy between actual and predicted energy use. Consequently, appraisers should take into account whether the energy-efficient features function as expected and determine whether there is a risk of underperformance resulting from misuse, improper installment, or a poor choice of materials and equipment. Further risk analysis should ascertain whether the implementation of the sustainability features conflicts with existing regulations (Bendewald et al., 2014).

Because environmental certification levels are often based on various combinations of sustainable features and outcomes, they are arduous to compare. It is also possible that properties without certifications have sustainable features that add value to the building. Consequently, it is important to ask whether there should be different types of valuation for buildings possessing various certifications, regardless of sustainable features present (Bendewald et al., 2014).

Retrofit valuation should include quantifiable measures of how sustainability features affect rental rates, volatility in operating costs, occupancy rates, space user demand, cap rates, absorption, maintenance costs, entitlement benefits, and discount rates. Changes in the durability and adaptability of a building and cost avoidance over the life of retrofits should be valued as well.

Valuations should include measures of occupant performance, such as health, productivity, satisfaction, and reduced resource use. For example, controlling moisture and pollution, improving ventilation, and efficiently managing temperatures can lower absenteeism and "presentism" (attending work while sick) and reduce litigation and future regulatory risk. Retrofits can also reduce insurance costs, increase space utilization due to equipment downsizing, decrease employee churn rates, decrease the frequency at which building occupants move, and reduce recruitment and retention costs (Bendewald et al., 2014). All of these non-energy benefits should contribute to a building's value.

Improved company reputation, increased company market value, and increased client or consumer loyalty for products and services may also play a role in tenants' or owners' decisions to demand more sustainable buildings. Public-sector benefits—such as reduced carbon output, landfill reduction, reduction in water and air pollution, acidification reduction, natural habitat preservation, and drought risk, deforestation, and desertification reduction—are difficult to assess but also offer evidence of a company's intentions to act responsibly. Carbon disclosure requirements demanded by most CSR reports also provide value if they are trending in a lower-emission direction. Public value through public grants, tax benefits, or favorable financing capital contribute to retrofit value, too.

However, few appraisers take into account these non-energy benefits, and without quantifying all benefits rather than merely energy cost savings, it can be difficult for building owners' desired hurdle rates to be achieved. As a result, fewer sustainability investments will be implemented than is likely optimal until such time as all appraisers recognize the benefits and solid returns on investments in better buildings.

8.6 Financing and insurance incentives

Since many companies use simple energy cost savings valuations, retrofits do not always appear to make good investments, if based upon only payback or the incremental impact on value as determined by an often naïve appraiser.[9] Building owners can therefore be motivated to conduct retrofitting through increased financing opportunities that are low cost and easy to obtain within limitations.

One form of available financing is through an energy service company (ESCO). The ESCO acts as a project developer; they assume responsibility for the project's design, financing, installation, and operational components. The most common financing options are shared-savings contracts and paid-from-savings contracts. The terms of the former contracts state that the dollar value of the energy savings is shared between the building owner and the ESCO. If there are no energy savings in a particular period, the owner continues to pay the energy bill, but there is no additional payment to the ESCO for that period. Under the latter contracts, the owner pays the ESCO a set amount each period, such as 85 percent of the predicted energy bill under the assumption of no improvements. Essentially, energy savings are guaranteed to cover the debt payments. Contract terms range between seven and 20 years, depending on the measures installed. In any ESCO financing, the ownership transfers to the building owner.

Property-assessed clean energy (PACE) programs, also known as tax-lien financing, are another form of financing available for residential and commercial buildings. The process begins with the state or local government validating the public purpose of a PACE program through statutes. Then they must establish a type of land or real property secured benefit district if there is no municipality already. The municipality provides funding by selling bonds that are secured by payments from participating property owners. These payments take the form of an increase in property taxes for up to 20 years. PACE assessments may be eligible for "expense pass-through" to tenants in the form of operating expenses, and payments are transferred to the new owner upon sale of the property.

Interest rates for PACE programs range between 6 to 7 percent. When up-front expenses are included, the real interest rate can actually range from 7 to 8 percent. However, if a city or state subsidizes the interest rate or provides a loan loss reserve, the interest rate will be lower (Managan and Klimovich, 2013). Examples of vendors of PACE products include the following:

- Figtree PACE Financing, within California, provides funding for all commercial property types in select counties at an interest rate of 4.5 to 6.99 percent.
- Set the PACE St. Louis provides financing for residential and commercial property owners at interest rates between 3 and 5 percent, the competitive market rates for land-secured loans.
- In Ann Arbor, Michigan, the interest rate is 1 percent above the rate being charged to the City of Ann Arbor.
- Milwaukee Energy Efficiency offers a fixed rate to homeowners between 4.5 and 5.25 percent for a maximum loan of $15,000.
- Efficiency Vermont provides financing at a rate of 7 to 8 percent, whereas a personal loan can carry an interest rate of 5 to 12 percent, a home equity loan 5.25 percent, and a 30-year mortgage 4 to 5 percent.

Other opportunities for sustainable real estate financing include internal and external debt financing, lease or lease-purchase agreements, energy service agreements, government loan programs, on-bill utility financing, tax credits, and bonds such as Qualified Energy Conservation Bonds (QECB).

Another way to encourage building owners to incorporate sustainable building features is through insurance discounts for buildings with green certifications. In 2010, Fireman's Fund Insurance Company offered policyholders with Energy Star buildings a 5 percent discount, recognizing the increased value and lower risk of energy-efficient buildings. Travelers Insurance Company also provides a 5 percent discount for LEED-certified homes.

With the mounting pressure for firms to become more sustainable and the increasing availability of financing for sustainability measures, retrofits are becoming a more desirable and feasible option. Firms must consider both the hard and soft costs of retrofitting and the non-energy benefits alongside energy cost savings. By improving valuation methods for retrofits and green buildings, firms can make more informed decisions about which sustainability measures to incorporate.

8.7 Chapter summary and conclusions

Only by retrofitting and converting existing buildings to more energy-efficient, greener buildings will we ensure that a reasonable percentage of the total stock is sustainable, since new build accounts for such a small percentage of the total stock each year.

Existing building retrofits are now surpassing new construction for LEED certifications and we expect this trend to continue. Workplace trends are also forcing the conversion of existing office stock, as new layouts and more natural light complement the goal of more efficient buildings.

There are a variety of other obsolete features in existing buildings from lighting and HVAC systems to building envelopes and windows. One can think of analyzing the returns from improved designs, features, or systems in terms of light, medium, and heavy retrofits. The types of energy retrofits can also be categorized by plug loads, lighting, ventilation, cooling, and heating. One can also put such investments into a cost–benefit framework. One of the benefits, savings in energy, depends on the local rates, which we observe vary greatly around the U.S. and the world. Those areas with expensive power, like Hawaii and California, will tend to observe faster payback from investments in such features as solar or wind-based electricity generation. At the same time, some areas need less heating and cooling, so the point is that local climate matters, which can be translated into average HDDs or CDDs.

The cost of certification also affects whether the owners of a more sustainable building bother engaging the services of third-party consultants to help with applications for LEED, Energy Star or any number of other scoring systems, such as BREEAM, Green Star, CASBEE, and so forth. These costs tend to be lower as a percentage of the building the larger it is. There are also portfolio and now experienced owner preferences being built into the LEED system so that it becomes more efficient and less costly.

One of the greatest potential benefits of more sustainable real estate is greater occupant productivity. Productivity is a function of design, layout, and also management and employee incentives. It is also something quite hard to measure since many tasks have long-term payoffs, that are not easily or quickly recognizable. For example, some projects might take years to construct and sell, and the people responsible for initiating a relationship might precede those executing it by several years. In such cases, it becomes very hard to know whether real productivity has been affected. Still, there is no choice but to try to measure productivity, and if productivity can be increased by several percent, then that benefit of better space can outweigh the higher rental costs of a better building.

It appears that valuing and financing a green building with added features is still a struggle, given that appraisers and lenders do not always recognize the benefits of better buildings in the valuation process. Even though insurance rates might be lower, these may not always be considered by a less enlightened analyst. Financing programs like PACE or using ESCOs described earlier, for upgrades, make sense in some but not all markets. Yet, the market is showing both rental and value premiums for green designs and features and certifications, generally in excess of the cost of the improvements. Certainly, attaining LEED certification up through Gold has been justified in most markets for existing building retrofits. Attaining Green Star in Australia, BREAAM in Europe, CASBEE in Japan, and DNGB in Germany is similar but in many cases the building regulations have changed to require better buildings. Even where they have not, the market is starting to recognize that value exceeds cost for many sustainable improvements and it is easy to find low-hanging fruit for improving most buildings.

Notes

1 This is based on CINCH data from HUD. See www.huduser.gov/portal/datasets/cinch.html.
2 Defined by Energy.Gov: https://energy.gov/eere/buildings/retrofit-existing-buildings.
3 Find the Guide to LEED Certification here: www.usgbc.org/cert-guide/commercial.
4 Lean enterprise or "lean thinking" is characterized by increased customer focus, elimination of "waste" throughout the organization, and reduced waste and inefficiencies in support functions.
5 See the work of Stephen Roulac on global places (Roulac, 2001).
6 When endogenous rent-setting policies are taken into account by measuring the dependent variable by the logarithm of effective rent (property owners can be expected to adopt differing asking rent strategies so, *ceteris paribus*, landlords who charge higher rents will experience higher vacancy rates), the results suggest that the effect of a green rating is even larger.
7 Conducted by the Pew Center on Global Climate Change and presented at the 2009 ACEEE Summer Study on Energy Efficiency in Industry Proceedings in 2009.
8 NCREIF (National Council of Real Estate Investment Fiduciaries) provide data on historic returns and a range of commentaries on the market. From details on their website (www.ncreif.org) at the time of writing most investment funds are looking for single-digit returns for seasoned core property, yet they want much higher returns for investments in sustainable features. The authors believe this is partially a result of simply not knowing what the risks are and whether technology will advance soon, and partially because there is little penalty for being slow to take advantage of incremental returns that require a fair amount of research and education on the part of property managers, developers, and investors.
9 According to surveys by Miller, appraisers are among the least informed on the impact of sustainable building features on value. Anecdotal evidence is not proof but as one example, David Gottfried, one of the founders of the U.S. Green Building Council, built a home with natural lighting, super insulation, and near net-zero energy consumption among other features. When the property was appraised for refinancing, the appraiser ignored all of the features that enhanced the energy efficiency of the home, including the savings on utilities. Such stories suggest that we have a long way to go before all appraisers learn how to recognize and account for the value impact on property of all types. As of 2015, fewer than 20 appraisers in the U.S. were LEED Accredited Professionals (APs), while tens of thousands of architects were certified as LEED APs.

References

Bendewald, M., Hutchinson, R., Muldavin, S., and Torbert, R. 2014. *How to Calculate and Present Deep Retrofit Value: A Guide for Owner-Occupants*. Rocky Mountain Institute. Available at: https://static1.squarespace.com/static/5175bc71e4b0ab734d253f1c/t/53c72ba3e4b03a362b345cd5/1405561783858/WinWerks+GSA+Retrofit+Value (accessed November 14, 2017).

Benson, A., Vargas, E., Bunts, J., Ong, J., Hammond, K., Reeves, L., Chaplin, M., and Duan, P. 2011. *Retrofitting Commercial Real Estate: Current Trends and Challenges in Increasing Building Energy Efficiency*. Technical report, UCLA Institute of the Environment and Sustainability.

Binkley, A. G. 2007. *Real Estate Opportunities in Energy Efficiency and Carbon Markets*. Diss. Massachusetts Institute of Technology.

Davis Langdon. 2007. *Cost of Green Revisited: Reexamining the Feasibility and Cost Impact of Sustainable Design in the Light of Increased Market Adoption*. Davis Langdon.
Eichholtz, P., Kok, N., and Quigley, J. M. 2010. "Doing well by doing good? Green office buildings." *The American Economic Review*. 100(5): 2492–2509.
Fuerst, F. and McAllister, P. 2009. "An investigation of the effect of eco-labeling on office occupancy rates." *The Journal of Sustainable Real Estate*. 1(1): 49–64.
Heerwagen, J., Kelly, K., and Kampschroer, K. 2005. *Changing Nature of Organizations, Work and Workplace*. Whole Building Design Guide.
Hunt, W. D. 2008. *Literature Review of Data on the Incremental Costs to Design and Build Low-Energy Buildings*. Pacific Northwest National Laboratory operated by Batelle for the US Department of Energy. Available at: www.pnl.gov/main/publications/external/technical_reports/PNNL-17502.pdf (accessed November 14, 2017).
Jackson, Jerry. 2009. "How risky are sustainable real estate projects? An evaluation of LEED and Energy Star development options." *The Journal of Sustainable Real Estate*. 1(1): 91–106.
Kats, G. H., Alevantis, L., Berman, A., Mills, E., and Perlman, J. 2003. *The Costs and Financial Benefits of Green Buildings: A Report to California's Sustainable Building Task Force*. Sustainable Building Task Force.
Kok, N., Miller, N. G., and Morris, P. 2012. "The economics of green retrofits." *The Journal of Sustainable Real Estate*. 4(1): 4–22.
Leonardo Academy Inc. 2008. *The Economics of LEED for Existing Buildings for Individual Buildings*. Leonardo Academy Inc.
Loutzenhiser Associates. 2004. *Final Evaluation Report: California Building Performance Contractors Association Comprehensive Whole House Residential Retrofit Program*.
Managan, K. and Klimovich, K. 2013. *Setting the Pace: Financing Commercial Retrofits*. Johnson Controls Institute for Building Efficiency, PACE*Now*, and the Urban Land Institute.
Miller, N. 2014. "Workplace trends in office space: Implications for future office demand." *Journal of Corporate Real Estate*. 16(3): 159–181.
Miller, N., Spivey, J., and Florance, A. 2008. "Does green pay off?" *Journal of Real Estate Portfolio Management*. 14(2): 385–401.
Miller, N., D. Pogue, D. Gough, Q.D., and Davis, S.M. 2009. "Green buildings and productivity". *The Journal of Sustainable Real Estate*. 1(1): 65–89.
Moomaw, W., and Johnston, L. 2008. "Emissions mitigation opportunities and practice in Northeastern United States." *Mitigation and Adaptation Strategies for Global Change*. 13(5–6): 615–642.
Muldavin, S. R. 2010. *Value Beyond Cost Savings: How to Underwrite Sustainable Properties*. Green Building Finance Consortium.
Muldavin, S., Torbert, R., and Bendewald, M. 2013. "The missing link: Transforming deep retrofits into financial assets." *Corporate Real Estate Journal*. 3(3): 244–259.
Navigant Consulting, Inc. and Rocky Mountain Institute (RMI). 2014. *NEEA Existing Building Renewal: Process Review Results*. Report #E14–292. Navigant Consulting, Inc. and RMI.
New Buildings Institute (NBI). 2010. *Deep Energy Savings in Existing Buildings: Home on the Range. NBI.*
Osser, R., Neuhauser, K., and Ueno, K. 2012. *Proven Performance of Seven Cold Climate Deep Retrofit Homes*. Building Science Corporation.
Roulac, S. 2001. *Stephen Roulac on Place and Property Strategy*. Property Press.
Steven Winter Associates, Inc. 2004. *GSA LEED Cost Study: Final Report*. U.S. General Services Administration.
Tatari, O., and Kucukvar, M. 2011. "Cost premium prediction of certified green buildings: A neural network approach." *Building and Environment*. 46(5): 1081–1086.
The David and Lucile Packard Foundation. 2002. *Building for Sustainability Report: Six Scenarios for the David and Lucile Packard Foundation Los Altos Project*. The David and Lucile Packard Foundation.
Thomsen, A. and van der Flier, K. 2011. "Understanding obsolescence: A conceptual model for buildings." *Building Research & Information*. 39(4): 352–362.
Tolkin, B. M., Blake, W., Bonanno, S., Conant, D., Mauldin, J., and Hoefgen, L. 2008. "How much more does it cost to build an Energy Star home? Incremental cost estimation process." 2008 ACEEE Summer Study on Energy Efficiency in Buildings.
U.S. Department of Energy (DOE). 2010. *Accelerating Adoption of Energy Efficiency and Renewable Energy. U.S. DOE.*
U.S. Energy Information Administration. 2014. *Commercial Buildings Energy Consumption Survey (CBECS)*. U.S. EIA. Available at: www.eia.gov/consumption/commercial/maps.php (accessed November 14, 2017).
U.S. Energy Information Administration State Electricity Profiles. Available at: www.eia.gov/electricity/state/ (accessed December 19, 2017).

9

Building sustainability into valuation and worth

Sarah Sayce

9.0 Introduction

There is a clear supra-national drive aimed at improving the sustainability of the building stock, notably in respect of its energy efficiency, but also in relation to the wider issues of climate change and social need. Successive reports, notably from the Intergovernmental Panel on Climate Change (IPCC), have identified buildings as major sources of carbon emissions (Bosteels and Sweatman, 2016). They are now a target for climate policy both in the UK (Green Construction Board, 2014a, 2014b) and the European Commission (EC) on behalf of the European Council and Parliament via the issuance of Directives such as that on energy efficiency (EC, 2012).

Within Europe, greenhouse gas (GHG) emissions are being tackled by carbon pricing through the European Union (EU)'s Emissions Trading System, but the best-known EU measure in this regard is the Energy Performance of Buildings Directive of 2002 (now updated in the Directive of 2010), which introduced Energy Performance Certificates (EPCs). However, the Directive did not immediately impose standards but sought to use market mechanisms to drive demand towards more 'sustainable' stock through greater transparency and data.

Whilst the revised Directive is more stringent in that new buildings must be nearly zero energy by 31 December 2020 (by 31 December 2018 for buildings occupied and owned by public authorities), there is no immediate easily defined translation to the use and demand for existing buildings, except where significant retrofits take place. Yet it is in relation to the need to upgrade existing stock that the position is critical, as the rate of building replacement is slow (and itself the cause of resource depletion) and will not allow for a move to a low-carbon economy along, via a 'road map',[1] the trajectory that has been adopted by the EU and notably by the UK via the Climate Change Act 2008,[2] which imposes strict accountability on the UK government for 80 per cent carbon reduction by 2050. Whilst this drive to fulfil the ambition of a more sustainable stock has been driven primarily by strong regulation in relation to new stock, there exists the hope (little more) that greater transparency and better data will lead to market transformation of the existing stock, based on stimulating demand for better, more resilient and healthier buildings. This is critical, as it is estimated that most of the building stock that will be in existence in 2050 in developed countries is already built (Kelly, 2009).

There is increasing recognition that markets are not responding as quickly as is required by governments so more stringent actions may be required. The strongest move aimed at upgrading the energy performance of existing buildings has been made by the UK, which is introducing restrictions on the lettings of buildings that fail to meet minimum energy standards from 2018,[3] though this will not apply to transactions or to owner-occupied buildings, despite the fact that such stock in the UK accounts for almost half of all commercial buildings and almost two-thirds of residential buildings (Property Industry Alliance, 2016). The clear intent is to both raise awareness and change the behaviour of market participants by stimulating a differential demand, though it is recognised that this could result in so-called stranded assets, whose value has fallen below any economic value in both use and redevelopment (Monasterolo, Battiston, Janetos, and Zheng, 2016).

In terms of demand, the expectation is that change within existing stock will come about through recognition of the benefits of 'green' buildings over those that are merely code compliant, leading to the emergence of a 'green' premium or 'brown' discount within property pricing (see, for example, World Green Building Council, 2013).

The questions of whether this has happened and what role valuers are playing in supporting transformation are explored in this chapter. Valuers advise on market value and investment worth[4] for market participants and, critically, those lending to facilitate purchase. The contention is that, if valuers do not take into account the cost savings and additional health and wellbeing benefits that flow from occupation of, and investment in, more climate-resilient stock, investment decisions by owner-occupiers and investors will remain sub-optimal. As valuers work on *evidence* of markets, they require evidence that market indicators have changed before reporting differently.[5]

The importance of 'value' to the sustainability agenda can be related to the notion of Elkington's triple bottom line (1997). For economic value to recognise notions of social and environmental improvements may be easy in principle, but in reality economics tends to takes precedence in decision-making, leading to a desire to 'do well whilst doing good'. Indeed this argument is put forward by, for example, Miller, Spivey, and Florance (2008) and Eichholtz, Kok, and Quigley (2010), who sought to argue that investing in so-called 'green' buildings is a good business commercial decision. The contention is that more resilient, more energy-efficient buildings will increase returns and reduce risk and obsolescence. According to the World Green Building Council (2013), the business case rests on a combination of value add and cost reduction. However, for Bosteels and Sweatman (2016: 4), representing the UNEP–FI's Property Working Group, it is a matter of fiduciary duty on the part of institutional investors "as a routine component of their business thinking, practices and management processes". To them, it is a social imperative which, if not adopted, leads to governance risk. It is not negotiable.

However, such investors, though influential, are in the minority and to encourage or speed up the transformation a variety of incentives schemes have been adopted across Europe, but many have failed or been short-lived.[6]

Valuers are concerned primarily with making judgements about the behaviours or likely behaviours of those market participants engaged in property management and transactions. If they find evidence that sustainable buildings are in higher demand and that is shown in transaction prices, they should report it. If they do not, they will not. It is with this proposition and the role of the valuer in supporting such claims or otherwise that this chapter now deals. It is only if valuers report differential valuations in the light of market evidence that the claim that 'green adds value' can be substantiated.

9.1 The purpose and bases of valuation

The process of valuing a property starts with the purpose for which the commission is being undertaken. This must be understood from the outset and governance on this is normally provided by the relevant professional body. Arguably the leading global professional body governing valuers is the Royal Institution of Chartered Surveyors (RICS). All RICS surveyors, have to adhere to the standards and processes laid out in the *RICS Valuation – Professional Standards 2014* (RICS, 2014), colloquially and hereafter known as the *Red Book*. This in turn embodies the standards of the International Valuation Standards Committee (IVSC). At the time of writing, the IVSC have issued a latest edition (IVSC, 2017), which will be followed by an update of the *Red Book*.

The *Red Book* aims to provide "[c]onsistency, objectivity and transparency [which] are fundamental to building and sustaining public confidence and trust in valuation" (RICS, 2014: 1). Whilst not a manual instructing valuers *how* to value, it sets out the processes, including due diligence and inspection, that a valuer practising anywhere around the world should complete when undertaking a valuation. Valuations can be undertaken for many purposes, notably for sale and to inform lenders as to the security of an asset. Furthermore, valuations are required for accounting purposes, in which the valuation placed in company accounts provides information on which both shareholders and property owners rely. It follows that it is imperative that valuations are accurate, reflective of market trends and based on sound data and judgement.

For all of these purposes, valuers are instructed to adopt the basis of market value (MV), defined as

> [t]he estimated amount for which an asset or liability should exchange on the valuation date between a willing buyer and a willing seller in an arm's length transaction, after proper marketing and where the parties had each acted knowledgeably, prudently and without compulsion (RICS, 2014: glossary).

In order to so do, valuers will normally rely on the analysis of other, comparable transactions (RICS, 2012). How this analysis is undertaken and against what criteria will fundamentally impact on the way that they then proceed with their valuation. For example, if the only details that they have of comparable transactions analysed comprise date, sizes, use and condition, those will be the variables factored into their value judgement of the property under consideration.

The *Red Book* also recognises, *inter alia*, investment value (IV) or worth. This is defined as "[t]he value of an asset to the owner or a prospective owner for individual investment or operational objectives" (RICS, 2014; glossary). This basis of valuation is prepared primarily to give strategic advice to investors in contemplation of purchase or for management purposes and was described by Mallinson (1994) as "the stuff of decisions" (as quoted in Sayce, 2011). Unlike MV, although comparable transactions may be used, the principal objective is to produce a cash flow of likely future net income from the asset, duly discounted to reflect both the investor's requirements and the risks inherent in the future performance.

It follows from the interpretation of these two bases that valuers may be involved in preparing calculations based on different premises: establishing MV essentially involves looking backwards at evidence and distilling this through a process of interpretation, often based on heuristics or mental 'short-cuts', which have been shown to reinforce or 'anchor' previous valuations (Gallimore, 1994; Diaz, Gallimore and Levy, 2002; Baum, Crosby, Gallimore, McAllister, and Gray, 2003; Levy and Frethey-Bentham, 2010; Lin and Chang, 2012). In contrast, the process of establishing IV is more forward looking and factors that have not yet entered the evidence base can be included within cashflow projections.

This raises two questions. To what extent is sustainability factored into the data streams that are captured within the valuation process? And to what extent is the data that is captured used for analysis?

Before considering these questions, the extant RICS advice in relation to sustainability to valuers is set out. It should be noted that, at the time of writing, there exists an intention to review and, if necessary, to update the guidance.

9.2 The current guidance in respect of sustainability and valuation

For sustainability to be reflected within any valuation, the first requisite is that valuers recognise what constitutes a 'sustainability attribute'; the second is that they collect such data; and the third is that they use it in their analysis of evidence and subsequent judgement. If they do not, they will not differentiate in their reported values between properties that are similar in terms of the 'normal' value attributes but that vary in sustainability terms. So there needs to be data – and it needs to be collected and used.

The RICS imposes globally mandatory requirements on valuers in relation to their due diligence and data collection processes, but these do not, currently, explicitly address any sustainability factors. The requirement is that "inspections and investigations must always be carried out to the extent necessary to produce a valuation that is professionally adequate for its purpose" (RICS, 2014: 38). The onus is simply to collect that which is 'necessary', although the requirement goes on to elaborate that this will normally include matters such as dimensions, accommodation, construction, age, 'apparent' condition, use, services and environmental issues such as contamination.

However the *Red Book* is not entirely silent on sustainability. From 2014, valuers globally have been advised (no stronger than this) to "collect appropriate and sufficient sustainability data, as and when it becomes available, for future comparability, even if it does not currently impact on value. Only where market evidence would support this, should sustainability characteristics be built into a report on value" (RICS, 2014: 59). The key comment here is that the valuer should look to market evidence to support any contention that sustainability is a material value factor.

In relation to advice for IV, the standards go further and recommend that "sustainability factors that could influence investment decision-making may properly be incorporated, even though they are not directly evidenced through transactions" (RICS, 2014: 59). Therefore, there is, in respect of sustainability, a clear distinction to be made between the role of the valuer in relation to MV and IV.

The issue that this reveals is that the expectation could be that market differentiation will be more likely to occur where the property being valued is one for which the prevalent market players are large-scale investors, such as those Bosteel and Sweatman (2016) were addressing.

In addition to mandated requirements, published guidelines for valuers in relation to both residential units (RICS, 2011) and commercial property (RICS, 2013) provide more fine-grained advice, but do not make definitive statements. Indeed, they state that sustainability issues can relate not only to the physical characteristics of buildings but also to "the impact of climate change on location, legislation, public policy and fiscal measures or the increasingly sustainability-aware attitudes of both occupiers and investors" (RICS, 2013: 6). It is open to the valuer to decide which aspects of a building may be regarded as sustainability related and which may not. These are now considered.

9.3 Building MV and IV into valuations: What are valuers doing?

As stated above, MV and IV are different concepts. In practice, which is being calculated influences both the evidence base and the approach to the calculation adopted. MV, based on market comparability, is inherently *backward looking* as the valuer seeks to analyse what *has* happened, which in turn can mean that valuations lag prices (Clayton, Geltner and Hamilton, 2001), although in theory the evidence provides a framework within which the valuer makes a judgement. However, in practice, not only anchoring (see, for example, Diaz and Hansz, 2001, for a discussion on how valuers use information supplied), but also the influence of clients (see, for example, Levy and Schuck, 2005 and Crosby, Devaney, Lizieri, and McAllister, 2015) and the fear of negligence, leading to the need to robustly support a valuation with evidence in the event of a valuation being challenged, can result in valuers being cautious in approach. This latter point is important, given that many valuations are prepared for the purposes of advising on secured lending and often only tested in the event of a market downturn (see, for example, De Silva, 2016).

In the case of IV, the appraisal is based on parameters that reflect the expectations of the specific buyer, seller or owner based on his/her own capabilities as well as on his/her own views and forecasts relating to the future; it may therefore be based on market research of future economic conditions (IPF/RICS, 1996). These may be above or below the market's expectations and can be expected to differ from the current average market situation. Therefore, MV is related to the current 'generalised' market understanding, whereas IV is influenced by factors and views specific to special players in the market and their forecast of future developments and new stimuli – such as some sustainability features represent.

It is possible, indeed probable, that MV may not be *provable* via evidenced transactions. However, as Michl, Lorenz, Lützkendorf, and Sayce (2016) argue, over time, IV and MV may converge as market understanding develops and the data available to valuers becomes more comprehensive. Such a view is not new. It was as long ago as 2006 that Boyd put forward a theoretical case for value differentiation. Ellison and Sayce (2006) and Sayce, Ellison and Parnell (2007) proposed that any market differentiation in terms of pricing would arise first from investors recognising market mispricing in relation to sustainable assets. They proposed, through modelling the characteristics that might impact on the market *in the future*, that it was possible to identify assets that were more resilient in terms of energy, waste, water and other sustainability features, and parameterise likely future value impacts. It must be emphasised that this argument was formulated in respect of institutional-grade property, not domestic or secondary units for which information on transactions may be more plentiful, due to greater volumes, but less fine-grained and, in the case of residential units, subject to other value drivers.

Whilst Ellison and Sayce concentrated on building sustainability into IV, much of the literature has concentrated on MV. Lützkendorf and Lorenz (2011) pointed to a need to change the processes of gathering, processing and presenting property-related information, suggesting that this could or should be led by professional bodies, whilst Warren-Myers (2012 and Chapter 7 in this book) concluded that it is not new methodology that is required, but for valuers to recognise their own knowledge shortfall on how to incorporate sustainability characteristics. This raises a concern about the role and competency of the valuer and the scope of their instructions to adequately look both backwards (MV) via market transaction analysis and forwards (IV) to assess the likely impact on cash flows of sustainability criteria.

The RICS, working with partners, has concluded that, at least in part, it is a skills agenda and through the RenoValue project[7] has developed a training toolkit for property valuation professionals aimed at enabling them to better factor energy efficiency and renewable energy issues into valuation practices in order to advise their clients. In particular, the project's final report argues that current valuation techniques have the capacity to capture sustainability but that there is "a lack of dedicated training for valuation professionals" in relation to user requirement and technologies (RenoValue, 2016: 10).

In summary, it would appear that the valuer is not under a mandated obligation to formally consider sustainability (though there is a strong recommendation so to do) and may not have the requisite technical valuation skills and appreciation of technologies. However, the valuer is not a building expert and for that reason may rely on the evidence supplied by others in order to form a view as to sustainability features. Notably, and as argued by Warren-Myers (2012) and Wilkinson, Sayce, and Christensen (2015), certifications such as BREEAM[8] and LEED[9] may be influential levers that are accessible to valuers to assist in their task; however, they tend to apply only to limited types of stock. To the valuer undertaking a secured-lending valuation of a secondary shop or a standard residential unit working in Europe, the limit of accessible sustainability data may be an EPC.

There is another dimension to this issue. Valuers must always work within the scope of their instructions. Therefore, as part of an RICS research project, some findings from which were reported in Michl et al. (2016), valuers were asked not only whether and how they built sustainability into their valuations, but whether they were under instruction from their clients so to do. The survey, to which there were over 300 respondents from European countries – most predominantly the UK, Germany and Switzerland (all countries in which there are significant numbers of RICS-regulated valuers) – sought to establish:

- how far the then extant guidance[10] was impacting on day-to-day practice;
- what sustainability factors were being requested; and, in consequence,
- what data was collected, analysed and used within their valuations.

In terms of the awareness and impact of guidance on day-to-day practice, the survey revealed little take-up, particularly in the UK. This in turn gives weight to the rationale both for RenoValue and for a tightening-up of requirements within professional guidance (Table 9.1).

The findings in relation to valuers who had been asked by clients to consider sustainability as part of their instructions are shown in Table 9.2.

Taken together, these findings point to a widespread lack of take-up by valuers, and to a wide variation in terms of instructions, with the majority of valuers never having been asked to consider sustainability. Whilst the client agenda appears to have been more developed within mainland European countries, penetration was not strong and unlikely to have been a major driver of value. The other 'stand-out' result in relation to instructions was that investor clients were more likely than other groups to ask about the impact of sustainability, which supports the proposition that investors, who might be seeking to establish worth, are concerned about future impacts.

It is acknowledged that this survey was conducted some years ago, but it remains probably the most comprehensive in terms of responses from valuers in relation to their instructions and

Table 9.1 The impact of extant guidance on valuers

	% of valuers using guidance on sustainability and valuation (rounded)		
	UK	*Germany*	*Switzerland*
Always	11	8	19
Sometimes	24	45	19
Seldom	27	19	23
Never	38	28	39
Total	100	100	100

Source: Adapted from Michl et al. (2016)

Table 9.2 The requirement by clients for valuers to report on sustainability in valuation reports

	% of client base requesting information regarding sustainability (rounded)		
	UK	*Germany*	*Switzerland*
Lenders	24	31	24
Investors	29	54	72
Lawyers	0	1	0
Owner/occupiers	13	26	48
No instruction to consider sustainability	60	29	12

Source: Adapted from Michl et al. (2016)

practice. However, feedback taken from unpublished sources available to the author[11] would indicate that, whilst the situation may have changed slightly, many valuers are still not considering sustainability explicitly within their valuations, as data is not supplied to them. Therefore, the integration of sustainability into the valuation process is thought to remain primarily implicit and hard to isolate as a value-determining factor.

If there is no change in either/both client requirements or/and valuer practice, then, in the light of the academic studies summarised below in Section 9.5, there would appear to be a disconnect between valuation practice and market pricing, at least in some sub-markets. Before considering the literature on pricing influences, valuers' understanding of what sustainability attributes are and how they may influence value is considered.

9.4 What do sustainability attributes look like?

One of the challenges in integrating sustainability attributes into property valuation is that many such attributes are already factored into a conventional valuation (e.g., location, access to public transport, structural flexibility, etc.), but additionally others, notably social and ecological attributes, are not. So the first challenge for the valuer is in relation to what evidence base should or could be collected and then how it is weighted. For this, there is no specific guidance, as it is up to valuers' judgement and their detailed knowledge of what factors might affect occupier or investor choice. In terms of the former, both Dixon, Ennis-Reynolds, Roberts and Sims (2009) and Levy and Peterson (2013) point to sustainability being only part of the decision process, whilst a literature review by Sayce, Sundberg and Clements (2010) found that the attributes placed under the general heading of 'sustainability' are extremely variable. Whilst most articles focused on energy efficiency, there was recognition of a range of contributory factors, from pollution to health and wellbeing to energy sources and biodiversity, and the literature revealed a lack of clarity of understanding.

Because of this variability of definition of both 'green' and 'sustainable', the tendency within the literature and, it is suggested, amongst practitioners is to use voluntary certification schemes such as LEED and BREEAM and other similar rating tools as surrogates for sustainability, yet both schemes embrace large, but not identical, numbers of factors assessed using differing weightings. As such, they are not precise surrogates. Furthermore, the market penetration of such voluntary schemes, whilst growing, is not complete and generally only relates to buildings of high specification[12] for which the costs of assessment are viable given the overall value of the building. They are not routinely found to apply to secondary and standard properties, which make up the majority of those transacted or used as loan security, purposes for which the majority of valuations are commissioned. As Schwartz and Raslan (2013) argue, these systems are largely market driven, relying on recognition of a value proposition attaching. However, whilst they may have a powerful role in influencing decision makers in some sub-markets, the technical differences between rating systems and their composition are not well understood by valuers (Warren, Bienert, and Warren-Myers (2009). Yet, without a nuanced understanding, it is hard to envisage how individual components can be isolated and used within valuations conducted using a comparative method.

This perhaps explains the concentration of recent research within Europe[13] on the isolation of energy efficiency as a value driver, as the number of properties with mandated EPCs is far greater and stretches across a wider range of assets in terms of built quality and value. Furthermore, as a single-factor rating, it provides focused information, which is freely accessible.

9.5 Incorporating sustainability into valuation: The academic evidence

The interest in exploring the linkage between transactional data and sustainability ratings has accelerated over the last decade; and over that period, the findings have become more positive. It was only in 2010 that the first major literature review of published work was undertaken (Sayce et al., 2010). This found only limited evidence of a direct link between sustainability factors and value, although the authors concluded that the case was beginning to develop. Similar findings of little linkage were reported by Fuerst and McAllister (2011), but by 2013 the World Green Building Council's review was more positive, and at the time of writing the evidence is widely accepted, although this is strongest in relation to 'prime' property, but this is a contested term (Atkinson-Baldwyn, Sayce and Smith, 2011).

In the majority of cases, hedonic regression analysis as designed by Rosen (1974) has been used as the tool for estimating the value of energy efficiency or sustainability rating based on actual market choices of individual purchasers (Fuerst, McAllister, Nanda, and Wyatt, 2015). The early studies predate both the introduction of EPCs in Europe and the increasingly widespread use of rating tools; as a consequence, they tended to suffer from small sample sizes and a lack of robustness (Laquatra, Dacquisto, Emrath, and Laitner, 2002). Although rating tools were more widespread even in 2009, McAllister argued that the results of studies are likely to be inconsistent given that each labelling system is different and that market conditions will also vary (McAllister, 2009).

The stream of research has resulted in findings of positive statistical evidence of value relationships between both rents achieved and capital values and sustainability characteristics. Amongst those examining links in the commercial sector, the largest part of the literature base has concentrated on US offices (see, for example, Fuerst and McAllister, 2009; Eichholtz et al., 2010; Das and Wiley, 2014). Whilst results vary, they tend to demonstrate a stronger link with Energy Star, which is a single-factor rating, than with LEED, which is a multi-factor rating. The literature base in Europe has been more recent and less extensive, possibly due to a lower concentration and number of certified commercial stock, and overall the evidence base is thin in relation to commercial properties other than city centre offices.

Within the residential field, the evidence in the US mirrors that of the commercial field (see, for example, Bond and Devine, 2016) but is less extensive. Within Europe, studies have primarily related to EPCs as other, voluntary, ratings systems tend not to be used except in large-scale new developments. Brounen and Kok's (2011) study of the relationship between EPC ratings and sales prices in the Netherlands was one of the first undertaken, but since then the evidence has increased (see, for example, Popescu, Bienert, Schützenhofer, and Boazu, 2012; Hyland, Lyon and Lyons, 2013; Cerin, Hassel, and Semenova, 2014; Stanley, Lyons and Lyons, 2015; Fuerst, McAllister, Nanda and Wyatt, 2016). However, although the evidence points to a connection between EPCs and higher transaction prices and/or rents, there are caveats to the findings. For example, Hyland et al. (2013) point to the connection with general market conditions, which can exercise a large influence, whilst Murphy (2014) found that many projections about the impact of the EPC have fallen short, with EPCs yielding only a weak influence on purchaser views and behaviours. Most of the above research concentrates on a particular country and/or sub-market. Within a European context, the most detailed and extensive report is that undertaken for the European Commission, which took household data from across selected EU countries and considered the role of energy labels, notably EPCs. This report is considered important within the context of influencing valuer judgement, as it concluded that "getting EPCs right is a necessary but not sufficient condition to ensure that markets value energy performance" (Mudgal, Lyons Cohen, Lyons, and Fedrigo-Fazio, 2013: 124).

In summary, studies reveal a strong general argument supporting differential values between energy-efficient/sustainable stock and other buildings, but results vary significantly across and even within national markets and cannot be automatically assumed to be applicable to other locations or building types. The evidence is strongest in central business districts (CBDs) of major cities with developed investment markets: here, energy efficiency may be part of the prime 'norm' specification and evidence extends beyond simple pricing to include duration to sell.

Commercial markets tend to be driven by rationality and increasing notions of corporate responsibility (Bosteels and Sweatman, 2016), and this should lead to accurate reflection in pricing of demand, with any premium resulting from a due consideration of costs, benefits and risks. However, there exists another potential and hitherto under-researched argument. Investors who are eco-champions may find that they are searching in a small pool of certified stock to match their requirements, resulting in *over*-pricing or what Fuerst, Gabrieli, and McAllister (2017) have dubbed the 'green winner's curse'. If this exists, then the price premium is less about a transformation of demand than about a lag in supply, which is a well-recognised phenomenon in property markets (Barras, 1994).

However, residential markets tend to operate differently and react to a different, more subjective range of value drivers (RICS, 2011), and the behaviour of market participants is more difficult to model and may be driven by factors other than those related to rational costs. This is important in potentially impacting on transaction prices.

Finally, the research recognises that market factors and average construction and quality standards vary across countries; therefore, the use of aggregated studies may be of limited applicability to the consideration of an individual unit or portfolio. From this, it is concluded that the extent to which energy efficiency and other sustainability features result in price differentiation does and perhaps always will depend on the conditions within a given local (sub-)market.

9.6 Reconciling the academic evidence with market practice

The evidence provided above points to sustainability being not currently considered explicitly within the valuation process. The requirement exists, under the *Red Book* (RICS, 2014), for valuers to seek to collect what data they can and to report on the risks and benefits of sustainability measures moving forwards even where, in their opinion, market evidence cannot support differential pricing. The argument then can be, and at times is, extended to a perceived lack of skills, knowledge and due diligence by valuers who are not accurately reflecting sustainability, particularly energy efficiency, in their reported values; in short, the implicit suggestion is that they are *undervaluing* in the light of the evidence provided by the hedonic analyses. The prevalence of this view is underscored by the very public ambitions of a range of projects, which imply that valuers are the 'problem'. For example, the EU is currently funding, or has recently funded, projects that seek to make explicit the link between energy efficiency and value.[14] All have a fundamental assumption implicit or explicit within them that the value is there but is just not being recognised. The premise is also that it may be due to a lack of education (RenoValue) or sufficiently strong instruction to valuers (Revalue) or issues in assessing the impacts (RentalCal).

However, if valuers do not have detailed evidence of individual transactions held in a way in which the individual component can be isolated and evaluated, it is unlikely that change will occur readily. And, with the exception of primarily prime commercial stock, the only readily available data is that contained in EPCs, the reliability of which has been questioned (Hobbs, 2013). Therefore, valuers assessing MV use heuristics to make judgements based on both the information that they have, which will be very limited in terms of sustainability, and on their experience gained from conducting past valuations (RICS, 2012). This is particularly the case where the valuation is of owner-occupied property and will provide a tendency to anchor their judgements in the past. Furthermore, the lack of explicit instructions, noted above, in respect of sustainability reporting reinforces the status quo.

Within the field of residential valuations, many valuers undertaking secured-lending valuations use comparable evidence obtained from automated valuation models (AVMs), but these typically do not include data in relation to EPCs, at least in the UK. If they did, it would immediately enable the impact of energy ratings to be considered explicitly within valuations. Until, or unless, the requirement for isolation and separate reporting of energy efficiency and other sustainability factors is made explicit, whether due to client instructions or professional body mandate, the resultant report is likely to remain a sealed 'black box' in which the judgement, revealed only in the valuer's file notes, is contained. One further issue that valuers will have to consider is the extent to which their reports, if challenged, can be supported in any dispute proceedings by firm evidence. This potential liability to professional negligence in itself is likely to engender a conservative approach and reliance back on traditional value factors, in which comparables are analysed by size, location, condition specification, tenure and market conditions. If sustainability is to be added to this list, explicit instructions, improved data collection and guidance may all be required.

In the case of preparing IV, a detailed cash flow (DCF) approach using a series of explicit inputs in which projections of future change are included is more likely to be the adopted methodology and reported explicitly. A DCF may include consideration of questions around the impact of rising environmental concerns, including projected energy costs and other risks. As an IV is less dependent on market evidence than an MV but can take account of external factors and developing agendas, the calculation is forward looking, though with reference to market evidence. It is therefore with IV that the actions of valuers will start to influence markets, leading to changes in MV.

Change is occurring and is likely to accelerate. One driver for this is the increasing awareness of banks and other lenders that properties with poor energy efficiency and other sustainability credentials may well present an increased risk profile. As this awareness grows, so some lenders are known to be actively considering whether energy efficiency and other sustainability criteria should be considerations that should be built into both/either the amount of lending that can be secured against the unit and/or the interest rate charged for the loan.[15] As the link with funding, either for development or, more importantly, for existing stock renovations, becomes explicit and banks and mortgage providers change their lending criteria, so clearer evidence will be presented to valuers in a way that they can more easily incorporate within their market appraisals.[16] This in turn may provide clear market evidence to assist valuers in their role.

9.7 Conclusions

The adage is that valuers reflect markets, they do not make them; this is built into valuer guidance. Valuers must work within the constraints imposed by the requirements of both clients and professional bodies. If there is no appetite by clients to be provided with valuations that pay explicit regard to sustainability, then it is unlikely that valuers will do so – but this does not necessarily mean that they have disregarded the factors. Most market valuations undertaken for transaction or lending purposes report a specific figure and do not specify the extent (if any) to which sustainability has influenced the judgement. Thus, they are a 'point estimate', with the actual process of arriving at a judgement remaining an item on the valuer's file. However, with IV, the explicit process may more easily reveal the extent to which energy efficiency and other factors are deemed influences on likely cash flows moving forwards; adjusted discount rates to reflect the risks of accelerated obsolescence may be discussed and used.

Returning to the two questions posed earlier: to what extent is sustainability factored into the data streams that are captured within the valuation process? Currently, there is not firm knowledge, but the most likely situation, in the absence of any current research on the point, is probably not a lot except in some sub-markets such as prime institutional buildings and those where the occupants are very cost conscious. And to what extent is the data that is captured used for analysis? Here, whilst limited for MV, the very process of IV more easily facilitates a move to better data capture.

The academic evidence of differential pricing is strong in some sub-markets but, if the values reported by valuers do not reflect differential values based on sustainability criteria but the hedonic regression studies do, there is an issue of credibility, requires addressing.

However, the hedonic regression findings are in aggregated form and, as such, are not necessarily 'de-composed' such that they can be separately analysed within the due diligence process, especially in the absence of strong instructions from clients. But does this mean that market participants and their advisors ignore energy efficiency and sustainability credentials in their purchase and rental decisions? The evidence is that some major investors consider the matter to transcend the issue of profit maximisation and risk reduction: it is a fiduciary concern and this appears consistent with larger pricing differentials for 'prime' city stock. To others, it is not a key driver. So are valuers taking note and building in modelling – or simply adjusting their implicit judgement process, possibly 'anchoring' backwards due to a lack of client drive and limited data?

The key, which has been strongly argued in the literature and supported by projects such as RenoValue, is to embrace DCF techniques, which require clarity and transparency of assumptions and value inputs. However, although written within a slightly different context, the words of Crosby and Henneberry resonate when they argue that, despite changes to the education of the last few generations of valuers, "explicit DCF has not supplanted conventional, comparison-based practice in investment valuation in this particular market" (Crosby and Hennebery, 2016: 14). If both the perception and the reality is that valuers are lagging behind the markets, this is potentially damaging to the profession and, more importantly, to their ability to assist in market transformation and fulfil Brundtland's (1987) ambitions.

The point has been reached at which a simple awareness of sustainability issues is insufficient; it is now time for valuers to take a more proactive stance, from their initial interface with their clients to their reporting process. One way forwards is to develop a deeper understanding of the process by which valuers are requesting data – or not – and using it in order to form their judgements. Only by opening the black box of their calculations will it become clear whether there exists a disparity between the weightings applied by the academic researchers and the heuristic judgements of practitioners.

Notes

1 www.roadmap2050.eu.
2 www.legislation.gov.uk/ukpga/2008/27/contents.
3 Energy Act 2011.
4 These terms are defined later in the chapter.
5 Under RICS (2014), valuers must undertake a process of due diligence in obtaining evidence. This is explored further.
6 An example of this was the UK's ill-fated Green Deal scheme, which sought to provide owners with a means of making capital improvements, payments for which would be made by instalments over time and related to cost savings. Launched in 2012, it was withdrawn in 2016 having failed to make impact (see www.gov.uk/green-deal-energy-saving-measures/overview).
7 http://renovalue.eu.
8 Building Research Establishment Environmental Assessment Method: a UK-based commercial voluntary rating scheme widely adopted for rating new builds and, increasingly, retrofitted stock but associated with prime stock.
9 Leadership in Energy and Environmental Design: a US-based scheme similar to BREEAM and used also in Europe.
10 The survey was undertaken prior to the introduction of the 2014 *Red Book*. At the time of the survey, the extant guidance, though similar to that contained in the 2014 *Red Book*, was not mandatory for valuers.
11 The author is part of the RICS Sustainability and Valuation Task Group and a European project on the integration of energy efficiency within valuation (http://revalue-project.eu) and as such has attended conferences and round tables with practising valuers.
12 BREEAM is the oldest scheme and as of 2014 was estimated to apply to 425,000 buildings (BRE Global, 2014), which is a small proportion of the stock. For a longer discussion of rating schemes, see Christensen and Sayce (2015).
13 EU projects such as Revalue, FINERPOL, LENDERS, RentalCal and EMF are all currently in progress and all have the aim of linking energy efficiency with valuation.
14 RenoValue: a training toolkit to integrate energy efficiency and renewable energies into property valuation practices (http://renovalue.eu/); Revalue: Recognising Energy Efficiency Value in Residential Buildings (http://revalue-project.eu/); RentalCal: a road map towards a sustainable housing stock by developing models and tools for assessing the commercial viability of energy efficiency retrofitting in rental properties (www.rentalcal.eu/).
15 See, for example, two current research projects exploring this: LENDERS, led by the UK's Green Building Council (www.ukgbc.org/event/lenders-project-launch) and EEMAP: Energy Efficient Mortgages Action Plan (http://globalabc.org/uploads/media/default/0001/01/33ca9d4a0801d5b51b699ea428cc5c461077a150.pdf).
16 For an example of a project driving innovation in funding for energy efficiency, see FINERPOL (www.interregeurope.eu/finerpol/).

References

Atkinson-Baldwyn, L., Sayce, S. and Smith, J. (2011) Time to define prime? A contested term within Western European office lease practice. Paper to European Real Estate Society Conference, Eindhoven, 15–18 June 2011

Barras, R. (1994) Property and the economic cycle: Building cycles revisited. *Journal of Property Research, 11*(3), pp.183–197

Baum, A., Crosby, N., Gallimore, P., McAllister, P. and Gray, A. (2003) Appraiser behaviour and appraisal smoothing: Some qualitative and quantitative evidence. *Journal of Property Research, 20*(3), pp.261–280

Bond, S. A. and Devine, A. (2016) Certification matters: Is green talk cheap talk? *The Journal of Real Estate Finance and Economics, 52*(2), pp.117–140

Bosteels, T. and Sweatman, P. (2016) Sustainable real estate investment: Implementing the Paris Climate Agreement: An action framework. UNEP–FI (United Nations Environment Programme – Finance Initiative), available at www.unepfi.org/publications/investment-publications/property-publications/sustainable-real-estate-investment-2/

Boyd, T. (2006) Evaluating the impact of sustainability on investment property performance. *Pacific Rim Property Research Journal, 12*(3), pp.254–271

BRE Global (2014) *The Digest of BREEAM Assessment Statistics Volume 01, 2014*, available from www.breeam.com/filelibrary/Briefing%20Papers/BREEAM-Annual-Digest---August-2014.pdf

Brounen, D. and Kok, N. (2011) On the economics of energy labels in the housing market. *Journal of Environmental Economics and Management, 62*(2), pp.166–179

Brundtland, G. H. (1987) *Our Common Future: Report of the 1987 World Commission on Environment and Development*. Oslo: United Nations, pp.1–59

Cerin, P., Hassel, L. G. and Semenova, N. (2014) Energy performance and housing prices. *Sustainable Development, 22*(6), pp.404–419

Christensen, P. and Sayce, S. (2015) Sustainable property reporting and rating tools, in Wilkinson, S., Sayce, S. and Christensen, P. (eds) *Developing Property Sustainably*. London: Routledge pp.203–233

Clayton, J., Geltner, D. and Hamilton, S. (2001) Smoothing in commercial property valuations: Evidence from individual appraisals. *Real Estate Economics, 29*(3), pp.337–360

Crosby, N. and Henneberry, J. (2016) Financialisation, the valuation of investment property and the urban built environment in the UK. *Urban Studies, 53*(7), pp.1424–1441, available at http://eprints.whiterose.ac.uk/91016/3/WRRO_91016.pdf

Crosby, N., Devaney, S., Lizieri, C. and McAllister, P. (2015) Can institutional investors bias real estate portfolio appraisals? Evidence from the market downturn. *Journal of Business Ethics*, pp.1–17

Das, P. and Wiley, J. A. (2014) Determinants of premia for energy-efficient design in the office market. *Journal of Property Research, 31*(1), pp.64–86

De Silva, C. (2016) Negligent valuation: Its development in the UK and the role of professional standards. Paper to COBRA, Toronto, RICS

Diaz III, J. and Hansz, J. A. (2001) The use of reference points in valuation judgment. *Journal of Property Research, 18*(2), pp.141–149

Diaz, J. III, Gallimore P. and Levy, D. (2002) Residential valuation behaviour in the United States, the United Kingdom, and New Zealand. *Journal of Property Research, 19*(4), pp.313–326

Dixon, T., Ennis-Reynolds, G., Roberts, C. and Sims, S. (2009) Is there a demand for sustainable offices? An analysis of UK business occupier moves (2006–2008). *Journal of Property Research, 26*(1), pp.61–85

Eichholtz, P., Kok, N. and Quigley, J. M. (2010) Doing well by doing good? Green office buildings. *The American Economic Review, 100*(5), pp.2492–2509

Elkington, J. (1997) *Cannibals with Forks: The Triple Bottom Line of 21st Century Business*. Oxford: Capstone

Ellison, L. and Sayce, S. (2006) *The Sustainable Property Appraisal Project: Final Report*, available at http://kueprints3.kingston.ac.uk/1435/1/Sustainable_Property_Appraisal_Project.pdf

European Commission (EC) (2012) *The Energy Efficiency Directive 2012/27/* available at https://ec.europa.eu/energy/en/topics/energy-efficiency/energy-efficiency-directive

Fuerst, F. and McAllister, P. (2009) New evidence on the green building rent and price premium, Conference paper, annual meeting of the American Real Estate Society, 9 April 2009, Monterey, CA

Fuerst, F. and McAllister, P. (2011) The impact of energy performance certificates on the rental and capital values of commercial property assets. *Energy Policy, 39*(10), pp.6608–6614

Fuerst, F., Gabrieli, T. and McAllister, P. (2017) A green winner's curse? Investor behavior in the market for eco-certified office buildings. *Economic Modelling, 61*, pp.137–146

Fuerst, F., McAllister, P., Nanda, A. and Wyatt, P. (2015) Does energy efficiency matter to home-buyers? An investigation of EPC ratings and transaction prices in England. *Energy Economics, 48*, pp.145–156

Fuerst, F., McAllister, P., Nanda, A. and Wyatt, P. (2016) Energy performance ratings and house prices in Wales: An empirical study. *Energy Policy, 92*, pp.20–32

Gallimore, P. (1994) Aspects of information processing in valuation judgment and choice. *Journal of Property Research, 11*(2), pp.97–110

Green Construction Board (2014a) *GCB 610 Mapping the Real Estate Lifecycle for Effective Policy Interventions*, available at www.greenconstructionboard.org/index.php/resources/valuation-demand-outputs

Green Construction Board (2014b) *GCB 620 Energy Efficiency Policies in the Domestic Real Estate Sector*, available at www.greenconstructionboard.org/index.php/resources/valuation-demand-outputs

Hobbs, D. (2013) So you thought the accuracy of EPCs was improving? *Building*, 13 November 2013, available at www.building.co.uk/so-you-thought-the-accuracy-of-epcs-was-improving?/5063502.article

Hyland, M., Lyons, R. C. and Lyons, S. (2013) The value of domestic building energy efficiency: Evidence from Ireland. *Energy Economics, 40*(c), pp.943–952

IPF/RICS (1996) *Calculation of Worth*. London: RICS Books/Investment Property Forum
IVSC (2017) *International Valuation Standards*, available at www.ivsc.org
Kelly, M. J. (2009) *Britain's Building Stock: A Carbon Challenge*. University of Cambridge/Department of Communities and Local Government
Laquatra, J., Dacquisto, D. J., Emrath, P. and Laitner, J. A. (2002) *Housing Market Capitalization of Energy Efficiency Revisited*. Summer Study on Energy Efficiency in Buildings. Washington, DC: American Council for an Energy-Efficient Economy
Levy, D. S. and Frethey-Bentham, C. (2010) The effect of context and the level of decision maker training on the perception of a property's probable sale price. *Journal of Property Research*, *27*(3), pp.247–267
Levy, D. and Peterson, G. (2013) The effect of sustainability on commercial occupiers' building choice. *Journal of Property Investment & Finance*, *31*(3), pp.267–284
Levy, D. and Schuck, E. (2005) The influence of clients on valuations: The clients' perspective. *Journal of Property Investment & Finance*, *23*(2), pp.182–201
Lin, T. C. and Chang, H. Y. (2012) How do appraisers absorb market information in property valuation? Some experimental evidence. *Property Management*, *30*(2), pp.190–206
Lützkendorf, T. and Lorenz, D. (2011) Capturing sustainability-related information for property valuation. *Building Research & Information*, *39*(3), pp.256–273
Mallinson, M. (1994) *The Mallinson Report: The Report of the President's Working Party on Commercial Property Valuations*. London: RICS
McAllister, P. (2009) Assessing the valuation implications of the eco-labelling of commercial property assets. *Journal of Retail and Leisure Property*, *8*(4), pp.311–322
Michl, P., Lorenz, D., Lützkendorf, T. and Sayce, S. (2016) Reflecting sustainability in property valuation: A progress report. *Journal of Property Investment & Finance*, *34*(6), pp.552–577
Miller, N., Spivey, J. and Florance, A. (2008) Does green pay off? Draft paper, Burnham Moores Center for Real Estate, University of San Diego
Monasterolo, I., Battiston, S., Janetos, A. and Zheng, Z. (2016) *Understanding Investors' Exposure to Climate Stranded Assets to Inform the Post-Carbon Policy Transition in the Eurozone*, available at https://papers.ssrn.com/sol3/papers.cfm?abstract_id=2766569
Mudgal, S., Lyons, L., Cohen, F., Lyons, R. and Fedrigo-Fazio, D. (2013) *Energy Performance Certificates in Buildings and Their Impact on Transaction Prices and Rents in Selected EU Countries*, Final Report to the European Commission (DG Energy), 19 April 2013, European Commission, available at https://ec.europa.eu/energy/sites/ener/files/documents/20130619-energy_performance_certificates_in_buildings.pdf
Murphy, L. (2014) The influence of the energy performance certificate: The Dutch case. *Energy Policy*, *67*, pp.664–672
Popescu, D., Bienert, S., Schützenhofer, C. and Boazu, R. (2012) Impact of energy efficiency measures on the economic value of buildings. *Applied Energy*, *89*(1), pp.454–463
Property Industry Alliance (2016) *Property Data Report 2016*, available at www.bpf.org.uk/sites/default/files/resources/PIA-Property-Report-2016-final-for-web.pdf
RenoValue (2016) *RenoValue: Final Report*, available from http://renovalue.eu/
RICS (2011) *Sustainability and Residential Property Valuation: Valuation Information Paper 22/2011*, available at www.rics.org/Global/Sustainability_residential_property_valuation_1st_edition_PGguidance_2011.pdf
RICS (2012) *Comparable Evidence in Property Valuation Information Paper 26/2012*, available at www.rics.org/Global/Comparable_evidence_in_property_valuation_1st_edition_PGguidance_2011.pdf
RICS (2013) *Sustainability and Commercial Property Valuation: A Guidance Note*. 2nd Ed, available at www.rics.org/Global/Sustainability_and_commercial_property_valuation_2nd_edition_PGguidance_2013.pdf
RICS (2014) *RICS Valuation: Professional Standards January 2014 (Red Book)*. London: RICS
Rosen, S. (1974) Hedonic prices and implicit markets. *Journal of Political Economy*, *82*(1), pp.34–55
Sayce, S. (2011) Valuing sustainability: The role of the valuer: From reflector to influencer. Presentation to SB2011, Helsinki October 2011
Sayce, S., Ellison, L. and Parnell, P. (2007) Understanding investment drivers for UK sustainable property. *Building Research & Information*, *35*(6), pp.629–643
Sayce, S., Sundberg, A. and Clements, B. (2010) *Is Sustainability Reflected in Commercial Property Prices: An Analysis of the Evidence Base*. London: RICS
Schwartz, Y. and Raslan, R. (2013) Variations in results of building energy simulation tools, and their impact on BREEAM and LEED ratings: A case study. *Energy and Buildings*, *62*, pp.350–359

Stanley, S., Lyons, R. C. and Lyons, S. (2015) The price effect of building energy ratings in the Dublin residential market. *Energy Efficiency*, *4*, pp.1–11
Warren, C. M., Bienert, S. and Warren-Myers, G. (2009) Valuation and sustainability: Are rating tools enough? Paper to 16th Annual European Real Estate Society Conference. ERES
Warren-Myers, G. (2012) The value of sustainability in real estate: A review from a valuation perspective. *Journal of Property Investment & Finance*, *30*(2), pp.115–144
Wilkinson, S., Sayce, S. and Christensen, P. (2015) *Developing Property Sustainably*. London: Routledge
World Green Building Council (2013) *The Business Case for Green Buildings: A Review of the Costs and Benefits for Developers, Investors and Occupants*, available at www.worldgbc.org/news-media/business-case-green-building-review-costs-and-benefits-developers-investors-and-occupants

10
Green REITs

David Parker

10.0 Introduction

For the purposes of this chapter, it is proposed to refer to those real estate investment trusts (REITs) that have made a significant commitment to sustainable property investment and that may hold a portfolio comprising a major proportion of sustainability-rated office, retail or industrial investment properties as "green REITs".

The history of green REITs spans little more than a decade, starting before the global financial crisis (GFC), with those entities that made a significant commitment to the precursor principles of corporate social responsibility (CSR) and sustainable development. In 2005, Pivo and McNamara (2005) cited eight publicly traded real estate investment companies and trusts around the world that had made such a commitment, including:

- UK: British Land, Land Securities;
- Europe: Klepierre, Wereldhave;
- Australia: Investa Property Group, Commonwealth Property Office Fund; and
- Asia: Mitsubishi Estate, Swire Pacific Ltd.

Newell (2008) added PruPIM and Hermes for the UK and Mirvac and GPT for Australia.

After the GFC, growth in the proportion of sustainability-rated office, retail or industrial properties within existing REITs was exponential, with Newell and Lee (2012) citing 19 leading property investors, including:

- UK PruPIM, Hermes, British Land, Land Securities;
- Europe Unibail-Rodamco, Klepierre, Wereldhave, Castellum;
- Australia GPT, Stockland, Dexus, Lend Lease, Mirvac;
- Asia Mitsubishi Estate, Mitsui Fudosan, CapitaLand, Swire Properties; and
- US Prologis, Hines.

Indeed, by 2012, Newell and Lee observed that:

> Increasingly, sustainability management by corporates is seen as being different from other corporate objectives, with stakeholders expecting it to be done even if it does not

> necessarily deliver enhanced financial performance. This seems to be the situation for AREIT [Australian real estate investment trusts] investors.

It was such that, by the time of writing, sustainable property investment has become embedded within most REITs, effectively having become "business as usual", such that it may be assumed that REITs will either be, or have a commitment to be, a green REIT.

Given that REITs are a massive component of global property investment, such a commitment to sustainable property investment is significant as it potentially comprises a large proportion of the world's prime office, retail and industrial investment property stock. For example, Newell and Lee (2012) note that AREITs are the largest property investment sector in Australia, accounting for 10 per cent of the global REIT market, holding 3,800 institutional-grade commercial properties with over $140 billion in property assets under management and holding over 50 per cent of the institutional-grade commercial property in Australia.

However, despite the vast body of research into sustainability and property, surprisingly little has been written specifically about sustainability in the context of REITs, with most contributions occurring in the post-GFC period. For the purposes of this chapter, that research focused on green REITs will be summarised as follows:

10.1 Green REIT risk–return performance

- 10.1.1 US REIT risk–return performance
- 10.1.2 European REIT risk–return performance
- 10.1.3 UK REIT risk–return performance
- 10.1.4 Asian REIT risk–return performance
- 10.1.5 Australian REIT risk–return performance
- 10.1.6 Global REIT risk–return performance

10.2 Green REIT value
10.3 Green REIT benefits
10.4 Green REIT future
10.5 Summary and conclusions

10.1 Green REIT risk–return performance

Following the GFC, there was a surge in interest in quantitative research into sustainable investment property performance generally, with three seminal papers, all published in 2008, finding generally higher rents and occupancy rates in sustainable investment properties and so providing quantitatively based evidence to support the qualitative benefits identified in the pre-GFC era for sustainable property investment. While not specific to REITs, the three papers comprised Miller, Spivey and Florance (2008), Fuerst and McAllister (2008) and Eichholtz REITs Kok and Quigley (2008), which essentially found higher occupancy rates, higher rents, lower operating expenses and lower capitalisation rates for sustainable buildings. These studies are considered in greater detail elsewhere in this book.

As such, these three seminal papers on sustainable investment property effectively comprised the "business case" for sustainable property investment and provided an understanding at the individual property level that contributed generally to an understanding at the portfolio level, including that of the REIT conceptualised as a portfolio, being supported by later papers including Pivo and Fisher (2010), Eichholtz, Kok and Quigley (2010), Fuerst and McAllister (2011), Szumilo and Fuerst (2013) and Fuerst, Oikarinen, Shimizu and Szumilo (2014). Again, these studies are considered in greater detail elsewhere in this book.

While some REITs held sustainable investment property before the GFC, growth in the proportion of sustainability-rated office, retail or industrial properties within existing REITs after the GFC was exponential, with Eichholtz, Kok and Yonder (2012) observing

> that the number of green-certified properties increased strongly during the recent downturn in the real estate markets, notwithstanding severe capital and liquidity constraints, in line with the findings of [Eichholtz et al. (2010)], who show that in the face of the historically severe downturn, the number of green-certified properties continued to grow exponentially, with no significant change in the financial premiums commanded by "green" properties. REITs seem to be persistent in pursuing the greening of their property portfolios.

In the specific context of green REITs, a relatively small number of published journal papers have investigated the impact of sustainable property investment on green REIT risk–return performance in the US, Europe, the UK, Asia, Australia and globally, which are considered further below.

10.1.1 US REIT risk–return performance

Eichholtz et al. (2012) undertook the first study at the REIT level for the US, investigating the effects of energy efficiency and sustainability of commercial properties on the operating and stock performance of a sample of around 70 US REITs, owning 708 LEED-registered and 919 Energy Star-certified properties as of August 2011, over the period 2000 to 2011, thus providing an insight into the benefits of green buildings in US REITs. Significantly, the authors found no ownership of green buildings by their REIT sample up to 2003, then very low levels up to 2005, then a minor increase up to late 2007, followed by a significant increase to late 2010.

Using longitude and latitude in GIS software, the authors matched data on LEED-registered and Energy Star-certified buildings with detailed information on REIT portfolios and calculated the share of green properties for each REIT, analysing operating performance to find that return on assets, return on equity and the ratio of funds from operations to total revenue are positively related to greenness. Furthermore, the authors found that, while greenness is not related to abnormal performance, those REITs with a higher proportion of green properties displayed significantly lower market betas (posited as relating to lower energy price fluctuation risk and lower occupancy risk and thus less exposure to the business cycle).

The authors noted that previous studies had focused on energy efficiency and sustainability at the level of individual property assets and on the relationship between green certification, cash flows and property valuation. This relationship had generally been found to be positive, with green buildings having higher rents, higher and more stable occupancy rates and higher prices than conventional buildings (Eichholtz et al., 2010; Fuerst and McAllister, 2011).

By focusing on REITs, the authors considered the impact of sustainability at a portfolio level, noting that such impact may be both at the property level (lower operating costs, higher and more stable occupancy levels, higher valuations) and at the entity level (better reputation, greater employee loyalty) contributing to improved financial performance.

In a further study of sustainable US REITs, Yonder (2013) intriguingly analysed the contributions of REIT CEOs to candidates during US federal elections, hypothesizing that those who contribute more to Democrats ("prone to follow environmental policies") are likely to have higher portfolio greenness than those who contribute to Republicans ("generally more conservative people ... reluctant to make new types of investments"). While the author did not find a relationship between the political preferences of REIT CEOs and their portfolio greenness as measured by LEED labels, CEO experience, firm size and locational greenness were found to significantly influence REIT portfolio greenness.

10.1.2 European REIT risk–return performance

Cajias, Geiger and Bienert (2012) undertook the first study at the entity level for Europe, examining 80 European listed real estate companies (including REITs) from 13 countries over the period 2006 to 2009, and found a positive linkage between a green agenda and green performance, including an increased ability to generate revenues and a decreased level of idiosyncratic stock volatility, concluding: "As a result, green commitments are not merely altruisms but are economically driven instead." The authors found a sharp increase in the representation of sustainability criteria in reporting from 30.8 per cent in 2006 to 48.1 per cent in 2007 to 51.9 per cent in 2009, with leading country performances by Sweden in 2006 and 2007, Finland in 2008 and the UK in 2009, and leaders including the UK's Land Securities and Hammerson.

Furthermore, the authors' econometric analyses found that European listed real estate companies that act responsibly, in line with international sustainability guidelines, experienced an increase in current financial performance and an ex post reduction in non-diversifiable risk, except in Greece and the UK, with the authors positing the latter to be a volatile market due to the number of REITs in the sample.

10.1.3 UK REIT risk–return performance

Newell (2009) undertook the first UK study and applied five socially responsible investment performance eligibility criteria to 11 identified UK property companies (rather than only REITs), with a total market capitalisation of £24.4 billion and representing 50 per cent of the UK listed property company market capitalisation, to establish a series of performance indices for the period 2003 to 2008, which were then used to assess the risk-adjusted performance and portfolio diversification benefits of UK socially responsible property companies.

The author found that UK socially responsible property companies delivered superior risk-adjusted returns compared to the overall UK property companies sector, with this performance being achieved with no loss of portfolio diversification benefits. Echoing the virtuous spiral of Lützkendorf (n.d.), Newell noted that establishing the performance aspects of sustainable investment will be a catalyst for further interest in sustainable investment.

10.1.4 Asian REIT risk–return performance

From an entity-level perspective in Asia, Ho, Rengarajan and Lum (2013) investigated the effect of sustainable properties, holding the Singapore Green Mark Award or Certification, on the operational and financial performance of two Singapore listed REITs, holding 100 per cent sustainable office properties and 50 per cent sustainable retail properties, respectively, and a public trust over the period 2007 to 2011.

The authors found that, for the office REIT, higher green ratings had a positive effect on both the operational and financial performance of the REIT, but that the findings were less conclusive for the retail REIT and for the public trust were "not of any significance". While the authors noted that their findings were generally not consistent with those of previous studies, such as Eichholtz, Kok and Quigley (2013), regard should be given to the nature and size of the study sample in Singapore.

10.1.5 Australian REIT risk–return performance

Newell, Peng and Yam (2011) undertook the first study at the REIT level for Australia, investigating CSR practices by AREITs across the environmental, social and governance (ESC) dimensions relative to other Australian Securities Exchange (ASX) sectors. The authors noted that many AREITs produced significant CSR reports and had clearly articulated CSR strategies with significant portfolios of highly rated NABERS and Green Star commercial properties, being key constituents of various global CSR performance measures including FTSE4Good Index, DJWSU and Global 100.

The authors developed three unique CSR AREIT performance indices (environmental, social and governance) and empirically assessed their risk-adjusted performance and portfolio diversification benefits for the period August 2005 to July 2010, finding no significant underperformance relative to conventional AREIT indices and portfolio diversification benefits. Using data for CSR ratings developed by the organisation Corporate Monitor, the authors identified seven AREITs for inclusion in an environmental index, six for inclusion in a social index and four for inclusion in a governance index, with the balance included in the non-environmental, non-social and non-governance AREIT indices, respectively.

Adopting individual AREIT returns for the period August 2005 to July 2010 (the timeframe when CSR was an increasingly important agenda item in AREIT corporate mandates), risk-adjusted returns and inter-asset correlations were assessed. For the environmental indices, the authors found average annual return to be lower (1.04 per cent versus 1.14 per cent indicating "no evidence of significant under-performance") and risk higher (24.88 per cent vs. 20.45 per cent) for the environmental index than for the non-environmental index with a correlation of r=0.74 indicating "some degree of portfolio diversification benefit". The authors acknowledged the limitations of sample size, with the analysis period including the turmoil of the GFC for AREITs.

In a subsequent study, Newell and Lee (2012) continued the analysis of CSR for AREITs by examining the impact of CSR factors and financial factors on the performance of the 16 CSR-rated AREITs in the ASX200 (accounting for 94.6 per cent of the AREIT sector market capitalisation) over the period 2005 to 2010. The authors again used CSR ratings developed by Corporate Monitor in the ESG dimensions with financial factors including size, book to market value, gearing and beta value.

The authors found that the ESG dimensions of CSR showed high correlations reflecting strong commitment by REITs, but were not separately priced by AREIT investors (though may be priced collectively), with corporate governance being the most significant CSR factor, but with most AREIT performance contributed by financial factors.

Alternatively, Siew (2015) continued the investigation of AREITs and the ESG dimensions over the period 2008 to 2014 using Markov chain analysis to predict AREIT behaviour, and found that price movements occurred in a random fashion and that sustainable AREITs did not necessarily show superior performance.

10.1.6 Global REIT risk–return performance

In addition to the various regional studies considered above, Kok, Eichholtz, Bauer and Peneda (2010) undertook a global study of the environmental performance of a sample of 688 listed property companies (including REITs) and private property funds from more than 20 countries in Europe, the US, Asia and Australia. While reporting a mixed response rate between sectors and regions, the authors found that listed property companies showed better environmental performance than private funds, with better environmental performance among larger listed property companies, though overall there was considerable room for improvement with Australian, listed Swedish and listed UK property companies outperforming the rest of the world.

Building on the findings of Kok et al. (2010), Bauer, Eichholtz, Kok and Quigley (2011) further developed the Global Real Estate Sustainability Benchmark (GRESB), finding the Australian REIT GPT to be the global leader with a score of 86 out of 100 and the best-performing US REIT to be Vornado Realty Trust with a score of 55 out of 100. While subsequent annual surveys have shown continuous improvement in scores, the 2015 GRESB results (GRESB, 2015) found the global average score to be only 56 out of 100.

A significant REIT-specific global study was undertaken by Fuerst (2015), analysing a sample of REITs from North America, Europe and Asia over the period 2011 to 2014 and finding that investing in sustainability enhances operational performance and lowers risk exposure and volatility. Interestingly, the author found evidence of "green talk" rather than "green walk", concluding that "Clearly, property companies do not necessarily practice what they preach when it comes to environmental management", with only 10 per cent of respondents classified as "green stars" and 67 per cent of the sample falling into the "green laggards" quadrant.

The author used GRESB – a quantified and multi-dimensional sustainability benchmark (that is more comprehensive than such "eco-labels" as LEED and Energy Star), including $8.9 trillion USD in assets, that has become standard practice for the world's largest real estate investment and asset management companies – to investigate whether the rating provided by GRESB is significantly associated with higher financial performance for a large sample of global REITs. The GRESB REIT universe includes 637 listed property company and private equity real estate company contributors covering 56,000 buildings across 12 countries with an aggregate value of 2.1 trillion USD, with the three largest contributors being the US (27 per cent), the UK (15 per cent) and Japan (12 per cent).

The author found a significant positive relationship between the REITs' overall GRESB scores and both returns on assets and returns on equity, but that the "evidence is less clear-cut" regarding absolute stock market performance; when adjusted for risk, though, a significant link between sustainability and stock market performance is revealed.

The author noted that, while a link exists between sustainability and financial performance, it may not be causal, concluding:

> Overall, the results of this global study suggest that investing in sustainability pays off for REITs both in terms of enhancing operational performance and lowering risk exposure and volatility. However, there remains significant room for improvement in the sustainability performance of REITs.

10.2 Green REIT value

In addition to positively impacting REIT performance as measured by risk and return, sustainable investment property may also positively impact the value of the REIT at the entity level, with such impact being quantitatively measurable in addition to the qualitative benefits identified in the pre-GFC era.

While there has been limited research into the positive impacts on value at the entity level for REITs, extensive research has been undertaken for entities in other sectors. For example, Hart and Ahuja (1996) found positive impacts from pollution abatement for a sample of manufacturing, mining and production entities and Konar and Cohen (2001) found a positive relationship between environmental performance and the intangible asset value of 321 publicly traded firms in the S&P500. Geiger, Cajias and Bienert (2013) noted the following findings from other previous studies:

- Lee and Faff (2009) and Hoepner, Yu and Ferguson (2010) discovered a reduction in sensitivity to market fluctuations for a portfolio with leading ESG stocks.
- Statman and Glushkov (2009) found higher alphas for companies ranking high in the fields of community, employee relations and the environment.
- Cho, Lee and Pfeiffer (2013) found that ESG ratings from indices reduce information asymmetry and that particularly well-informed investors (such as institutional investors) benefit from their information advantage.
- Derwall, Koedijk and Horst (2011), Borgers, Derwall, Koedijk and Horst (2012) and Bebchuk, Cohen and Wang (2013) explained the existence and disappearance of superior returns on the basis of a learning hypothesis, where investors learn to price the difference between indices listing sustainable and non-sustainable companies.

Accordingly, while a link between environmental performance and financial performance at the entity level has been identified for entities in various sectors, relatively limited research has been undertaken to investigate the link between environmental performance and value at the entity level in the REIT sector of the stock market.

Sah, Miller and Ghosh (2013) were the first to investigate the entity-level benefits from strategic initiatives aimed at increasing the ownership of green buildings, using REITs as investors/owners to test if management initiatives result in higher firm value. As recently as 2013, the authors noted that, at the firm level, the question of whether pursuing a green strategy as part of CSR provides any benefits at the operational level resulting in higher yields of capitalised value was yet to be investigated.

Using quarterly data over the period 2009–2010, the authors investigated 18 REITs that voluntarily participated in the US Energy Star partnership program, requiring such REITs to commit to measure, track and benchmark energy performance, to develop a plan to improve energy performance and to educate staff and the public about their partnership status. Using Tobin's q to measure the ratio of a firm's intangible-to-tangible value, the study sought to have regard to such intangible benefits as better corporate practices, improved productivity for employees and superior management at the firm level, in addition to tangible benefits such as lower operating costs and consequential net operating income.

The authors found that partner REITs held a higher proportion of green buildings than non-partner REITs with a higher return on average assets, a positive relationship between an REIT being green and its value (as represented by its Tobin's q) and higher returns from partner REITs than from non-partner REITs.

Cajias, Fuerst, McAllister and Nanda (2014) undertook a wider study comprising 34 US real estate groups, of which 43 per cent were REITs, in ESG dimensions over the period 2003 to 2010. The authors also found positive relationships between Tobin's q and ESG ratings, with REITs often among the weaker of the real estate groups, but driven by ESG concerns rather than strengths.

10.3 Green REIT benefits

In addition to positively impacting REIT performance as measured by risk and return and positively impacting the value of the REIT at the entity level, sustainable investment property may also provide REITs with further benefits including takeover defence, lower cost of capital and separate asset allocation potential.

Zochling and Phipson (2011) investigated sustainable investment property within Australian REITs as a takeover defence post-GFC, observing that significant trading discounts usually prompt takeover opportunities but that such activity was not occurring in the time period considered. From a literature review, the authors identified potential barriers to include obtaining shareholder support, managerial control, concentration of ownership, Foreign Investment Review Board (FIRB) approval and "poison pill" provisions. Interview-based research added principal shareholders, pre-emptive rights, perception of management and access to finance as further barriers. Significantly, the interview-based research into perception of management further identified market perception of the acquirer's inability to manage "A grade and premium grade assets", which presumably included environmentally sustainable office buildings in the takeover target considered, Principal Office Fund, rendering the competence to manage sustainable investment property within an REIT as a potential takeover defence.

Cajias, Fuerst and Bienert (2014a) investigated whether investing in CSR can lower a company's cost of capital or lead to higher market capitalisation for a company, through the analysis of 2,300 listed US companies over the period 2003 to 2010. The authors found that customer-oriented companies (such as telecommunications and automobile) outperformed asset-driven sectors (such as real estate or chemical companies), with "high-responsible firms" diminishing their cost of capital significantly across all sectors.

As sustainable investment properties become increasingly common, their performance becomes measurable and more clearly understood and REITs comprising solely sustainable investment properties emerge, the role of such REITs in portfolio diversification through separate asset allocation comes into focus. De Francesco and Levy (2008) identified asset allocation as a potential issue early in the REIT sustainability research timeline, considering both existing portfolios and new portfolios and suggesting that allocation rules other than popular measures such as risk-adjusted returns may be required.

Geiger et al. (2013) developed a sustainable real estate index comprising listed real estate companies (rather than REITs) following a CSR agenda over the period 2004 to 2010. The authors identified a unique risk–return pattern for sustainable real estate as an asset class, enabling it to be allocated across all portfolios ranging from low to high risk across the efficient frontier. In low-risk portfolios, sustainable real estate was found to act as a diversifier, whereas in medium-to high-risk portfolios, sustainable real estate became the main allocated asset class.

Furthermore, Geiger, Cajias and Fuerst (2014) analysed the effects of socially responsible investing within a multi-asset portfolio optimisation model using listed real estate companies (rather than REITs) with an active sustainability agenda, identified through the MSCI ESG database, to represent the sustainable real estate asset class over the period January 2003 to December 2010. The authors established empirically that, while sustainable real estate does not appear to yield high returns, it is also not exposed to excessively high risk as measured by variance, exhibiting low positive correlations with equities and low negative correlations with bonds and cash, rendering sustainable real estate an attractive asset class for diversification purposes.

Constructing a range of optimal portfolios, the authors found a significant role for sustainable real estate within multi-asset portfolio optimisation with a high allocation percentage in low-return portfolios, but with high-yielding portfolios unlikely to incorporate large shares of sustainable real estate investments. Furthermore, they found that risk-averse investors may benefit from including sustainable real estate as a distinct asset class in their investment decisions with investors seeking medium returns benefitting from sustainable real estate as a diversifier, stating:

> the analysis of [sustainable real estate] leads to the conclusion that a distinct asset class has evolved over the last decade which broadens the investment opportunity set for US and international market participants.

10.4 Green REIT future

The investment by REITs in green office, retail and logistics properties is now commonplace and expected by investors, with the creation of green REITs foreshadowed by Rohde and Lützkendorf (2009), who noted that "the untapped market potential for publicly offered sustainable property investment products is immense", with such REITs as Hines and GPT now fulfilling some of that market potential. The creation of green REITs has the capacity, as Lützkendorf (n.d.) notes, to create a virtuous spiral whereby the development of green REITs creates demand for further sustainable buildings, which leads the green REITs to grow and require further sustainable buildings to grow further and so forth.

While the possibility now exists to extend the REIT as a financing vehicle to other sustainable investment property sectors (such as hospitals and prisons), current REIT regulation and leasing structures in some property sectors create challenges. For example, in the context of business properties such as hotels, various regulatory regimes preclude REITs from operating a business such that the ownership of hotels subject to management contracts to hotel operators is a common occurrence within REITs around the world.

Melissen, van Ginneken and Wood (2016) considered the challenge to the implementation of sustainability in the evolving hotel industry where an "asset-light" approach to the adoption of hotel management contracts is emerging which favours the control-and-manage paradigm, adds additional stakeholders, encompasses stakeholder detachment and compromises alignment, so separating ownership and operation, which hinders sustainable development.

The authors noted that the hotel industry currently remains focused on only those sustainability measures that are needed or financially profitable in the short term with a focus, albeit inconsistently, on energy management and water management. Conversely, the authors contended that REITs are a clear example of the challenge of not only geographical distance from the actual location of the hotel but also of "the mental distance characteristic of certain 'detached' owners" from the actual business of the hotel, which causes the property to be regarded "as 'just' a real estate object generating lease income rather than as an actual business".

Accordingly, it may be contended that, while those REITs owning property housing operating businesses may still have the opportunity to further address sustainability through reshaping their leasing structures, global changes to REIT regulation may be more challenging, which could result in constraints to green REITs in some property sectors.

In addition to green REITs in individual property sectors investing in individual green buildings in different geographies, Mahoney (2012) advocated green REIT investment in residential communities developed on environmentally sustainable principles and characterised by walkability, mixed uses and multiple modes of transit, which, in the long term, will provide investors with the opportunity to earn both a competitive return on investment and help to produce sustainable communities:

> a new sustainable cities REIT, built upon the SRI model, is the most logical vehicle to achieve profit for companies, competitive returns for investors, stronger finance for municipalities, better health for individuals, and a more liveable environment for all.

McKinley (2014) considered the confluence of environmentally positive projects and REITs through the financing for commercial and utility-scale solar power installations. The author posited that REITs may offer a viable public financing structure allowing large-scale investment to take place and the cost of capital to fall through large-scale public capital raising with the benefits of pass-through taxation, established regulation and widespread acceptance by public investors, citing Power REIT, a publicly traded infrastructure REIT, which owns and leases land to entities generating solar power.

Taking the use of REITs to fund environmentally positive projects further, Ladewig (2011) applied traditional real estate principles to the design of a self-sustainable lunar habitat for an 80-person crew for an indefinite period on the moon. The author posited that an REIT would be an effective method for funding such a lunar colony with the habitat projected to be 100 per cent pre-leased prior to completion, based on the demand for research and positioning on the moon, with tenants signing a minimum five-year lease with the option to renew after the third year if their research needs to be continued.

The significant growth of sustainable investment property in existing REITs has taken little over a decade to achieve, which may be contended to be a remarkably brief period of time for such a major change in the property investment industry. With Lützkendorf's virtuous spiral now evident in green REITs, continued increase in sustainable property investment appears assured and may be supplemented by both investments in further property sectors and the combination of REITs as financing vehicles with other environmentally worthwhile property-related projects.

10.5 Summary and conclusions

10.5.1 Summary

By 2016, most of the major property companies and REITs in the UK, Europe, Australia, Asia and the US held portfolios comprising significant proportions of green investment properties, with sustainable property investment being embedded within REIT investment decision making and REIT portfolio construction.

The various studies concerning risk–return performance are summarised in Table 10.1, with the US study finding return on assets, return on equity and the ratio of funds from operations to total revenue to be positively related to greenness and green REITs displaying significantly lower market betas. Similarly, the European study found a positive linkage between a green agenda and green performance with a decreased level of idiosyncratic stock volatility, while the UK study also found superior risk-adjusted returns relative to the overall property companies sector. While the Asian study comprised only a very small sample in one country, the Australian study found the strong commitment to the ESG dimensions of CSR by REITs was not separately priced by AREIT investors, though may be priced collectively. Global studies generally found

Table 10.1 Summary of studies

Region	*No of entities*	*Period*	*Authors*
US	70	2000–2011	Eichholtz et al. (2012)
Europe	80	2006–2009	Cajias et al. (2012)
UK	11	2003–2008	Newell (2009)
Asia (Singapore)	3	2007–2011	Ho et al. (2013)
Australia	16	2005–2010	Newell and Lee (2012)

an increasing trend towards green portfolios with risk–return benefits, though the findings are constrained by the broad profile of the study samples and the study periods.

In addition to risk–return performance benefits for REITs from holding a significant proportion of sustainable investment properties in their portfolios, a positive impact on the value of the REIT at the entity level has also been identified by recent research, together with other benefits such as takeover defence, lower cost of capital and separate asset allocation potential.

Concerning the future, with the role of the green REIT clearly established, though with misalignment constraining sustainable investment in some sectors such as hotels, the potential arises to use the REIT structure as a financing vehicle for environmentally positive projects, potentially overlapping into those areas traditionally considered as infrastructure.

10.5.2 Conclusions

While the GFC was a watershed point for the greening of REITs, following which REIT investment in sustainable investment property grew exponentially, surprisingly little research has been undertaken into green REITs. Studies are found to exist in each major region of the world and at a global level, but these are generally not recently undertaken and are generally for relatively short time periods, sometimes overlapping the GFC, which may have a significant effect on findings.

However, the research that has been undertaken generally finds green to be good for REITs with positive risk–return performance outcomes, positive REIT value outcomes and other positive benefits for REITs. As it is now almost a decade since the advent of the GFC and with most of the major property companies and REITs in the UK, Europe, Australia, Asia and the US now holding portfolios comprising significant proportions of green investment properties, the scope for further research into green REITs in the future continues to increase.

From the research conducted to date, there is no logical investment reason for REITs not to invest in sustainable investment properties. Rather than seeing the creation of specialist green REITs occur, as often mooted in the past, it may be anticipated that existing REITs will continue to acquire or develop further sustainable investment properties as a matter of normal business, with non-green properties either sold or redeveloped to become green. In less than a decade, the REIT sector has made massive progress in becoming green, which may be expected to at least continue, if not accelerate, into the future.

References

Bauer, R, Eichholtz, P, Kok, N and Quigley, J 2011, "How green is your property portfolio? The global real estate sustainability benchmark", *Rotman International Journal for Pension Management*, vol. 4, pp. 34–43.

Bebchuk, L, Cohen, A and Wang, C 2013, "Learning and the disappearing association between governance and returns", *Journal of Financial Economics*, vol. 108, no. 2, pp. 323–348.

Borgers, A, Derwall, J, Koedijk, K, and Horst, J 2012, "Stakeholder relations and stock returns: on errors in investors' expectations and learning", PRI-CBERN Academic Conference working paper.

Cajias, M, Fuerst, F and Bienert, S 2014, "Can investing in corporate social responsibility lower a company's cost of capital?", *Studies in Economics and Finance*, vol. 31, no. 2, pp. 202–222.

Cajias, M, Geiger, P and Bienert, S 2012, "Green agenda and green performance: empirical evidence for real estate companies", *Journal of European Real Estate Research*, vol. 5, no. 2, pp. 135–155.

Cajias, M, Fuerst, F, McAllister, P and Nanda, A 2014, "Do responsible real estate companies outperform their peers?", *International Journal of Strategic Property*, vol. 18, no. 1, pp. 11–17.

Cho, S, Lee, C and Pfeiffer, R 2013, "Corporate social responsibility performance and information asymmetry", *Journal of Accounting and Public Policy*, vol. 32, no. 1, pp. 71–83.

De Francesco, AJ and Levy, D 2008, "The impact of sustainability on the investment environment", *Journal of European Real Estate Research*, vol. 1, no. 1, pp. 72–87.
Derwall, J, Koedijk, K and Horst, J 2011, "A tale of values-driven and profit-seeking social investors", *Journal of Banking and Finance*, vol. 35, pp. 237–247.
Eichholtz, P, Kok, N and Quigley, J 2008, *Doing well by doing good? Green office buildings*, Working Paper, UC Berkeley, Berkeley, CA.
Eichholtz, P, Kok, N and Quigley, J 2010, *The dynamics of green building*, Working Paper, UC Berkeley, Berkeley, CA.
Eichholtz, P, Kok, N and Quigley, J 2013, "The economics of green building", *The Review of Economics and Statistics*, vol. 95, no. 1, pp. 50–63.
Eichholtz, P, Kok, N and Yonder, E 2012, "Portfolio greenness and the financial performance of REITs", *Journal of International Money and Finance*, vol. 31, no. 7, pp. 1911–1929.
Fuerst, F 2015, *The financial rewards of sustainability: A global performance study of real estate investment trusts*, University of Cambridge, Cambridge.
Fuerst, F and McAllister, P 2008, "Pricing sustainability: an empirical investigation of the value impacts of green building certification", ARES Conference, April 2008.
Fuerst, F and McAllister, P 2011, "Green noise or green value? Measuring the effects of environmental certification on office values", *Real Estate Economics*, vol. 39, no. 1, pp. 45–69.
Fuerst, F, Oikarinen, E, Shimizu, C and Szumilo, N 2014, *Measuring "green value": An international perspective*, RICS Research, London.
Geiger, P, Cajias, M and Bienert, S 2013, "The asset allocation of sustainable real estate: a chance for a green contribution?", *Journal of Corporate Real Estate*, vol. 15, no. 1, pp. 73–91.
Geiger, P, Cajias, M and Fuerst, F 2014, *A class of its own? The role of sustainable real estate in a conditional value at risk multi-asset portfolio*, International Real Estate Business School, Regensburg and University of Cambridge, Cambridge.
GRESB 2015, "2015 GRESB results", www.gresb.com/2015-gresb-results-global-real-estate-industry-significantly-reduces-carbon-footprint, accessed 3 August 2016.
Hart, SL and Ahuja, G 1996, "Does it pay to be green? An empirical examination of the relationship between emission reduction and firm performance", *Business Strategy and the Environment*, vol. 5, no. 1, pp. 30–37.
Ho, KH, Rengarajan, S and Lum, YH 2013, "'Green' buildings and real estate investment trust (REIT) performance", *Journal of Property Investment & Finance*, vol. 30, no. 6, pp. 545–574.
Hoepner, A, Yu, PS and Ferguson, J 2010, *Corporate social responsibility across industries: When can who do well by doing good?*, SSRN working paper.
Kok, N, Eichholtz, P, Bauer, R and Peneda, P 2010, *Environmental performance: A global perspective on commercial real estate*, European Centre for Corporate Engagement, Maastricht.
Konar, S and Cohen, MA 2001, "Does the market value environmental performance?", *The Review of Economics and Statistics*, vol. 83, no. 2, pp. 281–289.
Ladewig, DM 2011, *Lunar real estate development*, ASCE Conference Paper, Fall 2011, Pasadena.
Lee, D and Faff, R 2009, "Corporate sustainability performance and idiosyncratic risk: a global perspective", *Financial Review*, vol. 44, no. 2, pp. 213–237.
Lützkendorf, T n.d. "How to break the vicious circle of blame? The contribution of different stakeholders to a more sustainable built environment", www/fec.unicamp.br/~parc, accessed 14 May 2016.
Mahoney, KP 2012, "An investment product for the times: the case for creating a real estate investment trust that fosters sustainable cities", unpublished Master's thesis, Georgia Institute of Technology, GA.
McKinley, B 2014, "Using REITs to invest in utility-scale solar projects", *Cornell Real Estate Review*, vol. 12, no. 10, pp. 60–70.
Melissen, F, van Ginneken, R and Wood, RC 2016, "Sustainability challenges and opportunities arising from the owner–operator split in hotels", *International Journal of Hospitality Management*, vol. 54, pp. 35–42.
Miller, N, Spivey, J and Florance, A 2008, "Does green pay off?", *Journal of Real Estate Portfolio Management*, vol. 14, no. 4, pp. 385–400.
Newell, G 2008, "The strategic significance of environmental sustainability by Australian listed property trusts", *Journal of Property Investment & Finance*, vol. 26, no. 6, pp. 522–540.
Newell, G 2009, "Developing a socially responsible property investment index for UK property companies", *Journal of Property Investment & Finance*, vol. 27, no. 5, pp. 511–521.

Newell, G and Lee, CL 2012, "Influence of the corporate social responsibility factors and financial factors on REIT performance in Australia", *Journal of Property Investment & Finance*, vol. 30, no. 4, pp. 389–403.
Newell, G, Peng, HW and Yam, S 2011, "Assessing the linkages between corporate social responsibility and A-REIT performance", *Pacific Rim Property Research Journal*, vol. 17, no. 3, pp. 370–387.
Pivo, G and Fisher, J 2010, "Income, value and returns in socially responsible office properties", *Journal of Real Estate Research*, vol. 32, no. 3, pp. 243–270.
Pivo, G and McNamara, P 2005, "Responsible property investing", *International Real Estate Review*, vol. 8, no. 1, pp. 128–143.
Rohde, C and Lützkendorf, T 2009, "Step-by-step to sustainable property investment products", *Journal of Sustainable Real Estate*, vol. 1, no. 1, pp. 227–240.
Sah, V, Miller, NG and Ghosh, B 2013, "Are green REITs valued more?", *Journal of Real Estate Portfolio Management*, vol. 19, no. 2, pp. 169–177.
Siew, RYJ 2015, "Predicting the behaviour of Australian ESG REITs using Markov chain analysis", *Journal of Financial Management of Property and Construction*, vol. 20, no. 3, pp. 253–267.
Statman, M and Glushkov, D 2009, "The wages of social responsibility", *Financial Analysts Journal*, vol. 65, no. 4, pp. 47–59.
Szumilo, N and Fuerst, F 2013, "The operating expense puzzle of US green office buildings", *Journal of Sustainable Real Estate*, vol. 5, no. 1.
Yonder, E 2013, "REIT investment decisions: governance, behaviour and sustainability", unpublished doctoral thesis, Maastricht University, Netherlands.
Zochling, H and Phipson, M 2011, *REIT takeovers: An evaluation of barriers to activity*, 17th Annual Pacific Rim Real Estate Society Conference, 16–19 January 2011, Gold Coast.

11

US green REITs

Vivek Sah and Norm Miller

11.0 Introduction

The investment category "socially responsible investment" pertains to investment criteria revolving around three fundamental factors, namely, environmental, social, and governance. Figure 11.1 shows the estimated size of these investments in 2016. As per the U.S. SIF Foundation's *Report on U.S. Sustainable, Responsible and Impact Investing Trends 2016*, these investments have grown by 33 per cent since 2014 (U.S. SIF Foundation, 2016). Part of this universe consists of real estate funds, both publicly listed and private, that invest in real estate assets that are "green". A label of "green" might suggest incorporating sustainable aspects of real estate in buildings that could vary from smart systems that save energy costs, to those that reduce carbon emissions or conserve water, to using materials in construction that are environmentally friendly. In the same vein, buildings around the world have been using various ratings systems designated to represent the sustainable aspects of the building and these also could be used to label a building as green. The Energy Star label, launched in 1992 by the Environment Protection Agency (EPA) in the U.S., is one of the oldest in the world.[1] Similarly, BREEAM, launched

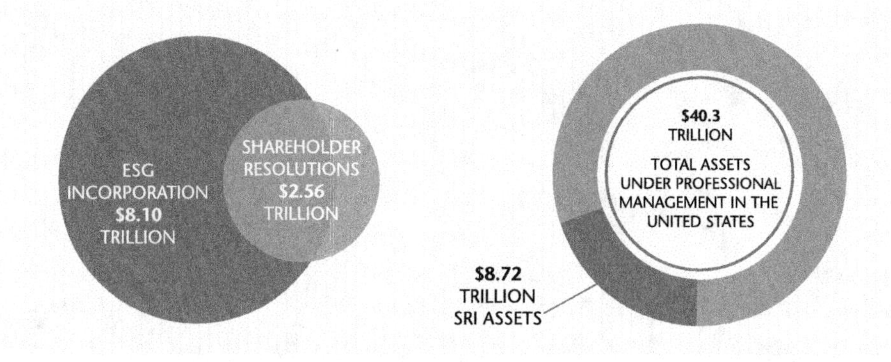

Figure 11.1 Size of sustainable, responsible, and impact (SRI) investing, 2016

Source: U.S. SIF Foundation (2016)

in 1990, is the world's first sustainability assessment method for buildings and enjoys 80 per cent market share in Europe (BREEAM, 2016).

However, in the last decade or so, the U.S. Green Building Council (USGBC) has promoted the Leadership in Energy and Environmental Design (LEED) certification, a more stringent requirement for buildings than Energy Star. As per the USGBC, "LEED is the most widely used third-party verification for green buildings, with around 1.85 million square feet being certified daily" (USGBC, 2016). U.S.-based real estate funds are increasingly moving towards LEED-certified properties, especially when it comes to new construction/development (Nelson, Rakau and Dörrenberg, 2010). However, ascertaining the volume and percentage of a portfolio of such green buildings for real estate investment trusts (REITs) is a big challenge. One of the problems in identifying public REITs as green is that they are striving to be "green" (higher proportion of sustainable properties, which could be any type of sustainable standard such as Energy Star or LEED), but using fuzzy reporting standards.

Table 11.1 List of REITs reporting sustainability measures

Company name	Industry	Priority	Energy	Water	Waste	GHG emissions
American Assets Trust, Inc.	Diversified	Moderate	✓	✓		
AvalonBay Communities, Inc.	Residential	High	✓	✓	✓	✓
Boston Properties, Inc.	Office	High	✓	✓	✓	✓
Cousins Properties, Inc.	Diversified	Moderate				✓
CubeSmart	Self storage	Low			✓	
DDR Corp.	Retail	Low	✓		✓	
DiamondRock Hospitality Company	Lodging	Low	✓	✓	✓	✓
Digital Realty Trust, Inc.	Diversified	Moderate	✓			
Douglas Emmett, Inc.	Diversified	Moderate		✓	✓	
Equity One, Inc.	Retail	Moderate	✓		✓	
Extra Space Storage, Inc.	Self storage	Moderate				✓
Federal Realty Investment Trust	Retail	Moderate				✓
First Potomac Realty Trust	Office	Moderate		✓		
Franklin Street Properties Corporation	Office	Moderate	✓		✓	
General Growth Properties, Inc.	Retail	High	✓	✓	✓	✓
HCP, Inc.	Health care	High	✓	✓	✓	✓
Welltower, Inc.	Health care	Moderate				✓
Hersha Hospitality Trust	Lodging	High	✓	✓	✓	
Kilroy Realty Corporation	Office	High	✓	✓	✓	
Kimco Realty Corporation	Retail	High	✓	✓	✓	✓
Macerich Company	Retail	High	✓	✓	✓	✓
Mack-Cali Realty Corporation	Diversified	High				✓
Parkway Properties, Inc.	Office	High	✓	✓	✓	✓
Prologis, Inc.	Industrial	Moderate	✓			✓
Public Storage	Self storage	Moderate	✓			
Regency Centers Corporation	Retail	High	✓	✓		✓
Simon Property Group, Inc.	Retail	Moderate	✓	✓	✓	✓
SL Green Realty Corporation	Office	Moderate	✓		✓	✓
UDR, Inc.	Residential	Moderate	✓			
Ventas, Inc.	Health care	Moderate	✓	✓	✓	
Vornado Realty Trust	Diversified	High	✓	✓	✓	✓
Weingarten Realty Investors	Retail	Moderate	✓	✓	✓	✓

Source: Vosilla, Behrendt and Hanson (2016)

As per a recent study titled *State of the Industry: Sustainability Reporting in the REIT Sector – 2016 Update* (Vosilla et al., 2016), the number of REITs providing quality, detailed information on green building practices and issues regarding sustainability is still small. The study was a follow-up to an earlier study, which was published in 2013. The 2013 study (Boundy and Poole, 2013) found a large amount of inconsistencies in reporting existing building certification standards, such as LEED or Energy Star. The 2016 study analysed 119 companies and found that only 55 REITs out of the sample published any information about their sustainability efforts. This represented 46 per cent of the overall sample studied in 2016, which was a significant increase from the 25.6 per cent reported in the first study. The study further found that reporting scope and quality differed widely from one REIT to the next. Lastly, the study found that multiple sources gave information on the REITs' sustainability effort. The authors found that the least amount of information was available from websites, followed by newsletters, and then reports, which were typically the most comprehensive. The 2016 study also reported disclosures included in different listed REITs' sustainability reports. Table 11.1 displays the list of the REITs as given in the study. The evolution of the green REIT sub-sector has been slow, like most other new developments in the financial markets. However, the growth is now strong, as reported in the above-mentioned study (Vosilla et al., 2016). As more REITs adopt strategies that promote sustainable measures both in development and operations, the reporting standards will become more transparent and streamlined across the sector. Going forward in the next five to ten years, we should expect to see 'green REITs' becoming the norm and not the exception as they are today.[2]

11.1 The benefits of green REITs

An obvious consideration for the management of a public REIT is the benefits of investing and/or developing sustainable properties. Similarly, the stakeholders may question the rationale behind such a strategy, especially given the low penetration of such investments/properties in the current stock of U.S. (and worldwide) commercial real estate. Given the relatively high cost of construction and/or acquisition of properties that are either some level of LEED-certified or are pursuing other smart management strategies to reduce operating costs and carbon emissions, owners are at the nascent stage of figuring out the perceived or actual benefits of sustainable investments in the commercial real estate sector. That is one of the main reasons why such initiatives are restricted to large public REITs that either have a significant portfolio of properties and/or have high-quality tenants that demand space in LEED-certified properties (which is now often part of their corporate social responsibility (CSR) initiative).

Two metrics for making such green investments would be returns and risk, as with any other investment. One can argue that over a longer holding period, such investment could systematically provide higher returns, specifically through higher income generation and reduced operating expenses. There could also be potential for higher valuations going forward, especially when rent premiums become significant and get capitalized in the value. However, such value-based returns would only be realized over longer investment periods due to the nature of the payoffs from green buildings.

Another potential benefit that would attract the investment community would be lower volatility of returns with such investments. If the profile of tenants renting space in green buildings, especially with higher sustainability ratings (such as LEED Silver and above), is mostly superior to that of an average tenant in a building, the risk profile of such buildings and hence of such a portfolio would be lower than average. If such a portfolio were to be rated, it can be argued reasonably that it may achieve a higher risk-based rating as compared to a non-green portfolio, assuming everything else is equal.

Lastly, some investors (especially socially responsible ones) may be willing to pay a premium for REITs that differentiate themselves from others due to their green strategy. This reputation or brand could warrant a premium in itself, as is the case, for example, for Nike, Coca-Cola, or Apple. This will be over and beyond the operational advantages/efficiencies of such REITs.[3]

The USGBC claims that LEED-certified buildings have lower operating costs while the Energy Star program claims that buildings with the Energy Star label tend to consume 35 per cent less energy and emit 35 per cent less carbon dioxide than average uncertified buildings. Various studies by Miller, Spivey, and Florance (2008), Wiley, Benefield, and Johnson (2010), Eichholtz, Kok, and Quigley (2010), and Fuerst and McAllister (2011) assessed the rents and transaction prices of LEED- and Energy Star-certified office buildings, relative to non-certified, comparable buildings in the U.S., and found rent premiums in the range of 5 per cent to 10 per cent for LEED-certified office properties. Such studies also found 3 per cent higher rents for Energy Star-rated buildings and reported increments for transaction prices of between 11 per cent and 19 per cent depending upon the rating standard (Energy Star or LEED). In a recent survey by the National Multifamily Housing Council, survey respondents said they would be willing to pay an extra $32.64 a month in rent to live in an apartment building that has earned a "green building" certification, such as an LEED (Anderson, 2016). Three-quarters of all respondents – 75 per cent – said they were "interested" or "very interested" in these "green" certifications. The certifications can even be the deciding factor for some potential renters. As investment strategies related to sustainable and green properties are relatively new, there is little empirical evidence to reveal the decision-making process. For the public REITs domain, only a handful of studies have tried to empirically prove the relevance and significance of green investing.

11.1.1 Empirical evidence of performance of green REITs

In 2012, a study titled "Portfolio greenness and the financial performance of REITs" (Eichholtz, Kok, and Yonder, 2012) examined the effects of the energy efficiency and sustainability of commercial properties on the operating and stock performance of a sample of U.S. REITs. The authors used data on LEED- and Energy Star-certified buildings and matched those with information on REIT portfolios. They then calculated the share of green properties for each REIT over the period from 2000 to 2011. Figures 11.2 and 11.3 show the share of green buildings in the REIT portfolios analysed in the study. The two figures show the two different metrics currently utilized in the U.S. property industry. In the sample, there were 708 LEED-registered properties owned by REITs as of August 2011, while 71 REITs owned an aggregate of 919 Energy Star-certified properties. This difference is expected due to higher costs, efforts, and stringent measuring standards required for LEED certification as compared to Energy Star certifications.[4]

The study found that green portfolios are positively related to operating performance. Specifically, if an REIT increases the share of green properties in its portfolio by 1 per cent, its return on assets increases by around 3.5 per cent for LEED-certified properties and by 0.31 per cent for Energy Star-certified properties. Also, the study found that by increasing the portfolio exposure to green properties by 1 per cent, the return on equity increases by 7.39 to 7.92 per cent for LEED-certified properties and by 0.66 per cent for Energy Star-certified properties. The significant difference between the value added by LEED-certified properties over Energy Star-certified buildings is largely due to the stringent measures and high cost (which gets captured in higher rents and/or higher valuations on sale) of attaining LEED certifications.

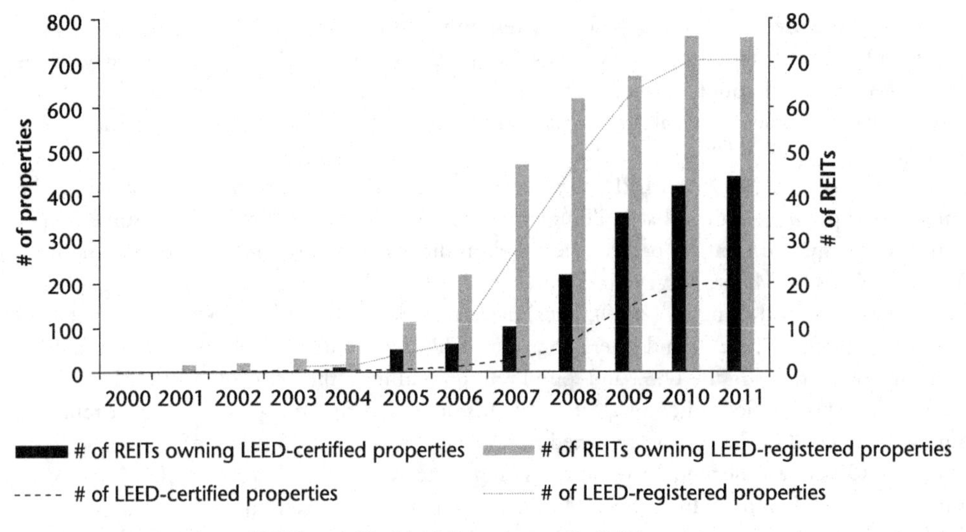

Figure 11.2 Number of LEED-certified buildings in public REITs

Source: Eichholtz et al. (2012)

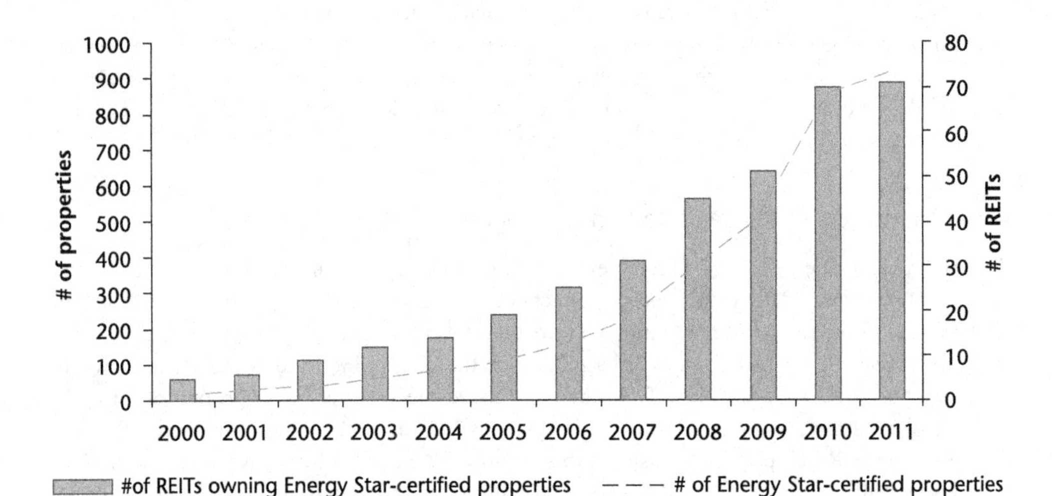

Figure 11.3 Number of Energy Star-certified buildings in public REITs

Source: Eichholtz et al. (2012)

For REITs, one of the important metrics followed by analysts is fund flow from operations (FFO). This is a measure of free cash flow, which is relevant for REITs as they have to distribute 90 per cent of their income as dividends to maintain their REIT status (largely meaning no corporate tax). The authors of this study also examined the impact of greenness on FFO. They found that a 1 per cent increase in certified properties within a portfolio raises the ratio between FFO and total revenue by 17 to 25 per cent for LEED-certified buildings and by 2 to 7 per cent for Energy Star-certified properties.

Lastly, the study also analysed the impact on REIT performance. Here, the authors did not find any significant relationship between predicted greenness and abnormal returns. However, they did find that predicted measures of portfolio greenness negatively affect market betas. Specifically, the study found that a 1 per cent increase in green properties decreases market beta by 0.14 and by 0.01 to 0.03 for LEED-certified properties and Energy Star-certified properties respectively. The authors attributed this to the fact that green properties are less exposed to energy price fluctuations and occupancy risks. This, to a large extent, affects the REITs that hold these properties in their portfolios and hence are less exposed to these impacts and idiosyncrasies of the business cycle.

The second documented evidence of superior performance comes from a study by Sah, Miller and Ghosh in 2013 titled "Are green REITs valued more?" The study was the first to analyse the link between following a sustainable strategy at the corporate level and the financial benefits that accrue from such initiatives (one of the CSR initiatives of a number of public REITs). As discussed previously, a formal and industry-acceptable-defined categorisation of green REITs does not exist. To get around this problem, the authors of this study used the Energy Star Partnership Program as a proxy for REITs that are targeting green real estate investments. This is sort of a signal to the investor community about the REITs' efforts to position themselves as a responsible corporate citizen and a "Green Fund". Under this program, REITs or other companies can become a partner by submitting a partnership letter. Partners and others are provided a set of valuable resources to improve the energy efficiency of their properties. Although the program is voluntary, the companies do commit to measure, track, and benchmark energy performance as well as to educate staff and the public about their partnership status. They also commit to developing a plan to improve energy performance. While there is no monitoring mechanism in the program, we can argue that the companies that participate would be more likely to follow a green strategy and proactively acquire and develop sustainable properties. This association with the program may also add to their investor image and result in enhanced value or maybe returns. While investing responsibly is an intangible benefit, it could add strength to the REITs' CSR case.

The objective of the REIT study was to find empirical evidence of any superior financial or operating performance of the by targeting such strategic initiatives at the firm level. The list of REITs that have signed up as partners under the program, which was launched in 1999, is available on Energy Star's website (Energy Star, 2016). There are over 100 REITs that have declared themselves as partners under this program since its launch. However, due to the absence of a monitoring system, REITs fall out of this programme as partners after their initial declaration/commitment. The study used data from 2009 to 2010 to analyse differences in key performance measures between partner and non-partner REITs. Due to unavailability of data and loss of partner REITs over the sample period, the final sample contained 18 partner REITs and 49 non-partner REITs. Table 11.2 shows the list of the partner and non-partner REITs used in this study. Key performance measures used in the study were Tobin's q, return on assets, and abnormal returns. Tobin's q is interpreted as the ratio of a firm's intangible to tangible value. The study tried to capture this intangible. The authors argued that the benefits arising from investing in "green buildings" are intangible, such as better corporate practices, improved productivity for employees, and superior management and strategic initiatives such as the Energy Star program, etc. It is essential that Tobin's q captures these intangibles. Such practices may result in lower operating costs, which directly affect the net operating income of REITs.

The study's results provide favorable evidence for pursuing a green strategy. The study found evidence of a positive relationship between green partner REITs and their value as represented by their Tobin's q. The authors also found that REITs that committed to the Energy Star Partnership Program have a higher return on assets than their less green peers. When the authors measured stock returns for the REIT's shareholders, they found that greener firms produced a

Table 11.2 List of partner and non-partner REITs

Non-Partner REITs	Partner REITs
Acadia Realty Trust	Brandywine Realty Trust
Agree Realty Corporation	Corporate Office Properties Trust
Alexander's, Inc.	Cousins Properties, Inc.
Alexandria Real Estate Equities, Inc.	Douglas Emmett, Inc.
AMB Property Corporation	Duke Realty Corporation
BioMed Realty Trust, Inc.	First Industrial Realty Trust, Inc.
Boston Properties, Inc.	Government Properties Income Trust
Brookfield Office Properties	Highwoods Properties, Inc.
Cedar Shopping Centers, Inc.	Kilroy Realty Corporation
Cogdell Spencer, Inc.	Liberty Property Trust
CommonWealth REIT	Mack-Cali Realty Corporation
DCT Industrial Trust, Inc.	Parkway Properties, Inc.
Developers Diversified Realty Corporation	Prologis, Inc.
Digital Realty Trust, Inc.	PS Business Parks, Inc.
DuPont Fabros Technology, Inc.	Simon Property Group, Inc.
EastGroup Properties, Inc.	SL Green Realty Corporation
Equity One, Inc.	Vornado Realty Trust
Essex Property Trust, Inc.	Washington Real Estate Investment Trust
Extra Space Storage, Inc.	
Federal Realty Investment Trust	
First Potomac Realty Trust	
Franklin Street Properties Corporation	
General Growth Properties, Inc.	
Gladstone Commercial Corporation	
Glimcher Realty Trust	
Gyrodyne Company of America, Inc.	
Healthcare Realty Trust Incorporated	
Inland Real Estate Corporation	
Investors Real Estate Trust	
Kite Realty Group Trust	
Lexington Realty Trust	
Mission West Properties, Inc.	
Monmouth Real Estate Investment Corporation	
MPG Office Trust, Inc.	
National Retail Properties, Inc.	
Nationwide Health Properties, Inc.	
Pacific Office Properties Trust, Inc.	
Pennsylvania Real Estate Investment Trust	
Public Storage	
Ramco-Gershenson Properties Trust	
Realty Income Corporation	
Regency Centers Corporation	
Saul Centers, Inc.	
Tanger Factory Outlet Centers, Inc.	
Taubman Centers, Inc.	
Urstadt Biddle Properties, Inc.	
U-Store-It Trust	
Weingarten Realty Investors	
Winthrop Realty Trust	

Source: Sah et al. (2013)

Table 11.3 REIT stock's return

	Abnormal returns	
Model	Partner REITs	Non-partner REITs
CAPM	0.018*	0.0001
Fama–French	0.014**	0.0026
Carhart	0.0093***	0.0045***

Source: Sah et al. (2013)

Notes:
*Significant at the 1% level.
**Significant at the 5% level.
**Significant at the 10% level.

higher annual return (5.68 per cent more) than their non-green peers from 2005 to 2010. Table 11.3 shows the REIT's performance based on different models used in the study.

The last and most recent study in this domain is titled "The financial rewards of sustainability: A global performance study of real estate investment trusts" (Fuerst, 2015). Unlike the previous two studies, which focused on the U.S. REIT sector, this study was the first to look at a sample of REITs from North America, Asia, and Europe. Just as each of the previous studies had a different way to measure the sustainability of a real estate portfolio, this study devised its own metric as well. Due to the lack of a uniform and universal standard for sustainability, and in order to better understand environmental, social, governance, and energy risks, several large pension funds launched the Global Real Estate Sustainability Benchmark (GRESB) in 2009. The role of GRESB is to provide a quantified and multi-dimensional sustainability benchmark, and it endeavors to become standard practice for the world's largest real estate investment and asset management companies. The GRESB rating system works as follows. GRESB gathers survey data at the portfolio level for both listed companies and private funds that invest directly in real estate. The survey is administered through its online portal to property companies and private funds between April and July every year and covers information on three aspects of the company: environmental, social, and governance. The information is reviewed continuously, and the results of the survey are published in September. To increase the participation rate of entities, survey respondents obtain a detailed scorecard, which allows them to gauge their individual sustainability performance against the GRESB universe. Some critics of GRESB argue that the system may be gamed, but the same can be said of LEED early on as well as BREEAM. One of the ways of gaming the system for such certifications could be to focus on improving those aspects of the scoring system that are least time- and resources-consuming but have a significant impact on the scores.

There are seven sustainability aspects on which the survey questions are based. Table 11.4 shows these seven aspects and their respective weights in the rating system. They make up the total GRESB score, which is scaled 0 to 100. The total GRESB score can be divided into two dimensions: (1) Implementation & Measurement, and (2) Management & Policy. These two dimensions reflect the role of sustainability in an organization's structure and portfolio of real estate assets. There is also an optional eighth aspect for participants with significant development activities (New Construction and Major Renovations), which is not included in the overall GRESB score to preserve the comparability of scores across all participants. By 2014, within five years of its launch, GRESB benchmarked companies with assets worth $8.9 trillion USD.[5]

Table 11.4 The GRESB rating's seven aspects

Survey aspect	*Absolute points*	*Weight (%)*
Management	12	9
Policy and Disclosure	14	10
Risks and Opportunities	16	12
Monitoring and EMS	13	9
Performance Indicators	33.5	24
Building Certification	15	11
Stakeholder Engagement	35	25

Source: GRESB (2016)

Each entity that is rated receives a sustainability rating based on a series of metrics and is placed in one of four possible performance categories: Green Starters, Green Walk, Green Talk, and Green Stars. The entities that receive the highest ratings are placed in the Green Stars quadrant. Figure 11.4 shows the dispersion of the entities in these quadrants.

Fuerst's study utilized these GRESB scores to test their relationship with return on assets (ROA) and return on equity (ROE) measures of REITs. Furthermore, as with its preceding studies, the author tested whether higher GRESB scores positively affected the stock performance as measured by total returns, alphas, and betas of REITs. The dataset used in the study contained 442 detailed sustainability ratings for REITs over the period from 2011 to 2014. Table 11.5 displays the GRESB scores of the REITs analysed in the study.

The results from this study show that ROA increased by roughly 1.26 per cent to 1.33 per cent for each 1 per cent increase in the GRESB score. ROE, on the other hand, increased by 3.29 per cent to 3.49 per cent for each 1 per cent increase in the GRESB score.[6] Lastly, on stock

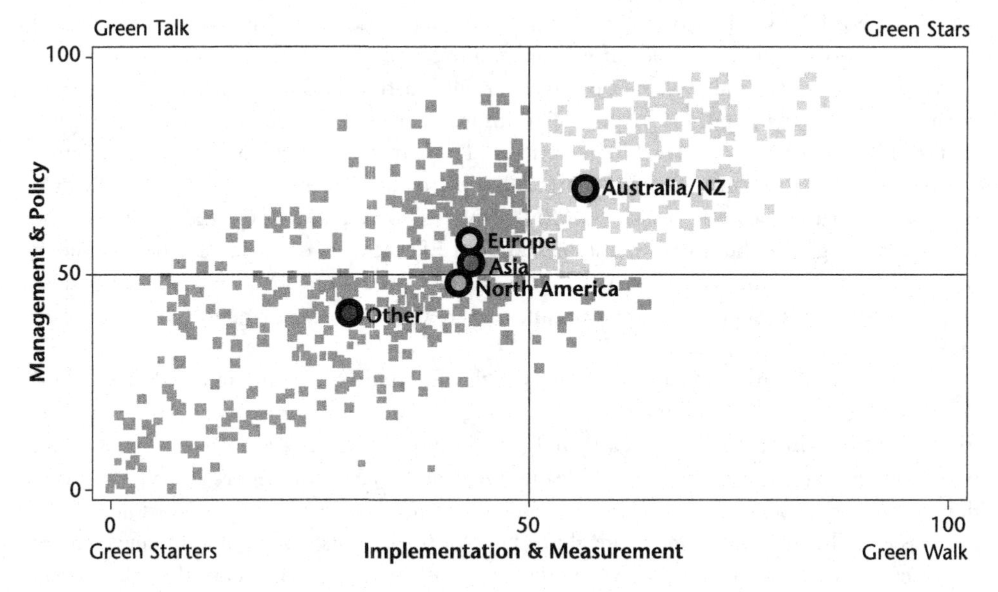

Figure 11.4 Performance categories of benchmarked firms

Source: GRESB (2016)

Table 11.5 Scores on GRESB variables of REITs

Variable	*Mean*	*Standard deviation*
Management	67.35	23.44
Policy Disclosure	51.13	26.99
Risks/Opportunities	64.40	21.92
Monitoring	50.77	27.14
Performance	33.76	23.61
Building	31.12	26.06
Stakeholder	48.17	18.78
New Construction	30.13	26.97
GRESB score	46.40	18.97
Management Policy	58.61	20.82

Source: Fuerst (2015)

performance measures, the study found that REITs with higher GRESB ratings consistently generated higher alphas. However, with stricter controls on the model, the rating results did not hold any significance. Further, there was no evidence of higher GRESB scores resulting in higher betas, a measure of higher risk.

It is evident from the handful of studies in this domain that there is a reasonable case for sustainable or green listed/public funds. Quite a few performance measures used in the studies discussed above show such funds' superiority over their ordinary/non-green counterparts. As more and more REITs increase their exposure to green properties, there will be more data at the fund/firm level to give researchers additional opportunities to make a case for their positive findings. These results will play a key role in determining whether these green REITs are here to stay in the long term or are just a passing fad, without much positive impact on either shareholder wealth or asset performance. So far, greener funds do have a point in their favor, though.

11.2 What are REITs doing to be green?

As mentioned previously, even though the number of REITs that are going green is still very low, there has been a significant increase in awareness among REITs towards implementing, maintaining, and developing sustainable properties. Cost and time remain the two big constraints, which is why the push to incorporate sustainability measures and goals to achieve greener standards has been restricted particularly to those that have a larger property footprint and/or need to attract large corporate clients that will otherwise not lease space in properties without high sustainable standards.[7] Some of the sustainable objectives are also part of the CSR guidelines laid by REITs. Kilroy Realty and AvalonBay are two of the larger REITs in the sector that publish an annual sustainability report (Nareit, 2017; Real Foundations/Nareit, 2014). The reports highlight not only their progress from the previous year but also include snapshots of sustainability measures in their existing or upcoming projects. Figure 11.5 shows the four stages of the sustainability strategy implemented by Kilroy Realty as given in their 2015 sustainability report (Kilroy Realty, 2016).

Large REITs are utilizing a three-pronged strategy for sustainability. These revolve around existing buildings, new development, and stakeholders. When it comes to existing

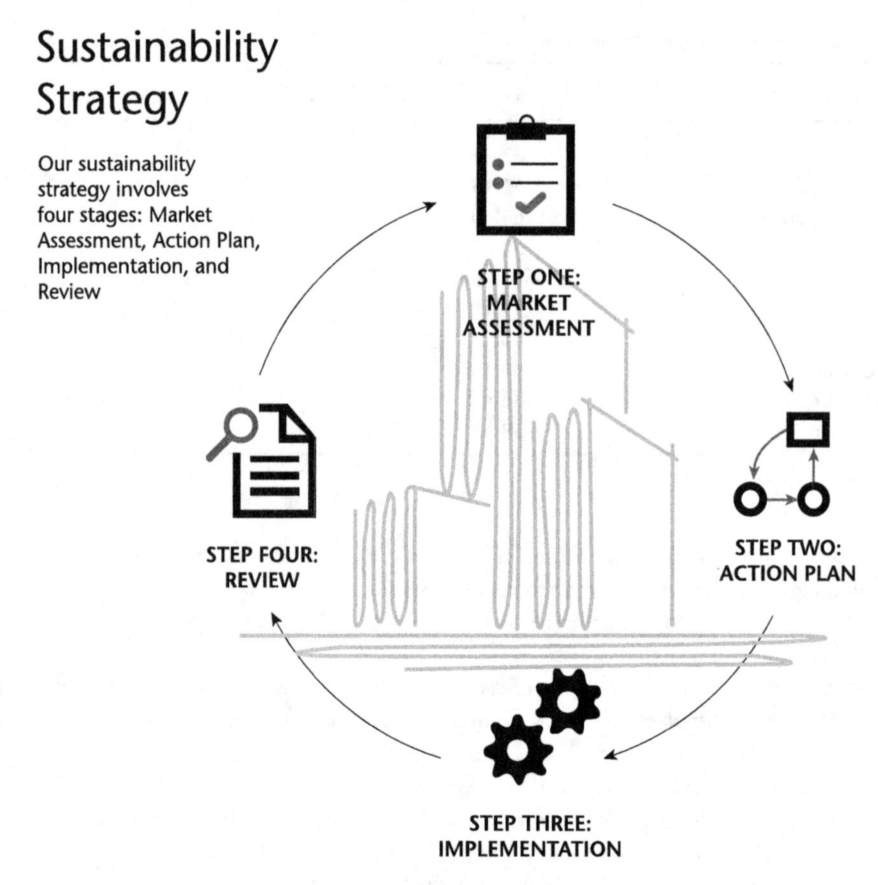

Figure 11.5 Kilroy Realty's strategy on sustainability

Source: Kilroy Realty (2016)

buildings, REITs are investing in systems and methods that lower energy cost, reduce waste and water consumption, improve quality of occupancy of tenants, and create awareness among their tenants. Additionally, the buildings in their portfolio that are either without any certification or have received lower certification standards are being prepared for achieving higher LEED certifications.

For projects that are under development or being conceptualized, emphasis is on the environmental impact of construction and adopting newer standards for building materials and technologies that are eco-friendly. An example of the latter is the WELL Building Standard, developed by the International Well Building Institute.[8] WELL is a new protocol that focuses on human wellness in the built environment. Larger and high-profile tenants are choosing to pursue WELL as it analyses the impact that buildings have on human health (which in turn increases productivity and lowers absenteeism in employees due to sickness). Kilroy Realty is one of the first REITs that is pursuing WELL certification on one of the first two WELL residential high-rise buildings in the U.S., at Hollywood Proper (the Columbia Square Residential Tower), having targeted completion in June 2016. Also, The Exchange on Sixteenth in San Francisco is one of the first projects to be pursuing WELL Ready certification on a shell and core building (Kilroy Realty, 2016).

We have a variety of programs to promote health and wellness for our tenants. They include:

PHYSICAL FITNESS	NUTRITION
• Access to stairwells • Discounted or free gym passes in properties with onsite gyms • Bike racks, showers, and bike concierge facilities • Bike rentals with available bike repair kits • Support of alternative transportation	• Hydration stations • Healthy food for building events • Healthy onsite food, such as an onsite GMO-free healthy bakery
REDUCED SICKNESS	COMMUNITY PROGRAMS
• Hand-sanitizing stations in lobbies and throughout buildings • Flu shots • Onsite dental	• Blood drives • Bike to Work Days • Wellness and Health Day fairs • Great Outdoors Month: recent prizes were annual passes to national parks

Figure 11.6 Programs for tenants at Kilroy Realty
Source: Kilroy Realty (2016)

Lastly, stakeholders of REITs include their investors, tenants, and the local communities where their projects are located. Information dissemination about their sustainability practices through periodic announcements through the media, at shareholder meetings, and in annual reports (or in specific sustainability reports like those of Kilroy Realty) helps investors learn about their green investments. Educating tenants in their buildings about following energy-efficient practices in both the wrok and the home environment and creating awareness in the local community through partnerships is another way of pursuing a green strategy. Figure 11.6 is an example of how Kilroy Realty is promoting health and wellness for its building tenants.

As with Kilroy Realty's efforts in the office market, an apartment REIT and another large REIT, AvalonBay Communities, has been making substantial progress in its sustainability efforts. AvalonBay currently owns 16 LEED and ten Energy Star-certified communities, and an additional 23 communities are pursuing certification. In 2016, it ranked highest among its peers and listed among the top third of the 500 largest U.S. publicly traded companies in the 2016 *Newsweek* Green Rankings. Launched in 2009, the *Newsweek* Green Rankings, created in partnership with Corporate Knights and HIP Investor, is one of the foremost corporate environmental rankings. It assesses the 500 largest publicly traded companies in the U.S. and the 500 largest publicly traded companies globally on overall environmental performance. Figure 11.7 shows some of the steps taken by AvalonBay to promote sustainability in its existing communities, which did not have any previous sustainability measures in place. At the larger corporate/firm level, the initiatives are shown in Figure 11.8.

Table 11.6 highlights the corporate social behavior of selected REITs.

11.3 Green REITs: Going forward

The sustainability platform is relatively new for the REIT sector. At present, larger REITs dominate the scene mainly due to the higher costs for such initiatives with minimum or no return

REDEVELOPMENT/REMERCHANDISING COMMUNITY HIGHLIGHTS

EAVES DUBLIN

NORTHERN CA

- Replaced lighting and fixtures with LED lamps in the apartment homes, leasing office and fitness center
- Replaced roofs with more efficient materials; recycled existing shingles
- Replaced condensers, hydronic heat coils and air handlers in all apartment homes
- Replaced all water heaters eight years or older with more efficient versions

Year Built	1989
Apartment Homes	204
Total Redevelopment Investment	$8,700,000

AVALON PLAYA VISTA

LOS ANGELES, CA

- Upgraded approximately 1,000 lamps to LEDs in common areas
- Enhanced landscaping to include drought-resistant plant material, converted irrigation to a drip system and installed a weather-based irrigation system
- Installed motion-detecting faucets and low-flow toilets in common-area restrooms

Year Built	2006
Apartment Homes	309
Total Redevelopment Investment	$1,500,000

AVA PACIFIC BEACH

SAN DIEGO, CA

- Replaced fixtures with LEDs in common areas and the pool, and installed model-minders and other motion sensors
- Replaced mechanical systems with more efficient equipment, including 196 through-wall air conditioning units, nine boilers and the entire HVAC system in the leasing office
- Replaced common-use washers and dryers with energy-efficient models and installed 728 low-flow toilets
- Replaced all refrigerators and dishwashers with Energy Star-rated appliances
- Enhanced landscaping to include drought-resistant plant material and built four new recycling centers

Year Built	1969
Apartment Homes	564
Total Redevelopment Investment	$23,600,000

Figure 11.7 Initiatives at some AvalonBay communities

Source: AvalonBay Communities (2015)

ENVIRONMENTAL INNOVATION

GOING PAPERLESS

AvalonBay launches project to go digital with all job site communications.

New digital-plan-room software eliminates hard-copy sets of drawings. At each AvalonBay construction site, project managers can review, mark-up and stamp all documents electronically, enhancing productivity, collaboration and communication among team members. This change will also facilitate collaboration with third-party architects and engineers.

AVALON WHARTON WINS TWO PRESTIGIOUS ENGINEERING AWARDS

Now a community of 247 apartment homes in Wharton, NJ, Avalon Wharton used to be an abandoned ore mine. Through complex engineering and geotechnical efforts, the mine was remediated and the site developed into a community. This project has received two prestigious awards:

Distinguished Award for mine remediation – awarded by the American Council of Engineering Companies, New Jersey chapter.

Award for Excellence for Best Engineering Design – site design for multifamily/mixed-use housing awarded by the Metropolitan Builders & Contractors Association of New Jersey.

33% fewer weekly trash pickups. Avalon First & M in Washington, DC, reduced their weekly trash by recycling clean cardboard. By taking advantage of the program, they saved their community over $14,500.

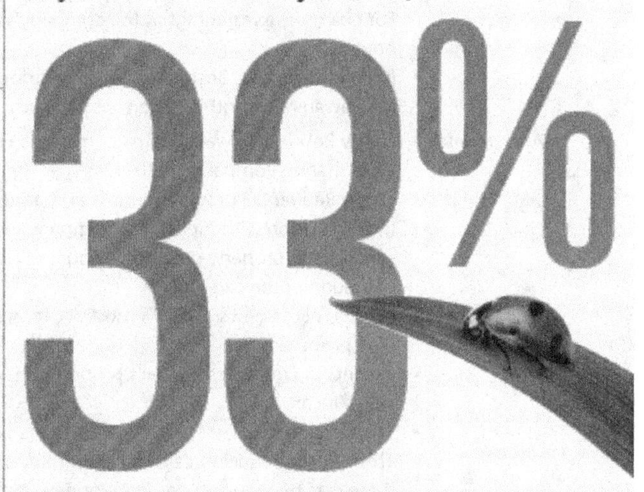

EXCELLENCE IN RECYCLING

Avalon at Grosvenor Station in Bethesda, MD, received an Excellence in Recycling Award from Montgomery County. This award is given to multifamily properties that display outstanding efforts in promoting recycling, reducing waste and buying recycled products and materials.

CERTIFIED SUSTAINABLE PROGRAM

In 2015 we launched the Certified Sustainable pilot program to award an internal green certification to existing communities that are not likely to achieve any third-party environmental certification. Eighteen communities participated in the program, achieving certification levels from Bronze to Platinum. As a result, we saw significant benefits in building performance – for example, at eaves Rancho Penasquitos in the San Diego area, we saw an 11% reduction in electricity consumption, a 9.2% reduction in gas consumption and a 10.6% reduction in water consumption.

Chief Investment Officer Matt Birenbaum awards the team at eaves Los Feliz with our highest internal green certification, Platinum.

Figure 11.8 Corporate-level initiatives at AvalonBay

Source: AvalonBay Communities (2015)

Table 11.6 Corporate social behavior of selected REITs

REIT name	*REIT focus*	*Sustainability initiatives*
Hersha Hospitality Trust	Hospitality	• Partnership with the nonprofit group Clean the World in 2011 • To reduce waste and help nations with hygiene-related deaths of children in developing countries, Hersha housekeepers collect partially used soap bars, which are then placed into prelabeled FedEx bins and shipped to Clean the World facilities. There, the soap is sanitized, packaged, and sent to developing countries. So far, Hersha has donated enough leftover soap to create 350,000 new bars
Lamar Advertising Co.	Billboards	• Partnership with outdoor gear manufacturer RAREFORM, to repurpose used billboards into backpacks, tote bags, surf bags, and other accessories • Lamar donates approximately 10,000 pounds of used billboard material per month • Lamar extends the life of old billboards by donating them to the community to be used as temporary roofs for storm victims and canvases for art projects, among other uses
Boston Properties	Office	• Partnership with Best Bees, a Boston-based beekeeping operation • Hives were installed at six Boston Properties rooftop locations, turning areas that were previously underutilized into active honey production zones • Boston Properties is producing more than 100 liquid pounds of honey per year, which in some cases are being utilized by restaurants in their office properties. For example, a restaurant at the company's Atlantic Wharf property utilizes the honey • In other locations, Boston Properties distributes the honey as gifts designed to raise awareness for this important program
Equity Residential	Apartments	• Equity Residential was the first company in its peer group to partner with Zipcar, a car-sharing company, in 2011 • Equity Residential provides onsite car sharing at 38 of the properties in its portfolio • In conjunction with Zipcar, the company tweaks the locations and numbers of cars at each property on an ongoing basis to ensure availability and efficient utilization of the cars • Boston is Equity Residential's most active market for Zipcar use, with 36 cars at nine properties • Tenants in Equity Residential's properties have reserved almost 250,000 hours with Zipcar • Their average cars get utilized eight hours per day • This helps in reducing carbon footprint in their communities due to sharing of these cars by reducing demand for new cars to be purchased
Simon Property Group	Retail malls	• Simon has reduced food waste in its properties by promoting food recycling programs • Food waste produces methane, which causes global warming • Simon is investigating a variety of food waste technologies, including digesters, dehydrators, and composters • At Simon's The Shops at Mission Viejo in California, they collect leftovers from food prep and guest plates and then transport it to the local recycling facility, where it is converted into engineered bio slurry. The slurry is then trucked to a county plant, where it is used to create electricity to power the facilities • In Massachusetts, Simon's Emerald Square and Southshore Plaza centers both have composting programs in place that divert about a ton of food waste weekly • Going forward, the company plans to expand its efforts to customer food court waste

Source: Borcherson-Keto (2016)

to the firm. However, the growth expected in this area is very high. With various stakeholders (investors, tenants, and owners) slowly and steadily getting better educated and informed about impact of sustainable investment strategies, almost no REIT in the public domain will remain unexposed to any or all aspects of sustainable real estate. Be it as a response to investor expectations or to demand driven by tenant requirements, the future of REITs is to go green. The changing demographics favoring millennials, who tend to understand, appreciate, and demand sustainable real estate products in their workplace or living environment, will drive REITs into such practice. More REITs are adapting to such changing real estate needs; those that are not will be forced to change or will be left to lose ground relative to their peers. The U.S. REIT industry is leading this change worldwide, with REITs in other countries set to follow in the next decade.

Notes

1 More than 1.6 million homes are Energy Star-certified out of a total stock of 111.1 million housing units as of 2005 and more than 25,000 buildings have earned certification.
2 Almost all of the green REIT evolution is restricted to the U.S. markets (Bourne, 2013).
3 Note that such a brand premium requires an REIT to be very much dominated by green properties, not just a small percentage of its portfolio.
4 Check the USGBC's website at www.usgbc.org to get the LEED certification requirements.
5 See the GRESB's website for a number of resources on profiles for this benchmark – for example, GRESB (2016b).
6 This depends on the model. (The study uses two models, the fixed-effects model and the Heckman model, to control for selectivity bias for estimating the impact of sustainability performance on financial performance as measured by return on equity.)
7 Some smaller firms that are not large REITs are also trying to be sustainable and green (Sledd and Stika, 2016).
8 The International WELL Building Institute (IWBI) is a public benefit corporation that is leading the movement to promote health and wellness in buildings and communities everywhere. The IWBI was launched in 2013 following a Clinton Global Initiative commitment made by founder Paul Scialla. See IWBI (2017).

References

Anderson, B., 2016 (March 21). Are Apartment Renters Willing to Pay More for Green Features? *National Real Estate Investor*. Available at: www.nreionline.com/multifamily/are-apartment-renters-willing-pay-more-green-features.

AvalonBay Communities, 2015. *Reaching Scale: Corporate Responsibility Report 2015*. Arlington, VA: AvalonBay Communities. Available at: www.avaloncommunities.com/~/media/Images/Corp/CorporateResponsibilityRevamp/2015-Corporate-Reponsibility-Report.pdf?la=en.

Borcherson-Keto, S., 2016 (July 25). REITs Gone Green. *REIT Magazine*. Available at: www.reit.com/news/reit-magazine/july-august-2016/reits-gone-green.

Boundy, E., and Poole, B., 2013. *State of the Industry: Sustainability Reporting within the REIT Sector*. Washington, D.C.: USGBC. Available at: www.usgbc.org/resources/state-industry-sustainability-reporting-within-reit-sector.

Bourne, K., 2013. How Green Is Your REIT? *IPE Real Assets*. Available at: https://realassets.ipe.com/how-green-is-your-reit/53975.fullarticle.

BREEAM, 2016. Why BREEAM? Available at: www.breeam.com/why-breeam.

Eichholtz, P., Kok, N., and Quigley, J. M., 2010. Doing Well by Doing Good? Green Office Buildings. *American Economic Review*, 100:5, 2492–2509.

Eichholtz, P., Kok, N., and Yonder, E., 2012. Portfolio Greenness and the Financial Performance of REITs. *Journal of International Money and Finance*, 31:7, 1911–1929.

Energy Star, 2016. www.energystar.gov.

Fuerst, F., 2015. The Financial Rewards of Sustainability: A Global Performance Study of Real Estate Investment Trusts. Available at: https://papers.ssrn.com/sol3/papers.cfm?abstract_id=2619434.

Fuerst, F., and McAllister, P. M., 2011. Green Noise or Green Value? Measuring the Price Effects of Environmental Certification in Commercial Buildings. *Real Estate Economics*, 39:1, 45–69.

GRESB, 2016a. www.gresb.com.

GRESB, 2016b. *2016 Global Snapshot*. Available at: https://gresb.com/wp-content/uploads/2017/07/2016_Global_Snapshot.pdf.

International WELL Building Institute (IWBI), 2017. www.wellcertified.com.

Kilroy Realty, 2016. *Sustainability Report 2015*. Los Angeles, CA: Kilroy Realty. Available at: http://kilroyrealty.com/sites/default/files/kilroy-realty-corporation-sustainability-report-2015v2.pdf.

Miller, N., Spivey, J., and Florance, A., 2008. Does Green Pay Off? *Journal of Real Estate Portfolio Management*, 14:4, 385–399.

Nareit, 2017. REITs & Sustainability. Available at: www.reit.com/nareit/reits-sustainability.

Nelson, A. J., Rakau, O., and Dörrenberg, P., 2010 (April 12). *Green Buildings: A Niche Becomes Mainstream*. Deutsche Bank Research.

Real Foundations/Nareit, 2014. *2014 Leader in the Light Award Entries & Trend Analysis*. Washington, D.C.: Nareit. Available at: www.reit.com/sites/default/files/media/PDFs/2014NAREITLeaderLightAward_Trend%20Analysis_RF.pdf.

Sah, V., Miller., N, and Ghosh, B., 2013. Are Green REITs Valued More? *Journal of Real Estate Portfolio Management*, 19:2, 169–177.

Sledd, A., and Stika, N., 2016 (May 20). Big Efficiency for Small and Medium Buildings. *Greenbiz*. Available at: www.greenbiz.com/article/big-efficiency-small-and-medium-buildings.

U.S. Green Building Council (USGBC), 2016. LEED. Available at: www.usgbc.org/leed.

U.S. SIF Foundation, 2016. *Report on U.S. Sustainable, Responsible and Impact Investing Trends 2016*. Washington, D.C.: U.S. SIF Foundation. Available at: www.ussif.org/files/SIF_Trends_16_Executive_Summary(1).pdf.

Vosilla, G., Behrendt, J., and Hanson, M., 2016. *State of the Industry: Sustainability in the REIT Sector – 2016 Update*. Washington, D.C.: USGBC. Available at: www.usgbc.org/sites/default/files/State%20of%20the%20US%20REIT%20Industry%20Sustainbility%20Reporting%20-%202016%20Update_0.pdf.

Wiley, J., Benefield, J., and Johnson, K., 2010. Green Design and the Market for Commercial Office Space. *Journal of Real Estate Finance and Economics*, 41:2, 228–243.

12

The 'green value' proposition in real estate

A meta-analysis

Ben Dalton and Franz Fuerst

12.0 Introduction

In recent years, mandatory government-led environmental rating systems have gained traction in several countries; for example, in Singapore the Building Control Act mandated that existing buildings achieve at least Green Mark Standard from 2012 onwards. At the same time, there has been a proliferation of voluntary eco-labels, such as BREEAM in the UK, Green Star in Australia and LEED in the US. The very existence of the voluntary labels is indicative of a market-led environmental agenda. Any voluntary initiatives that exceed regulatory requirements and national building codes could potentially create 'green value', which should, at least hypothetically, be capitalised into prices and rents of both commercial and residential property. The existence of a green premium would also reflect consumer willingness to pay, which studies have found to be primarily related to increased energy efficiency; therefore, a premium may also indicate the ability to successfully and credibly convey a property's energy efficiency.

Amongst the issues that may hinder energy-efficient investment include those that stem from principal–agent problems and a 'vicious circle of blame' (Cadman, 2000) between developers/landlords and tenants. Estimating the green premium, and analysing its dynamics within a transaction setting, may be able to provide a clearer understanding of the incentives available to stakeholders.

Evidence-based policy depends on reliable and robust analytical results, particularly in innovative areas such as green real estate finance. However, the growing body of literature on the green premium is disjointed and at least partly inconclusive. The general incentives and disincentives of energy efficiency and broader sustainability are now widely researched but the empirical studies are often limited in terms of geography and time periods analysed. Practitioners and academics also debate whether 'brown discounts' (i.e., discounts to the rents and prices of non-sustainable buildings) are more relevant in practice than green premiums due to their larger proportion in the existing stock (Runde and Thoyre, 2016). The present study acknowledges this point but empirical evidence in the peer-reviewed literature appears to be thus far limited to green premium studies. Whilst there have been previous attempts to consolidate the extensive green premium literature (e.g., Sayce, Sundberg and Clements, 2010; World Green Building Council, 2013), very few have attempted to synthesise the evidence using statistical methods and to place the individual studies in a larger context. This chapter attempts to achieve this objective via a meta-analysis of green premium studies in a real estate context and illustrate the implications arising from the green premium consensus on the analysis of property investment.

12.1 Background

The real estate sector revolves around reshaping the environment, and thus has an inevitable impact on it. Given that buildings reportedly account for 32 per cent of final energy consumption (International Energy Agency, 2016), there should be an impetus for property owners and investors to improve the energy efficiency and environmental performance of their assets. Prior to 2010, the majority of the literature concerning the environmental performance of property emerged from engineering disciplines and focused on construction systems and technology, rather than on the implications for financial stakeholders (Eichholtz, Kok and Quigley, 2010; Sayce, Ellison and Parnell, 2007). Although investors suspected that premiums for environmental performance were available (Sayce et al., 2007), the evidence was largely theoretical.

Voluntary eco-labels, sometimes also referred to as environmental labels (Fuerst and McAllister, 2011a) or green ratings (Eichholtz et al., 2010), aim to reduce harmful emissions through communicating an asset's environmental impact, thereby influencing consumer preferences, supplier production and the overall market supply and demand (Fuerst and McAllister, 2011a, 2011b; Cole, 2005). Since the inception of BREEAM in the UK in 1990, commercial and residential eco-labels have gained significant momentum, with many countries following suit (see Figure 12.1). Many regulators and government bodies (e.g., the Welsh Assembly Government) now require attainment of certain eco-labels. As such, the idea of a 'green building' is becoming increasingly institutionalised and has provided eco-labels with what Fuerst and McAllister (2011a, 2011b) describe as a 'quasi-compulsory' status. However, they are not without criticism. For example, there is ongoing debate concerning the attribution of weightings to environmental impacts (Lee, 2013), from which a clear tension between actual measures and theoretical measures emerges. The rating is usually based on the theoretical energy efficiency of the building's services and fabric, rather than its actual efficiency. Large differences between theoretical and actual efficiency have been reported, mainly due to intervening behavioural factors that are not taken into account in many eco-certification schemes (Wedding and Crawford-Brown, 2008; Ingle, Moezzi, Lutzenhiser and Diamond, 2014).

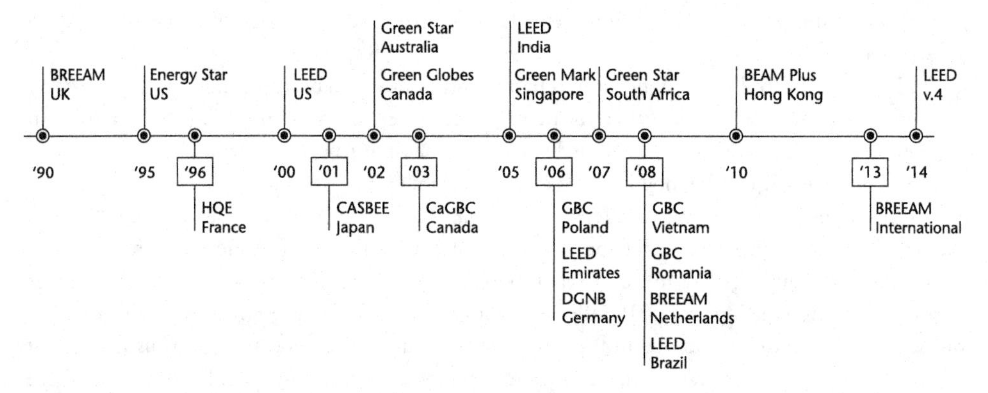

Figure 12.1 An international timeline of eco-label development

Source: Adapted from Arup (2014)

12.2 Research strategy

To our knowledge, only one meta-analysis of the green premium has been undertaken in a real estate context. Brown and Watkins (2015) analyse the green premium for environmentally certified homes, through a review sample of 20 studies primarily focused on the US residential market. The present study focuses on both residential and commercial real estate, primarily office buildings, in a large number of countries. Whilst there are several systematic review databases for medicine and health, such as the Cochrane Database of Systematic Reviews (CDSR), no such database exists for finance and economics. A search within Google Scholar reveals only two previous systematic reviews of the green premium literature, both undertaken by McAllister (2013; n.d.); however, they are not within peer-reviewed journals, nor were they intended to be (hence the title of the 2013 paper, 'An "off the record" record'). As such, these reviews lack an explicit search and selection strategy, which is central to the validity of a systematic review.

The other peer-reviewed papers that undertake a systematic review of green real estate often do so in a far broader context. A paper by Zhang (2015) employs broad search terms to draw general themes from the literature rather than narrow search terms to focus on a particular aspect of green real estate or to perform a meta-analysis. Moreover, the review is focused on the Chinese market, which the author notes to have predominantly government-led sustainability and environmental initiatives; by contrast, Western commercial and residential markets such as the US and UK, as identified in the preliminary literature review, are largely market-led. Another recent systematic review, by Olubunmi, Xia and Skitmore (2016), explores the incentives and disincentives for green building and green procurement. However, their broad search terms produce broad results, and the green premium is therefore overlooked.

The systematic review within this chapter follows the procedure proposed by Klewitz and Hansen (2014), as illustrated in Table 12.1.

Steps 1–4: Search process

Step 1: The research question necessitates a literature search comprising keywords related to four key components (search groups): energy-efficient, real estate, price, premium. An overall total

Table 12.1 The six-step review process

Overall process	*Individual steps*	*Analysis*	*Resulting no. articles*
Search process	Step 1: Identify keywords (17 keywords).	Previous research and reviews	NA
	Step 2: Develop exclusion/inclusion criteria.	NA	NA
	Step 3: Specify relevant search engines.	Title and abstracts (automated based on keywords)	21267
	Step 4: Develop A, B, C, list		
	C-list	NA	21,267
	B-list	Title and abstracts (manual)	299
	A-list	Full content	42
Meta-analysis	Step 5: Code A-list for their methodology, effects and errors.		
	Step 6: Aggregate the study effects. Estimate overall and subgroup effects.		

Source: Klewitz and Hansen (2014, p.60)

Table 12.2 Overview of search keywords

Search groups				*Search syntax*
Energy-efficient	*Real estate*	*Price*	*Premium*	
green, environ*, eco, responsible, RPI, sustainab*, energy-efficient, energy efficient, energy rating, energy performance, energy certificat*, environ* rating, environ* performance, environ* certificate, EPC, performance certificat*, eco-label, BREEAM, LEED, Energy Star	real estate, property, properties, building$	price$, value$, rent$, transaction$, sale$	premium$, capitali?ed, capitali?e	("green" OR environ* OR "eco" OR "responsible" OR "RPI" OR sustainab* OR "energy-efficient" OR (("energy" OR environ*) AND (efficien* OR "rating" OR "performance" OR certificat*)) OR "EPC" OR performance certificat* OR "eco-label" OR "BREEAM" OR "LEED" OR "Energy Star") AND ("real estate" OR "property" OR "properties" OR building$) AND ("price" OR "value" AND "premium" OR capitali?ed OR capitali?e)

Note: Wildcards:? = one character (e.g., capitalise, capitalize); * = zero or more characters (e.g., environment, environmental); $ = zero or one character

of 32 keywords were considered to describe the four search groups (see Table 12.2). The exemplary search syntax in Table 12.2 will only return studies that contain at least one word from each search group.

Step 2: Borenstein, Hedges, Higgins and Rothstein (2009) emphasise that a systematic review does not rid the review process of subjectivity and bias; rather, there are different biases that threaten the review's validity (Hopewell, Loudon, Clarke, Oxman and Dickersin, 2009). Ensuring methodological quality at this step in the process can be crucial in alleviating systematic bias (Bennett et al., 2012), particularly publication bias, whereby studies with results that favour a certain direction, with higher statistical significance and perceived importance, are more likely to be published (Hopewell et al., 2009). As such, the exclusion of the non-peer-reviewed or 'grey' literature from a meta-analysis may lead to overestimation of an effect (Borenstein et al., 2009; Hopewell, McDonald, Clarke and Egger, 2007) as positive and large findings may have a better chance of getting published. Whilst the inclusion of grey literature should reduce publication bias, it will also provide exposure to industry publications that may present first-hand insights into the incentives/disincentives for green property.

Steps 3 and 4: The review, which was undertaken in June 2016, digitally searched the content of studies within several large complementary academic and general databases: Web of Science, Wiley Online Library, Taylor & Francis Online, Science Direct, Emerald Insight, Sage Journals, Google, Business Source Complete. The general databases, Google and Business Source Complete, should capture the grey literature. It must be noted that the search syntax varies for some databases; the eight databases required three moderate adaptations of the exemplary search syntax in Table 12.2. Following Klewitz and Hansen (2014), the initial search results formed a 'C-list' of 21,267 articles that may have been largely irrelevant. The article titles and abstracts were downloaded into *Endnote* and were manually reviewed in order to further categorise the relevant articles into a 'B-list' of 299 articles. After removing 29 duplicate studies, a review of the full text of these articles, based on the inclusion and exclusion criteria in Table 12.3, formed the 'A-list' of the 42 most relevant articles from 2007 to 2016 (Figure 12.2).

Table 12.3 Review of the full text of digitally searched articles, based on the inclusion and exclusion criteria

Inclusion	*Exclusion*
Studies by government, academics, business and industry (commercially and non-commercially published).	Studies not available in English.
Studies that estimate the relationship between the energy efficiency of individual properties, and their rental and capital values (based on transaction data/proxies).	Studies prior to 2007.
Studies where the price effect is provided as a percentage relative to the expected value (based on comparables).	Summary articles that lack a detailed methodology.
Studies that exhibit adequate methodological quality and transparency.	
Studies that analyse different data. When encountering studies using the same data, the study that demonstrates the most statistically robust model estimates, whilst controlling for the most factors (excluding interaction terms – see Appendix A1) will be used. Preference will also be given to the study that reports standard errors (essential for the meta-analysis), and estimates the effect size as an elasticity coefficient.	

Note: Due to the Energy Performance of Building Directive 2002/91/EEC mandating EPCs for property transactions, studies in which energy efficiency is measured using EPCs are benchmarked against other levels of EPC. As a result, this meta-analysis will only include the highest group of EPC levels provided, i.e., A–C (similar to Brown and Watkins, 2015).

Steps 5 and 6: Aggregating studies and estimating an effect size

Step 5: The final sample comprises 42 unique studies that have been coded by: measure of energy efficiency; methodology; sample market; sales and rental premiums; and estimate standard errors (dataset available upon request from the authors). All of the selected studies were published in journals; a search of the grey literature did not return any non-commercially published studies that met the inclusion criteria. Of the 29 journals in which the studies were published, the most influential were *Energy Policy* (four studies), *Journal of Real Estate Finance and Economics* (four studies) and *Regional Science and Urban Economics* (three studies). The publication of green premium studies peaked during 2013 and 2014, and may be in decline (see Figure 12.3).

The literature on green premiums appears to be predominantly US-focused, particularly on commercial properties (Figure 12.4). This does not come as a surprise – the data available on the US office market has been acknowledged to exhibit the highest quality, in terms of its sample size and number of variables (McAllister, n.d.), not to mention that the US is a fitting subject due to its widespread certification schemes, LEED and Energy Star (Eichholtz, Kok and Quigley, 2013), in comparison to other countries such as China where government-led green building initiatives have been met with resistance (Zheng, Wu, Kahn and Deng, 2012). In some cases, authors have circumvented this through constructing indices based on properties that are marketed as 'energy efficient' as a proxy for 'greenness' (Zheng et al., 2012; Sánchez-Ollero, García-Pozo and Marchante-Mera, 2014; Aroul and Hansz, 2012; Shewmake and Viscusi, 2015). As such, these studies are prone to error from the energy efficiency of properties being over-understated or inaccurate. This highlights the well-recognised issue of obtaining quality data in real estate.

Step 6: See methodology section in the Appendix.

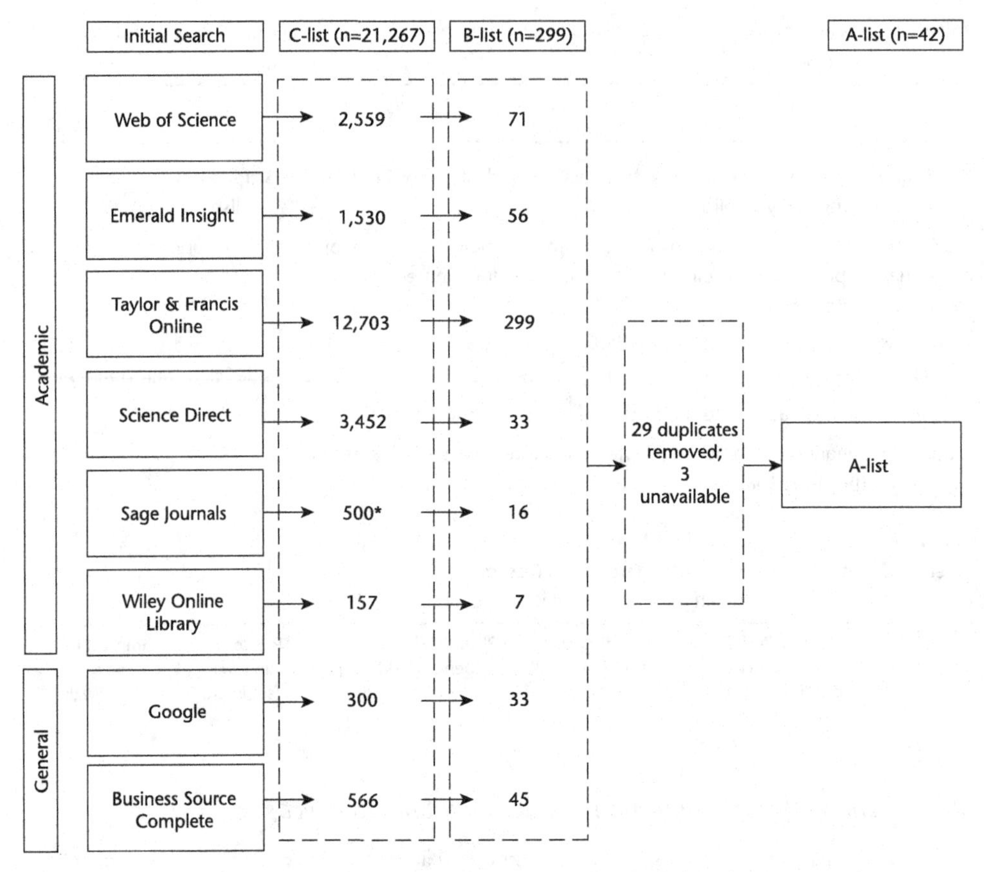

Figure 12.2 Step 4: Developing the A-list of studies to be included in the meta-analysis

Note: *Due to the limited search string-length in the SAGE search engine, 73,321 results were returned. Therefore, the results were ordered by relevance and the first 500 studies were taken.

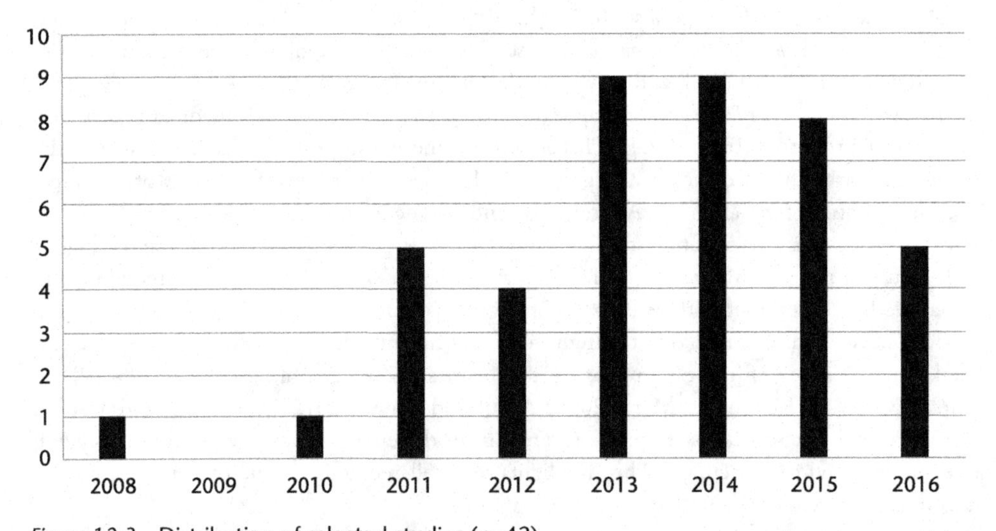

Figure 12.3 Distribution of selected studies (n=42)

		Property type				Energy efficiency measure								
						Environmental certifications						Other		
	Market	Commercial	Residential	Hotels	TOTAL	LEED	ES	EPC	Dual: LEED & ES	BREEAM	Other cert.schemes	Marketing	Actual energy consumption	Mixed
Rental value observations	US	16	3	0	19	6	6	0	3	0	1	1	0	2
	UK	4	0	0	4	0	0	2	0	2	0	0	0	0
	Japan	0	0	0	0	0	0	0	0	0	0	0	0	0
	Sweden	0	0	0	0	0	0	0	0	0	0	0	0	0
	Australia	2	0	0	2	0	0	0	0	0	2	0	0	0
	Singapore	0	0	0	0	0	0	0	0	0	0	0	0	0
	Spain	0	0	1	1	0	0	0	0	0	0	1	0	0
	Canada	1	0	0	1	1	0	0	0	0	0	0	0	0
	China	0	1	0	1	0	0	0	0	0	0	1	0	0
	France	1	0	0	1	0	0	0	0	0	0	0	1	0
	Germany	0	1	0	1	0	0	1	0	0	0	0	0	0
	Hong Kong	0	0	0	0	0	0	0	0	0	0	0	0	0
	Netherlands	0	0	0	0	0	0	0	0	0	0	0	0	0
	Switzerland	0	1	0	1	0	0	0	0	0	0	0	1	0
	TOTAL	24	6	1		7	6	3	3	2	3	3	2	2
Capital value observations	US	12	5	0	17	6	5	0	2	0	0	1	0	3
	UK	2	1	0	3	0	0	2	0	1	0	0	0	0
	Japan	0	3	0	3	0	0	0	0	0	3	0	0	0
	Sweden	1	2	0	3	0	0	3	0	0	0	0	0	0
	Australia	1	0	0	1	0	0	0	0	0	1	0	0	0
	Singapore	0	2	0	2	0	0	0	0	0	2	0	0	0
	Spain	0	1	0	1	0	0	1	0	0	0	0	0	0
	Canada	0	0	0	0	0	0	0	0	0	0	0	0	0
	China	0	1	0	1	0	0	0	0	0	0	1	0	0
	France	1	0	0	1	0	0	0	0	0	0	0	1	0
	Germany	0	1	0	1	0	0	1	0	0	0	0	0	0
	Hong Kong	0	1	0	1	0	0	0	0	0	0	0	0	1
	Netherlands	0	1	0	1	0	0	1	0	0	0	0	0	0
	Switzerland	0	0	0	0	0	0	0	0	0	0	0	0	0
	TOTAL	17	18	0		6	5	8	2	1	6	2	1	4

Figure 12.4 Literature on green premiums by property type and energy efficiency measure

Note: Darker shading indicates a higher number of studies for each combination

12.3 Results and discussion

The primary analyses of rental and sales premiums produced weighted mean effects of 6.0 per cent and 7.6 per cent respectively, indicating positive price effects for energy efficiency and environmental certification (Table 12.4). However, the confidence intervals suggest that the true mean effects may be between 4.3 per cent and 7.8 per cent and between 5.9 per cent and 9.4 per cent respectively, which are similar to the range anticipated by Morri and Soffietti (2013).

These estimates are highly statistically significant. The confidence intervals signify a range outside of which the true value is improbable ($p < 0.05$); and, in this case, the ranges have been estimated with high statistical power (p-values of < 0.0001) to be narrow and positive. This indicates a confident rejection of the null hypotheses of a zero mean.

The results of the primary analyses of the rental and sales premiums are visualised as forest plots (Figures 12.5 and 12.6 respectively). The horizontal lines depict the 95 per cent confidence intervals for the observed effect sizes. The central point of each of these lines corresponds to the size of a respective study's observed effect, encompassed by a box providing a size representation of its weighting in the analysis. The diamond represents the estimated overall effect size, and its width represents a confidence interval, the range outside of which the true value is improbable.

Upon visual inspection, it is apparent that the majority of the studies included within the analyses observed a positive price effect, with the exception of three rental observations (Fuerst and McAllister, 2011c; Gabe and Rehm, 2014; Zheng et al., 2012) and four sales observations (Fuerst and McAllister, 2011c; Nappi-Choulet and Décamps, 2013; Yoshida and Sugiura, 2015; Cerin, Hassel and Semenova, 2014). As the confidence intervals are essentially normal distributions of where the true effect may lie, absence of overlapping intervals between studies, *ceteris paribus*, suggests a significant between-study variance (t^2) of the true effects (Borenstein et al., 2009; Cumming and Finch, 2005). A significant t^2 illustrates statistical heterogeneity, and thus a rejection of the null hypothesis of homogeneity (i.e., studies estimating the same effect). Whilst heterogeneity in a meta-analysis is inevitable, measuring the magnitude is of importance (Higgins, 2008). The *a priori* assumption of unexplained heterogeneity between the studies was a key reason for employing a random-effects model, thus separating the total variance to also account for t^2. The forest plots show little overlap between the studies, which prompts a further test for statistical heterogeneity.

As alluded to in the methodology section (Appendix), systematic reviews – and therefore meta-analyses – are prone to publication bias. We investigated publication bias using the Light and Pillemer (1984) 'funnel plots' (Figures 12.7 and 12.8). The scatter plots of the observed effects, with their size on the x-axis and their standard errors on the y-axis, should show that the larger studies with lower standard errors appear towards the top in a narrow spread and smaller studies

Table 12.4 Summary statistics from the primary analysis of rental and capital value green premiums

	Magnitude and significance				*Heterogeneity*		
	Effect size	*Std. error*	*k*	*Z-value*	τ^2	*Q*	I^2
Sales	0.0761*** [0.0586; 0.0936]	0.0179	35	8.53	0.0017	1564.75***	97.8% [97.5; 98.1]
Rental	0.0602*** [0.0430; 0.0775]	0.0176	31	6.84	0.0017	574.05***	94.8% [93.5; 95.8]

Note: 95 per cent confidence intervals are shown in square brackets. *** *pp*< 0.0001

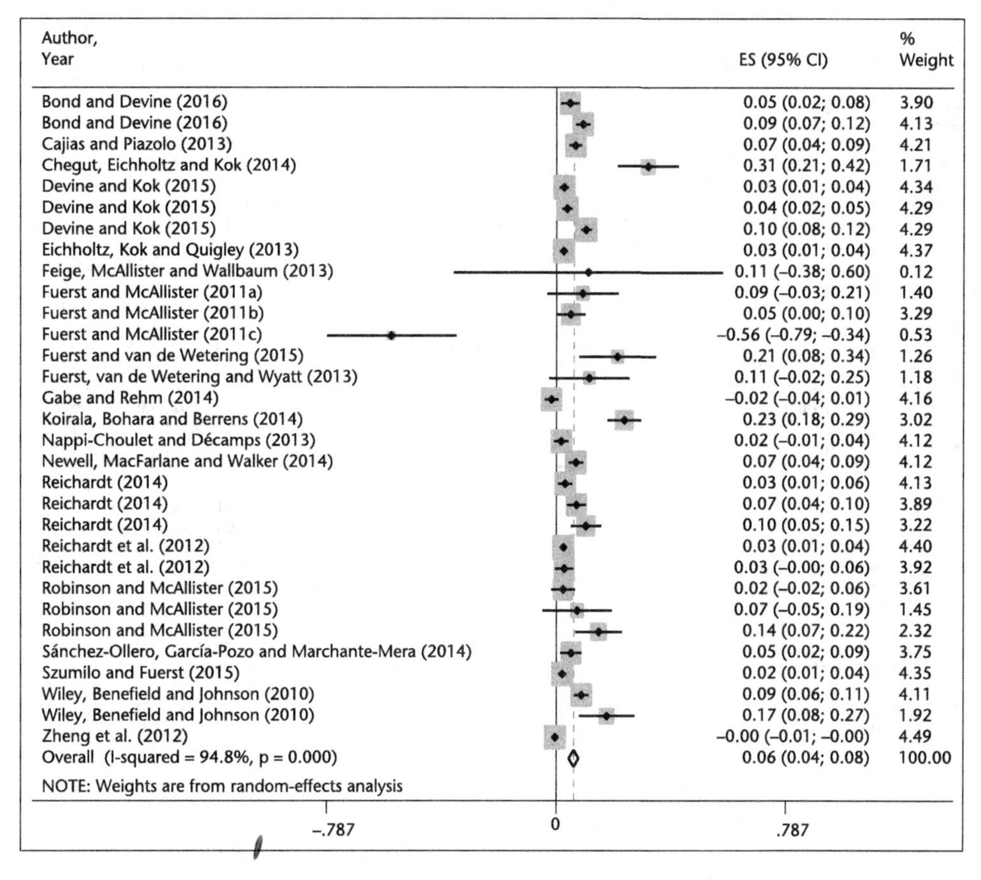

Figure 12.5 Primary rent premium forest plot

Note: In some cases different data has been extracted from the same articles – hence the article may be included in more than one line of the table.

at the bottom in a wider spread. The funnel plots for both the rental and sales premiums display asymmetry towards smaller studies on the x-axis where publications may be missing (Sterne, Gavaghan and Egger, 2000). The lack of smaller studies estimating negative premiums is an indication, but not an accurate test (Lau, Ioannidis, Terrin, Schmid and Olkin, 2006), of possible publication bias or studies that have not been undertaken. The widely used test proposed by Egger et al. (1997) is used to test for funnel plot asymmetry, indicating significant bias at the 5 per cent level (Table 12.5), although this test is known to have low power (Sterne, Egger and Moher, 2011). Whilst the detection of bias prompts consideration, there is no set solution to the problem (Sterne et al., 2011).

Table 12.5 Results from the Egger, Smith, Schneider and Minder (1997) test

	t	*p*	*95% CI*
Sales	2.58	0.015	0.6779; 5.7723
Rental	5.53	0.000	2.2871; 4.9694

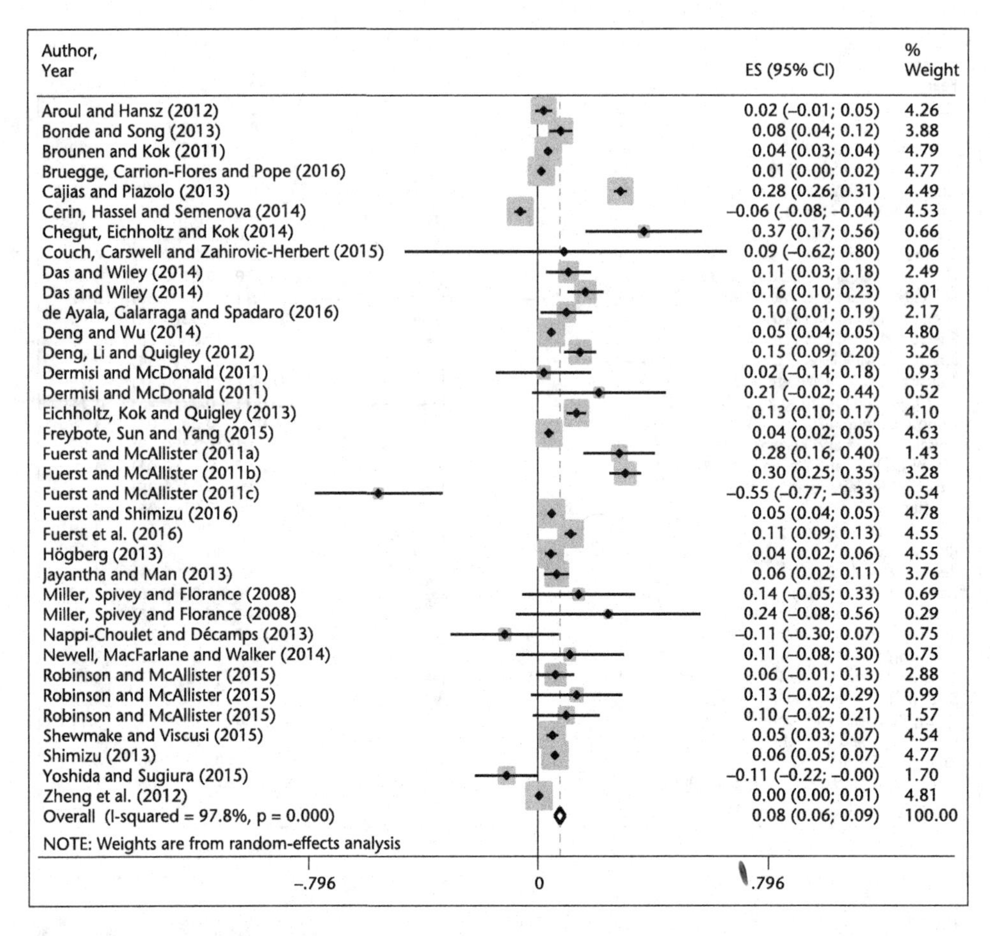

Figure 12.6 Primary sales premium forest plot

Note: : in some cases different data has been extracted from the same articles – hence the article may be included in more than one line of the table

12.4 Heterogeneity and subgroup analysis

t^2 is an absolute measure dependent on the scale of effect sizes (Borenstein et al., 2009). The magnitude of heterogeneity in the random-effects model can be quantified independent of scale using the I^2 index, a descriptive statistic proposed by Higgins, Thompson, Deeks and Altman (2003):

$$I^2 = \left(\frac{Q - df}{Q}\right) \times 100\% \qquad (1)$$

where Q is Cochran's Q as estimated in the methodology section and *df* is the degrees of freedom ($k-1$). Thus, I^2 is a percentage estimate of the total variance ($\tau^2 + \upsilon_{y_i}$) that is attributable to the between-study variance (t^2).

Higgins et al. (2003) categorise, albeit tentatively, the I^2 index percentages of 25 per cent, 50 per cent and 75 per cent as low, moderate and high levels of heterogeneity respectively. However, it must be stressed that a high level of heterogeneity does not indicate a wrong result; if that was the case, real estate meta-analyses, such as this one, that are *a priori* heterogeneous would be redundant. A later commentary on the I^2 index by Higgins (2008) emphasised that a high level of heterogeneity warrants exploration.

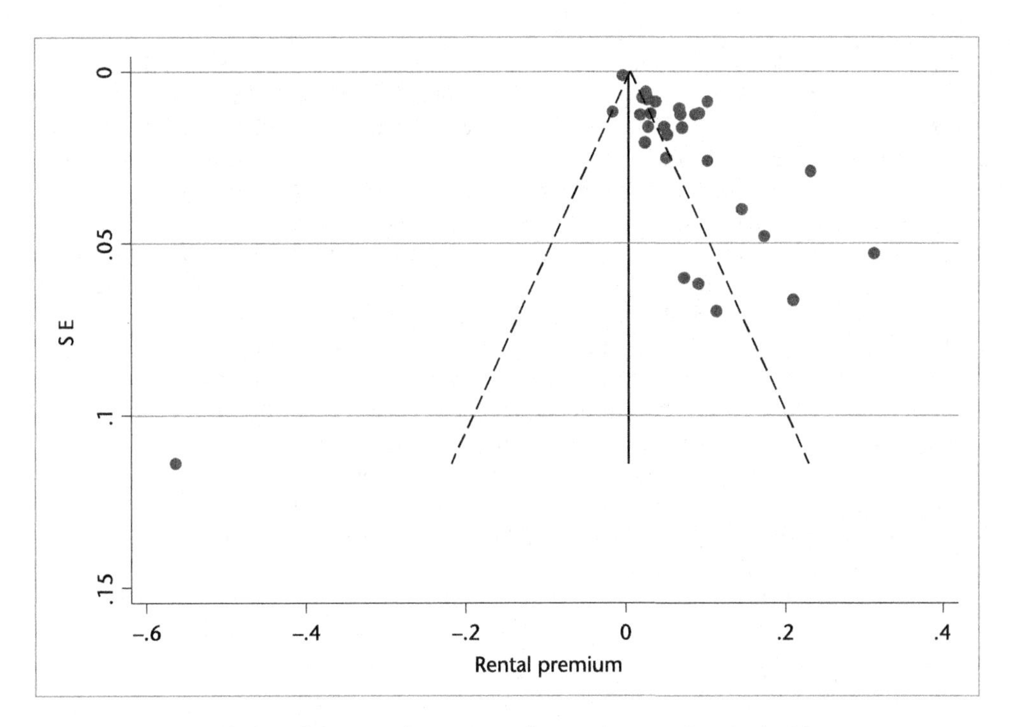

Figure 12.7 Funnel plot of the rental premium observations, with pseudo 95 per cent confidence limits

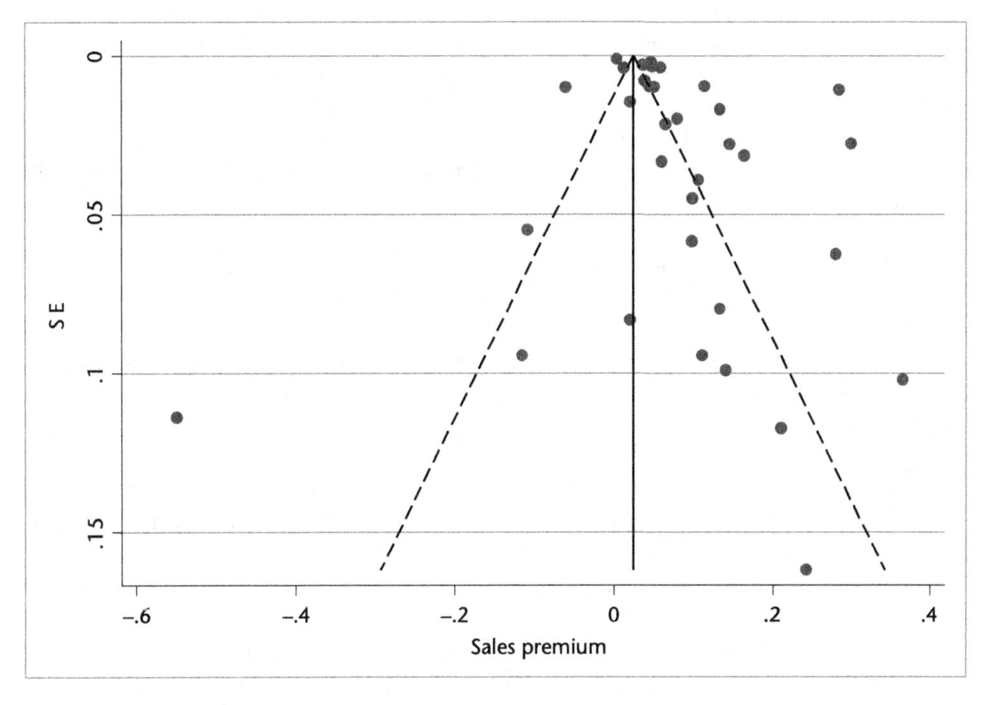

Figure 12.8 Funnel plot of the sales premium observations, with pseudo 95 per cent confidence limits

As a function of Cochran's Q, the power to predict a reliable I^2 is dependent on the number of studies included within the analysis; thus, a small number of studies will estimate heterogeneity with low power and often underestimate I^2 (Thorlund et al., 2012), providing an erroneous sense of precision (Von Hippel, 2015). In this case, the I^2 is estimated with high statistical power due to the large number of observations attributing to a significant Q ($p < 0.0001$).

As expected, the primary analyses for the sales premium and rental premium exhibit high levels of heterogeneity at an I^2 of 97.9 per cent and 95.1 per cent respectively. Subgroup analysis may present further explanation for the magnitude of the heterogeneity. The selection of the subgroups for further analysis should be made *a priori*; in this case, the groups are categorical, discrete, study-level variables of market (geography), measure of energy efficiency and property type.

The heterogeneity between groups is analysed through conducting separate meta-analyses for each group (forest plots available upon request from the authors) and testing the null hypothesis that the effect size is not dependent on inclusion within a subgroup (Borenstein et al., 2009). Testing heterogeneity between the subgroups follows the same logic as that in the primary analysis; however, the total variance is partitioned into within-subgroup variance and between-subgroup variance (Borenstein et al., 2009). The variance between subgroups is then tested for statistical significance:

$$Q_b = Q - \sum_{j=1}^{k} Q_j \tag{2}$$

where Q_b is the weighted between-subgroup variance around the overall effect; Q is the total weighted variance around the overall effect for k subgroups; Q_j is the weighted variance of the subgroup j.

Due to the small sample sizes within the studies, it is difficult to make accurate inferences from subgroups with a small number of studies. As alluded to above, the number of studies must also be taken into consideration when estimating Q as a test for heterogeneity; subgroups with few studies will estimate Q_j with low power and subgroups in which there is only one study will estimate Q_j as 0 as there will be 0 degrees of freedom.

The market, property type and energy efficiency measure groups for both the sales and rental premiums exhibited highly significant Q_b values ($p < 0.001$ as a chi-square distribution) (Tables 12.6 and 12.7): a rejection of the null hypothesis that all subgroups observe the same effect size. Thus, all groups are heterogeneous and attribute to the statistical heterogeneity of the primary analysis, although, as noted above, subgroups with only few observations will significantly limit the power of Q_b.

For both the sales and rental premiums, the US market subgroup has a sufficient number of observations (17 and 19) to make significant inferences. The US sales premiums produced a weighted mean average of 10.5 per cent, which is markedly higher than the overall average premium of 7.6 per cent estimated in the primary analysis. This is estimated with a confidence interval (CI) of 7–14 per cent, which is a significant rejection of the null hypothesis of a zero mean. The US rental premiums produced an average of 5.9 per cent (CI 4.3; 7.5 per cent), which is similar in size and precision to the overall average premium of 6 per cent (CI 4.3; 7.8 per cent). This is not surprising, as 61 per cent of the observations for the rental premium are from the US market.

The residential and commercial property type subgroups also have sufficient observations to make inferences. The commercial property subgroups produced an average sales premium of 11.5 per cent (CI 5.8; 17.3 per cent) and an average rental premium of 5.4 per cent (CI 3.7; 7.2 per cent). The premiums are similar in size to those from the US market subgroup as there is a

Table 12.6 Sales premium subgroup heterogeneity

Sales (Q =1564.75)		*1564.8*			
Subgroup	Q_b	*n*	*ES*	*95% CI*	Q_j
Market	1225.31***				
US		17	0.105	0.070; 0.140	202.85
Sweden		3	0.020	−0.063; 0.103	70.94
Netherlands		1	0.036	0.030; 0.042	0
Germany		1	0.284	0.263; 0.305	0
UK		3	−0.016	−0.406; 0.378	39.92
Spain		1	0.098	0.010; 0.186	0
Singapore		2	0.092	−0.005; 0.189	12.42
Japan		3	0.049	0.032; 0.066	13.31
Hong Kong		1	0.064	0.022; 0.106	0
France		1	−0.114	0.299; 0.071	0
Australia		1	0.110	0.295; 0.750	0
China		1	0.004	0.001; 0.006	0
Property	132.02***				
Commercial		17	0.115	0.058; 0.173	106.25
Residential		18	0.055	0.036; 0.075	1326.48
Energy measure	741.98***				
Marketing		2	0.005	−0.005; 0.015	1.25
EPC		8	0.047	−0.025; 0.118	673.1
Energy Star		5	0.075	−0.002; 0.151	26.23
BREEAM		1	0.365	0.165; 0.565	0
LEED		6	0.083	0.026; 0.141	7.78
Other cert. schemes		6	0.053	0.041; 0.065	29
Mixed		4	0.134	0.045; 0.223	80.85
Dual: LEED & ES		2	0.187	0.008; 0.367	4.56
Energy consumption		1	−0.114	−0.299; 0.071	0

Notes: *** $p < 0.001$, as a chi-square distribution with j–1 degrees of freedom, for j number of subgroups. Q_b is the weighted between-subgroup variance around the overall effect, n is number of studies within the subgroup and Q_j is the weighted variance of the subgroup around its effect.

considerable overlap between the observations as 71 per cent of commercial sales premiums and 60 per cent of commercial rental premiums are observed in the US market. Interestingly, the residential sales premiums, which are less biased towards the US market (only 27 per cent of observations), average at 5.5 per cent (CI 3.6; 7.5 per cent). This may suggest that either the green premium is generally lower for residential capital values or, because this subgroup primarily captures the effects of different markets, the premium for capital values may actually be less in countries outside of the US. This would explain why the average sales premium for the US market was substantially higher than the overall average from the primary analysis. The residential rental premium is estimated at 8.2 per cent (CI 2.4; 14 per cent), which is higher than both the commercial property rental premium (above) and the overall rental premium of 6 per cent from the primary analysis. However, this is estimated with only six observations, which is reflected in the wide confidence interval; thus, it may not be appropriate to infer that a higher rental premium is obtained for residential property.

LEED and Energy Star had the most observations for the rental premium. The subgroup analysis of the rental premium by energy efficiency measure estimates a significant positive premium for

Table 12.7 Rental premium subgroup heterogeneity

Rental (Q = 574.05)		574.1			
Subgroup	Q_b	*n*	*ES*	*95% CI*	Q_j
Market	381.28***				
US		19	0.059	0.0430; 0.0750	119.84
Germany		1	0.066	0.0450; 0.0870	0
UK		4	0.033	–0.2500; 0.3160	49.55
Canada		1	0.102	0.0840; 0.1200	0
Switzerland		1	0.110	–0.3850; 0.6050	0
Australia		2	0.026	–0.0560; 0.1080	23.38
France		1	0.018	–0.0070; 0.0430	0
Spain		1	0.052	0.0150; 0.0880	0
China		1	–0.004	–0.0060; –0.0010	0
Property	204.13***				
Commercial		24	0.054	0.0370; 0.0720	197.87
Residential		6	0.082	0.0240; 0.1410	172.05
Hotels		1	0.052	0.0150; 0.0880	0
Energy measure	357.92***				
Marketing		3	0.029	–0.0140; 0.0720	18.6
LEED		7	0.073	0.0400; 0.1050	71.52
EPC		3	–0.104	–0.3760; 0.1680	30.81
BREEAM		2	0.268	0.1690; 0.3680	1.47
Energy Star		6	0.036	0.0170; 0.0550	20.97
Mixed		2	0.028	0.0140; 0.0410	0.84
Energy consumption		2	0.018	–0.0060; 0.0430	0.13
Dual: LEED & ES		3	0.112	0.0720; 0.1530	0.95
Other cert. schemes		3	0.091	–0.0140; 0.1960	70.84

Notes: *** $p < 0.001$, as a chi-square distribution with j–1 degrees of freedom, for j number of subgroups. Q_b is the weighted between-subgroup variance around the overall effect, n is number of studies within the subgroup and Q_j is the weighted variance of the subgroup around its effect.

LEED at 7.3 per cent (CI 4; 10.5 per cent) and Energy Star at 3.6 per cent (CI 1.7; 5.5 per cent). With a similar number of observations, both primarily in the US, the evidence suggests that LEED certification provides the optimal rent premium. It is worth noting that whilst 69 per cent of the US commercial rental premiums were estimated using LEED and Energy Star ratings, a further 19 per cent were estimated from dual LEED and Energy Star certification at 11.2 per cent (CI 7.2; 15.3 per cent). With only three observations for dual certification, it is difficult to draw any conclusions in regards to its influence on the overall premium; however, inferences can be made about the US rental market premium, given that it is primarily based on commercial observations (84 per cent), 88 per cent of which are LEED, Energy Star or dual certified. The US market rental premium is a lower 5.9 per cent (CI 4.3; 7.5 per cent), in comparison to LEED and dual certification, which suggests that the Energy Star rating may be lowering the overall US rental premium.

Whilst LEED and Energy Star attribute to the most observations for rental premiums, the sales premium is also largely influenced by EPCs and other certification schemes, such as Green Mark and NABERS. The subgroup analysis of the sales premium by energy efficiency measure estimates a significant positive premium for LEED at 8.3 per cent (CI 2.6; 14 per cent) and other certification schemes at 5.3 per cent (CI 4.1; 6.5 per cent); however, the confidence intervals for EPCs (CI –2.5; 11.8 per cent) and Energy Star (CI –0.2; 15.1 per cent) include zero, indicating that their estimates are not a significant rejection of the null hypothesis of zero means. Whilst the EPC subgroup exhibits high heterogeneity, which may suggest that the premium varies throughout Europe, the Energy Star subgroup was observed entirely in the US and does not exhibit as much heterogeneity. As such, Energy Star certification may not result in a sales price premium, although the confidence interval is largely positive. Again, this signifies that LEED is the optimal choice of certification.

12.5 Implications for investments in sustainable real estate

The evidence of positive price premiums reflects a consumer willingness to pay (WTP) for green real estate and, perhaps the ability to convey its energy efficiency. Subject to this study's limitations, the significant premiums obtained from the primary and subgroup meta-analyses may be directly input into cash-flow analyses to inform investment and policy-making decisions. The estimations within Table 12.8, with positive confidence intervals and sufficient observations, may be employed.

A discernible application of the green premiums would be within real estate valuation. Usually, the value of a real estate asset is based on what an investor would expect to pay for its projected net operating income in perpetuity. The viability of an investment over a given time period can be modelled using a discounted cash flow (DCF) – the present value of future cash flows minus the initial investment:

$$NPV = \sum_{t=1}^{n} \frac{NCF_t \times (1+g)^t}{(1+r)^t} - NCF_0 \tag{3}$$

where NPV is the asset's net present value, t is the holding period, NCF_t is the net cash flow at period t, NCF_0 is the initial investment at the project outlay, g is the market growth rate and r is the rate of return. In practice, the fund's weighted average cost of capital (WACC) is often employed as the rate of return to solve for NPV; this is the rate required to cover the fund's debt obligations, and to meet its return on equity. The NPV is therefore the excess investment return at the fund's cost of capital, expressed as a monetary value. An NPV of zero indicates that the fund is 'breaking even' at its cost of capital. An NPV greater than zero – the excess return – could be added to the purchase price in a competitive bidding process.

In using a DCF to assess the viability of an energy-efficient real estate investment, practitioners can anticipate higher NCF_t throughout the entire holding period, driven by a higher rental premium and presumably a reduction in operating costs. A sales premium would also increase NCF at its sale. Both premiums result in a higher NPV, indicating that their inclusion within a DCF may permit funds to bid more competitively. These premiums can be applied in both

Table 12.8 Significant positive rental and sales premiums

	Overall	*US*	*Commercial*	*Residential*	*LEED*	*Energy Star*
Rental premium	6	5.9	5.4	8.2	7.3	3.6
	[4.3; 7.5]	[4.3; 7.5]	[3.7; 7.2]	[2.4; 14]	[4; 10.5]	[1.7; 5.5]
Sales premium	7.6	10.5	11.5	5.5	8.3	NS
	[5.9; 9.4]	[7; 14]	[5.8; 17.3]	[3.6; 7.5]	[2.6; 14]	

Note: NS = not significant to reject the null hypothesis of a zero mean.

development and acquisition scenarios, and factored in alongside other variables associated with energy-efficient real estate – notably, construction cost premiums, operating expenses and structural vacancy rates. The base case green premiums can be taken as the weighted mean effect sizes for the property's respective subgroup, whilst the downside and upside scenarios can be taken as the bounds of its confidence interval. As the statistically significant premiums are all positive, providing that the cost of implementing environmental certification does not outweigh the premiums, the split-incentive problem, which agency theorists such as Jaffe and Stavins (1994) posit, does not arise; the owners would benefit from investing in energy efficiency every time.

Previous studies have focused mostly on the premiums available to owners, rather than on the net savings available to tenants; however, studies such as Morri and Soffietti (2013) and Zalejska-Jonsson (2014), have indicated that the tenant's willingness to pay a premium depends on perceived savings from reduced operating expenses. In other words, at least part of the observed green premiums may stem from cost savings for tenants who in turn pay higher rents. The premiums could also be analysed from the standpoint of a tenant within a cost–benefit analysis, to analyse whether a perceived reduction in operating expenses outweighs a rental premium. If the rental premiums are not outweighed by savings to the tenant, it may also be the case that there is no 'vicious circle of blame'; in fact, the developers' perception of insufficient demand for energy-efficient properties may hold some truth if a large number of buyers and tenants are unconvinced of the savings. Exploring this area is a recommendation for future research.

12.6 Limitations of the meta-analysis

Although the price premiums reported above are highly statistically significant, the limitations in the estimation must be realised before informing investment or policy decisions. One of the limitations of the underlying studies, as raised by Das and Wiley (2014), is that the estimated green premium coefficients are stationary; most of the studies do not report how the premium is changing over time and may omit an 'attrition effect'. Indeed, in one of the few studies that has provided coefficients over time, Reichardt, Fuerst, Rottke and Zietz (2012) did find that the Energy Star premium has been decreasing over time, albeit using a relatively short time series, perhaps reflecting dynamic changes in local demand and supply conditions and underlying energy prices.

Figure 12.9 plots the variance-weighted averages of the premiums by their year of publication, but no clear trend is visible. The time series is too short and the observations are too few to draw any conclusions on whether there is an underlying attrition effect. However, if an attrition effect does exist, later publications that have employed a longer time series would be prone to overestimating the green premium due to the bias caused by the inclusion of earlier studies. This prompts an investigation for future studies.

This chapter examines the premium for energy efficiency, on the premise that WTP for environmental certification is related to reduced operating expenses; however, only three observations directly measured energy consumption. As the majority of observations were premiums for environmental certification, there are undoubtedly other factors besides energy efficiency that influence WTP.

The WTP for energy efficiency may not be financially driven, as this chapter has assumed. For example, corporate tenants may also be seeking to satisfy corporate social responsibility (CSR) requirements, which have received increased attention and response in the past two decades. There is also a bias towards the US market, particularly LEED and Energy Star commercial properties.

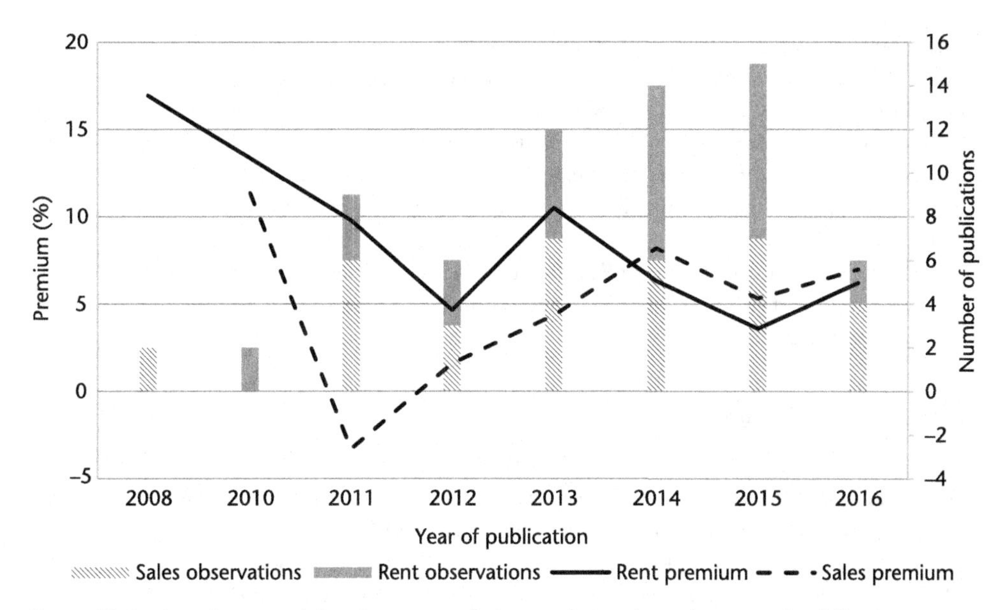

Figure 12.9 A variance-weighted average of observed premiums by year of publication

12.7 Conclusions

The random-effects meta-analysis aggregated 42 studies that examine the relationship between energy efficiency and property prices. The studies were identified through a systematic review, searching two general databases and six academic databases. The primary analysis included all observations to estimate an overall weighted mean premium for rental and sales values. The rental premium was estimated as 6.02 per cent (CI 4.30; 7.75 per cent) and the sales premium as 7.61 per cent (CI 5.86; 9.36 per cent). Although the estimates were both highly significant, the analyses revealed considerable statistical heterogeneity and potential publication bias.

Further subgroup analyses were conducted as a means of explaining the high degree of heterogeneity. This involved analysing the effect sizes and between-subgroup variances of different markets, energy efficiency measures and property types. All were found to significantly contribute to heterogeneity. From the subgroup analyses, it was also determined that the US market, commercial property, residential property, LEED and Energy Star subgroups contained a sufficient number of studies to estimate significant premiums. The Energy Star rating appears inferior to the LEED rating, particularly with regard to the sales premium, which was not significantly positive. As such, the Energy Star rating is believed to be reducing the estimated premiums for the US market, due to the large number of Energy Star observations underlying the subgroup.

The potential publication bias was identified by using the Egger et al. (1997) test to measure funnel plot asymmetry. The test indicated significant bias, most likely attributable to the exclusion of smaller studies estimating negative premiums that may or may not exist.

The green premiums can be applied within a DCF model to demonstrate their implications for investors and stakeholders. If the premiums outweigh the cost of their implementation/construction, the DCF would yield a higher *NPV* and therefore the principal–agent problem could be rejected as a barrier to energy efficiency. Based on previous studies, such as Morri and Soffietti (2013) and Zalejska-Jonsson (2014), the tenant's WTP for energy-efficient property primarily related to a perceived reduction in operating expenses. The quantification of a green premium prompts an interesting discussion about the tenant's WTP, and whether the energy savings justify the rental premium. Widespread tenant and buyer scepticism of the savings may also provide an alternative to the 'vicious circle of blame' as an explanation for the low development of green properties (Andelin, Sarasoja, Ventovuori and Junnila, 2015).

This chapter contributes towards the body of evidence on real estate green premiums to inform policy-making and investment decisions. There has been little prior research on the green premium from the tenant's perspective, and on the net savings available to them. The majority of studies also report a 'static' premium, which has been identified as a limitation due to a potential underlying attrition effect. These are both areas for future research.

Appendix: Methodology

This chapter contains a meta-analysis of the overall rental and sales value green premiums and includes all observations recorded in the dataset, besides those that are considered 'supplementary' to another effect reported within the same study, e.g., the LEED and Energy Star estimates in Eichholtz et al. (2013) when an LEED and/or Energy Star estimate is also provided. The primary analysis is expected to exhibit large heterogeneity due to the variety of the studies, in terms of the property type, country and energy efficiency measure analysed, although it will be biased towards the US commercial market and LEED/Energy Star certification (Figure 12.4). Further estimations of market, property type and certification subgroup effect sizes will be undertaken to explain heterogeneity between studies. For the purpose of this meta-analysis, the single study on hotel green premiums is included as a rental green premium, although the short-term nature of a hotel stay will invariably have an impact on consumer WTP.

All but one of the selected studies (Freybote, Sun and Yang, 2015) estimate the price effect of the green premium through a hedonic log-linear model, which originates from the work of Rosen (1974). The model assumes that the price of a good is a function of its attributes (see Equation 4). In this case, the rental or sales price P of a building is a function of energy efficiency φ, and other attributes – commonly those concerned with its physical (ζ), locational (ψ) and sales/lease (η) characteristics.

$$P = f(\varphi, \zeta, \psi, \eta) \tag{4}$$

Thus, the regression can be notated in log-linear form as:

$$\text{in}\, P = X\beta + \varepsilon \tag{5}$$

where ln P is the natural log of rental/sales price, X is a matrix of attribute vectors (e.g., φ, ζ, ψ, η), β is a matrix of respective vectors to be estimated and ε is the stochastic error term. As such, the result is reported as an elasticity coefficient ($\%\Delta P/\%\Delta\varphi$): an intuitive ratio that represents the price differential, *ceteris paribus*, between a 'treatment' sample of energy-efficient properties and a non-energy-efficient 'control' sample. This ratio is taken to directly represent the green premium as a percentage, and will serve as the observed effect size for each study. In the case of Freybote et al. (2015), who employed a second-order parsimonious spatio-temporal autoregressive (2PSTAR) model, the estimate was still provided as an elasticity coefficient. Within each study, there are usually multiple model iterations that estimate the green premium, usually controlling for different factors or interaction terms. It is important to acknowledge that the green premium may not be solely attributable to energy efficiency and the variables denoted in Equation 4; rather, it may be dependent on an interaction between energy efficiency and another variable (e.g., age), or confounded by selection bias from uncontrolled systematic differences between the treatment and control samples.

Whilst a few studies have introduced instrumental variables to mitigate selection bias from unobservables (e.g., Wiley, Benefield and Johnson, 2010; Szumilo and Fuerst, 2015; Brounen and Kok, 2011), the studies have more commonly addressed selection bias from observables – through the Rosenbaum and Rubin (1983) propensity score matching methods (e.g. Reichardt, 2014; Chegut, Eichholtz and Kok, 2014; Yoshida and Sugiura, 2015). More often than not, the studies explore various interactions between energy efficiency and other variables through the inclusion of interaction terms (e.g., $\delta E_{it} \times \delta AGE_{it}$); by doing so, they can identify more precise and informative effect sizes for different subgroups (e.g., the green premium for energy-efficient buildings in different age categories). However, the terms are inconsistent throughout the selected studies. Therefore, to ensure consistency and simplicity, the green premium will be taken from the most statistically robust model estimates that exclude interaction terms, as suggested by Stanley and Doucouliagos (2012), but control for the most factors. In cases where the authors only provided the interaction terms of their models, a 'crude' summary effect size was estimated by adding an equal-weighted average of the subgroup coefficient estimates.

As the intention of this research is to inform policy-making and investment decisions, it is of utmost importance that a suitable model is used to estimate the overall effect and to alleviate publication bias, which can lead to its overestimation (Van Assen, Van Aert and Wicherts, 2015). Whilst the inclusion of grey literature within the systematic review has sought to mitigate bias at the editorial level, it is equally important to include observed effect sizes from all studies that met the inclusion criteria; this is regardless of the direction of the effect and its statistical significance (see Hedges and Olkin, 1985, for a discussion on the importance of including non-significant results).

A common issue within meta-analyses is missing standard errors within studies (Higgins, Deeks and Altman, 2011). Whilst it is common practice for medical journals to report standard errors, the real estate literature is inconsistent in its reporting; 25 of the studies did not directly report standard errors. Algebraic and approximate algebraic recalculation of the standard errors from other statistical information provided, such as exact P-values and T-statistics, was used in 22 of the studies (Stevens, 2011). In the remaining three studies where this was not possible, values were imputed from other studies within the meta-analysis – a technique that Furukawa, Barbui, Cipriani, Brambilla and Watanabe (2006) demonstrated to yield accurate results.

In estimating an overall effect, a meta-analysis should address the weighting of significance that it assigns to the results of its underlying studies. There are predominantly two meta-analytic models: 'fixed effects' and 'random effects'. A fixed-effects model assigns weighting based on the inverse variance within the studies (similar to the size of the studies), assuming a common effect size (Borenstein, Hedges and Rothstein, 2007); therefore, the methodology followed by each study should be identical. However, as Hedges and Vevea (1998) note, model selection is not only a question of study homogeneity; rather, it is also the inferences that can be made. In the case of the fixed-effects model, by limiting the analysis to assign weights based solely upon the variance within studies, inference is also limited to the observed studies (conditional inference). This model does not lend itself to the heterogeneity of the studies selected for the meta-analysis, which exhibit variation in their samples, methodologies and control factors; this is expected due to the inherent heterogeneity of real estate. Furthermore, a conditional inference is only pertinent to those studies analysed; therefore, in order to apply the estimated effect to other transactions or cash flows, it is imperative that unconditional inferences – generalised observations that can extend beyond the observed studies – can be made (Hedges and Vevea, 1998). Unconditional inferences can be made through the application of the random-effects model. This is because the model does not assume a common effect size and accounts for the variance between studies (tau-squared), as well as the variance within studies (Borenstein et al., 2007) (see Equation 7). Incorporating the variance between studies also accounts for heterogeneity, such as different market conditions between studies.

As noted above, the random-effects model does not assume a common effect size; rather, it seeks to estimate a distribution of true effects, y_i, to estimate an overall mean effect, μ (Borenstein et al., 2007). The true effect size of a study is essentially a projection of its observed effect θ_i for an infinite sample; thus, sampling error is inevitably present (Borenstein et al., 2009). With this in mind, a study's observed effect should be equal to the deviation of its true effect from the overall mean effect, δ_i, adjusted for sampling error ε_i (Borenstein et al., 2009):

$$y_i = \mu + \delta_i + \varepsilon_i \tag{6}$$

d_i is derived from the standard deviation of the true effects distribution, t^2, which is essentially the between-study variance; its estimation can be observed in Equation 8.

However, a meta-analysis must start with the observed effects in order to arrive at an estimation of μ. Considering k number of studies with index ($i = 1, 2, \ldots, k$), the random-effects model assigns weights to studies based on both within-study and between-study variance:

$$w_i^* = \frac{1}{\tau^2 + v_{yi}} \tag{7}$$

where w_i^* is the inverse of the sample variance in the ith study (σ_i^{-2}), t^2 is the variance of parameter effect size between studies and v_{y_i} is the variance within study i. Whilst v_{y_i} is observable, t^2 is estimated following the DerSimonian and Laird (1986) non-iterative method-of-moments estimator:

$$\tau^2 = max\left(0, \frac{Q - (k-1)}{\Sigma w_i - \dfrac{\Sigma w_i^2}{\Sigma w_i}}\right) \tag{8}$$

where Q is Cochran's Q, the observed weighted sum of squares, estimated by:

$$Q = \sum_{i=1}^{k} w_i (y_i - \bar{y}_w)^2 \tag{9}$$

where $\bar{y}_w$ is the weighted mean effect estimator $\left(\sum_{i=1}^{k} w_i y_i \big/ \sum_{i=1}^{k} w_i\right)$; and y_i is the estimated effect of the ith study.

The weighted mean (summary) effect, $\hat{\mu}$, can then be estimated:

$$\hat{\mu} = \frac{\sum_{i=1}^{k} w_i^* y_i}{\sum_{i=1}^{k} w_i^*} \tag{10}$$

Following Borenstein et al. (2009, p.74), the estimated summary effect is then used to calculate the variance, standard error and a 95 per cent confidence interval of the summary effect:

$$\upsilon_{\hat{\mu}} = \frac{1}{\sum_{i=1}^{k} w_i^{\star}} \tag{11}$$

$$SE_{\hat{\mu}} = \sqrt{\upsilon_{\hat{\mu}}} \tag{12}$$

$$Lower_{\hat{\mu}} = \hat{\mu} - 1.96 \times SE_{\hat{\mu}}, \tag{13}$$

$$Upper_{\hat{\mu}} = \hat{\mu} + 1.96 \times SE_{\hat{\mu}}$$

References

Andelin, M., Sarasoja, A.L., Ventovuori, T. and Junnila, S., 2015. Breaking the circle of blame for sustainable buildings: evidence for Nordic countries. *Journal of Corporate Real Estate*, 17(1), pp.26–45.

Aroul, R.R. and Hansz, J.A., 2012. The value of 'green': evidence from the first mandatory residential green building program. *Journal of Real Estate Research*, 34(1), pp.27–49. Available at: <Go to ISI>://WOS:000301320900002.

Arup, 2014. *International Sustainability Systems Comparison*, Available at: http://publications.arup.com/~/media/Publications/Files/Publications/I/International_Sustainability_Systems_Report.ashx.

de Ayala, A., Galarraga, I. and Spadaro, J.V., 2016. The price of energy efficiency in the Spanish housing. *Energy Policy*, 94, pp.16–24.

Bennett, D.A., Eliasz, T.K., Forbes, A., Kiszely, A., Khosla, R., Petrinic, T., Praveen, D., Shrivastava, R., Xin, D., Patel, A. and MacMahon, S., 2012. Study protocol: systematic review of the burden of heart failure in low- and middle-income countries. *Systematic Reviews*, 1(59), pp.1–5.

Bond, S.A. & Devine, A. J., 2016. Real Estate Finance and Economics, 52, p.117. Available at: https://doi.org/10.1007/s11146-015-9499-y.

Bonde, M. and Song, H. S., 2013. Is energy performance capitalized in office building appraisals? *Property Management*, 31(3), pp.200–215.

Bonde, M. and Song, H.S., 2014. Does greater energy performance have an impact on real estate revenues? *Journal of Sustainable Real Estate*, 5(1), pp.171–182.

Borenstein, M., Hedges, L.V. and Rothstein, H.R., 2007. *Meta-Analysis: Fixed Effect vs. Random Effects*, Available at: www.meta-analysis.com/downloads/Meta-analysis fixed effect vs random effects.pdf.

Borenstein, M., Hedges, L.V., Higgins, J. and Rothstein, H.R., 2009. *Introduction to Meta-Analysis*, Chichester: John Wiley and Sons, Ltd.

Brounen, D. and Kok, N., 2011. On the economics of energy labels in the housing market. *Journal of Environmental Economics and Management*, 62(2), pp.166–179. Available at: www.sciencedirect.com/science/article/pii/S0095069611000337.

Brown, M.J. and Watkins, T., 2015. The 'green premium' for environmentally certified homes: a meta-analysis and exploration. Research Gate Working Paper.

Bruegge, C., Carrión-Flores, C. and Pope, J. C., 2016. Does the housing market value energy-efficient homes? Evidence from the Energy Star program. *Regional Science and Urban Economics*, 57, pp.63–76.

Cadman, D., 2000. The vicious circle of blame. *Upstream*. Available at: www.upstreamstrategies.co.uk.

Cajias, M. and Piazolo, D. (2013). Green performs better: energy efficiency and financial return on buildings. *Journal of Corporate Real Estate*, 15(1), pp.53–72. Available at: https://doi.org/10.1108/JCRE-12-2012-0031.

Cerin, P., Hassel, L.G. and Semenova, N., 2014. Energy performance and housing prices. *Sustainable Development*, 22(6), pp.404–419. Available at: http://dx.doi.org/10.1002/sd.1566.

Chegut, A., Eichholtz, P. and Kok, N., 2014. Supply, demand and the value of green buildings. *Urban Studies*, 51(1), pp.22–43. Available at: <Go to ISI>://WOS:000328600300002.

Cole, R.J., 2005. Building environmental assessment methods: redefining intentions and roles. *Building Research & Information*, 33(5), pp.455–467.

Couch, C., Carswell, A.T. and Zahirovic-Herbert, V., 2015. An examination of the potential relationship between green status of multifamily properties and sale price. *Housing and Society*, 42(3), pp.179–192. Available at: https://doi.org/10.1080/08882746.2015.1121675.

Cumming, G. and Finch, S., 2005. Inference by eye. *American Psychologist*, 60(2), pp.170–180.

Das, P. and Wiley, J.A., 2014. Determinants of premia for energy-efficient design in the office market. *Journal of Property Research*, 31(1), pp.64–86. Available at: http://dx.doi.org/10.1080/09599916.2013.788543.

de Ayala, A., Galarraga, I. and Spadaro, J.V., 2016. The price of energy efficiency in the Spanish housing market. *Energy Policy*, 94, pp.16–24.

Deng, Y. and Wu, J., 2014. Economic returns to residential green building investment: the developers' perspective. *Regional Science and Urban Economics*, 47, pp.35–44.

Deng, Y., Li, Z. and Quigley, J., 2012. Economic returns to energy-efficient investments in the housing market: Evidence from Singapore. *Regional Science and Urban Economics*, 42(3), pp.506–515.

Dermisi, S. and McDonald, J., 2011. Effect of 'green' (LEED and Energy Star) designation on prices/sf and transaction frequency: the Chicago office market. *Journal of Real Estate Portfolio Management*, 17(1), pp.39–52.

DerSimonian, R. and Laird, N., 1986. Meta-analysis in clinical trials. *Controlled Clinical Trials*, 7(3), pp.177–188. Available at: http://linkinghub.elsevier.com/retrieve/pii/0197245686900462.

Devine, A. and Kok, N., 2015. Green Certification and Building Performance: Implications for Tangibles and Intangibles. *Journal of Portfolio Management*, 41(6), pp.151–163.

Egger, M., Smith, G.D., Schneider, M. and Minder, C., 1997. Bias in meta-analysis detected by a simple, graphical test. *British Medical Journal*, 315(7109), pp.629–634.

Eichholtz, P., Kok, N. and Quigley, J., 2010. Doing well by doing good? Green office buildings. *American Economic Review*, 100(5), pp.2492–2509.

Eichholtz, P., Kok, N. and Quigley, J.M., 2013. The economics of green building. *Review of Economics and Statistics*, 95(1), pp.50–63. Available at: http://search.ebscohost.com/login.aspx?direct=true&db=bth&AN=86185656&site=bsi-live.

Feige, A., McAllister, P. and Wallbaum, H., 2013, Rental price and sustainability ratings: which sustainability criteria are really paying back? *Journal of Construction Management and Economics*, 31(4), pp.322–334.

Freybote, J., Sun, H. and Yang, X., 2015. The impact of LEED neighborhood certification on condo prices. *Real Estate Economics*, 43(3), pp.586–608. Available at: http://search.ebscohost.com/login.aspx?direct=true&db=bth&AN=108675017&site=bsi-live.

Fuerst, F. and McAllister, P., 2011a. Green noise or green value? Measuring the effects of environmental certification on office values. *Real Estate Economics*, 39(1), pp.45–69.

Fuerst, F. and McAllister, P., 2011b. The impact of Energy Performance Certificates on the rental and capital values of commercial property assets. *Energy Policy*, 39(10), pp.6608–6614. Available at: www.sciencedirect.com/science/article/pii/S0301421511006021.

Fuerst, F. and McAllister, P., 2011c. Eco-labeling in commercial office markets: do LEED and Energy Star offices obtain multiple premiums? *Ecological Economics*, 70(6), pp.1220–1230.

Fuerst, F. and Shimizu, C., 2016. Green luxury goods? The economics of eco-labels in the Japanese housing market. *Journal of the Japanese and International Economies*, 39, pp.108–122.

Fuerst, F. and van de Wetering, J., 2015. How does environmental efficiency impact on the rents of commercial offices in the UK? *Journal of Property Research*, 32(3), pp.193–216.

Fuerst, F., McAllister, P., Nanda, A. and Wyatt, P., 2016. Energy performance ratings and house prices in Wales: an empirical study. *Energy Policy*, 92, pp.20–33.

Fuerst, F., Oikarinenm E. and Harjunen, O., 2016. Green signalling effects in the market for energy-efficient residential buildings. *Applied Energy*, pp.560–571.

Fuerst, F., van de Wetering, J. and Wyatt, P., 2013. Is intrinsic energy efficiency reflected in the pricing of office leases? *Building Research and Information*, 41(4), pp.373–383.

Furukawa, T.A., Barbui, C., Cipriani, A., Brambilla, P. and Watanabe, N., 2006. Imputing missing standard deviations in meta-analyses can provide accurate results. *Journal of Clinical Epidemiology*, 59(1), pp.7–10.

Gabe, J. and Rehm, M., 2014. Do tenants pay energy efficiency rent premiums? *Journal of Property Investment & Finance*, 32(4), pp.333–351. Available at: www.emeraldinsight.com/doi/abs/10.1108/JPIF-09-2013-0058.

Hedges, L.V. and Olkin, I., 1985. *Statisical Methods for Meta-Analysis*, London: Academic Press Inc.

Hedges, L.V. and Vevea, J.L., 1998. Fixed- and random-effects models in meta-analysis. *Psychological Methods*, 3(4), pp.486–504. Available at: http://personal.psc.isr.umich.edu/yuxie-web/files/pubs/Articles/Hedges_Vevea1998.pdf.

Higgins, J.P.T., 2008. Commentary: heterogeneity in meta-analysis should be expected and appropriately quantified. *International Journal of Epidemiology*, 37(5), pp.1158–1160.

Higgins, J.P.T., Deeks, J.J. and Altman, D.G., 2011. Special topics in statistics. In J.P.T. Higgins and S. Green, eds. *Cochrane Handbook for Systematic Reviews of Interventions*. The Cochrane Collaboration. Available at: handbook.cochrane.org.

Higgins, J.P., Thompson, S.G., Deeks, J.J. and Altman, D.G., 2003. Measuring inconsistency in meta-analyses. *British Medical Journal*, 327(7414), pp.557–560.

Högberg, L., 2013. The impact of energy performance on single-family home selling prices in Sweden. *Journal of European Real Estate Research*, 6(3), pp.242–261. Available at: https://doi.org/10.1108/JERER-09-2012-0024.

Hopewell, S., McDonald, S., Clarke, M.J. and Egger, M., 2007. *Grey Literature in Meta-Analyses of Randomized Trials of Health Care Interventions*. The Cochrane Library.

Hopewell, S., Loudon, K., Clarke, M.J., Oxman, A.D. and Dickersin, K., 2009. *Publication Bias in Clinical Trials Due to Statistical Significance or Direction of Trial Results*. The Cochrane Library.

Ingle, A., Moezzi, M., Lutzenhiser, L. and Diamond, R., 2014. Better home energy audit modelling: incorporating inhabitant behaviours. *Building Research & Information*, 42(4), pp.409–421. Available at: www.tandfonline.com/action/journalInformation?journalCode=rbri20.

International Energy Agency, 2016. Energy efficiency. Available at: www.iea.org/aboutus/faqs/energyefficiency.

Jaffe, A.B. and Stavins, R.N., 1994. The energy paradox and the diffusion of conservation technology. *Resource and Energy Economics*, 16(2), pp.91–122. Available at: http://scholar.harvard.edu/files/stavins/files/theenergyparadox.ree1994.pdf?m=1440012158.

Jayantha, W.M. and Man, W.S., 2013. Effect of green labelling on residential property price: a case study in Hong Kong. *Journal of Facilities Management*, 11(1), pp.31–51. Available at: https://doi.org/10.1108/14725961311301457.

Klewitz, J. and Hansen, E.G., 2014. Sustainability-oriented innovation of SMEs: a systematic review. *Journal of Cleaner Production*, 65, pp.57–75.

Koirala, B. Bohara, A. and Berrens, R., 2014. Estimating the net implicit price of energy efficient building codes on U.S. households. *Energy Policy* 73, pp.667–675.

Lau, J., Ioannidis, J., Terrin, N., Schmid, C. and Olkin, I., 2006. The case of the misleading funnel plot. *British Medical Journal*, 333(1), pp.597–600.

Lee, W.L., 2013. A comprehensive review of metrics of building environmental assessment schemes. *Energy and Buildings*, 62, pp.403–413. Available at: http://dx.doi.org/10.1016/j.enbuild.2013.03.014.

Light, R.J. and Pillemer, D.B., 1984. *Summing Up: The Science of Reviewing Research*, Cambridge: Harvard University Press.

McAllister, P. n.d., *Handle with Care? An Evaluation of Empirical Research on the Financial Returns from Investing in Real Estate Assets with Superior Environmental Performance*. LefargeHolcim Foundation for Sustainable Construction.

McAllister, P., 2013. Studies of price effects of eco-labels in real estate markets: an 'off the record' record, Available at: www.reading.ac.uk/web/FILES/REP/Table_of_studies_(May_2013).pdf.

Miller, N., Spivey, J. and Florance, A., 2008. Does green pay off? *Journal of Real Estate Portfolio Management*, 14(4), pp.385–400.

Morri, G. and Soffietti, F., 2013. Greenbuilding sustainability and market premiums in Italy. *Journal of European Real Estate Research*, 6(3), pp.303–332.

Nappi-Choulet, I. and Décamps, A., 2013. Capitalization of energy efficiency on corporate real estate portfolio value. *Journal of Corporate Real Estate*, 15(1), pp.35–52. Available at: www.emeraldinsight.com/doi/abs/10.1108/JCRE-01-2013-0005.

Newell, G., MacFarlane, J. and Walker, R., 2014. Assessing energy rating premiums in the performance of green office buildings in Australia. *Journal of Property Investment & Finance*, 32(4), pp.352–370. Available at: https://doi.org/10.1108/JPIF-10-2013-0061.

Olubunmi, O.A., Xia, P.B. and Skitmore, M., 2016. Green building incentives: a review. *Renewable and Sustainable Energy Reviews*, 59, pp.1611–1621.

Reichardt, A., 2014. Operating expenses and the rent premium of Energy Star and LEED certified buildings in the Central and Eastern U.S. *Journal of Real Estate Finance and Economics*, 49(3), pp.413–433. Available at: http://search.ebscohost.com/login.aspx?direct=trueanddb=bthandAN=97875167andsite=bsi-live.

Reichardt, A., Fuerst, F., Rottke, N. and Zietz, J., 2012. Sustainable building certification and the rent premium: a panel data approach. *Journal of Real Estate Research*, 34(1), pp.99–126. Available at: <Go to ISI>://WOS:000301320900005.

Robinson, S. and McAllister, P., 2015. Heterogeneous price premiums in sustainable real estate? An investigation of the relation between value and price premiums. *Journal of Sustainable Real Estate*, 7(1), pp.1–20.

Rosen, S., 1974. Hedonic prices and implicit markets: product differentiation in pure competition. *The Journal of Political Economy*, 82(1), pp.34–55.

Rosenbaum, P.R. and Rubin, D.B., 1983. The central role of the propensity score in observational studies for causal effects. *Biometrika*, 70(1), pp.41–55.

Runde, T. and Thoyre, S., 2010. Integrating sustainability and green building into the appraisal process. *Journal of Sustainable Real Estate*, 2(1), pp.221–248.

Sánchez-Ollero, J.L., García-Pozo, A. and Marchante-Mera, A., 2014. How does respect for the environment affect final prices in the hospitality sector? A hedonic pricing approach. *Cornell Hospitality Quarterly*, 55(1), pp.31–39. Available at: http://cqx.sagepub.com/cgi/content/abstract/55/1/31.

Sayce, S., Ellison, L. and Parnell, P. 2007. Understanding investment drivers for UK sustainable property. *Building Research & Information*, 35(6), pp.629–643.

Sayce, S., Sundberg, A. and Clements, B., 2010. *Is Sustainability Reflected in Commercial Property Prices: An Analysis of the Evidence Base*. London: RICS. Available at: http://eprints.kingston.ac.uk/15747/1/Sayce-S-15747.pdf.

Shewmake, S. and Viscusi, W.K., 2015. Producer and consumer responses to green housing labels. *Economic Inquiry*, 53(1), pp.681–699. Available at: <Go to ISI>://WOS:000345350200039.

Shimizu, C., 2013. Sustainable measures and economic value in green housing. *Open House International Journal*, 38(3), pp.57–63.

Stanley, T.D. and Doucouliagos, H., 2012. *Meta-Regression Analysis in Economics and Business*. London: Routledge.

Sterne, J.A.C., Egger, M. and Moher, D., 2011. Addressing reporting biases. In J.P.T. Higgins and S. Green, eds. *Cochrane Handbook for Systematic Reviews of Interventions*. The Cochrane Collaboration. Available at: handbook.cochrane.org.

Sterne, J.A.C., Gavaghan, D. and Egger, M., 2000. Publication and related bias in meta-analysis: power of statistical tests and prevalence in the literature. *Journal of Clinical Epidemiology*, 53, pp.1119–1129.

Stevens, J.W., 2011. A note on dealing with missing standard errors in meta-analyses of continuous outcome measures in WinBUGS. *Pharmaceutical Statistics*, 10(4), pp.374–378.

Szumilo, N. and Fuerst, F., 2015. Who captures the 'green value' in the US office market? *Journal of Sustainable Finance and Investment*, 5(1-2), pp.65–84. Available at: http://dx.doi.org/10.1080/20430795.2015.1054336.

Thorlund, K., Imberger, G., Johnston, B.C., Walsh, M., Awad, T., Thabane, L., Gluud, C., Devereaux, P.J. and Wetterslev, J., 2012. Evolution of heterogeneity (I2) estimates and their 95% confidence intervals in large meta-analyses. *PLoS ONE*, 7(7), p.e39471.

van Assen, M.A.L.M., van Aert, R.C.M. and Wicherts, J.M., 2015. Meta-analysis using effect size distribution of only statistically significant studies. *Psychological Methods*, 20(3), pp.293–309.

von Hippel, P.T., 2015. The heterogeneity statistic I2 can be biased in small meta-analyses. *BMC Medical Research Methodology*, 15(35).

Wedding, C.G. and Crawford-Brown, D., 2008. Improving the link between the LEED green building label and a building's energy-related environmental metrics. *Journal of Green Building*, 3(2), pp.85–105.

Wiley, J.A., Benefield, J.D. and Johnson, K.H., 2010. Green design and the market for commercial office space. *Journal of Real Estate Finance and Economics*, 41(2), pp.228–243. Available at: <Go to ISI>://WOS:000280074400006.

World Green Building Council, 2013. *The Business Case for Green Building: A Review of the Costs and Benefits for Developers, Investors and Occupants*. Available at: www.worldgbc.org/files/1513/6608/0674/Business_Case_For_Green_Building_Report_WEB_2013-04-11.pdf.

Yoshida, J. and Sugiura, A., 2015. The effects of multiple green factors on condominium prices. *Journal of Real Estate Finance and Economics*, 50(3), pp.412–437. Available at: <Go to ISI>://WOS:000351100800006.

Zalejska-Jonsson, A., 2014. Stated WTP and rational WTP: willingness to pay for green apartments in Sweden. *Sustainable Cities and Society*, 13, pp.46–56.

Zhang, X., 2015. Green real estate development in China: state of art and prospect agenda – a review. *Renewable and Sustainable Energy Reviews*, 47, pp.1–13.

Zheng, S., Wu, J., Kahn, M.E. and Deng, Y., 2012. The nascent market for 'green' real estate in Beijing. *European Economic Review*, 56(5), pp.974–984. Available at: www.sciencedirect.com/science/article/pii/S001429211200030X.

13

Sustainability and housing value in Victoria, Australia

Neville Hurst and Sara Wilkinson

13.0 Introduction

Shelter for humans, in the form of housing, is essential for survival and has, over the centuries, evolved into the shapes and forms seen today. It is the case that in developed countries, it is often seen both as a place of habitation and of wealth creation, providing security subsequent to retirement from work (Logan and Molotch 2007). However, in the use of houses as well as their construction, renovation and/or extension to accommodate people's changing needs and desires, there is a significant impact upon the environment (Bardhan, Jaffee, Kroll and Wallace 2014). In Australia, and particularly Victoria, housing contributes around 20 per cent to greenhouse gas (GHG) emissions (Your Home 2013). These impacts comprise largely of GHG emissions as a result of utilising fossil fuel for energy generation; however, excessive water consumption and landfill waste are increasingly becoming a concern (Ghaffarian Hoseini 2012). With predictions of irreversible environmental impact, the need for housing to become more energy efficient and for occupants to behave in a more environmentally sustainable way has never been more urgent (Garnaut 2011). The question is: how can we achieve this, when the housing market is made up of such disparate elements, and the value drivers are a meld, often comprising personal, social and pragmatic attitudes and beliefs? This chapter considers these issues from the perspective of real estate markets and, as market facilitators, real estate agents. It does so in an Australian context, and in particular, in one of its most populated states: Victoria. Given the access to the housing market that real estate agents have, it is quite likely that they would be aware of the importance of sustainability measures in housing. In addition, given the position of influence that real estate agents possess, it is conceivable that they can influence the market in a positive way. To achieve this, they must first have the knowledge of sustainability's importance if they do not already.

Globally, governments have been challenged to encourage the uptake of more energy-efficient housing (UNECE 2013). Some governments, such as in Australia, have adopted mandatory approaches through regulatory frameworks with minimum building energy performance standards for new and homes and those that are substantially refurbished or altered (VBA 2014). Other authorities have added reporting requirements; for example, the UK and EU constituencies require energy ratings to be provided in the form of Energy Performance Certificates (EPCs) at the point of sale (UK Government 2016; BPIE 2016). In response to international calls to reduce GHG emissions during the early 2000s, Australian federal and state governments initially planned the introduction of mandatory energy efficiency reporting at the point of sale. Queensland was the only state to implement the policy in 2009 (Bryant and Eves 2012); however, it has since been repealed. Contemporary Australian governments have adopted a neo-liberal perspective favouring 'free-market forces' to generate demand for energy-efficient housing with the legislative framework only requiring new homes and extensions to comply with minimum energy efficiency standards. Considering this perspective, the question therefore becomes: how effective is this approach when the majority of housing stock was constructed prior to the introduction of such legislation?

In considering the topic of energy-efficient housing, it must be acknowledged that there are numerous global, regional and local consequences and impacts resulting from the occupation of housing and domestic activities. It is therefore necessary to consider, first, what exists in order to understand how efficiencies and improvements can be made. There has been extensive research into the effects of anthropological activity upon the environment, and the broader scientific community has concluded that significant action is needed to reduce the existing harmful and negative impacts of human activity upon the environment if significant global warming and associated extreme weather events are to be mitigated (Stern 2007; Garnaut 2011; UN 2015). The next section considers the imperatives for energy-efficient housing. This discussion is provided in the context of detached housing. Detached housing in Australia remains the most popular form of housing and choices regarding energy-efficient technologies can be made with regard to personal benefits. Such a situation provides rich research opportunities.

13.1 The need for energy-efficient housing

Across Australia, housing is estimated to contribute up to 20 per cent of the nation's total GHG emissions (Department of Industry 2014; EPA 2014). In Victoria, housing releases around 20 per cent of the state's total GHG emissions due to its reliance on coal for electrical power generation (Environment Victoria 2014). This total percentage is lower than the UK, where in 2012 25.3 per cent of total GHG emissions were derived from the residential sector (DECC 2013). Arguably, the need to avoid these undesirable changes to global climate should be sufficient in itself; however, acceptance of this reality and the call for action appears to remain in the purview of the informed to date, rather than the broader population where real change can be effected (Pereira 2012). The question of how to encourage and embed the uptake of environmentally friendly behaviours and more energy-efficient housing has challenged, and is still challenging, governments worldwide (Westley et al. 2011).

The financial benefits of reduced energy costs, the sense of contributing to a greater societal or community benefit and the opportunity for innovative housing design are just some examples of how we can profit from energy-efficient housing. However, these benefits are often typically not taken into account when buying a house (Henning 2008; Hurst 2012). In regards to house owner-occupiers, research to date has shown that whilst buyers like the idea of an energy-efficient house, they are typically not willing to pay for it (Eves and Kippes 2010; Yu and Tu 2011). Conversely, some researchers argue that this is not case (Högberg 2013; Fuerst, McAllister, Nanda and Wyatt 2015). Recent evidence is suggesting that buyers of established housing typically make no specific allowance for energy-efficient technologies but rather apply more traditional paradigms of location, etc. (Bruegge, Carrión-Flores and Pope 2016). Disagreement in regard to willingness to pay for energy-efficient housing characteristics is possibly the result of climate variances in the geographic locations of the research.

In Australia, much of the existing housing stock has poor thermal performance, with an energy efficiency performance rating of two stars or less when measured using the Nationwide Home Energy Rating System (NatHERS) (Environment Victoria 2014). The NatHERS system adopts a scale of zero to ten, where ten stars is considered thermally comfortable without the need for artificial heating and cooling (Department of Industry 2014). Notably, and unlike Europe, Australia does not have a means of conveying any information about the energy performance of the house in question to an interested buyer at the point of sale. In Europe, residential sales information is accompanied with an EPC, which is displayed in a prominent position on the advertisement, although there is debate about whether the availability of the EPC information is having an effect on the European market. So how and what, if anything, is likely to cause the Australian housing market to include energy-efficient housing characteristics as part of the choice criteria? In terms of governance, Australia is a federation of states and territories, and under the constitution, it is these state governments that hold much of the decision-making powers (Australian Government 2015). These include powers to govern the land title system and housing policy (DELWP 2016). In terms of housing, this can be considered as a positive attribute, because each state is able to respond appropriately to the unique climate in which it governs. For example, Queensland is well known for its tropical climate and Victoria for its cool climate with ski resorts in the north of the state. These disparate climates dictate a need for differing approaches to building design and how optimum energy efficiency is realised and ought to be interpreted locally.

In relation to environmental objectives, the unique characteristics of such a vast country as Australia need to be factored into any decisions about housing policy; to date, some believe this has not been done well (Ben-David 2012). The dominant political paradigm in the developed world is that of neo-liberalism, and when it comes to structural market change for energy-efficient housing in the established housing market, Australia, like many countries, has embraced a market-led doctrine for the adoption of energy-efficient housing and governments have been reluctant to interfere with traditional market processes (Australian Government Climate Change Authority 2016). There have been many cases posited for the implementation of information concerning the energy performance of houses at the point of sale (Boza-Kiss, Moles-Grueso and Urge-Vorsatz 2013; Golubchikov and Deda 2012) and globally these have varied in form and shape. The rebuttal to the case for information at the point of sale is more recent research suggesting that it is unlikely that homebuyers will respond to general information alone when it comes to developing sustainable practices (McKenzie-Mohr and Smith 1999; Henning 2008). This leaves us in a conflicted position. On the one hand, buyers are not responding to information presented in regard to house energy performance and on the other, market forces alone are likely to adhere to traditional paradigms unless something extraneous to the current status quo emerges.

If the current doctrine of reliance upon housing markets to innately drive energy efficiency continues, change is likely to be slow, as housing markets exhibit significant inertia when considering characteristic changes (Case and Shiller 1988; Walker, Karvonen and Guy 2015). Following this line of thought, it is therefore probable that only market externalities, such as legislative imperatives and/or increasing energy prices, would become the catalyst for change. The variable that is most likely to impact on buyer choice is that of reduced energy costs, as it will affect the household budget and must come from residual monies after mortgage payments are made.

Electrical energy costs in Victoria have risen considerably in recent years; over the period from 2008 to 2014, they increased by approximately 62 per cent for the average household (Essential Services Commission Victoria 2015). These increases will undoubtedly continue, although the rate of increase may abate – but even this is not guaranteed. The most recently available data from the Australian Bureau of Statistics (ABS) shows as at September 2013, the average Victorian household was spending around 2.4 per cent of their income on electrical energy (ABS 2013). Whilst not large by international standards, electrical energy costs are gaining public attention. On the face of it, one would imagine that this cost increase almost alone would create a case for people to consider more energy-efficient housing in their home search. In Victoria, this does not appear to be the case. Therefore, when considering the introduction of new technologies into any sector, impediments to the implementation must be considered and this also includes existing housing styles and designs.

Typical housing types in Victoria exhibit low energy efficiency standards (Environment Victoria 2014). With regard to their design and character, Victorian houses have followed a rather traditional pattern of being detached and sited on varying sized allotments. Whilst early forms of subdivisions typically created allotments in the order of $1000m^2$, more recent subdivisions are around $400–500m^2$. In regard to buildings constructed between 1984 and 2003, the average-sized Victorian new house increased 36 per cent from $163.6m^2$ to $222.4m^2$ (ABS 2005). In terms of the building fabric, kiln-fired brick or timber weatherboard are almost exclusively the choice offered (DELWP 2014). Outside of suburban regions, occasionally mud brick, local stone or hay bale may also be found (ibid). Examples of typical Victorian housing can be seen in Figures 13.1–13.4.

More recently in Victoria, a rapidly increasing population has placed considerable pressure on the provision of housing and housing forms are evolving, with medium- and high-density developments appearing on the urban and suburban landscapes. In summary, the trend over the twentieth century of increased affluence no doubt contributed to an increase in homebuyer aspirations for larger housing with additional spaces for leisure and car parking, all of which

Figure 13.1 Italianate homestead circa 1900

Figure 13.2 Bungalow circa 1920

Figure 13.3 Post-war bungalow circa 1960

increased the energy and water consumption per metre squared and per capita, which has a direct impact on energy consumption (Clune, Morrissey and Moore 2012; McKinlay, Baldwin and Stevens 2016). These typologies are likely to be found to some extent in many different regions and countries globally.

Figure 13.4 Neo-modern house circa 2000

13.2 Barriers for energy-efficient housing

When considering housing markets in developed or developing countries, the composition of the housing stock is likely to be diverse in terms of age, design and building fabric. In Australia, approximately 86 per cent of the current housing stock existed prior to the energy rating standards introduced in the Building Code of Australia (BCA) in 2005 (Sustainability Victoria 2014). As these houses were built prior to the introduction of energy efficiency standards, the energy efficiency of the majority of Australian housing stock is likely to be substandard (Environment Victoria 2014). The inclusion of energy efficiency in the building codes lagged significantly behind other European countries, where in the UK, for example, energy efficiency was mandated in the Building Regulations in 1984, some 21 years earlier (Wilkinson 2014). This proportion of housing stock with poor energy efficiency may be similar to that of some countries, though possibly greater than that of others throughout the world. In order to reduce GHG emissions from housing, the proportionally small number of houses equipped to meet prevailing energy standards must be recognised and the resulting issues addressed. The question is: how can we compel homeowners to invest in energy-efficient technologies when they are not confident about recovering the capital expenditure? Herein lies the dilemma for Australia. If a market-led policy is allowed to prevail, to lead the way towards energy-efficient housing becoming a 'line' item on a homebuyer's shopping list, there needs to be a sufficiently large selection of houses from which the buyers can choose; otherwise, it is likely that such housing will remain the exception rather than the norm.

In most countries, the majority of the housing stock was built prior to the introduction of stringent energy efficiency standards. Hence, retrofitting these homes is generally considered a more viable strategy for achieving higher average energy efficiency than replacement and new construction. Thus, it would appear sensible to target this activity and encourage homeowners to seek such technologies.

Three key barriers in the Australian housing market have been identified. The first is the payback or return on investment of energy-efficient technologies. The installation of post-construction energy-efficient measures must be economically viable for homeowners to be encouraged to outlay capital funds (Pellegrini-Masini, Bowles, Peacock, Ahadzi and Banfill 2010), with viability often measured in terms of homeowners' ability to recover their capital expenditure and obtain the future benefits of reduced energy bills prior to selling (Bardhan et al. 2014). Households in Australia are relatively mobile, with only 27 per cent living in their current home for more than 15 years (ABS 2010). This in part is perceived to have reduced the opportunity of gaining an economic advantage from investing in energy-efficient measures (Hurst and Wilkinson 2015; Bardhan et al. 2014; Amstalden, Kost, Nathani and Imboden 2007). This high rate of mobility may impede future investment into energy-efficient technologies, unless there is a premium paid for energy efficiency in house purchases or, conversely, there is a discount for houses that have poor energy efficiency thereby enabling retrofit shortly after occupation.

The second issue is to know which measure, or measures, will provide maximum benefit when improving housing energy efficiency (Willrath and Logic 1997; Ren, Chen and Wang 2011). Without professional help, this decision can be daunting and may result in homeowners preferring to do nothing rather than waste money on ineffective technologies (Frederiks, Stenner and Hobman 2015; Hurst and Wilkinson 2015). There is a role for government support and education programs to mitigate this possibility (Mlecnik, Visscher and Van Hal 2010; Hepburn 2010), and also for trusted, knowledgeable tradespeople.

The third potential barrier that must be considered is the market facilitator–namely, the real estate agent. This cohort must be able to sense a market appetite for energy-efficient technologies if they are to be promoted in advertisements and the sales process (Bridge 2001; Arndt, Harrison, Lane, Seiler and Seiler 2013; Hurst and Wilkinson 2015). If real estate agents do not understand and value the benefits to occupants, they are unlikely to make such references in advertisements. At worst, the technologies may even be referred to as being irrelevant to the 'bigger picture' of the house purchase decision process. As the key professionals interacting at the point of sale, inability or reluctance on the part of agents to positively promote energy-efficient characteristics could act as an impediment to long-term market acceptance of energy-efficient technologies.

Having discussed the drivers and barriers for residential energy efficiency, it is appropriate to review the context in which this case is considered. Climate influences lives and behaviours in many ways. The lifestyle of a family in the 'snowbelt' of Canada, for example, will differ from that of a similar family in Melbourne, Australia. A cold country with significant snowfall will undoubtedly influence how people build and occupy housing and what they expect from it regarding thermal comfort and performance, whereas people dwelling in temperate places, such as Melbourne, will require something quite different. For example, a family in Melbourne are unlikely to think about the effects of extreme cold upon their thermal comfort in the house and may have only one centrally located internal heater and an air-conditioning unit to attenuate extreme summer temperatures. A Canadian family, on the other hand, would be likely to require triple-glazed windows and whole-of-house heating by a furnace that is continuously working throughout the winter period. Thus, in having a discussion about energy-efficient housing, it is necessary to consider the effects of the prevailing climate of the house's location.

13.3 The influence of climate on housing design

The objective for designers of sustainable houses is to create a house that has minimal, if any, reliance on external energy resources to produce a thermally comfortable environment (Marszal et al. 2011; Miller and Buys 2012). This objective needs to be achieved with regard to the climate of the place where the house will be located. The optimum goal is the creation of a house that can be occupied in a 'free-running' state – that is, a passive house that is independent of external energy sources for its functioning (Hernandez and Kenny 2010). This must also be achieved with

Table 13.1 Description of climate zones

Climate zone	*Climate*
Zone 1	High humidity summer, warm winter
Zone 2	Warm humid summer, mild winter
Zone 3	Hot dry summer, warm winter
Zone 4	Hot dry summer, cool winter
Zone 5	Warm temperate
Zone 6	Mild temperate
Zone 7	Cool temperate
Zone 8	Alpine

regards to occupant needs and stylistic preferences. Thus, it is logical to conclude that, for these objectives to be achieved successfully, the designer must be cognisant of the climate in which the house will be located (Willrath and Logic 1997; Ren et al. 2011).

Figure 13.5 illustrates the range of climates within Australia recognised by the Australian Building Codes Board (ABCB). There are several different climates nationally, from tropical in the northern regions to cool temperate in the southern mainland regions and Tasmania.

As can be observed from Figure 13.5, Victoria has four recognised climate zones: 'Hot', 'Mild', 'Cool' and 'Alpine'. Figure 13.6 provides a more detailed version of this chart for Victoria.

The names of the individual zones provide insight into the climate characteristics that the ABCB is labelling. It is evident from these names that the design and building fabric of a house in, for example, the Alpine zone, would need to be very different from that of a house in, say, the Mild zone. These differences, however, may be taken for granted by homebuyers within the climate zones and may only be discernible if the homebuyer were considering options across the zones. The large geographic expanse would render this an unlikely event for most purchasers.

The ABCB is responsible for determining the minimum building standards for energy efficiency in housing, and more generally, all construction standards throughout Australia (ABCB 2016c). This also includes extensions and renovations of existing dwellings that go beyond certain criteria. These criteria and others are included in the BCA and are interpreted by the local municipalities for compliance and enforcement. Similar systems operate in other countries; for example, in the UK, the Building Regulations set out minimum standards in respect of new and existing buildings, which are interpreted by building control officers in local authorities (UK Government 2016). In Australia, regional authorities such as the Victorian Building Authority and their state equivalents also influence and exert measures of control over building codes but discussion of their relevance is beyond the scope of this chapter. One of the common regulatory requirements in Victoria is that when 50 per cent or more of the dwelling floor area is extended, the whole house must be brought up to current standards. When the extension is below 50 per cent of the floor plate, only the new part is required to comply. When the renovation does not alter the existing roofline no adherence to contemporary building codes is required, other than to normal safety standards such as electrical wiring and gas plumbing (VBA 2016). This situation leads to a considerable variation in the standards of energy-efficient housing that are required in the Victorian housing market. Unless an extensive alteration to the original house is made, these arrangements allow the individual homeowner to decide the extent to which he or she wishes to engage with the building codes for energy efficiency. Even where compliance is mandatory, the actual type of technology employed is discretionary; it is meeting the performance standard that is important (VBA 2014).

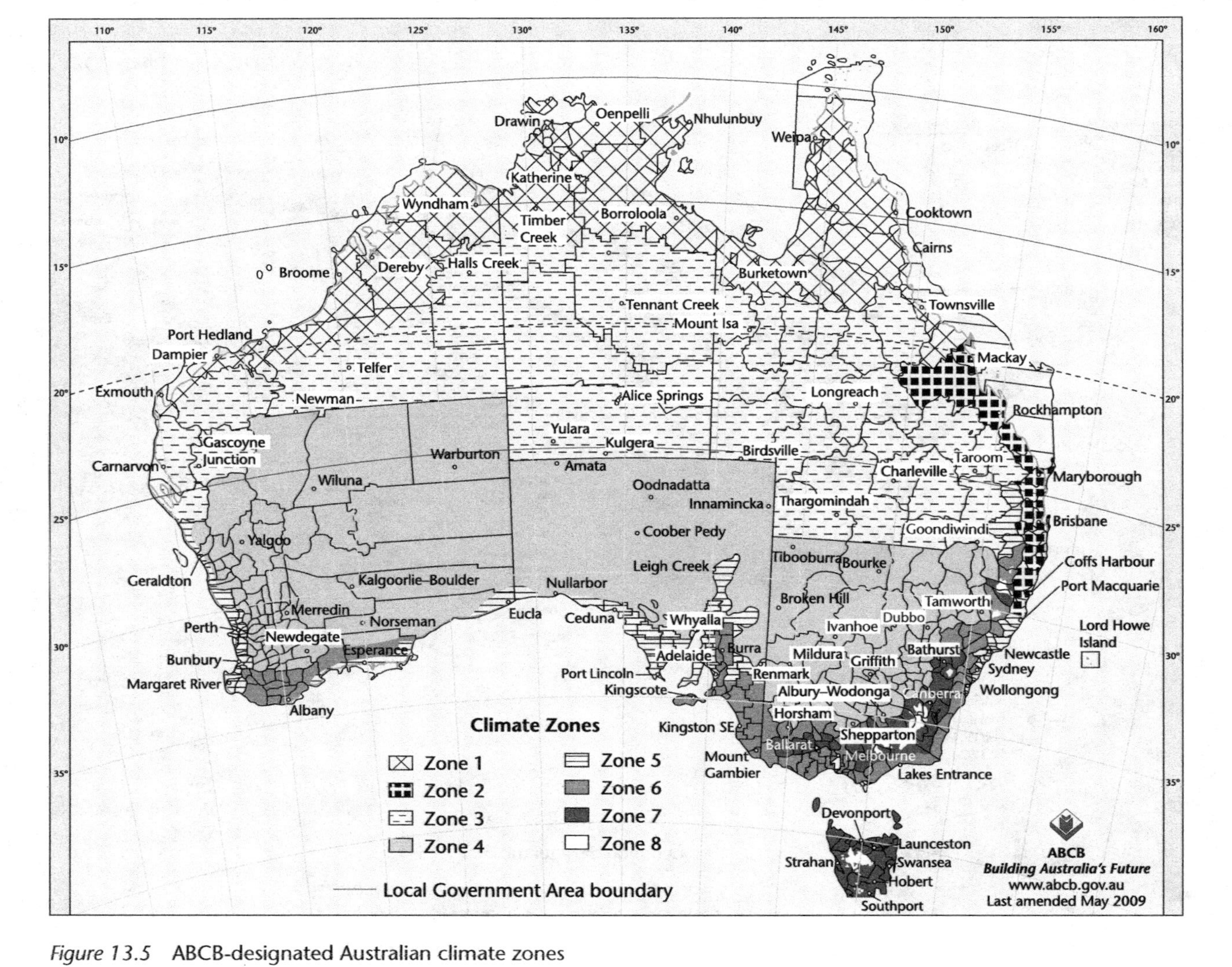

Figure 13.5 ABCB-designated Australian climate zones

Source: ABCB (2016a)

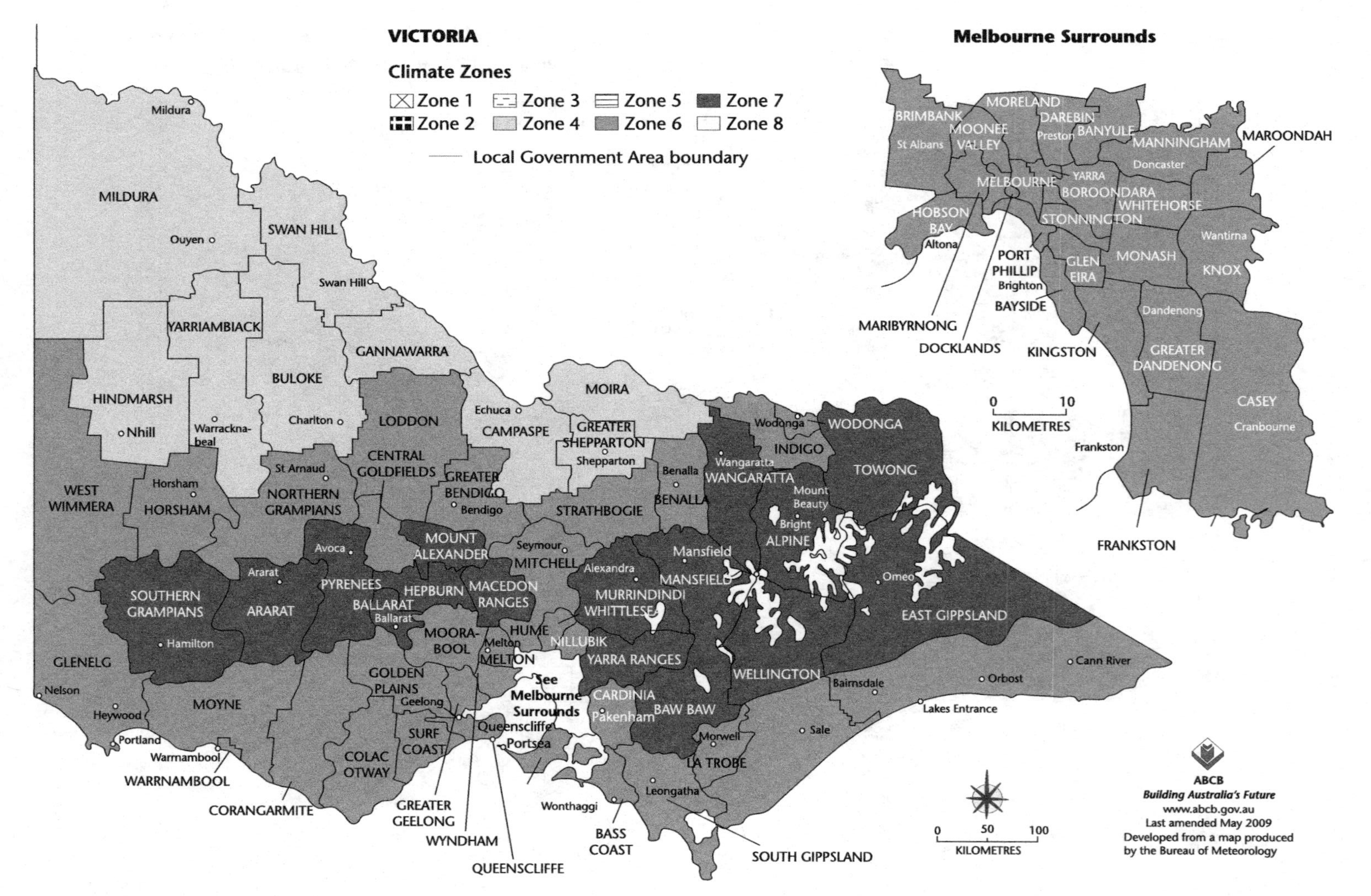

Figure 13.6 ABCB-designated Victorian climate zones

Source: ABCB (2016b)

Given this situation, how are housing markets responding to the need for energy-efficient housing? Furthermore, what are real estate agents, as market facilitators, doing in terms of promoting such housing when it has energy-efficient technologies within it? This discussion now turns to the role that real estate agents play and how they are promoting energy-efficient characteristics when marketing a house for sale.

13.4 Real estate agency practice in Victoria

Due to specific national and regional laws and cultural customs, each country that has an operational real estate agency industry will function differently in terms of practice (Li and Wang 2006; Brinkmann 2009). Therefore, the discourse here focuses on agency practice in Victoria, Australia.

Real estate agents are market facilitators and their role is therefore a dichotomous one in that they typically act on behalf of the seller and engage with the buyer in an effort to sell the house (Patron and Roskelley 2008). In the course of this role, agents also prepare advertising material, often in consultation with the seller. Victorian real estate agents must first undertake and pass a State Government-prescribed course of study that introduces them to the various consumer laws affecting property transactions and dealing with members of the public. These laws can be onerous and breaches can carry significant penalties (CAV 2015). Each state within Australia has similar requirements for real estate agents operating within their respective jurisdictions.

To undertake their responsibilities, real estate agents must be skilful in negotiations, be able to explain rudimentary building construction and possess a sound knowledge of marketing principles and apply these within the laws affecting property transactions. These skills are initially acquired via formal training and further refined through experience. They must also have an in-depth understanding of the area/region in which they work (Arndt et al. 2013) and be able to interpret the impacts of macro occurrences, such as fluctuations in home lending interest rates, on housing markets in order to advise their clients. All of these functions must be achieved within the frame of law. In Victoria, real estate agents are usually remunerated by a commission, which is earned only when the house is sold (Department of Justice 2012). This arrangement motivates both sellers and real estate agents towards a mutually successful outcome. The higher the sale price, the more return on investment for the seller and the higher the commission for the agent. A corollary of this structure is that the marketing campaign should focus greatly upon the property's most desirable attributes to attract the most suitable buyer (Bridge 2001; Perkins, Thorns and Newton 2008).

Within Victoria, real estate agents must firstly obtain written authority from the seller before performing any tasks that could be interpreted by a buyer as having listed a property for sale (Estate Agents Act 1980 [Vic] s49A). The reason for this is simply to make explicit the agent's legal authority to offer the property as available for sale. Once the authority is signed by the seller, the skills of the agent are directed towards presenting the house to the market in the best possible light. The agent is expected to align the characteristics of the house with the appetite that the market is exhibiting and promote those that are most likely to attract the buyer who is willing to pay the highest price (Pryce and Oates 2008). This is achieved usually through advertising.

Research has shown, and real estate agents have acknowledged, that location is one of the main drivers of value (Archer, Gatzlaff and Ling 1996; Kiel and Zabel 2008). To address this aspect, real estate agents will generally make known the suburb and any other specific but desirable locational information in the advertising caption. This is intended to secure the buyer's attention. Whether knowingly or not, the agent is applying one of the long-standing advertising principles, namely AIDA (Attention-Interest-Desire-Action) (Wischmeyer 2015). The way in which advertisements are formed and how they can be used to interpret social trends and values is the subject of the remainder of the chapter.

Advertising is central to selling and a properly constructed marketing campaign will invariably contain advertisements designed to target identified buyers. As real estate agents typically earn their commission on a success basis, they ought to be 'in sync' with prevailing market and fashion trends. They will therefore make every reasonable and legally permissible effort to construct a marketing campaign, including advertising script, to attract suitably qualified buyers (Perkins et al. 2008). These advertisements often attempt to 'position' the property in the marketplace through the use of 'persuasive textual' formation (Bruthiaux 2000).

The real estate agent will often use emotive words and phrases to construct images in the minds of buyers (Pryce and Oates 2008) and will manipulate linguistic patterns to achieve favourable outcomes (Beangstrom and Adendorff 2013; Perkins et al. 2008; Schöllmann, Perkins and Moore 2001). The textual variances found in real estate advertisements can be useful when examining how society is responding to change or innovations – in this instance, the need for more energy-efficient houses. A useful example of how advertisements can be used to examine housing trends is found in the study by Rodriguez and Siret (2009). They reviewed real estate advertisements that had been published over a 20-year period and observed that housing preferences had altered in respect of the use of space and the concept of comfort. They concluded that advertisements are a compact description of the characteristics and qualities that dwellers and real estate agents give to a house in order to make the best sale (Rodriguez and Siret 2009). In summary, advertisements therefore should be able to inform researchers about how housing markets are responding to the proliferation of energy-efficient technologies.

Energy-efficient technologies in the home come in many forms and range from the inexpensive to the very expensive (Pellegrini-Masini et al. 2010). For example, weather seals around doors and windows are an example of the former, whereas smart house systems are an example of the latter. As previously discussed, the performance and suitability of individual technologies are, to a significant extent, determined by the particular climate zone and physical design of the individual house (Yilmaz 2007; Ren et al. 2011; Miller and Buys 2012).

Research findings now presented in this chapter reveal that real estate agents are including words and phrases that describe energy-efficient technologies within advertisements, but only at a low level (Hurst and Wilkinson 2015; Hurst and Halvitigala 2016). Considering the earlier comments made concerning the relationship between successful marketing and agent remuneration, this suggests that real estate agents believe that buyers are not factoring energy efficiency into their house purchase decision to any great extent. Traditional locational value drivers such as proximity to transport, preferred schools and occupant accommodation needs appear to remain the key factors affecting choice (Levy and Lee 2004, 2011; Semeraro and Fregonara 2013).

13.5 Research design and methodology

The research presented in this chapter analysed just over 123,000 real agent advertisements (after data cleaning had been completed) that were used to successfully market a property resulting in an enforceable contract between 2008 to 2015 in Victoria. These advertisements were provided by Victoria's peak real estate agency industry body, the Real Estate Institute of Victoria (REIV). The files provided required considerable cleaning for use to remove corrupted data and were then coded using SPSS V 23.0 to identify house sales, land sales and apartment/townhouse sales. The particular focus of this research was on detached houses and therefore apartment and land sales were removed from the dataset. The rationale for this separation was that homeowners are free to make choices about what, if any, energy-efficient technologies they wish to install, whereas owners in building complexes such as apartments do not enjoy such freedoms. This is because they must abide by the collective decisions of the owners corporation, an entity that governs the common areas of the building and occupants' standards of conduct within the building (CAV 2016). Australia's most recently published census showed that 79 per cent of Australians live in detached housing, with 10 per cent in semi-detached and the remaining 11 per cent in flats, apartments (medium to high-rise) and single-or two-story units (ABS 2012).

Before reviewing the advertisements, grey literature was reviewed and advertisements from commercial vendors of energy-efficient technologies were examined to identify typical words/phrases used to describe technologies such as solar panels, double glazing and insulation (for examples, see Your Home 2013; IEA 2015; Willrath and Logic 1997). Over 40 terms relating to energy-efficient technologies were found to be commonly used and Table 13.2 below illustrates a sample of these words/phrases. The words/phrases were then grouped based upon synergies. For example, words/phrases that related to the building design and fabric, such as 'double glazed' and 'orientation' (northern is the preferred orientation in Victoria), were grouped as 'building design and fabric', whereas words/phrases that related to the sun's ability to provide energy were grouped as 'solar'.

These groupings were then coded using SPSS V 23.0 to enable analysis. The aim was to review the advertisements within the database to understand how real estate agents were advertising energy-efficient characteristics, the premise being that real estate agents would advertise such technologies if they perceived a market appetite for them and if they existed in the house. Furthermore, the primacy/recency theory, developed in the field of psychology, posits that readers of text are more likely to remember the first component (primacy) and

Table 13.2 Advertising word/phrase grouping for analysis

Word/phrase category	*Sample of words/phrases found in advertisements*	
Solar	solar	solar heating
	solar panel	solar home
	solar power	solar electricity
	solar-power	solar electric
	solar HWS	solar boosted
	solar hot water	solar enhanced
	solar heated hot water	solar heated
	solar energy	solar-heated
	solar-energy	solar passive
	solar system	solar design
		solar principles
Building design and fabric	double glazing	brick veneer
	d/glazed	split system
	double glazed	water tank
	miglas windows	north facing
	double-glazed	northern aspect
	smart glass	facing north
	efficient design	north-facing
	sustainable design	north sun
	passive design	northerly aspect
	eco design	northern sun
	BV (brick veneer)	

(continued)

Table 13.2 Advertising word/phrase grouping for analysis *(continued)*

Word/phrase category	*Sample of words/phrases found in advertisements*	
Environment	grey water	eco-
	insulation	eco-waste
	greenhouse gas	eco features
	certified green builder	eco home
	green technology	eco conscious
	environmentally green	eco-sustainable
	eco friendly	eco efficient
	eco-friendly	eco technologies
Energy efficiency	energy efficient	energy-efficient
	energy-efficiency	energy conscious
	energy efficiency	energy cost
	energy saving	low-energy
	energy-saving	energy consumption
	energy save	energy rated
	energy bill	energy rating
	low energy	energy-rated
	energy conservation	star energy
	energy reports	passive energy

the last component (recency) of the advertisement message (Ohanian and Cunningham 1987; Forgas 2011). Thus, applying this primacy/recency theory to the advertisements, it was suggested that if real estate agents believed energy-efficient technologies were important and ought to be considered by buyers, the words/phrases relating to these technologies would be located early or late in the advertisements. Similarly, when considering the AIDA principle, if real estate agents believe these words/phrases are of considerable interest to house buyers, they would be seen early in the advertisements.

Another significant variable is the relevance of demographic profiles to the engagement with environmentally friendly practices and the uptake of environmentally friendly technologies. Research has shown that people who are more educated and wealthier are more likely to engage with environmentally friendly behaviours and invest in energy-efficient technologies (Mills and Schleich 2012). Age also has relevance, but not in the way that many would assume, as older people are tending to engage with energy-efficient technologies, possibly in order to reduce expenses post retirement (Valkila and Saari 2012; Vassileva and Campillo 2014).

To evaluate the data, a statistical logit model was developed to conduct a content analysis of the advertisements in regard to the four categories of words/phrases (see Pampel 2000 for further reading regarding logit models).

The research was quantitative, sharing the characteristics of quantitative research set out by Silverman (2010). The data was converted into the form of numbers and not words. Quantitative research relies on the interpretation of large amounts of data and is objective rather than subjective (Silverman 2010). Its aim is to quantify (1) the size of the relationship between variables within the model, and (2) the probability that the results are generalisable. Thus, with regard to this research, the analysis of the data was conducted to reveal how, if at all, energy-efficient technologies were being advertised and, utilising logistic regression, the likelihood (probability) that the independent variables within the model will influence the appearance of such words/

phrases. The results show that energy-efficient technologies did not feature highly in buyers' purchasing decisions in Victoria, Australia between 2008 and 2015.

The statistical techniques in the research design applied to evaluate the data were undertaken in four distinct stages. The first stage comprised a high-level descriptive statistical analysis to examine the quantum of advertisements containing the identified words/phrases and an evaluation of where the words/phrases appeared in the textual string (the advertisement). The second stage of the analysis was a logistic regression. Following best practice, this regression technique was adopted because the dependent variable was dichotomous (Hair, Black, Babin and Anderson 2014). Logistic regressions enable researchers to determine the probability of an event, in this case the appearance of the identified words/phrases, occurring (Aldrich and Nelson 1984). The coefficients of the logistic regression model provide information of the probability of a change in the dependent variable for a unit change in the independent variable. The third stage was an analysis of variance (ANOVA). The ANOVA was used to compare the mean of the transfer prices of houses advertised without mention of energy-efficient technologies to that of houses advertised with a reference to energy-efficient technologies. However, the findings here need to be treated with some caution as the model is unable to be sufficiently detailed to discern the effect of other important considerations such as allotment size or individual room size or quality. For purposes of comprehension, 'allotment' in the Australian context refers to the parcel of land upon which the house is built. Various countries have different appellations for this. For example, in New Zealand the land would be referred to as a 'section' whilst in the UK it would be a 'plot'. It is nonetheless a useful indicator of whether or not there appears to be a price premium for more energy-efficient houses. The fourth and final stage of the analysis was to forecast possible future growth/decline of the appearance of the key words/phrases used within this project. To undertake this task, a secondary database was created. This database utilised the original data and extended time variables, omitting the other dependent variables, to 'crystal ball' into the future for the dependent variable of interest, such as, projected increase in the appearance of words/phrases referencing solar technologies.

13.6 Findings and interpretation: The disappointing reality

First, the overall results showed that less than 5 per cent of all advertisements had any reference to energy-efficient technologies, although certain regions exhibited up to 12 per cent. The variable that distinguished these geographic regions from the other regions was the demographic profile. The regions that displayed a greater likelihood of energy-efficient technologies being advertised were those where more educated, professionally qualified people resided. Typically, these regions also tended to exhibit higher land values. Areas that were less developed with more natural settings also exhibited a greater number of the identified words/phrases in advertisements. This finding supports previous research concerning the relationship between education and environmentally friendly behaviours (Tan 2012; Yu and Tu 2011).

In terms of where the words/phrases appeared within the advertisements, which provides an indication of the emphasis given to the energy efficiency message when considering primacy/recency theory, it appears that the environmental message is being lost. Figure 13.7 presents four histograms of the frequency of appearances plotted against the location of the words/phrases within the advertising text. The x-axis indicates the number of characters into the text script before the first word/phrase appears.

Examination of the histograms in Figure 13.7 reveals that, with the exception of words/phrases relating to building design and fabric, the environmental message is largely embedded within the body of the advertisement but nonetheless tending towards the earlier portion of the advertisement. This suggests that the message is likely to be lost relative to information at

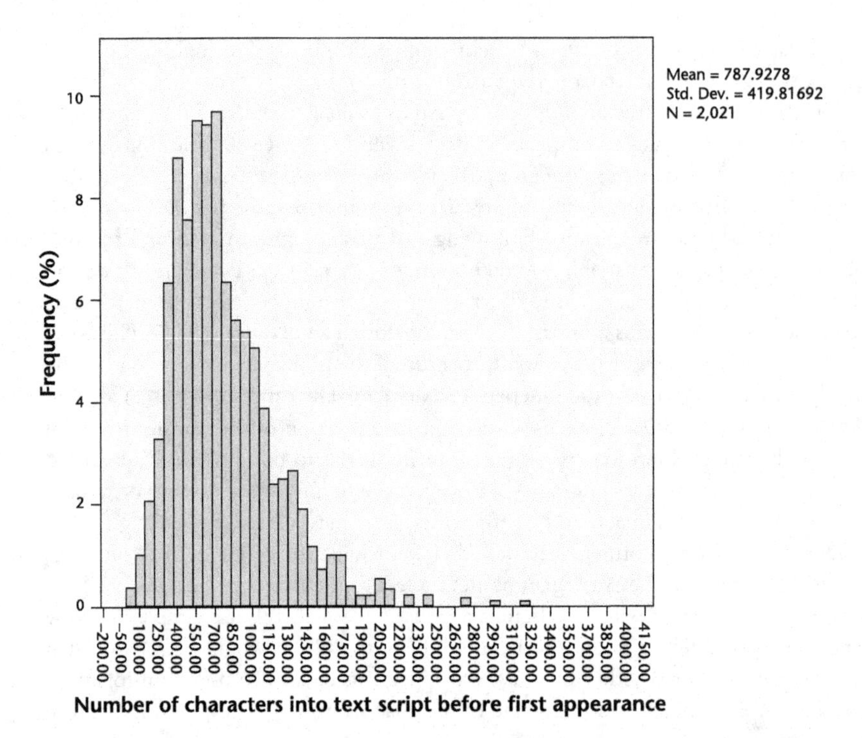

Figure 13.7a Frequency of words/phrases from the solar category plotted against text location

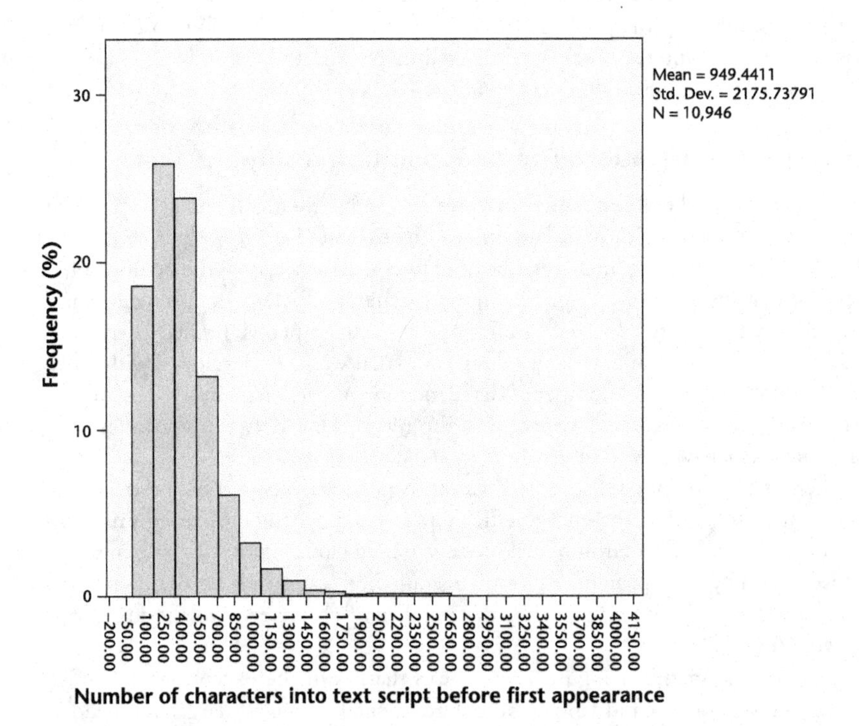

Figure 13.7b Frequency of words/phrases from the building design and fabric category plotted against text location

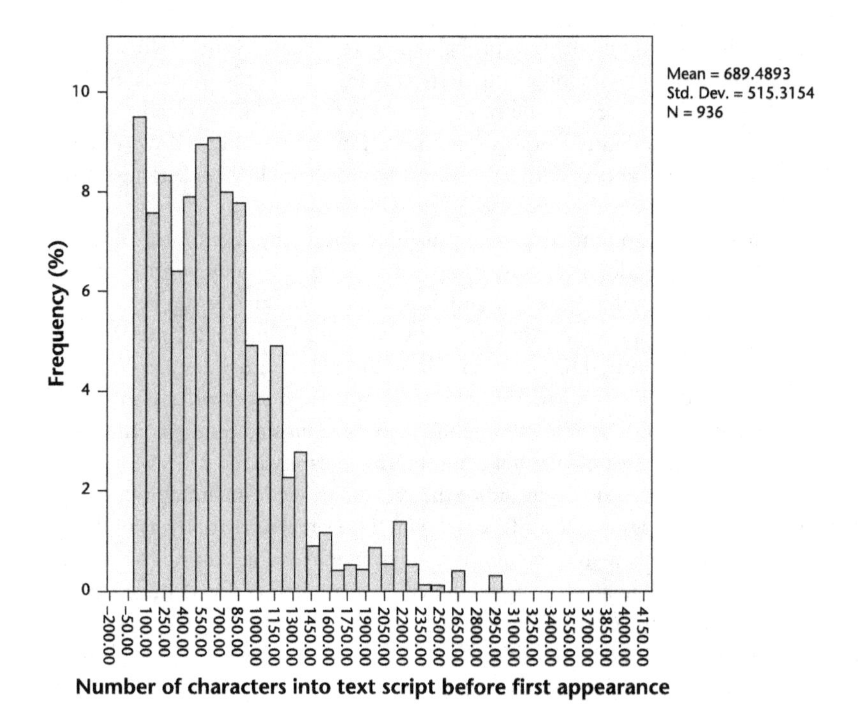

Figure 13.7c Frequency of words/phrases from the environment category plotted against text location

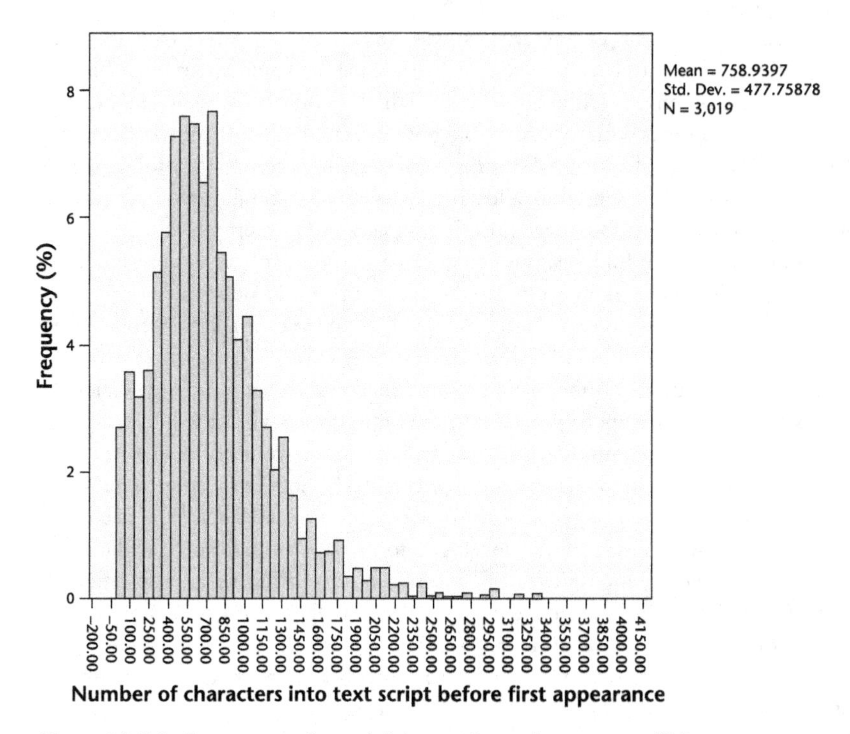

Figure 13.7d Frequency of words/phrases from the energy efficiency category plotted against location

the beginning and end of the advertisement. It is probable that real estate agents have written about what they believe to be attention-grabbing and interesting features for the potential buyers and are now just 'listing' the other features in the house. If this is the case this is clearly not an ideal situation, to be relying on established housing markets, rather than direct government policy, to encourage buyers to factor energy-efficient measures into their choices (Parliament of Australia 2015).

The logistic regressions, which sought to provide analysis about the likelihood of each independent variable of interest appearing in an advertisement, revealed that only words/phrases relating to solar technologies and building design and fabric were statistically significant for household wealth and education, again supporting previous literature (Miller, Colantuoni and Crago 2014). Regions with well-educated residents and those located within wealthier suburbs were found to include more references to solar technologies in advertisements when compared to areas where the demographic consisted of lower-income and less educated residents. In other words, in areas where income levels are relatively higher and persons have achieved post-secondary qualifications, there is a greater likelihood of the real estate agent including reference to environmentally friendly technologies in the advertisement. This suggests that buyers in these markets are expressing more interest in such technologies and agents are attempting to make those buyers aware of their existence in an effort to attract an enquiry. It could be deduced from this evidence that hope exists for house energy-efficient technologies to become part of everyday house purchase decisions, although when considered with reference to the forecasts shown below, this may be quite some time into the future.

The ANOVA is the next stage for consideration. As stated previously, the ANOVA output must be viewed with caution, but it is useful in providing an indication of current attitudes to the pricing of energy-efficient technologies. For the data evaluated, the mean transfer for properties without energy-efficient characteristics was $837,221 compared to $885,860 for properties with energy-efficient characteristics. At first glance, it would appear that buyers are willing to pay more for energy-efficient houses, which would positively encourage homeowners to retrofit their homes with such technologies to attract higher prices when selling. This market characteristic is typical of capitalistic markets, creating a temptation for governments to adopt a 'hands-off' approach, thus allowing markets to naturally evolve towards energy-efficient housing at all levels. However, this finding cannot be interpreted in isolation. When reviewed with the regression findings, it is evident that houses that have energy-efficient technologies installed are also more likely to be sited in areas of greater personal wealth, and therefore likely to be in more expensive locations. So again evidence of market acceptance of energy-efficient technologies is veiled by other realities, if it does exist.

One final evaluation would be to assume that a growing market appetite for energy-efficient technologies does exist and then attempt to project into the future to see what it will look like in five years' time. Why five years? Because scientists are telling us that we need to make radical change now if we are to avert environmental disaster (Garnaut 2011). So if real change is not seen over the next five years, then it is likely to be too slow to be of relevance, at least as far as detached housing is concerned. Forecasting models were developed to provide an indication of the time it will take before we see regular appearances of words/phrases relating to energy-efficient technologies in advertising. These models were created by developing a database that considered current appearances of the four keywords of interest over the eight-year period of the research and then, using the SPSS V 23.0 forecasting function, projecting five years hence. The graphs in Figure 13.8 show the outputs for each category of words/phrases.

Only words/phrases in the solar category seem to be increasing in likelihood of appearances (Figure 13.8a). In Victoria, as is the case throughout Australia, solar PV cells attract a rebate from the government at point of installation and electricity feed-in tariff rates into the grid (Sustainability Victoria 2015). However, the extent of the rebate has been significantly reduced, decreasing from $0.60 per kw hour in 2009 to $0.05 per kw hour in 2016 (Essential Services Commission 2016). This no doubt will have had a negative impact upon the projection of the uptake of solar technologies as government rebates have been shown to positively impact upon public engagement with new technologies (Boza-Kiss et al. 2013).

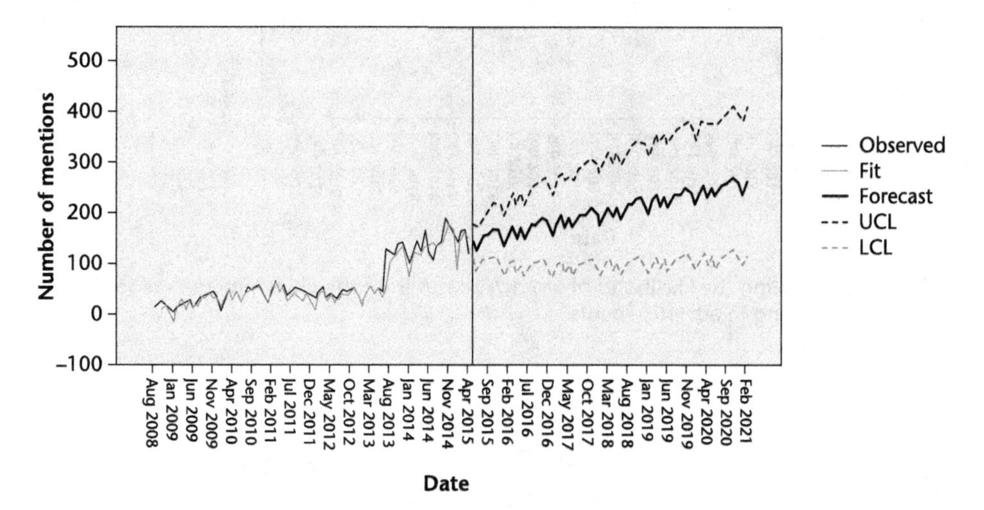

Figure 13.8a Forecasting the likelihood of words/phrases from the solar category appearing in advertisements

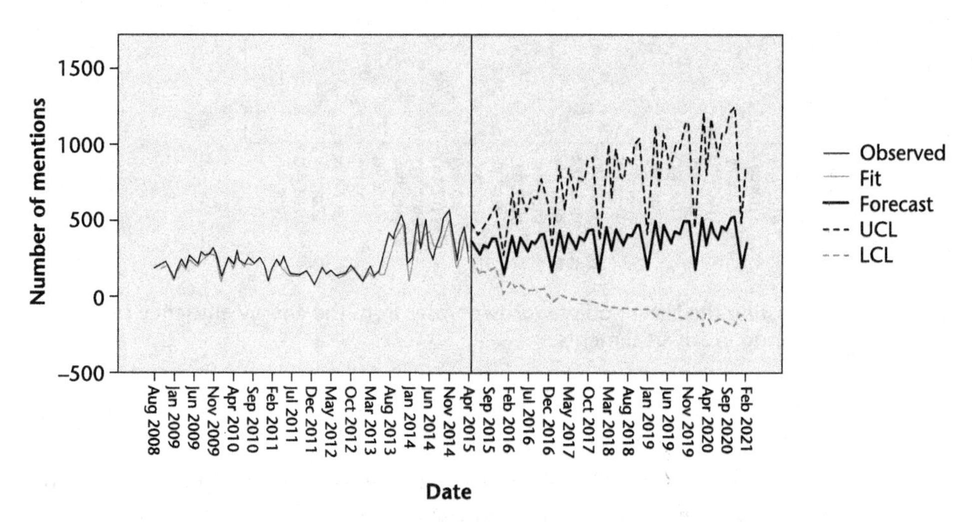

Figure 13.8b Forecasting the likelihood of words/phrases from the building design and fabric category appearing in advertisements

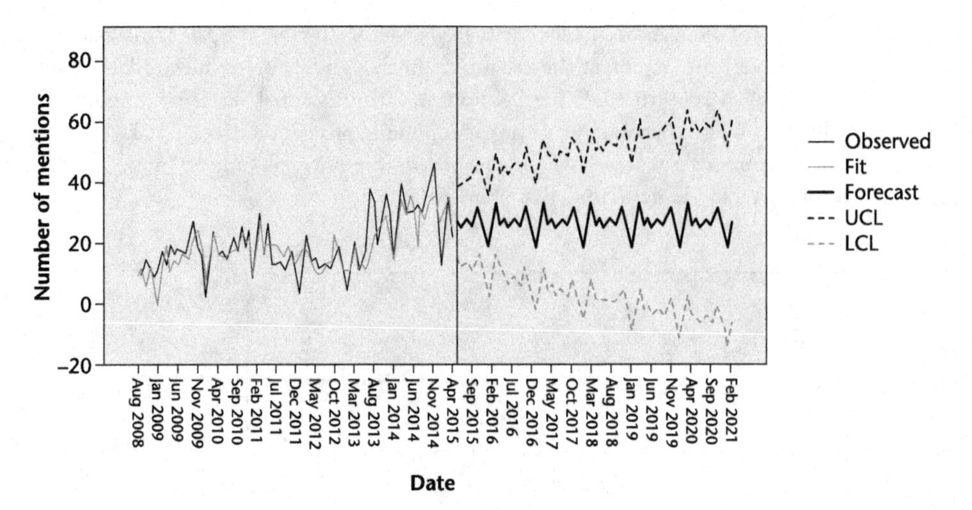

Figure 13.8c Forecasting the likelihood of words/phrases from the environment category appearing in advertisements

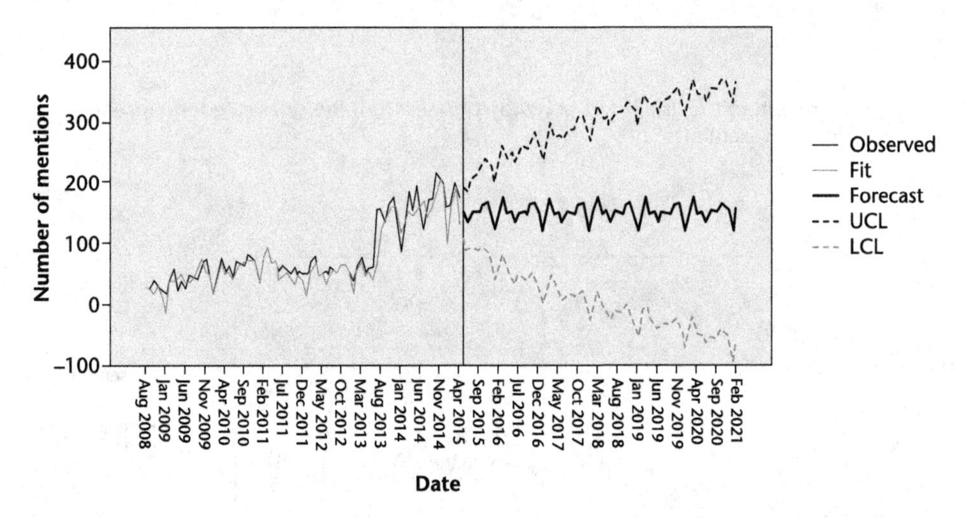

Figure 13.8d Forecasting the likelihood of words/phrases from the energy efficiency category appearing in advertisements

13.7 Conclusions

Housing is a significant contributor to GHG emissions and therefore warrants attention to investigate ways of reducing its impact on the environment. This chapter has highlighted the importance of integrating environmental considerations into the house purchase decision, through real estate agents' advertisements for owner-occupied detached dwellings. Agents were shown to be important influencers of the market as well as being acutely tuned into features that buyers consider valuable. The results have highlighted the failings of a laissez-faire approach that relies upon the market to stimulate and promote the uptake of energy-efficient technologies in the housing sector. Such an outcome supports the views of some who argue for mandatory interventions as opposed to voluntary methodologies for energy efficiency gains in the housing sector (Millward-Hopkins 2016; Janda, Bright, Patrick, Wilkinson and Dixon 2016; Van der Heijden 2015). The lack of active and positive government intervention with regard to the uptake of energy-efficient technologies in the housing sector has led to a cool reception by homebuyers. Traditional value drivers such as location are likely to remain 'front of mind' for buyers and the challenge continues for governments to more actively engage in market processes in Victoria. Models such as the EU's energy disclosure requirements in the marketing process are posited as being appropriate means of ensuring that buyers are constantly reminded of the importance of energy reduction, particularly at a time of consistent rising energy costs.

International summits, such as the recent Paris Summit of 2015, have highlighted the need for real and positive action to avoid catastrophic and irreversible environmental damage. It has been stated here that housing is a major contributor to GHG emissions and increasing buyer demand for energy-efficient housing can bring at it real change. It is imperative that we monitor the market acceptance of energy-efficient technologies in order to ensure progress, and the speed of acceptance must be rapid if we are to avert further severe negative impacts. A longitudinal study, using a similar methodology, would be useful to monitor acceptance and inform community of progress made. Clearly, every effort that assists in reducing our individual environmental footprint cannot be considered wasted.

This study has highlighted the need for more positive action from governments, industry bodies, researchers and educators. Market-led approaches alone, whilst perhaps convenient, are not likely to produce the much-needed change in the time remaining before irreversible damage is done. Much has been achieved and there is still more to be done. Any efforts to influence the creation of a market appetite for cleaner, more energy-efficient housing can only aid preservation of our lifestyle and planet.

Acknowledgement

The authors wish to acknowledge the generous contribution of the Real Estate Institute of Victoria (REIV) for allowing access to their database, the use of which has provided a very useful and unique insight into how real estate agents and homebuyers are responding to changes in energy-efficient technologies.

References

Aldrich, J. H. and Nelson, F. D. (1984). *Linear probability, logit, and probit models.* Thousand Oaks, CA: Sage.

Amstalden, R. W., Kost, M., Nathani, C. and Imboden, D. M. (2007). 'Economic potential of energy-efficient retrofitting in the Swiss residential building sector: The effects of policy instruments and energy price expectations.' *Energy Policy* vol. 35, no. 3, pp. 1819–1829.

Archer, W. R., Gatzlaff, D. H. and Ling, D. C. (1996). 'Measuring the importance of location in house price appreciation.' *Urban Economics* vol. 40, no. 3, pp. 327–353.

Arndt, A., Harrison, D. M., Lane, M. A., Seiler, M. J. and Seiler, V. L. (2013). 'Can agents influence property perceptions through their appearance and use of pathos?' *Housing Studies* vol. 28, no. 8, pp. 1105–1116.

Australian Building Codes Board (ABCB) (2016a). 'Climate zone map: Australia wide.' Accessed 20 January 2016. www.abcb.gov.au/Resources/Tools-Calculators/Climate-Zone-Map-Australia-Wide.

Australian Building Codes Board (ABCB) (2016b). 'Climate zone map: Victoria.' Accessed 20 January 2016. www.abcb.gov.au/Resources/Tools-Calculators/Climate-Zone-Map-Victoria.

Australian Building Codes Board (ABCB) (2016c). 'Homepage.' Accessed 11 March 2016. www.abcb.gov.au.

Australian Bureau of Statistics (ABS) (2005). 'Year Book Australia 2005'. Accessed 9 February 2016. www.abs.gov. au/ausstats/abs@.nsf/0/609e28eb4ba28e14ca256f7200832ff6?OpenDocument.

Australian Bureau of Statistics (ABS) (2010). 'Australian social trends: moving house.' Accessed 1 February 2016. www.abs.gov.au/AUSSTATS/abs@.nsf/Lookup/4102.0Main+Features30Dec+2010.

Australian Bureau of Statistics (ABS) (2012). 'Year Book Australia 2012.' Accessed 4 October 2016. www.abs.gov.au/ausstats/abs@.nsf/Lookup/by%20Subject/1301.0~2012~Main%20Features~Types%20of%20 Dwellings~127.

Australian Bureau of Statistics (ABS) (2013). 'Household energy consumption survey, Australia: Summary of results 2012'. Accessed 4 October 2016. www.abs.gov.au/AUSSTATS/abs@.nsf/Lookup/4670.0Explanatory%20Notes12012?OpenDocument.

Australian Government. (2015). 'How government works.' Accessed 23 February 2016. www.australia.gov.au/about-government/how-government-works.

Australian Government Climate Change Authority. Targets and progress review. Accessed 23 February 2016. www.climatechangeauthority.gov.au/reviews/targets-and-progress-review/part-b.

Bardhan, A., Jaffee, D., Kroll, C. and Wallace, N. (2014). 'Energy efficiency retrofits for U.S. housing: Removing the bottlenecks.' *Regional Science and Urban Economics* vol. 47, no. 1, pp. 45–60.

Beangstrom, T. and Adendorff, R. (2013) 'An APPRAISAL analysis of the language of real estate advertisement.' *Southern African Linguistics and Applied Language Studies* vol. 31, no. 3, pp. 325–347.

Ben-David, R. (2012). 'Climate change and the importance of shouting, context and economics', transcript. Presented at Financial Counselling Australia Conference, 16 May 2012. Accessed 25 October 2017. www.esc.vic.gov.au/wp-content/uploads/esc/40/40faa29b-a9d3-4174-9a58-9876d7868766.pdf

Boza-Kiss, B., Moles-Grueso, S. and Urge-Vorsatz, D. (2013). 'Evaluating policy instruments to foster energy efficiency for the sustainable transformation of buildings.' *Current Opinion in Environmental Sustainability* vol. 5, no. 2, pp. 163–176.

Bridge, G. (2001). 'Estate agents as interpreters of economic and cultural capital: The gentrification in the Sydney housing market.' *International Journal of Urban and Regional Research* vol. 25, no. 1, pp. 87–101.

Brinkmann, J. (2009). 'Putting ethics on the agenda for real estate agents.' *Journal of Business Ethics* vol. 88, no. 1, pp. 65–82.

Bruegge, C., Carrión-Flores, C. and Pope, J. C. (2016). 'Does the housing market value energy efficient homes? Evidence from the energy star program.' *Regional Science and Urban Economics* vol. 57, pp. 63–76.

Bruthiaux, P. (2000). 'In a nutshell: Persuasion in the spatially constrained language of advertising.' *Language and Communication* vol. 20, no. 4, pp. 297–310.

Bryant, L. and Eves, C. (2012). 'Home sustainability policy and mandatory disclosure: A survey of buyer and seller participation and awareness in Qld.' *Property Management* vol. 30, no. 1, pp. 29–51.

Building Performance Institute (BPIE). 'Energy performance certificates across the EU'. Accessed 21 September 2016. http://bpie.eu/publication/energy-performance-certificates-across-the-eu/.

Case, K. E. and Shiller, R. J. (1988). 'The efficiency of the market for single-family homes.' *The American Economic Review* vol. 79, no. 1, pp. 125–137.

Clune, S., Morrissey, J. and Moore, T. (2012). 'Size matters: House size and thermal efficiency as policy strategies to reduce net emissions of new developments.' *Energy Policy* vol. 48, pp. 657–667.

Consumer Affairs Victoria (CAV) (2015). 'Penalties: estate agents'. Accessed 24 February 2016. www.consumer.vic.gov.au/businesses/licensed-businesses/estate-agents/penalties.

Consumer Affairs Victoria (CAV) (2016). 'What is an owners corporation?' Accessed 24 February 2016. www.consumer.vic.gov.au/housing-and-accommodation/owners-corporations/buying-into-an-owners-corporation/what-is-an-owners-corporation.

Department of Energy and Climate Change (DECC) (2013). '2013 UK greenhouse gas emissions, provisional figures and 2012 UK greenhouse gas emissions, final figures by fuel type and end-user.' Accessed 24 January 2016. www.gov.uk/government/uploads/system/uploads/attachment_data/file/295968/20140327_2013_UK_Greenhouse_Gas_Emissions_Provisional_Figures.pdf.

Department of Environment, Land, Water and Planning (DELWP) (2014). 'What house is that?' Accessed 21 March 2016. www.dtpli.vic.gov.au/heritage/research-and-publications/what-house-is-that.

Department of Environment, Land, Water and Planning (DELWP) (2016). 'Property and land titles.' Accessed 23 February 2016. www.propertyandlandtitles.vic.gov.au.

Department of Industry (2014). 'Nationwide House Energy Rating Scheme (NatHERS), Star Rating.' Accessed 13 January 2016. www.nathers.gov.au/accredited-software/how-nathers-software-works/star-rating-scale.

Department of Justice (2012). 'Real estate guide for buyers and sellers'. Accessed 24 February 2016. www.consumer.vic.gov.au/library/publications/housing-and-accommodation/buying-and-selling-property/real-estate-a-guide-for-buyers-and-sellers-word.doc.

Environmental Protection Authority (EPA) (2014). 'Household and greenhouse gas emissions'. Accessed 16 March 2016. www.epa.vic.gov.au/agc/research.html.

Environment Victoria (2014). One million homes. Accessed 23 February 2016. http://environmentvictoria.org.au/onemillionhomes.

Essential Services Commission (2016). 'Electricity standing offer tariffs,1994–95 to current.' Accessed 17 January 2016. www.esc.vic.gov.au/Publications/Search-Results?sector=Energyandkeywords=electricity percent20standing percent20offer.

Estate Agents Act 1980 (Vic).

Eves, C. and Kippes, S. (2010). 'Public awareness of "green" and "energy efficient" residential property: An empirical survey based on data from New Zealand.' *Property Management* vol. 28, no. 3, pp. 193–208.

Forgas, J. P. (2011). 'Can negative affect eliminate the power of first impressions? Affective influences on primacy and recency effects in impression formation.' *Journal of Experimental Social Psychology* vol. 47, no. 2, pp. 425–429.

Frederiks, E. R., Stenner, K. and Hobman, E. V. (2015). 'Household energy use: Applying behavioural economics to understand consumer decision-making and behaviour.' *Renewable and Sustainable Energy Reviews*, vol. 41, pp. 1385–1394.

Fuerst, F., McAllister, P., Nanda, A. and Wyatt, P. (2015). 'Does energy efficiency matter to home-buyers? An investigation of EPC ratings and transaction prices in England.' *Energy Economics* vol. 48, pp. 145–156.

Garnaut, R. (2011). *The Garnaut Review 2011: Australia in the global response to climate change*. Melbourne: Cambridge University Press.

Ghaffarian Hoseini, A. (2012). 'Ecologically sustainable design (ESD): theories, implementations and challenges towards intelligent building design development.' *Intelligent Buildings International* vol. 4, no. 1, pp. 27–48.

Golubchikov, O. and Deda, P. (2012). 'Governance, technology, and equity: An integrated policy framework for energy efficient housing.' *Energy Policy* vol. 41, pp. 733–741.

Hair, J. F., Black, W. C., Babin, B. J. and Anderson, R. E. (2014). *Multivariate data analysis*, 7th ed. Harlow: Pearson.

Henning, A. (2008). 'Heating Swedish houses: A discussion about decisions, change and stability.' *Anthropological Notebooks* vol. 14, no. 3, pp. 53–66.

Hepburn, C. (2010). 'Environmental policy, government, and the market.' *Oxford Review of Economic Policy* vol. 26, no. 2, pp. 117–136.

Hernandez, P. and Kenny, P. (2010). 'From net energy to zero energy buildings: Defining life cycle zero energy buildings (LC-ZEB).' *Energy and Buildings* vol. 42, no. 6, pp. 815–821.

Högberg, L. (2013). 'The impact of energy performance on single-family home selling prices in Sweden.' *Journal of European Real Estate Research* vol. 6, no. 3, pp. 242–261.

Hurst, N. (2012). 'Energy efficiency rating systems for housing: An Australian perspective.' *International Journal of Housing Markets and Analysis* vol. 5, no. 4, pp. 361–376.

Hurst, N. and Halvitigala, D. (2016). 'The influence of climate zones on energy efficiency advertising of residential properties.' 22nd Annual Pacific-Rim Real Estate Society Conference, Sunshine Coast, Queensland, Australia, 17–20 January 2016.

Hurst, N. and Wilkinson, S. (2015). 'Housing and energy efficiency: What do real estate agent advertisements tell us?' RICS COBRA Conference, UTS, Sydney, 8–10 July 2015.

International Energy Agency (IEA) (2015). *'Energy technology perspectives 2015.'* Accessed 14 March 2016. www.iea.org/bookshop/710-Energy_Technology_Perspectives_2015.

Janda, K. B., Bright, S., Patrick, J., Wilkinson, S. and Dixon, T. J. (2016). 'The evolution of green leases: Towards inter-organizational environmental governance.' *Building Research & Information*, vol. 44, no. 5–6, pp. 1–15.

Kiel, K. A. and Zabel, J. E. (2008). 'Location, location, location: The 3L approach to house price determination.' *Journal of Housing Economics* vol. 17, no. 2, pp. 175–190.

Levy, D. S. and Lee, C. K. C. (2004). 'The influence of family members on housing purchase decisions.' *Property Investment & Finance* vol. 22, no. 4, pp. 320–338.

Levy, D. and Lee, C. K. C. (2011). 'Neighbourhood identities and household location choice: Estate agents' perspectives.' *Place Management and Development* vol. 4, no. 3, pp. 243–263.

Li, L. H. and Wang, C. (2006). 'Real estate agency in China in the information age.' *Property Management* vol. 24, no. 1, pp. 47–61.

Logan, J. R. and Molotch, H. L. (2007). *Urban fortunes: The political economy of place*, 20th anniversary ed. London: University of California Press.

Marszal, A. J., Heiselberg, P., Bourrelle, J. S., Musall, E., Voss, K., Sartori, I. and Napolitano, A. (2011). 'Zero Energy Building: A review of definitions and calculation.' *Energy and Buildings* vol. 43, no. 4, pp. 971–979.

McKenzie-Mohr, J. and Smith, W. (1999). *Fostering sustainable behavior: An introduction to community-based social marketing*. Gabriola Island, BC: New Society Publishers.

McKinlay, A., Baldwin, C. and Stevens, N. J. (2016). 'Influences on dwelling size in Australia.' 22nd Annual Pacific-Rim Real Estate Society Conference, Sunshine Coast, Queensland, Australia, 17–20 January 2016.

Miller, K. H., Colantuoni, F. and Crago, C. L. (2014). 'An empirical analysis of residential energy efficiency adoption by housing types and occupancy.' Selected Paper prepared for presentation at the Agricultural and Applied Economics Association's 2014 AAEA Annual Meeting, Minneapolis, MN, 27–29 July 2014.

Miller, W. and Buys, L. (2012). 'Anatomy of a sub-tropical Positive Energy Home (PEH).' *Solar Energy* vol. 86, no. 1, pp. 231–241.

Mills, B. and Schleich, J. (2012). 'Residential energy-efficient technology adoption, energy conservation, knowledge, and attitudes: An analysis of European countries.' *Energy Policy* vol. 49, no. C, pp. 616–628.

Millward-Hopkins, J. T. (2016). 'Natural capital, unnatural markets?' *Wiley Interdisciplinary Reviews: Climate Change* vol. 7, no. 1, pp. 13–22.

Mlecnik, E., Visscher, H. and van Hal, A. (2010). 'Barriers and opportunities for labels for highly energy-efficient houses', *Energy Policy* vol. 38, no. 8, pp. 4592–4603.

Ohanian, R. and Cunningham, I. C. V. (1987). 'Application of primacy-recency in comparative advertising.' *Current Issues and Research in Advertising* vol. 10, no. 1, pp. 99–121.

Pampel, F. C. (2000). *Logistic regression: A primer.* Thousand Oaks, CA: Sage.

Parliament of Australia (2015). 'Australian climate change policy: A chronology (2013)'. Accessed 20 January 2016. www.aph.gov.au/About_Parliament/Parliamentary_Departments/Parliamentary_Library/pubs/rp/rp1516/ClimateChron.

Patron, H. E. and Roskelley, K. D. (2008). 'The effect of reputation and competition on advice of real estate agents.' *The Journal of Real Estate Finance and Economics* vol. 37, no. 4, pp. 387–399.

Pellegrini-Masini, G., Bowles, G., Peacock, A. D., Ahadzi, M. and Banfill, P. F. G. (2010). 'Whole life costing of domestic energy demand reduction technologies: Householder perspectives.' *Construction Management and Economics* vol. 28, no. 3, pp. 217–229.

Pereira, T. (2012). 'The transition to a sustainable society: A new social contract.' *Environment, Development and Sustainability* vol. 14, no. 2, pp. 273–281.

Perkins, H. C., Thorns, D. C. and Newton, B. M. (2008). 'Real estate advertising and intraurban place meaning: Real estate sales consultants at work.' *Environment and Planning* vol. 40, no. 9, pp. 2061–2079.

Pryce, G. and Oates, S. (2008). 'Rhetoric in the language of real estate marketing.' *Housing Studies* vol. 23, no. 2, pp. 319–378.

Ren, Z., Chen, Z. and Wang, X. (2011). 'Climate change adaptation pathways for Australian residential buildings', *Building and Environment* vol. 46, no. 11, pp. 398–412.

Rodriguez, G. and Siret, D. (2009). 'The future of houses: What real-estate ads tell about the evolution of single-family dwellings.' *International Journal of Architectural Research* vol. 3, no.1, pp. 92–100.

Schöllmann, A., Perkins, H. C. and Moore, K. (2001). 'Rhetoric, claims making and conflict in touristic place promotion: The case of central Christchurch, New Zealand.' *Tourism Geographies* vol. 3, no. 3, pp. 300–325.

Semeraro, P., and Fregonara, E. (2013). 'The impact of house characteristics on the bargaining outcome.' *Journal of European Real Estate Research*, vol. 6, no. 3, pp. 262–278.

Silverman, D. (2010). *Qualitative research*, 3rd ed. London: Sage.

Stern, N. (2007). *The economics of climate change: The Stern review.* Cambridge: Cambridge University Press.

Sustainability Victoria (2014). *'Victorian households energy report.* Accessed 2 February 2016. www.sustainability.vic.gov.au/services-and-advice/households/energy-efficiency/toolbox/reports.

Sustainability Victoria (2015). 'Solar photovoltaic (PV) systems: What is a solar PV system?' Accessed 2 February 2016. www.sustainability.vic.gov.au/services-and-advice/households/energy-efficiency/at-home/solar-photovoltaic-pv-systems

Tan, T. H. (2012). 'Predicting homebuyers' intentions of inhabiting eco-friendly homes: The case of a developing country.' International Real Estate Symposium (IRERS) 2012: Globalization of Real Estate: Transforming and Opportunities (6th), National Institute of Valuation (INSPEN), Ministry of Finance, Malaysia, 24–25 April 2012.

UK Government (2016). 'Policy: Building Regulation.' Accessed 11 March 2016. www.gov.uk/government/policies/building-regulation.

United Nations Economic Commission for Europe (2013). 'Good practices for energy-efficient housing in the UNECE region.' UNECE Information Service, Palais des Nations, CH-1211 Geneva, Switzerland.

United Nations. (2015) 'Framework Convention on Climate Change.' Accessed 21 April 2016. http://newsroom.unfccc.int/unfccc-newsroom/finale-cop21/.

Valkila, N. and Saari, A. (2012). 'Consumer panel on the readiness of Finns to behave in a more pro-environmental manner.' *Sustainability* vol. 4, no. 7, pp. 1561–1579.

van der Heijden, J. (2015). 'Interacting state and non-state actors in hybrid settings of public service delivery.' *Administration and Society* vol. 47, no. 2, pp. 99–121.

Vassileva, I. and Campillo, J. (2014). 'Increasing energy efficiency in low-income households through targeting awareness and behavioral change.' *Renewable Energy* vol. 67, pp. 59–63.

Victorian Building Authority (VBA) (2014). *Residential sustainability measures.* Practice Note 2014–55. Accessed 24 February 2016. www.vba.vic.gov.au/__data/assets/pdf_file/0003/20397/PN-55-2014-Residential-Sustainability-Measures.pdf.

Victorian Building Authority (VBA) (2016). 'Energy efficiency performance requirement for residential buildings'. Accessed 24 February 2016. www.vba.vic.gov.au/consumer-resources/building/pages/energy-efficiency-performance-requirement-for-residential-buildings.

Walker, G., Karvonen, A. and Guy, S. (2015). 'Zero carbon homes and zero carbon living: Sociomaterial interdependencies in carbon governance.' *Transactions of the Institute of British Geographers* vol. 40, no. 4, pp. 494–506.

Westley, F., Olsson, P., Folke, C., Homer-Dixon, T., Vredenburg, H., Loorbach, D., Thompson, J., Nilsson, M., Lambin, E., Sendzimir, J., Banerjee, B., Galaz, V. and van der Leeuw, S. (2011). 'Tipping toward sustainability: Emerging pathways of transformation.' *AMBIO: A Journal of the Human Environment* vol. 40, no. 7, pp. 762–780.

Wilkinson, S. J. (2014). 'Office building adaptations and the growing significance of environmental attributes.' *Journal of Corporate Real Estate* vol. 16, no. 4, pp. 252–265.

Willrath, H. and Logic, S. (1997). 'Thermal sensitivity of Australian houses to variations in building parameters.' 35th Annual Conference of the Australian and New Zealand Solar Energy Society, Canberra, 1–3 December 1997.

Wischmeyer, A. (2015). 'Humor in house advertising: Positive effects of wordplay.' *Business and Economic Policy* vol. 2, no. 3, pp. 208–213.

Yilmaz, Z. (2007). 'Evaluation of energy efficient design strategies for different climatic zones: Comparison of thermal performance of buildings in temperate-humid and hot-dry climate.' *Energy and Buildings* vol. 39, no. 3, pp. 306–316.

Your Home 'Energy.' (2013). Accessed 24 February 2016. www.yourhome.gov.au/energy.

Yu, S. M. and Tu, Y. (2011). 'Are green buildings worth more because they cost more?' IRES Working Paper Series IRES 2011-023.

Part 3
Management

14

Corporate real estate management

The missing link in sustainable real estate

Christopher Heywood

14.0 Introduction

Sustainability, arguably, is now core to corporate real estate management (CREM) practice. A 2009 CoreNet Global/JLL survey (CoreNet Global, 2015) showed that sustainability was highly important to 70 per cent of CRE managers with very high numbers (approximately 90 per cent) considering sustainability aspects in their CREM – for example, location decisions, green building certification and energy labelling. More recently, a survey showed that approximately 52 per cent of UK corporates had 'well-underway' or 'mature' sustainability agendas, with another 14 per cent finding the sustainability agenda 'transformational' (JLL, 2015b). Within those agendas, CREM aspects – green building/workplaces (≈69 per cent) and resource efficiency and waste (≈67 per cent) – figured highly. Despite these encouraging signs, the same survey showed substantial immaturity, with approximately 14 per cent 'starting out', approximately 16 per cent having an 'ad-hoc' agenda and even 4 per cent with 'declining' maturity. So progress is occurring in CREM and its associated corporate world, but with more to do and more transformations to come (JLL, 2015a).

By nomenclature, CREM places real estate into 'corporate' contexts where real estate is not core business. For some, 'corporate' presumes private-sector, 'for-profit' organisations' real estate; for others, including the author, CRE includes real estate used by 'for-profit' or 'not-for-profit' and private and public sectors – essentially, all organisations. The latter conception is not universal as some distinguish between 'corporate' and 'public' real estate (Evers, Van der Schaaf and Dewulf, 2002), suggesting different management imperatives. However, if CRE is managed to meet occupiers' strategic business ends, then the need to meet those objectives is consistent, with only the strategic objectives differing. 'Occupiers', as used here, is one of several terms, such as 'tenant', 'user' or 'CRE organisation', that denote types of and nuances to relationships. 'Occupier' also conflates occupying organisations and their people. It also allows for the distinction between ownership, where real estate as an investment is central, and occupation, where the real estate's use is important. 'Organisations' also includes everything from large, global corporations to sole traders and micro businesses. The principles are common but the management task's scale differs.

With CREM encompassing the management of real estate of organisations where real estate is incidental to and not their core business (CoreNet Global, 2015) – essentially, all 'real estate you have to do business' – CREM then offers a different perspective on real estate's purposes, which is that real estate exists to meet space occupiers' business ends. Here, real estate's monetary value (be that market, investment or development value), which is to the fore in general real estate, is de-emphasised. Instead, the way the real estate is used to meet occupiers' strategic business objectives comes to the fore, where CRE's five roles in organisational economies – as a factor of production, a corporate investment, a corporate asset, a real estate commodity and a contribution to the public realm (Heywood and Kenley, 2013) – need recognition and management.

It is possible to likely, with real estate incidental to a core business, that in focusing on that business the space occupier becomes oblivious to its real estate or just sees it as a business cost – rent or utilities to be paid and a management impost when it has to be maintained. Many CRE managers, when advocating more proactive CREM, have encountered their senior management saying 'We are not in real estate', with the implication that CREM should focus on business needs rather than trying to 'win' in real estate deals that are incidental to the core business; however, this is changing. The last 25 to 30 years have seen a shift towards not just more active management of CRE but also conceptions of it as a 'strategic resource' based on a greater understanding of how properties strategically enable (Joroff, Louragand and Lambert, 1993; McGregor and Then, 1999) or add value (Jensen, Van der Voordt and Coenen, 2012) to space-occupying organisations. This ethos of CRE as a strategic resource requires CREM to have business competencies to augment CRE managers' existing real estate ones (Kenley, Heywood and McGeorge, 2000; McCarty, Hunt and Truhan, 2006). When considering sustainability and real estate, there are several implications:

- It is not just the environmental sustainability virtues of real estate objects ('green' buildings or 'green' leases) that matter, but also how the real estate objects connect with the occupiers' stance on sustainability. Indeed, corporate social responsibility (CSR) positions where there are corporate concerns about corporations' ethical and human impacts on society (after Torrecchia, 2015) may be the point of leverage rather than just relying on any altruistic organisational desires to 'save the planet'.
- CREM, because of its focus on property use, represents the commercial real estate system's demand side (Heywood and Kenley, 2010). This conceptualisation is far from usual and has some consequences for the circles of 'blame' or 'virtue' that occur in discussions of sustainable real estate. Both the demand side and the two circles are discussed further below.
- CREM's intersection with business invokes all Brundtland's sustainability dimensions (WCED, 1987), not just building energy efficiency (resource or environmental dimensions) and market or investment values (economic dimension), which are often general real estate's focus. Recently, CoreNet Global (2015) identified that a CREM environmental dimension was evident in concerns about waste's effect on the natural environment. The economic dimension was found in operating cost reductions (presumably from energy efficiency), sustainable products and services offered, productivity (presumably from a more sustainable property) and optimised CRE (presumably throughout its life cycle economic performance and in alignment with business strategy). The social dimension was evident in occupant comfort and health, real estate's aesthetics, externalities and overall quality of life. While arguably incomplete, this list summarises current thinking about the multiple dimensions across both business and real estate concerns.

This means that when considering CREM and sustainability, it is not just the energy efficiency of the real estate that matters. How the real estate fits an occupier's sustainability agenda or whether they have one or not can be influential in how CREM engages with sustainability.

To develop this argument, this chapter reviews the state of the art for CREM and sustainability using three lenses:

1 Cadman's (2000) 'vicious circle of blame' (Keeping, 2000) and its 'virtuous circle' antidote (Hartenberger and Lorenz, 2008) locates real estate occupiers and their CRE managers as important actors in sustainable and non-sustainable real estate.
2 A consumption-based perspective locating CREM as the ultimate demand point in the CRE system makes it critical in the operation of that system.
3 CREM as a field operates at the intersection of business and real estate systems. As such, CREM leverages both business and real estate intentions with regards to sustainability.

These lenses are addressed first before several 'provocations' are also considered based on a CREM view as a way of advancing sustainability considerations in real estate. These include occupiers' 'perpetual' search for new premises being complicit in wasteful real estate practices and choosing CRE locations on 'exo-property' considerations like carbon in business supply chains. This extends CREM's real estate business nexus from the individual property focus that is otherwise so evident.

This chapter focuses on the aspect of property in use for business ends with a slant towards the physical artefact's fit-out and workplace configuration. The related aspects of sustainable building operations more usually fall into facilities management, which is outside this chapter's scope.

14.1 The 'circle of blame'

Cadman (2000) advanced the 'vicious circle of blame' to explain why more sustainable buildings are not created – new or refurbished. An antidote, a 'virtuous circle' where sustainability is advanced, was put forward by Hartenberger and Lorenz (2008).These circles show occupiers as actors whether real estate is sustainable or not. In the blame circle, occupiers are an object of shame (as are all actors in the circle). Various authors have addressed these actors, such as Newell, MacFarlane and Kok (2011) and Wilkinson, Sayce and Christensen (2015). Warren-Myers (2013) adds valuers (appraisers), while Hodges (2005) examines facility managers.

When occupiers receive attention in this circle, driven by a stance where economic wealth derives from property ownership, the attention is often in terms of their willingness to pay a 'green' premium, or perhaps to receive a 'brown' discount, for less sustainable property (Newell et al., 2011). Another significant body of occupier-focused work, often with a building science stance, considers indoor environmental qualities (IEQ) in more sustainable buildings (Abbaszadeh, Zagreus, Lehrer and Huizenga, 2006), where IEQ is virtuous in its own right (it is enough to just be a better environment for the occupying humans). Often, productivity benefits as an economic sustainability benefit derived from higher IEQ are sought (Seppänen and Fisk, 2006), or at least inferred, which presumably increases the willingness to pay.

14.1.1 Occupiers' roles in sustainable real estate

The CREM perspective allows fuller understanding of sustainable real estate through an examination of occupiers' roles in such real estate. A most obvious role is as the income stream that creates market, investment or development value for any real estate. As noted above, the sustainable real estate discourse manifests this as occupiers' presumed willingness to pay more for green (sustainable) real estate, less for brown (unsustainable) real estate or the same regardless.

Whether CREM does any of these can depend on several things. These include real estate practitioners' propensity to negotiate a 'deal'. From the occupier side, least cost in the real estate deal is a traditional, high-order concern. Combined with the deal-making propensity, there is a predisposition to negotiate to pay as little as possible, regardless – though this is changing, as discussed below.

It can also include the occupiers' tenure practices and commitment to 'manage' the CRE. Occupiers' tenure practices include owner-occupancy, various lease forms and sale and leasebacks. Depending on the practice, the issue of split incentives to invest in sustainable real estate, which underpins the circle of blame, will variously manifest. In owner-occupied CRE, the split is minimised, with Lowe and Gereffi (2008) suggesting the incentive is maximised; however, the CREM perspective distinguishing between use and ownership still has opportunities for splits, even here. With non-green leases, the split is larger. Green leases attempt to deal with the split through 'green' clauses (Bright and Dixie, 2014). A sale and leaseback for a seller of sustainable CRE, presumably, will seek a post-sale sustainability commitment. A purchaser's ability to deliver that should be a key ingredient in the sale.

It has been persistently noted throughout CREM's post-1980 professionalisation that the percentage of owned CRE is declining. For example, Brounen (2014) noted that between 1983 and 2013, Dow Jones Global 1000 companies' real estate assets declined from 22 per cent to 14 per cent of total assets. Reasons for this include occupiers' search for more flexible tenure, 'getting out of real estate' and off-balance-sheet treatment of leases to improve corporate financial ratios. This suggests that the split-incentive problem is increasing. Whether this trend will continue is unclear because impending changes to the International Financial Reporting Standards require lease values to be reported on balance sheets. This may change occupiers' attitudes to tenure methods (Brounen, 2014).

Commitment to manage CRE may be a more practically significant role in achieving sustainable CRE than any theoretical degree of split incentives. It is conceivable that the least split tenure method (owner-occupied) with low commitment to manage the CRE will produce less sustainable real estate than the more split leases coupled with higher commitment to manage CRE. The increase in leasehold CRE noted above can also be attributed to the rise in strategic CREM. As CREM emerged as a distinct real estate discipline, two things happened. First, an awareness of the amount of CRE used for business purposes contributed to greater commitment to manage organisations' CRE; metaphorically, 'We *are* in real estate.' Related was the identification of levels of evolved practice by Joroff et al. (1993), who posited evolutionary levels discernible by approaches to: real estate and particularly business information in decision making; the treatment of real estate capital costs; the internal pricing of real estate costs; the degree of CREM involvement with organisational senior management; and approaches to real estate products. The five levels are:

1. taskmaster (least evolved)
2. cost controller
3. dealmaker
4. intrapreneur
5. business strategist (most evolved).

The sustainable real estate implications are that it is likely, though empirically untested, that different evolved levels will engage differently with sustainability. For example, a taskmaster with a technical focus on the physical stuff of real estate and construction might appreciate the technical virtuosity of the sustainable building for its own sake without connecting with real estate's relevant business objectives. By contrast, the business strategist, while also appreciating the technical virtuosity, understands the CRE's strategic role in business strategy and seeks to align both. At this level of practice, the justification of capital and operational expenditures and how they advance the strategic business objectives are much more common. Also at this level, the traditional least-cost approach to real estate is replaced with a 'right-cost' approach that takes as its mantra, 'Pay the right cost for the required real estate quality, quantity, location, technology and management practices' (Osgood Jr, 2004). With this approach, a greater willingness to pay for sustainability may be more evident because of strategic business benefits – perhaps productivity gains, corporate positioning or meeting CSR commitments.

Finally, occupiers have an important role in achieving sustainable real estate over the course of occupation–what might be called 'occupants behaving properly'. Occupants are often blamed for the discrepancy between property's designed performance and actual performance (D'Oca, Corgnatia and Hong, 2015; Turner and Hong, 2014). Examples of behaviour thought to be blameworthy include:

- leaving lights and office equipment running overnight (Masoso and Grobler, 2010);
- 'interfering' with thermostats in seeking thermal comfort from building services;
- 'incorrectly' operating mixed-mode building systems by not operating windows to suit the outdoor climate (D'Oca et al., 2015);
- not using available public transport and bicycle facilities where these have been provided for and scored by the sustainability rating scheme; and
- 'incorrectly' sorting waste streams where recycling is available.

Undoubtedly, there are bases to these claimed poor behaviours. However, it can equally be said that design modelling may not adequately account for occupant behaviour, or that systems may be designed with inadequate knowledge of how people actually work or interact with built environments (Turner and Hong, 2014).

14.1.2 Occupiers' propensity to pay

Occupiers' propensity to pay seems to crucially underpin thinking about more sustainable real estate – certainly so for the economics of more sustainable real estate development and investment. There is an assumption that there is higher value from a premium, green-based income stream, and/or lower risk. A CREM perspective challenges this assumption through a disconnect that is evident in prioritising sustainability in CRE decisions. This flows through to, and is augmented by, other factors in occupiers' willingness to pay for sustainability.

First, surveys of occupiers show that sustainability is a lower-order priority in selecting CRE: Dixon, Ennis-Reynolds, Roberts and Sims (2009), Gensler (2006) and Van de Wetering and Wyatt (2011) are typical. Higher-order priorities include accessibility, flexibility and lease terms (Van de Wetering and Wyatt, 2011).This contrasts with green CRE being to the forefront of corporate sustainability agendas (JLL, 2015b) and occupiers seeing themselves as drivers of change towards sustainability (Van de Wetering and Wyatt, 2011). Reducing this disconnect is unlikely because CREM's priority is aligning CRE with business needs. The higher-order priorities noted are stronger alignment mechanisms, unless an occupier's business is sustainability. Therefore, it is likely that sustainability will continue as a not unimportant CRE decision factor, but one below other, more business-enabling criteria.

Second, whether occupiers are willing to pay remains an open question. Surveys certainly show their willingnes; for example, Van de Wetering and Wyatt (2011) found approximately 42 per cent would pay up to 10 per cent more. However, they also found that the majority would not want to pay more and those that were willing often had qualifications, such as a return on any extra cost. Given the relatively low prioritisation of green CRE noted above, a movement by occupiers to pay more seems less than certain. Any movement at all appears to hinge on:

- The mechanics of price setting in negotiating the occupancy transaction. This may be reliant on the understanding of sustainability of those in the negotiation, be they real estate agents, valuers (appraisers) (Warren-Myers, 2013), tenant representatives, or employee CREM executives. Why would, or should, an occupier pay more if a valuer takes as the most comparable evidence for setting the rent the non-sustainable property next door to the subject sustainable property? From a CREM perspective, you would not, nor should you.
- Occupiers' attitudes to real estate costs in the business and whether the attitude is one of least cost, right cost (for a strategic resource), rent cost or total cost of occupancy. As noted above, a right-cost approach can strengthen arguments for occupying more sustainable real estate. CREM's historically prevalent least-cost approach can focus on only the deal's rent cost. More evolved CREM considers total cost of occupancy or the inter-relationship between real estate costs and other infrastructure resource costs (after Materna and Parker, 1998); and
- The sophistication in CREM, as noted above in the evolutionary level and ability to build an argument for real estate that includes sustainability.

14.2 The demand–supply conceptualisation of the real estate system

Within the vicious circle of blame and even its virtuous antidote, every actor has equivalence, which is why the blame and virtue circle endlessly around them. To break this circularity, a different perspective is necessary – one that is provided by CREM and that supports an argument that CREM is the 'missing link'. Two ideas are relevant:

1. CREM is consumption-based perspective on the purpose of the commercial real estate economy (Heywood and Kenley, 2010); and
2. CREM's operation at the intersection of the real estate and business systems.

CREM's consumption-based perspective takes it as axiomatic that property exists to be useful ('consumed') and that real estate wealth derives from that use. CREM pays attention to that usefulness – that is, how the property meets the occupying organisation's business purposes. Under this conception, the occupant constitutes the system's ultimate point of demand (Figure 14.1). This contrasts with other perspectives about sustainability in real estate, which are

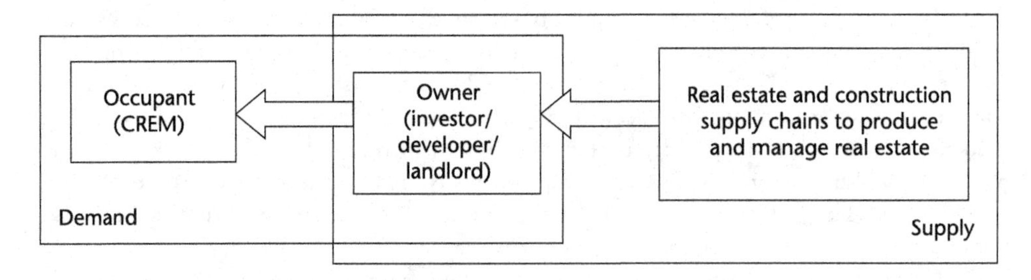

Figure 14.1 The real estate demand–supply system

Sources: Adapted from Heywood and Kenley (2010) and Warren-Myers and Heywood (2016)

supply-side led or pushed from there. This is not to say that there is no role for supply-side initiatives or for landlords or occupiers working together (Andelin, Sarasoja, Ventovuori and Junnila, 2015), but, without demand as the circuit breaker, the circle remains endless. This suggests that CREM occupies *the* critical point in making real estate sustainable and that CREM's motivations for business real estate warrant further consideration. This conception also assumes that the CREM supply relationship is independent of real estate and economic cycles that change the balance of power towards or away from occupiers in negotiations. At the model's level of abstraction, this assumption is acceptable because the CRE is still supplied to meet the demand. What the cycles affect is the capacity of CREM to demand the most favourable CRE objects and terms for objects' supply.

A second demand consequence results from CREM operating at the intersection of real estate and business systems where the inter-relationship between them is important. This allows CREM to positively leverage the occupiers' sustainability agenda such as is expressed as CSR obligations. The corollary is that without such occupier business concerns, demand for sustainable real estate will diminish. The CREM perspective shows that for all the supply-side effort towards more sustainable real estate, the demand for it is critical for its take-up and the delivery of supply-side intentions. Studies showing CREM prioritising decision-making criteria ahead of sustainability (Dixon et al., 2009) reiterate the scale of the challenge.

14.3 Changing workplace practices and sustainable real estate

Recently, for a variety of reasons, workplace designs and their management requirements have changed, with implications for sustainable real estate. These workplaces, variously called 'new-ways-of-working' or 'activity-based working', are made possible by mobility-enabling information and communication technologies (ICT) whereby work is no longer fixed in space (Schneider Electric, 2016). With various configurations, these workplaces are characterised by smaller footprints for individual, flexible workpoints and larger collaborative spaces. Various distributed working modes like teleworking and serviced offices are comparable manifestations of the phenomenon (Schneider Electric, 2016). Reasons for these changes include:

- Strategic business reasons such as competitive advantage from innovation or flexibility; and
- Seeking efficiencies in response to persistent evidence of workplace under-utilisation of fixed workstations dedicated to individuals. Vacancies at any time of the working day greater than 30 per cent were not uncommon, attributable to people being on leave or working elsewhere, permanently or temporarily. New configurations are showing allocations of, say, 130 people per 100 workpoints (National Australia Bank, 2013) though with not all present given the structural vacancy noted above.

Such workplace practices could enable a reduction in the emissions per employee assigned to a building as a way of addressing greenhouse gas (GHG) emissions. However, the intensification in occupancy can increase the energy use and related GHG emissions per square metre of floor area (Surmann and Hirsch, 2016). There can also be challenges inserting these new workplaces into existing properties where, due to the more intense occupation, HVAC systems and lifts work harder than they were designed to do, with consequences for energy use and building obsolescence.

Where these workplace practices reduce real estate emissions, these may be compensated for by increasing ICT-related emissions. For instance, between 3.6 and 6.2 per cent of all electricity use worldwide (approximately 2 per cent of all energy) is now attributable to operating the internet (Strickland, 2012). This points towards needing an integrated workplace management approach where CRE, ICT and human resources are managed holistically to optimise sustainability, as suggested by the Corporate Infrastructure Resource concept (Materna and Parker, 1998).

It is in the social dimension that new workplaces' sustainability has been challenged. Critiques of poor privacy, loss of territoriality and poor acoustics have been made (Kim and de Dear, 2013). In extremity, these new workplaces seem portrayed as the twenty first-century knowledge workers' 'bright satanic offices' (Baldry, Bain and Taylor, 1998), cruelly foisted on unsuspecting workers by 'evil', cost-minimising bosses and real estate managers. As they are alternatively portrayed as places for people to achieve their best and to innovate and as sites for WELL-ness (International WELL Building Institute PBC, 2015), it is clear that more research is required into their social sustainability.

Despite the prominence of the new workplace designs, it should be noted that they remain a small (though growing) percentage of the world's workplaces and that much more work is required to increase the sustainability of all workplace designs and their management.

14.4 Occupier churn and wasteful real estate practices

There has been a long-term trend towards shorter lease (occupancy) periods. This has been particularly promises in the UK, where 25-year leases were a norm (Crosby, Gibson and Murdoch, 2003) but leases are now closer to an average of seven years (MSCI Inc with Strutt & Parker, 2015). The trend is driven by many of the strategic CREM forces noted above where more dynamic business conditions require more dynamic CREM, such as shorter leases and new workplace designs. Balancing occupiers' needs are developer-investors' needs for commitments to support their activities. Combined with an established 'let's do a deal' mentality prevalent in all real estate, an 'unholy alliance' is created for wasteful real estate practices.

As occupiers are attracted to new developments, there is an investor need to backfill the vacated space. This creates adaptation requirements (Wilkinson, Remøy and Langston, 2014) where the physical fabric is reconfigured to suit new occupiers. Depending on the lease requirements, this potentially means that the exiting occupier strips out their old fit-out and makes good. The new occupier then strips out some of that make-good – say some or all of the carpets, ceilings and lighting (Figure 14.2). In addition to the waste streams created, it is evident that the embodied energy waste can be substantial when aggregated across the entire commercial real estate sector, but research is necessary to quantify this waste.

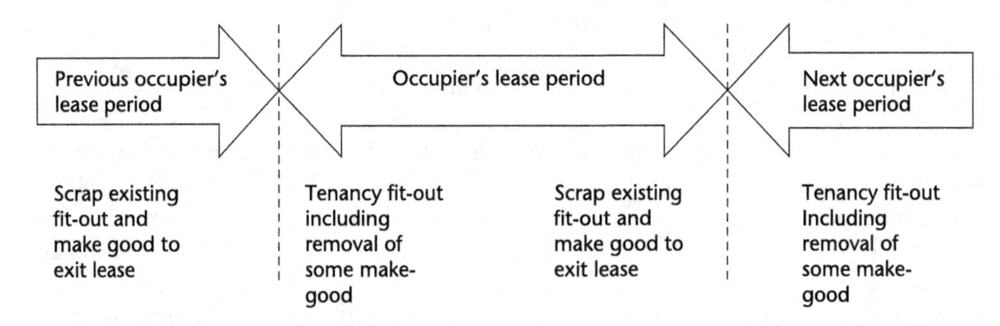

Figure 14.2 The wasteful fit-out cycle

14.5 CRE location selection from an exo-property perspective

CRE location selection is one of the longest-standing CRE domains of practice. It has also been of interest to the general business literature in terms of, usually, manufacturing plant location, for example, Schniederjans (1999). Concerns from time to time include human resources issues (workforce availability and skill levels), real estate availability and infrastructure support, cost and the business environment in the location (Rabianski, Delisle and Carn, 2001).

This literature generally focuses on making a single property's location decision. While decision-making guides such as Rabianski et al. (2001) account for externality and locality factors, by and large the focus is within the bounds of a single property decision, so-called 'endo-property' considerations ('endo-' means internal, inner or inside); this is true of most property activities, even external intelligence gathering like market research.

Three themes have emerged that suggest an 'exo-property' perspective ('exo-' means outer or external) might be useful in CRE location selection from a sustainability perspective. These are:

1. Globalisation of CRE portfolios, both as a more complex CREM task and as a part of reducing production costs – real estate and particularly human resource costs. This incorporates CRE into businesses' physical global supply chains;
2. Buildings' embodied energy (Crawford, 2014), to be added to their operational energy; and
3. 'Green supply chains', where carbon figures in their logistics and production and procurement trade-offs (Benjaafar, Li and Daskin, 2013; Sundarakani, De Souza, Goh, Wagner and Manikandan, 2010).

This convergence brings together sustainable CREM's more usual operational-energy concerns and to a lesser extent embodied-energy concerns and the physical global supply chain's carbon consequences as modelled in Figure 14.3.

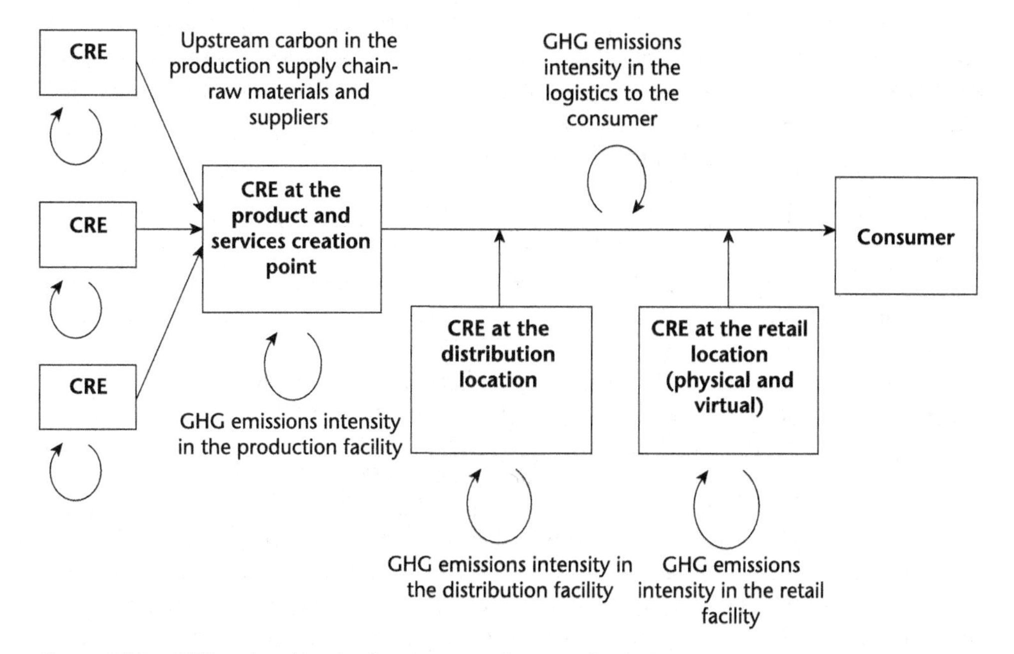

Figure 14.3 CRE and carbon in the consumption supply chain

Implications are then identifiable for the supply chain and potentially location decision making in minimising carbon in the consumed product. With this exo-property perspective, real estate's location in the logistics system becomes important – inbound from suppliers and outbound to the point of consumption, including distribution and retailing CRE locations. In full carbon accounting, this could also capture upstream carbon from raw materials or suppliers.

Calculating this whole-of-supply-chain carbon is complicated. It is difficult enough to calculate embodied energy at the endo-property level; calculating operational energy looks easy in comparison. But if the world is serious about reducing the GHG emissions related to economic activities, then thinking of this sort is necessary.

Consequently, depending on each step's GHG intensity, it may be better to embed less carbon in consumed products and services by locating production in higher-GHG locations and CRE closer to the consumption point rather than further away in low-GHG locations and CRE. Obviously, it is better to reduce GHG throughout the consumption supply chain, but, for a unit of embodied carbon in a product or service, then this argument can hold.

The question is how to apply this to individual property decision making? Perhaps a first step that CREM could initiate is to embed the CRE carbon work into the existing green supply chain work. There is some reference to the facility and location, but, with the logistics and operational research focus often evident in the green supply chain work, there is an obvious need for CREM inputs.

The CREM perspective, in which the real estate system and the space-using business system are connected, allows for (requires) real estate to be placed into a broader systemic context when considering sustainability. It redraws the system boundaries from the individual property level to CRE's contribution into the broader business productive system that might be necessary if true advances in the sustainability of CREM are to be achieved.

14.6 Conclusions

This chapter argues that occupiers are key participants in real estate sustainability without which the blame circles endlessly. A CREM consumption-based consideration interrupts that circularity, arguably making CREM the 'missing link' in sustainable real estate. If there is little demand, then little rationale for supply exists, other than for society's public good. This does not mean that no supply-side effort is warranted but the penetration of sustainable real estate can depend on:

- organisations taking sustainability into their business model and practices – CREM's role is then to align and enable or add value; and
- greater CREM professionalisation (necessary to extract optimum value from CRE) to strengthen arguments for sustainable real estate's enabling or adding value.

This indicates a need to change demand for sustainable real estate. CREM has a role to play in this, with survey results showing, with advanced CREM practice, that this is occurring, but that there is more to do to achieve a sustainability mindset.

A challenge is that occupiers' focus is on the business, not the real estate per se, other than when CREM argues for how CRE and CREM enables or adds value to the business. Real estate sustainability can be or is ignored, like many real estate considerations. However, the business connection can be used as leverage (or not) depending on organisational sustainability agendas.

Finally, because of CREM's business connectivity to sustainability, there are untapped opportunities to connect sustainable real estate with wider sustainable business concerns like CRE in green supply chains that warrant future research.

References

Abbaszadeh, S., Zagreus, L., Lehrer, D. and Huizenga, C. 2006. Occupant satisfaction with indoor environmental quality in green buildings. Available: www.escholarship.org/uc/item/9rf7p4bs.

Andelin, M., Sarasoja, A. L., Ventovuori, T. and Junnila, S. 2015. Breaking the circle of blame for sustainable buildings: Evidence from Nordic countries. *Journal of Corporate Real Estate*, 17, 26–45.

Baldry, C., Bain, P. and Taylor, P. 1998. 'Bright satanic offices': Intensification, control and team Taylorism. *In:* Thompson, P. and Warhust, C. (eds.) *Workplaces of the future*. Basingstoke: Macmillan.

Benjaafar, S., Li, Y. and Daskin, M. 2013. Carbon footprint and the management of supply chains: Insights from simple models. *IEEE Transactions on Automation Science and Engineering*, 10, 99–116.

Bright, S. and Dixie, H. 2014. Evidence of green leases in England and Wales. *International Journal of Law in the Built Environment*, 6, 6–20.

Brounen, D. 2014. *Corporate real estate acceptance*. Brussels: European Public Real Estate Association (EPRA).

CoreNet Global 2015. Sustainability. *In:* CoreNet Global (ed.) *The essential guide to corporate real estate*. Atlanta, GA: CoreNet Global.

Crawford, R. H. 2014. Embodied energy: What is it and how can it be optimised across the building life cycle? *In:* Clarke, D. (ed.) *How to rethink building materials: Creating ecological housing for the designer, builder and homeowner*. Empire Bay: CL Creations Pty Ltd.

Crosby, N., Gibson, V. and Murdoch, S. 2003. UK commercial property lease structures: Landlord and tenant mismatch. *Urban Studies*, 40, 1487–1516.

D'Oca, S., Corgnatia, S. and Hong, T. 2015. Data mining of occupant behavior in office buildings. *Energy Procedia*, 78, 585–590.

Dixon, T., Ennis-Reynolds, G., Roberts, C. and Sims, S. 2009. Is there a demand for sustainable offices? An analysis of UK business occupier moves (2006–2008). *Journal of Property Research*, 26, 61–85.

Evers, F., van der Schaaf, P. and Dewulf, G. 2002. *Public real estate management challenges for government: An international comparison of public real estate strategies,* Delft: Delft University Press Science.

Gensler 2006. *Faulty Towers: Is the British office sustainable?* London: Gensler.

Hartenberger, U. and Lorenz, D. P. 2008. Breaking the vicious circle of blame: Making the business case for sustainable buildings. *RICS FiBRE*. London: RICS.

Heywood, C. and Kenley, R. 2010. An integrated consumption-based demand and supply framework for corporate real estate. *In:* Wang, Y., Yang, J., Shen, G. Q. P. and Wong, J. (eds.) *2010 International Conference on Construction and Real Estate Management, 1–3 December 2010, Brisbane*. China Architecture and Building Press, 380–385.

Heywood, C. and Kenley, R. 2013. Five axioms for corporate real estate management: A polemical review of the literature. *In:* Callaghan, J. (ed.) *19th Pacific Rim Real Estate Society Conference*. Melbourne: Pacific Rim Real Estate Society. Available: www.prres.net.

Hodges, C. P. 2005. A facility manager's approach to sustainability. *Journal of Facilities Management*, 3, 312–324.

International WELL Building Institute PBC. 2015. *WELL Building Standard* International WELL Building Institute PBC. Available: www.wellcertified.com/well.

Jensen, P. A., van der Voordt, D. J. M. and Coenen, C. (eds.) 2012. *The added value of facilities management: Concepts, findings and perspectives*. Lyngby: Polyteknisk Forlag.

Jones Lang LaSalle (JLL) 2015a. The big eight for real estate: Understanding critical sustainability trends and what they mean for the future of your business. London: JLL.

Jones Lang LaSalle (JLL) 2015b. *Corporate real estate sustainability survey 2015*. London: JLL. Available: www.jll.co.uk/united-kingdom/en-gb/Documents/JLL_Sustainability-survey-PDF_March-2015.pdf.

Joroff, M., Louragand, M. and Lambert, S. 1993. *Strategic management of the fifth resource: Corporate real estate*. Atlanta, GA: IDRC.

Keeping, M. 2000. *What about demand? Do investors want 'sustainable buildings'*? The RICS Research Foundation, Available: www.rics.org/Practiceareas/Builtenvironment/Sustainableconstruction/what_about_the_demand_20000101.html.

Kenley, R., Heywood, C. and McGeorge, D. 2000. Australian corporate real estate management: Identification of the hybridisation of property and financial knowledge in practice. *6th Pacific Rim Real Estate Society Conference*. Sydney: Pacific Rim Real Estate Society. Available: www.prres.net

Kim, J. and de Dear, R. 2013. Workspace satisfaction: The privacy–communication trade-off in open-plan offices. *Journal of Environmental Psychology*, 36, 18–26.

Lowe, M. and Gereffi, G. 2008. *An analysis of the US real estate value chain with environmental metrics*. Durham, NC: Center on Globalization, Governance and Competitiveness, Duke University.

Masoso, O. T. and Grobler, L. J. 2010. The dark side of occupants' behaviour on building energy use. *Energy and Buildings*, 42, 173–177.

Materna, R. and Parker, J. R. 1998. *Corporate infrastructure resource management: An emerging source of competitive advantage*. Atlanta, GA: IDRC.

McCarty, T. D., Hunt, R. and Truhan, J. E. 2006. Transforming CRE value through relationship management. *Journal of Corporate Real Estate*, 8, 4–18.

McGregor, W. and Then, D. S.-S. 1999. *Facilities management and the business of space*. London: Arnold.

MSCI Inc with Strutt & Parker 2015. *IPD UK Lease Events Review: November 2015*. London: British Property Federation.

National Australia Bank 2013. *700 Bourke Street Building fact sheet*. Melbourne: National Australia Bank.

Newell, G., MacFarlane, J. and Kok, N. 2011. *Building better returns: A study of the financial performance of green office buildings in Australia*. Sydney: Property Council of Australia and Property Funds Association.

Osgood Jr, R. T. 2004. Translating organisational strategy into real estate action: The strategy alignment model. *Journal of Corporate Real Estate*, 6, 106–117.

Rabianski, J. S., Delisle, J. R. and Carn, N. G. 2001. Corporate real estate site selection: A community-specific information framework. *Journal of Real Estate Research*, 22, 165–197.

Schneider Electric 2016. *Activate to collaborate: The evolution of the smart workplace*. Kingston upon Thames: Schneider Electric.

Schniederjans, M. J. 1999. *International facility acquisition and location analysis*. Westport, CT: Quorum Books.

Seppänen, O. A. and Fisk, W. 2006. Some quantitative relations between indoor environmental quality and work performance or health. *HVAC&R Research*, 12, 957–973.

Strickland, J. 2012. How much energy does the internet use? *HowStuffWorks.com*. Available: http://computer.howstuffworks.com/internet/basics/how-much-energy-does-internet-use2.htm.

Sundarakani, B., De Souza, R., Goh, M., Wagner, S. M. and Manikandan, S. 2010. Modeling carbon footprints across the supply chain. *International Journal of Production Economics*, 128, 43–50.

Surmann, M. and Hirsch, J. 2016. Energy efficiency: Behavioural effects of occupants and the role of refurbishment for European office buildings. *Pacific Rim Property Research Journal*, 22, 77–100.

Torrecchia, P. 2015. Corporate social responsibility. *In:* Idowu, E. I. C., Samuel, O., Capaldi, N., Fifka, M. S., Zu, L. and Schmidpeter, R. (eds.) *Dictionary of corporate social responsibility: CSR, sustainability, ethics and governance*. Cham: Springer International Publishing.

Turner, W. and Hong, T. 2014. *A technical framework to describe occupant behaviour for building energy simulations*. Berkeley, CA: Ernest Orlando Lawrence Berkeley National Laboratory.

van de Wetering, J. and Wyatt, P. 2011. Office sustainability: Occupier perceptions and implementation of policy. *Journal of European Real Estate Research*, 4, 29–47.

Warren-Myers, G. 2013. Is the valuer the barrier to identifying the value of sustainability? *Journal of Property Investment & Finance*, 31, 345–359.

Warren-Myers, G. and Heywood, C. 2016. Investigating demand-side stakeholders' ability to mainstream sustainability in residential property. *Pacific Rim Property Research Journal*, 22, 59–75.

Wilkinson, S. J., Remøy, H. and Langston, C. 2014. *Sustainable building adaptation: Innovations in decision-making*. Chichester: John Wiley & Sons.

Wilkinson, S. J., Sayce, S. L. and Christensen, P. H. 2015. *Developing property sustainably*. London: Routledge.

World Commission on Environment and Development (WCED) 1987. *Our common future*. Oxford and New York: Oxford University Press.

15

The burgeoning influence of sustainability in managing UK retail property

Jessica Ferm and Nicola Livingstone

15.0 Introduction

There can be no doubt that sustainability is becoming an increasingly relevant and pertinent consideration for landlords and occupiers, which introduces burgeoning, continually evolving implications for real estate management. At the same time, retail property is undergoing significant changes, particularly with the shift from in-store to online sales. In 2012, global retail sales saw an annual growth rate of 2 per cent, whereas non-store internet retailing grew by 17.7 per cent year on year and represented 4 per cent of total retail sales (Cushman & Wakefield, 2013). In the UK, retail represents 5 per cent of national GDP (BRC, 2012), which is low compared to other countries globally and reflects the degree of domestic market saturation (Luce, 2013). However, the growth of online retailing has been strong. Today, UK internet sales account for approximately 13.2 per cent of all retail spending and this figure has been consistently growing year on year for 35 consecutive months, up to March 2016 (ONS, 2016). As the retailing market continues to adjust and respond to changing consumption patterns and the growing online marketplace, what emergent approaches have been adopted by landlords and occupiers to integrate sustainability into real estate management strategies and decision making? This chapter reflects on the extent to which sustainability is influencing decision making for retailers in the UK currently, and the anticipated effects for real estate management in the medium term, over the next five to ten years.

Until recently, the majority of work relating to sustainability in real estate has been concentrated on the office market (Oyedokun, Jones and Dunse, 2015; Warren-Myers, 2012), in line with topics such as sustainable retrofitting (Morrissey, Dunphy and MacSweeney, 2014; Wilkinson, 2012), valuation (Lorenz and Lützkendorf, 2011; Fuerst and McAllister, 2011), rent premiums (Kok and Jennen, 2012; Eichholtz, Kok and Quigley, 2010), green leases (Janda, Bright, Patrick, Wilkinson and Dixon, 2016) and corporate social responsibility (Jones, Hillier, Comfort and Clarke-Hill, 2009; Ang and Wilkinson, 2008), linking these issues directly to the 'triple bottom line' (TBL) (Elkington, 1997), of economic, social and environmental outcomes. In terms of their interaction with sustainability in comparison to office occupiers, retailers appear to be more complex and dynamic when considering the breadth of stakeholders involved and the very different user interactions that retailers foster as gateways to consumption. Due to the variation in retail offering, there is likely to be diversity in the way in which sustainability is managed and interpreted, relating to the size of the retailer and the type of goods they sell. Retail is the largest commercial sector in the UK, accounting for 43 per cent of the real estate market (PIA, 2015), and across Europe retailing is responsible for approximately 23 per cent of the total energy consumption of non-domestic real estate (Doak, 2009). Despite these facts, retailers and their responses to sustainability have received far less attention than office occupiers.

In the academic literature on retailing, the topic of sustainability remains fairly underexplored, and Wiese, Kellner, Lietke, Toporowski and Zielke (2012a) suggest that, compared to other industries, retailing has a time lag of ten years in terms of integrating sustainable practices. However, there is a burgeoning literature emerging. Most of the recent literature focuses on the drivers and barriers for retailers in adopting sustainable practices more generally, in their selection of sustainable products and brands, promotion of sustainable consumer behaviour and adoption of sustainable business processes (Wiese, Zielke and Toporowski, 2015). In the latter category, there has been some focus on the so-called 'sustainable store' (ibid.: 438) promoted by the large retailers: IKEA, Marks & Spencer (in the UK) and Whole Foods (in the US) (Robaton, 2012; Wiese et al., 2012a; Wilson, 2015). There has also been some discussion of the extent to which retailers' choice of store location (out of town versus town centre) is becoming more sustainable (Doak, 2009) and of the relative carbon footprint of online versus traditional retailing (Edwards, McKinnon and Cullinane, 2010, 2011; Wiese, Toporowski and Zielke, 2012b). Although previous literature has clearly recommended that retailers would 'be wise to continue to try and ensure that their strategic management thinking and decision making is consistently informed, and perhaps increasingly driven, by the need to move to a more sustainable long-term future' (Jones, Comfort and Hillier, 2008: 568), there are suggestions that such movements are merely 'genuflections' to adopting truly sustainable approaches (ibid.).

With sustainable real estate in retailing established as an important area of further research and attention, the next section reviews the more specific literature in this field, focusing on the drivers and barriers for retailers in implementing sustainable real estate strategies. The remainder of the chapter then sets out the research design and findings from an initial scoping study, compiled to assess UK retailers' current adoption of sustainable approaches to their real estate, and their views on the question 'What is next for sustainable retail real estate?'

15.1 Retailers and sustainable real estate: Drivers of and barriers to progress

The customer-facing nature of retailing as an industry, and fierce competition, means that retailers are under increasing pressure to be seen to be addressing sustainability, particularly as consumer awareness and media attention rises (Whitson and Crawford, 2013). Consumers can be cynical, are suspicious of 'greenwashing' and tend to be critical of retailers who seek publicity for sustainability initiatives before they are embedded (Ipsos MORI, 2008). Retailers' claims of their sustainability credentials are often contested by pressure groups who argue for the disproportionate impact of retailers on the environment, communities and local economies; therefore, retailers are working harder than ever to demonstrate their commitment to sustainable development (Jones, Comfort and Hillier, 2007). This can be difficult for large retailers, since there are inherent tensions between sustainable development's emphasis on 'local' and 'small' and their operational requirements, which rely on large-format stores and global supply chains (Jones, Comfort and Hillier, 2008). There has been some criticism that the unspecific nature of the sustainable development debate allows retailers to pick and choose practices that are both profitable and socially acceptable (Lehner, 2015), and that many of the environmental initiatives adopted by retailers – especially those relating to energy consumption – also help to reduce costs and are primarily 'business driven' and therefore represent only a 'weak' model of sustainability (Jones, Comfort and Hillier, 2011). However, from a cost/profit perspective, even though leading retailers may be better placed to integrate sustainability features into their business strategies, the paradox of small and local versus large and global could be a significant barrier to sustainable development, which may not be easily overcome.

This notion of 'weak' sustainability may persist, but the competitive nature of retailing means that as sustainable strategies are adopted by leading retailers, this drives change in the industry more broadly. Wilson (2015) argues that Marks & Spencer – through implementing their sustainability strategy 'Plan A' and opening the first eco-store in the UK – could be considered to operate a 'strong' model of sustainability. This pioneering or leading role of key retailers was also shown in the US (Robaton, 2012). Whereas there are relatively few LEED-certified stores in the US, compared to offices, those that do exist are driven by retailers (rather than investors). One such example is Whole Foods, who in the early 2000s rolled out LEED-certified stores in the US, but found few shopping centre developers that shared their commitment to sustainability. This is important because there is – as yet – no significant sustainability effect on the investment portfolios of retail properties (Vlasveld, 2012), and sustainability effects on rent premiums in a Dutch retail portfolio of 128 properties were found to be insignificant after controlling for other factors (Op't Veld and Vlasveld, 2014). However, the development of corporate social responsibility (CSR) initiatives has incentivised and influenced retailers to adopt more sustainable approaches (Jones, Comfort and Hillier, 2005), and today retail real estate investment trusts (REITs) have made sustainability an integral part of their CSR platforms (Robaton, 2012). This is consistent with observations that retail property companies are taking a leadership role in driving the sustainability agenda (Newell, 2009). Nevertheless, retailers are increasingly considered to have a pivotal role as they are a critical link in the supply chain, and their actions are considered to have broader influence through socially constructed networks on (say) manufacturers, who are thought of as bigger direct polluters (Doak, 2009). This illustrates, argue Fernie, Sparks and McKinnon (2010), the transformation that has occurred over the last 20 to 30 years, where retailers have moved from being passive recipients of products to controlling the product supply in relation to consumer demand so that they now 'control, organize and manage the supply chain from production to consumption' (ibid.: 895). Developing this idea more broadly, Jones et al. (2008: 569) suggest that retailers are now 'actively constructing and disseminating sustainable development agendas, which are driven by their own commercial goals', which include efficient supply chain management, CSR, responding to consumer demands and adopting environmental certifications.

Real estate and occupancy concerns in relation to sustainability are becoming an increasingly important area of research as retailers concentrate their efforts on the physical operation of their stores, methods of distribution and in-store operations – aspects of their operations that are essentially within their control (Jones, Comfort and Hillier, 2011). This is supported by a survey of retailers by the Retail Industry Leaders Association (RILA, 2015), which showed that retailers typically begin with a focus on their own operations and then turn to addressing the product and supply chain impacts. Within their own operations, 'retail facilities often represent the main environmental impacts that are directly under the company's control. Stores are a major touch-point for customers, and corporate offices for the headquarter employees' (RILA, 2015: 31). The British Retail Consortium (BRC, 2012) also distinguishes between retailers' 'direct impact' (buildings, waste, transport, refrigeration, climate change adaptation), 'customer impact' (green products, packaging, carrier bags) and 'supply chain impact' (sourcing products responsibly, regulatory frameworks). The BRC estimates that the retail sector is directly responsible for around 3.5 per cent of UK greenhouse gas (GHG) emissions, and that energy use in buildings contributes over half the direct carbon footprint of the retail sector. In a further survey of European (primarily food) retailers (Forum for the Future, 2009), it was shown that carbon management and reduction through building or selecting sustainable buildings and reducing in-store energy use is becoming an increasingly important focus for retailers; whereas packaging and waste have been top of their sustainability agenda, carbon reduction is now at the top, or near the top, influenced by expectations of further legislation in this area. This was confirmed by an Ipsos MORI survey (2008), which identified climate change as the most important issue facing retailers currently. These changing priorities are supported by retailer surveys (RILA, 2015) showing that, whereas waste and recycling currently ranks as one of the most mature dimensions for the industry in terms of sustainability impact, retailers' retail operations are ranked as 'standard' but are identified as an area where they expect to be excelling in the near future.

Although there is evidence that prime retailers are driving much sustainable practice in real estate strategies, and that property is becoming a more important focus for retailers, there remain a number of challenges to implementation and future progress, particularly in the sustainable management and occupation of buildings. The uptake of green leases in the UK commercial sector has been slow compared to other countries such as Australia and Canada (Sayce, Sundberg, Parnell and Cowling, 2009). Green leases can be categorised as either light green and voluntary, or dark green and enforceable (Bright, 2008). Most green leases fall into the light green category, particularly in the UK (Bright and Dixie, 2014). Internationally, the use of green leases remains low in the retail sector, compared to the office sector, but the UK is starting to take the lead in retail (Bright et al., 2015). Change is being driven by larger companies with public environmental commitments; leading this turn are Marks & Spencer, who in 2012 adopted an environmental leasing strategy, informed by the BBP 'green lease toolkit', implementing light-green clauses in leases for all their existing and new stores (Janda et al., 2016; Bright et al., 2015). This is consistent with findings from studies of office occupiers where large corporate tenants are more likely to adopt sustainable practices (Miller and Buys, 2008). Research on barriers to implementing sustainable practices in multi-tenanted office buildings may be applicable to retailers in shopping centre environments. For example, Van de Wetering and Wyatt (2011) – in their study of a multi-tenanted office building in Bristol, UK – found that there were few financial incentives for tenants to manage their buildings sustainably, particularly in relation to energy consumption, either because there were fixed service charges, infrequent energy bills, bills paid by headquarters, or a lack of information available on operational energy consumption, all of which undermine the feedback loop required to incentivise behaviour. Lack of motivation to implement sustainable operations by tenants may also be related to wider structural changes within the real estate market, specifically the shortening of lease lengths. The average lease length is now 6.8 years, compared to 9.6 years in 1999, with the retail market experiencing the most pronounced change (PIA, 2015). If a tenant is signing a lease for a short time frame only, are they generally less likely to be concerned with the day-to-day functionality of a building if they are not a large corporate occupier who prioritises sustainable practices?

Research on retailers' sustainable practices points to some specific challenges for retailers. One of the barriers faced by many retailers is the problem of 'split incentives' between the landlord and the tenant (retailer) in implementing energy-efficient upgrades or engaging in better energy management (Janda et al., 2016). In a report on sustainability practices in UK shopping centres, Whitson and Crawford (2013) argue that this is exacerbated by a lack of trust between retailers and landlords, where retailers fear 'rogue' landlords and where there is often poor communication and the demarcation of responsibilities is unclear. In shopping centres, they found that the problem is magnified further as landlords lack control over much of the property and facilities managers and/or branch managers of stores are inexperienced or lack awareness of sustainable practices. Other barriers for retailers highlighted in the report include the high staff turnover and high numbers of part-time workers in retail (relative to other sectors), which makes it difficult to train staff in environmental practices. Another key difference between retailers and office occupiers is that whereas cost savings tend to be a key driver for office occupiers to implement sustainable measures in-use, in retail there is an inherent conflict as optimum heating, cooling and lighting is essential for the retail experience of customers and links to retailers bottom-line sales (Thompson, 2007; Whitson and Crawford, 2013). Jones Lang LaSalle (JLL, 2012) also discuss tensions relating to split incentives, suggesting that collaboration between landlords and occupiers is essential to foster best-practice approaches and overcome barriers to create mutually beneficial outcomes for all parties, founded on clear communication and close working relationships.

15.2 Research design

This chapter now presents results from an initial scoping study, compiled to assess UK retailers' current adoption of sustainable approaches to their real estate. In order to ascertain where the drive for change is coming from and to account for broader experiences of real estate sustainability, commercial agents, landlords and retailer representatives were consulted, as well as retailers.

The focus of the research was based around the following five key questions, which were developed from the literature reviewed and compiled to offer perspective into the influences on the integration of sustainable real estate features into the market:

1. What is currently driving the sustainability agenda for retailers?
2. How might this change in the future?
3. How is sustainable real estate being implemented by retailers?
4. What do you perceive to be the most important sustainability issues for retailers?
5. What are the main barriers for retailers to the adoption of sustainable real estate practices?

For Question 4, respondents were asked to rank the following list and comment on their rationale:

- sustainably sourced products;
- waste management;
- energy consumption of building;
- carbon footprint of retail practices more generally (consider the importance of online retailing, staff travel to work patterns);
- good access to public transport (for staff and customers); and
- staff working environment and well-being.

The structured questionnaire was disseminated via email to 18 UK-based participants in early 2016. Seven responses were received, representing a 40 per cent response rate. Respondents have not been directly identified to preserve individual and corporate anonymity; however, all contacts are employed by major industry leaders, either in retail (all multiple international retailers, primarily offering clothing) or real estate (landlords and commercial agents). Any direct quotes are identified by a particular abbreviation and number for reference – for example, 'L1' for 'landlord', 'A1' for 'commercial agent' and 'R1' for 'retailer'.

Overall trends in responses were identified across the five themed questions, and also from the three areas of industry expertise (commercial agents, landlords and retail experts), to offer insight into the nuanced relationships inherent in developing and managing sustainable retail strategies. From these responses, an indicative snapshot of the current influence of sustainability in retailing is presented and developed, to offer insight into future research in this topical area.

15.3 Retailers' sustainable practices and the role of real estate

The questionnaire asked retailers and their agents/landlords/representatives to rank and comment on the most important sustainability issues for the sector and to describe the sustainable real estate initiatives that retailers are engaged in.

On the most important sustainability issues, there was little consensus on the ranking amongst this sample, but there was a strong message that initiatives under the banner of sustainable real estate were already being tackled by retailers. So, for example, 'good access to public transport for staff and customers' was considered very important but 'is already being done through store design and site selection' (R2) or 'we regard [these] as fundamental factors of location planning' (R1). Similarly, 'energy consumption of buildings' is already being addressed so it is no longer so high on retailers' sustainability agendas (R2). However, as the retailers contacted were market leaders, the issue of energy consumption is likely to be experienced very differently by smaller or independent retailers in more secondary locations, occupying older buildings; there is more opportunity for further research across this spectrum of retailing in the future.

One agent (A1) commented that sustainable real estate is less of a priority in retail than in other commercial sectors since '[in] retail, sustainability is more "customer facing" than in other sectors, so sourcing of products and waste management are probably dominant'. This customer-focused nature of retailing also meant that, whereas staff working environments and well-being were key considerations driving office occupiers' sustainability agendas (Andelin, Sarasoja, Ventovuori and Junnila, 2015; Armitage, Murugan and Kato, 2011), for retailers, 'this is true for consumer well-being as well as staff' (R2). One retailer (R1) emphasised that they prioritise initiatives that they see as 'important from a sustainability perspective, make sound business sense ... add to the integrity of our business, and typically save money'. This has led to the following list of priorities (in no particular order), which go beyond the list included in the questionnaire to include aspects of social sustainability:

- human rights and working conditions in our supply chains;
- sustainably sourced materials;
- safe products;
- waste management in our operations and in our customer services;
- energy efficiency;
- health and well-being of our partners; and
- local communities, including training and recruitment and local jobs for local people.

From the perspective of a retailer, sustainably sourced products and materials were considered 'to be the most important due to reputational aspects and the reach of retail across very diverse and large supply chains', followed by the carbon footprint of retailer practices more generally, since 'understanding their carbon footprint addresses multiple sustainability aspects and gives ... a more comprehensive view across the whole business' (R2). The increasing importance of carbon management and reduction for retailers supports predictions on changing priorities by the Forum for the Future report (2009).

One landlord (L1) suggested that priorities came down to 'how much money can be saved', and therefore the focus was more on waste management (in particular recycling) and energy consumption. Agents suggested that '[if] a landlord can provide sustainable buildings with low energy costs, that will be a bonus but not a deal breaker' (A3); they provided some examples of on-site renewable energy, use of LEDs, reducing the impact of the fit-out, green walls/roofs and green lease clauses, but '[on] the whole I don't believe many are doing much' (A2). This perspective may reflect the priorities of the majority of retailers as occupiers, whereas large retailers (particularly owner-occupiers) who see themselves as market leaders may be more interested in developing and/or leasing sustainable buildings that meet the benchmark standards of BREEAM or LEED. One retailer (R1) has developed a Responsible Development Framework for informing their approach to capital projects, be they new premises or refurbishments, where they seek early engagement with landlords/investors/developers and design teams to find the optimum solution for the investor and occupier. Other retailers see the acquisition of buildings that meet BREEAM standards more as part of their risk assessment and due diligence process (R2). However, the acquisition of BREEAM-rated properties is not the main focus for most retailers; as one landlord commented, 'despite many London businesses opting for new developments, there may be a turn towards retrofitting old or existing buildings' (L1). Another retailer explained why there is more emphasis on fit-out than on new build in retailing:

> The problem really is that retail is successful based on footfall and therefore location is the driving force in site selection. This means retailers have focused a great deal on fit-out to implement their sustainability strategy rather than seek high environmentally performing buildings (R2).

15.4 Drivers and barriers in the implementation of sustainable real estate in retailing

The influence of the consumer emerged as a dominant consideration from respondents reflecting on the key drivers of and barriers to implementing sustainable retail practices, as well as themes such as costs, split incentives and legislation.

The visibility of retailer branding and CSR coupled with advances in technology and social media have created a marketplace where the consumer can access key information about retail practices easily and efficiently. Although this is also linked to the evolution of online shopping, the fact that 'consumers are becoming increasingly well informed through mobile and digital technology' means that retailer information is now more accessible and therefore should be transparent (R1). 'Word-of-mouth' communication is also still recognised as an important influential factor for consumers, with their feedback driving changes on sustainability from 'retailers at shop level and upwards', but also from the top down, 'by head office, who try to feed sustainability through all departments, from leasing to property management, to development' (L1). These social drivers are encouraging retailers not only to become more sustainable, but also to ensure that consumers are aware of their 'green practices' to remain competitive and relevant:

> Competition is fierce, and as consumers increasingly factor sustainable performance into purchasing decisions (especially at the more prestige end of the market), then retailers ... have to respond in order to remain successful (A1).

On the flip side, negative publicity is never welcome and retailers also need to be conscious that 'social media allows bad news stories to spread quickly' (A1) and that 'public image' is a key driver of consumer decision making (A3). This is very much in line with work published by Hammerson (2014), which discussed 'the considered consumer', who typically takes significantly longer to make decisions not only about what to buy but also where to buy it today, in a retail market awash with competitors. For retailers today, there is 'great social expectation to improve their environmental and social impacts' (A2). How retailers effectively communicate their 'green' or 'sustainable' operating principles is of paramount importance to developing and maintaining successful consumer relations. There is a requirement for retailers to 'demonstrate a good appreciation of the issues, as CSR and sustainability agendas assume a higher profile' (R1) and 'consumers have generally become more savvy to "greenwashing", so retailers need to be a lot more robust now in their sustainability claims' (A1).

The need to communicate clear sustainable principles to consumers is essential, not only for retailers to remain competitive, but also to 'future-proof' and save costs going forward (A1). There is an awareness of cost pressures as both a barrier to and driver of change, from landlords', agents' and retailers' perspectives. Improving energy efficiency in line with burgeoning legislation, such as the Minimum Energy Efficiency Standards (MEES) coming into operation in the UK in 2018, is identified as a driver of change (A1, R2), but a change that is beneficial to retailers in the long term. The key reasons for improving energy management are linked to cost savings and risk management (A1, A3, R1). Sustainability drives the desire to consume less energy, due to retailer competitiveness, efficiency and burgeoning legislative requirements, but as a result, by reducing energy consumption, costs and risk are more effectively managed:

> Developing sustainability strategies has moved from being seen as purely an environmental good, to encompassing risk management, building resilience to future costs and legislation, and delivering greater efficiencies across operations and the supply chain: effectively maximising profitability (R2).

It is essential that the design and implementation of such sustainability strategies is appropriate to the usage and purpose of the building, for both owner-occupiers (R1) and tenants (R2): 'retailers need to buy into the tangible benefits of sustainable practices' (L1). However, the cost of implementing such practices can be prohibitive and although owner-occupiers and landlords may see significant returns over the life cycle costs of buildings, for occupiers the situation can be a little more challenging and limited by costs. Put simply, 'retailers are free to be as sustainable as they wish, but often it's expensive. Where it is cost efficient to do so, retailers will do it; where it isn't, they won't' (A3). One of the central barriers to adopting sustainable real estate practices is the issue of the split incentive, and the 'key to success is collaboration … to ensure that respective objectives are aligned' (R1), from the perspectives of landlords, investors, developers and occupiers. The lack of 'value proposition' is problematic for tenants (A2). However, the MEES are seen as a counter-balance to the split-incentive problem, as they will impose

> requirements to improve properties on landlords. This also addresses the issue of shortening lease lengths and tenants being unwilling to invest in efficiency initiatives where they may not see payback within their remaining lease term (A1).

It was suggested that the split incentive was the 'main blockage' to successfully growing a sustainable approach to retail real estate (R2), with suggestions for remedying this situation proposed, such as joint investments between landlords and tenants where there was sufficient trust between parties. However, it was also noted that currently there is 'a huge opportunity in joint sustainability strategies that is being missed through a lack of engagement on those issues between landlord and tenant' (R2).

15.5 What is next for sustainable retailing?

Changing consumer demands and changing legislation and taxation were the two main drivers of change mentioned by respondents.

Consumer demands are changing for a number of reasons. Improvements in mobile and digital technology mean that consumers are increasingly well informed about sustainability, but awareness is more around the sustainable sourcing of products and the working conditions and human rights of workers in the supply chains than the sustainability credentials of the buildings that retailers operate (R1). Improved technology is also fuelling the rise of online retail, which 'means tackling different issues to traditional bricks-and-mortar retail as there are distribution and delivery as well as data centre operations to consider, as well as potentially office and (some) store challenges' (R2). This was confirmed by an agent, who suggested that retailers would shift from consumer-facing shops to warehousing and distribution and thus 'become more akin to logistics firms in this respect' (A1). This means that retailers are increasingly aware of the need for the flexibility of their buildings in the future:

> We need our buildings to be increasingly flexible to enable us to change our proposition to adapt to the changing demands of our customers – for example, to change part of a shop from the display of products to a beauty spa, or simply to surrender part of a building back to the landlord for an alternative third party use (e.g., leisure) (R1).

This is consistent with the landlord perspective: 'Shopping centre lettings will be geared more to items that you cannot buy online; food, beverage, entertainment and leisure pursuits. Shopping areas will change to leisure areas' (L1). The increasing focus on deliveries to support online retailing and changing policy and taxation associated with this is likely to have an impact on retailers. One agent (A1) suggested that retailers are likely to make more significant investments into hybrid or electric vehicles as councils roll out low-emission zones in urban centres, to avoid both increasing taxation on their distribution vehicles and associated tolls for vehicles not classed as low emission (e.g., the London congestion charge). However, retailers emphasised the uncertainty and inconsistency surrounding government policies and legislation around carbon reduction and the efficiency of buildings. The recent removal of Carbon Reduction Commitments by the UK Government meant that one retailer went so far as to say that 'legislation is not a key driver' (R1). Another suggested, however, that retailers will need to focus on building resilience to those changes and that employing alternative energy sources and managing emissions 'in the face of rising energy and carbon costs is good for the long-term survival of their business' (R2).

15.6 Conclusions

Despite the significant contribution of retailing to carbon emissions internationally, the dominance of retail in the commercial real estate market and the sector's changing property requirements, retailers have received little explicit attention in the academic research and professional literature on sustainable real estate to date. This chapter has started to address this gap and reveals some of the specific drivers and challenges facing UK retailers, which frame their specific response to addressing sustainability. The customer-facing nature of retail, fierce competition and increasing consumer awareness and media attention mean that large international retailers are working hard to demonstrate their commitment to sustainable development and are emerging as leading drivers of sustainability in the real estate market. Real estate and occupancy concerns are important for retailers, as they concentrate on operations that are under their control. The types of initiatives engaged in are driven by retailer branding and a desire to be a market leader on the one hand, and risk management and future-proofing on the other. However, the actions of retailers in terms of occupancy are limited to aspects that are within their control, rather than that of their landlord or head office, and by initiatives that do not undermine the customer experience, which ultimately drives their bottom-line sales. Split incentives, problematic landlord–tenant relationships, shortening leases and concern about consumer well-being are currently barriers to more widespread implementation. Changing consumer preferences, improvements to digital and mobile technologies and the changing legislative landscape are expected to impact retailers in the future, with an increased emphasis on the flexibility and resilience of buildings, and a shift away from a focus on the 'sustainable store' to warehousing, distribution and delivery and carbon management and reduction across all their operations.

This research is necessarily limited by its nature as an initial scoping study focused on the UK. More research, with larger samples, extending internationally, needs to be undertaken in order to understand in greater depth retailers' motivations, landlord–tenant relationships and leasing issues, and whether there is a particular process involved in becoming more sustainable by moving from 'weak' to 'strong' models of sustainability as suggested by Jones, Comfort and Hillier (2007). We know that retailers seem to start by prioritising certain sustainable features in their business plans, which can be integrated with a low cost margin and in a more straightforward way – for example, waste management and in-store operations. In some cases, the adoption of specific retailing practices may be merely attempts to 'greenwash' and are truly 'genuflections' to sustainability (Jones, Comfort and Hillier, 2008); however, such practices could also indicate shifts towards a deeper, more holistic sustainability strategy. Is there a larger process at play, and if so, how does it differ between retailers and in terms of resilience? How can the split incentive be effectively managed and where are best-practice approaches emerging? How can smaller, more secondary retailers be encouraged to adopt sustainable practices? In the Australian office market, there is a documented environmental performance gap between the top-quality and lower-grade office stock (Wilkinson, 2014) and we might speculate that a similar phenomenon could occur in the retail sector. Rydin (2016) points to the lack of attention given to the non-prime market in the promotion of sustainable commercial real estate, which is particularly relevant to retailers and merits further research. Finally, given the increasing importance of online retailing, research on sustainability in retail needs to explicitly extend to other domains of the commercial market and operations, such as warehousing and logistics. As Wilson (2015: 443) argued, sustainable retailing is indeed a 'journey not a destination' and our research efforts need to continue to reflect this moving forward.

References

Andelin, M., Sarasoja, A., Ventovuori, T. and Junnila, S. (2015) 'Breaking the circle of blame in sustainable buildings: Evidence from Nordic countries'. *Journal of Corporate Real Estate*, Vol.17, No.1, pp.26–45.

Ang, S. L. and Wilkinson, S. J. (2008) 'Is the social agenda driving sustainable property development in Melbourne, Australia.' *Property Management*, Vol.25, No.5, pp.331–343.

Armitage, L., Murugan, A. and Kato, H. (2011) 'Green offices in Australia: A user perception survey'. *Journal of Corporate Real Estate*, Vol.13, No.3, pp.169–180.

Bright, S. (2008) 'Drafting green leases'. *Conveyancer*, Vol.6, pp.498–517.

Bright, S. and Dixie, H. (2014) 'Evidence of green leases in England and Wales'. *International Journal of Law in the Built Environment*, Vol.6, Nos.1/2, pp.6–20.

Bright, S., Patrick, J., Thomas, B., Janda, K.B., Bailey, E., Dixon, T. and Wilkinson, S. (2015) 'The evolution of "greener" leasing practices in Australia and England'. Paper presented at RICS COBRA

AUBEA 2015, 8–10 July, Sydney, Australia. Available at: www.rics.org/uk/knowledge/research/conference-papers/the-evolution-of-greener-leasing-practices-in-australia-and-england (accessed 3 March 2016).

British Retail Consortium (BRC) (2012) *A better retailing climate: Towards sustainable retail*. British Retail Consortium (BRC), London.

Cushman & Wakefield (2013) *Global perspective on retail: Online retailing*. July 2013. Cushman & Wakefield, London. Available at: www.cushmanwakefield.co.uk/en/research-and-insight/2013/global-perspectives-on-retail (accessed 3 March 2016).

Doak, J. (2009) 'An inspector calls: Looking at retail development through a sustainability lens'. *Journal of Retail and Leisure Property*, Vol.8, No.4, pp.299–309.

Edwards, J., McKinnon, A. and Cullinane, S. (2010) 'Comparative analysis of the carbon footprints of conventional and online retailing: A "last mile" perspective'. *International Journal of Physical Distribution and Logistics Management*, Vol.40, Nos.1/2, pp.103–123.

Edwards, J., McKinnon, A. and Cullinane, S. (2011) 'Comparative carbon auditing of conventional and online retail supply chains: a review of methodological issues'. *Supply Chain Management: An International Journal*, Vol.16, No.1, pp.57–63.

Eichholtz, P., Kok, N. and Quigley, J. (2010) 'Doing well by doing good: green office buildings'. *American Economic Review*, Vol.100, No.5, pp.2494–2511.

Elkington, J. (1997) *Cannibals with forks: The triple bottom line of 21st century business*. Capstone, Oxford.

Fernie, J., Sparks, L. and McKinnon, A. (2010) 'Retail logistics in the UK: Past, present and future'. *International Journal of Retail and Distribution Management*, Vol.38, Nos.11/12, pp.894–914.

Forum for the Future (2009) *Sustainability trends in European retail*. September 2009. Available at: www.forumforthefuture.org/sites/default/files/images/Forum/Documents/Sustainability_trends_in_European_retail_Sept09.pdf (accessed 7 May 2016).

Fuerst, F. and McAllister, P. (2011) 'Green noise or green value? Measuring the effects of environmental certification on office values'. *Real Estate Economics*, Vol.39, No.1, pp.45–69.

Hammerson (2014) *The considered consumer*. Hammerson, London.

Ipsos MORI (2008) *Sustainability issues in the retail sector*. February 2008. Available at: www.ipsos-mori.com/researchpublications/publications/1297/Sustainability-Issues-in-the-Retail-Sector.aspx (accessed 7 May 2016).

Janda, K., Bright, S., Patrick, J., Wilkinson, S. and Dixon, T. (2016) 'The evolution of green leases: Towards inter-organizational environmental governance'. *Building Research & Information*. DOI: 10.1080/09613218.2016.1142811.

Jones Lang LaSalle (JLL) (2012) *A tale of two buildings*. JLL/Better Building Partnership, London.

Jones, P., Comfort, D. and Hillier, D. (2005) 'Corporate social responsibility and the UK's top ten retailers'. *International Journal of Retail and Distribution Management*, Vol.39, No.4, pp.256–271.

Jones, P., Comfort, D. and Hillier, D. (2007) 'Sustainable development and the UK's major retailers'. *Geography*, Vol.92, No.1, pp.41–47.

Jones, P., Comfort, D. and Hillier, D. (2008) 'UK retailing through the looking glass'. *International Journal of Retail and Distribution Management*, Vol.36, No.7, pp.564–570.

Jones, P., Comfort, D. and Hillier, D. (2011) 'Sustainability in the global shop window'. *International Journal of Retail and Distribution Management*, Vol.33, No.12, pp.882–892.

Jones, P., Hillier, D., Comfort, D. and Clarke-Hill, C. (2009) 'Commercial property investment companies and corporate social responsibility'. *Journal of Property Investment & Finance*, Vol.27, No.5, pp.522–533.

Kok, N. and Jennen, M. (2012) 'The impact of energy labels and accessibility on office rents.' *Energy Policy*, Vol.46, pp.489–497.

Lehner, M. (2015) 'Translating sustainability: The role of the retail store'. *International Journal of Retail and Distribution Management*, Vol.43, Nos.4/5, pp.386–402.

Lorenz, D. and Lützkendorf, T. (2011) 'Sustainability and property valuation: Systematisation of existing approaches and recommendations for future action'. *Journal of Property Investment & Finance*, Vol.29, No.6, pp.644–676.

Luce, S. (2013) *Global retail report*. October 2013. UNI Global Union, Nyon. Available at: http://blogs.uniglobalunion.org/commerce/wp-content/uploads/sites/7/2013/10/Global-Retail-Report-EN.pdf (accessed 16 June 2016).

Miller, E. and Buys, L. (2008) 'Retrofitting commercial office buildings for sustainability: Tenants' perspectives'. *Journal of Property Investment & Finance*, Vol.26, No.6, pp.552–561.

Morrissey, J., Dunphy, N. and MacSweeney, R. (2014) 'Energy efficiency in commercial buildings: Capturing added-value of retrofit'. *Journal of Property Investment & Finance*, Vol.32, No.4, pp.396–414.
Newell, G. (2009) 'The significance of sustainability best practice in retail property'. *Journal of Retail and Leisure Property*, Vol.8, No.4, pp.259–271.
Office for National Statistics (ONS) (2016) 'Retail sales: March 2016'. ONS Statistical Bulletin, March 2016. Available at: www.ons.gov.uk/businessindustryandtrade/retailindustry/bulletins/retailsales/march2016 (accessed 8 May 2016).
Op't Veld, H. and Vlasveld, M. (2014) 'The effect of sustainability on retail rents, values and investment performance: European evidence'. *Journal of Sustainable Real Estate*, Vol.6, No.1, pp.163–185.
Oyedokun, T. B., Jones, C. A. and Dunse, N. (2015) 'The growth of the green office market in the UK'. *Journal of European Real Estate Research*, Vol.8, No.3, pp.267–284.
Property Industry Alliance (PIA) (2015) *Property data report*. October 2015. PIA, London.
Retail Industry Leaders Association (RILA) (2015) *Retail sustainability report*. RILA, Arlington, VA.
Robaton, A. (2012) 'Retail REITs seeing green'. *REIT Magazine*, July/August 2012, pp.34–36. Available at: www.reit.com/news/reit-magazine/july-august-2012 (accessed 2 May 2016).
Rydin, Y. (2016) 'Sustainability and the financialisation of commercial property prices: An analysis of the evidence base'. *Environment and Planning D. Online first*. Available at: http://epd.sagepub.com/content/early/2016/02/23/0263775816633472.abstract (accessed 8 May 2016).
Sayce, S., Sundberg, A., Parnell, P. and Cowling, E. (2009) 'Greening leases: Do tenants in the United Kingdom want green leases?' *Journal of Retail and Leisure Property*, Vol.8, No.4, pp.273–284.
Thompson, B. (2007) 'Green retail: Retailer strategies for surviving the sustainability storm'. *Journal of Retail and Leisure Property*, Vol.6, No.4, pp.281–286.
van de Wetering, J. and Wyatt, P. (2011) 'Office sustainability: Occupier perceptions and implementation of policy'. *Journal of European Real Estate Research*, Vol.4, No.1, pp.29–47.
Vlasveld, M. (2012) *Sustainable retail performance*. MRE Master's Thesis. Amsterdam School of Real Estate. Available at: www.ivbn.nl/viewer/file.aspx?fileinfoID=560 (accessed 7 May 2016).
Warren-Myers, G. (2012) 'Sustainable management of real estate: Is it really sustainability?' *Journal of Sustainable Real Estate*, Vol.4, No.1, pp.177–197.
Whitson, M. and Crawford, H. (2013) *UK shopping centres and the sustainability agenda: Are retailers buying?* CEM (College of Estate Management) Occasional Paper Series, May 2013.
Wiese, A., Kellner, J., Lietke, B., Toporowski, W. and Zielke, S. (2012a) 'Sustainability in retailing: A summative content analysis'. *International Journal of Retail and Distribution Management*, Vol.40, No.4, pp.318–335.
Wiese, A., Toporowski, W. and Zielke, S. (2012b). 'Transport-related CO_2 effects of online and brick-and-mortar shopping: A comparison and sensitivity analysis of clothing retailing'. *Transportation Research Part D: Transport and Environment*, Vol.17, No.6, pp.473–477.
Wiese, A., Zielke, S. and Toporowski, W. (2015) 'Sustainability in retailing: Research streams and emerging trends'. *International Journal of Retail and Distribution Management*, Vol.43, Nos.4/5.
Wilkinson, S. (2012) 'Analysing sustainable retrofit potential in premium office buildings', *Structural Survey*, Vol.30, No.5 pp.398–410.
Wilkinson, S. J. (2014) 'Office building adaptations and the growing significance of environmental attributes'. *Journal of Corporate Real Estate*, Vol.16, No.4. pp.252–265.
Wilson, J. P. (2015) 'The triple bottom line: Undertaking an economic, social and environmental retail sustainability strategy'. *International Journal of Retail and Distribution Management*, Vol.43, Nos.4/5, pp.432–447.

16

Sustainable facilities management

Paul Appleby

16.0 Introduction

This chapter outlines the key areas of policy and legislation that have driven the sustainability agenda in the management of non-residential and multi-residential property, with particular attention to UN and EU initiatives and their impact on countries such as the UK. This is followed by an examination of the main areas of sustainable facilities management, including modern practices in the integrated management of energy, water, waste, air quality, indoor environment, hygiene, health and safety, fire and ecology and how these relate to maintenance and purchasing strategies.

As defined by CEN (the European Committee for Standardisation) and ratified by BSI British Standards in BS EN 15221 1:2006, '[f]acility management (FM) is the integration of processes within an organisation to maintain and develop the agreed services which support and improve the effectiveness of its primary activities' (BSI 2006). The term 'facilities management' is widely used in the UK, whilst internationally the term 'facility management' is common, although both terms tend to be used interchangeably with each other and with 'property management', all of which may be applied to the management of individual buildings and of multi-building portfolios. In the UK, however, 'property management' is normally used in the context of leased buildings and frequently multi-residential properties. 'Facilities management', on the other hand, usually covers both 'hard' and 'soft' elements of building management, where hard FM includes maintenance and repair of fabric and building services and soft FM includes cleaning, catering, waste management, reception and such like.

16.1 Policy and legislation

Since humans first sought shelter in caves and dwellings, an element of facilities management entered into their lives. From managing waste, cleaning and repairing the fabric to ensuring catering arrangements are maintained, there has always been a maintenance element to ensuring the health, comfort and wellbeing of building occupants. As populations grew and sought security in numbers, clusters of buildings evolved into towns and cities, resulting in the development of buildings that had to cater for more than simply providing accommodation for sleeping and eating. Increased population density and prosperity resulted in greater volumes of waste and consequent odour and epidemic risk. The Greeks recognised the problem as far back as 500 BC when they introduced a law that required all waste to be dumped no less than one mile from a city boundary (Chandrappa and Bhusan Das 2012).

Remarkably, it took until the fourteenth century for the rapidly growing London to introduce a similar law: the 1309 City of London Regulation, which banned the dumping of rubbish into streets or lanes and stated that people 'ought to have it carried to the Thames or elsewhere out of town'. A 1357 Royal Order forbade the dumping of rubbish into the Thames and Flete, and the 1388 First English Sanitary Act 'forbade its citizens from dumping animal waste in any waterways or ditches'.[1] However, until the end of the fifteenth century, the Thames remained effectively an open sewer, resulting in widespread cholera and culminating in the 'Great Stink' of 1858 and the construction of the network of sewers and pumping stations by Joseph Bazalgette that followed, much of which has survived into modern times. Meanwhile, Paris was experiencing similar problems to London, although as early as 1370 closed sewers were built under the Rue Montmartre, but not extended to most of Paris until the early fifteenth century. However, this system did not have sufficient capacity to cope with the increased flows from the recently developed flushing toilets, which were implicated in a cholera epidemic that killed 20,000 people in 1832 (Appleby 2011).

Alongside the growth of urban populations in the Middle Ages was the adoption of coal as a fuel for heating, which in London in particular led very quickly to poor air quality, resulting in the burning of so-called 'soft coal' being prohibited in 1273, whilst in 1306 King Edward I introduced a law stating that 'no coal was to be burned by industry and artisans during Parliament' (Griffin 2016).

Urbanisation also resulted in increased risk of fire spread, such that after a major fire in London in 1212, thatched roofs were banned. However, since buildings continued to be built of highly combustible materials and so close together that fire could easily spread across streets, this did not prevent 80 per cent of the city being destroyed by the Great Fire of London in 1666.[2]

It was this history of reacting to epidemic and disaster that led to the building, fire, hygiene, health and safety regime that exists today – not just in the UK, of course, but across most of the planet. This regime not only shapes the way buildings are built, but how communities are planned, how infrastructure is developed and how processes and property are managed.

It was the global oil crisis in the early 1970s that focused the minds of building operators on energy consumption, mostly for economic reasons. The greenhouse effect and global warming had been mooted by the Swedish scientist Svante Arrhenius as far back as 1896, but no formal connection was made with energy generation and consumption until the 1970s, when the world's leading climate scientists gathered in Geneva in 1979 for the first World Climate Conference. However, it took another 18 years for an international commitment to action to be agreed at the 1997 UN Framework Convention on Climate Change in Kyoto, since when energy management of the existing building stock has increased in importance exponentially.[3]

Out of Kyoto grew the EU's Emissions Trading Scheme (EU ETS) along with a raft of policies and legislation introduced by the signatory nations (which did not include the US), such as the Climate Change Levy (CCL) in the UK, launched in 2001 (Seely, 2009). This targeted the larger energy consumers (using more than 1,000 kWh electricity per month or more than 4,397 kWh of gas) whilst initially exempting electricity from a renewable source, although this exemption was withdrawn in July 2015, to much consternation from environmental pressure groups, third-party intermediaries (such as energy brokers) and energy companies.[4] Specified 'energy-intensive' industries (Environment Agency 2015) were able to obtain a rebate of 90 per cent on electricity CCL and 65 percent on gas CCL if they signed up to an energy reduction commitment, as set out in an umbrella Climate Change Agreement (CCA) established for their industry sector (Environment Agency 2017).

The EU has introduced a suite of directives focusing on improving the energy performance of buildings, including the Energy Performance of Buildings Directive (EPBD), revised in 2010 to push member countries to achieve 'nearly zero-energy' new buildings by 2021, and the 2012 Energy Efficiency Directive (EED) (which is discussed at greater length in Section 16.2).

Wherever they are operating in the world, modern facilities managers have to comply with literally hundreds of pieces of legislation. Those that are applicable will depend, to a large extent, on the activities that are carried out in and around the property and the operational risks that fall within the ambit of the facilities management function. In a factory, for example, these responsibilities may be divided between process management and facilities management: process managers being responsible for the impacts associated with the processes, facilities managers for buildings and associated plant and infrastructure. Sometimes, these two responsibilities overlap: for example, where the building services are an important component of the process control, such as ventilation to minimise occupational exposure to toxic airborne emissions from a process.

The techniques involved in facilities management are applicable across the headings discussed below, and are frequently applied to the same plant, equipment or building components. For example, energy management includes the operation, maintenance and monitoring of domestic hot water and cooling towers, which also require water management and hygiene management within the context of health and safety risk mitigation.

There is a series of international, EU and British standards covering facilities management strategy and systems, including the ISO 41000 series, which is under development at the time of writing. This is a high-level document entitled *Facilities Management: Integrated Management Systems*, which will link with the current ISO 15221 series, such as BS ISO EN 15221 5:2011 and 7:2012, which deal with FM processes and benchmarking respectively. The latter provides a comprehensive approach to benchmarking that includes environmental, energy, water and waste metrics, as well as guidance on financial, spatial, service quality, satisfaction and productivity benchmarking, thus providing a framework for tracking the performance of the facilities management process across a wide range of parameters.

Some organisations may wish to establish an environmental management system (EMS), for which certification is possible through a number of routes, such as the European Eco-Management and Audit System (EMAS) or the International Standard ISO 14000 series, which is based on the same template as ISO 41000, as well as the energy management and quality standards, ISO 50001 (see below) and ISO 9001 respectively. Annex IV of EMAS sets out the requirements for the reporting of Environmental Performance Indicators (EPI) including energy efficiency, material efficiency, water, waste, biodiversity and emissions (EC 2009).

There is not space in this chapter to examine multiple industrial processes, so the focus in the following sections is on the role of facilities management in relation to offices and multi-residential properties.

It is possible that a facilities manager will need to consider retrofitting and refurbishment in order to improve the operational efficiency of a property. For more information, refer to *Sustainable Retrofit and Facilities Management* (Appleby 2013).

16.2 Energy, greenhouse gas and water management

Good energy management can have a major impact on the energy use in a building and consequent carbon dioxide emissions. The carbon dioxide associated with burning fossil fuels and electricity use is usually the primary source of greenhouse gas (GHG) emissions associated with office and residential use, although leakage of refrigerant gases may also make a significant, if unpredictable, contribution. Common refrigerants used in air-conditioning and refrigeration systems are fluorinated gases, such as hydrofluorocarbons (HFCs), which may have global warming potentials (GWPs) many thousands of times greater than carbon dioxide, although globally they contribute only 2 per cent (±0.4 per cent) of total GHG emissions (Blanco et al. 2014). EU fluorinated gas (F-gas) regulations are quite onerous and demand a programme of leakage checks depending on the mass of refrigerant and the carbon equivalent emissions (EU 2014). The US, on the other hand, which has never signed up to the Kyoto Protocol, allows for a loss of refrigerant of up to 20 per cent of system contents per year for large commercial and industrial refrigeration systems and 10 per cent for comfort cooling before action is required under Section 668 of the Clean Air Act.[5] These 'trigger repair leak rates' were lowered from 35 per cent and 15 per cent respectively through a proposed rule change out for consultation at the time of writing. This demonstrates a very large difference in the allowable leakage rates for these gases between the US, where refrigeration and air conditioning are far more widespread, and the EU.

Energy management is 'the systematic use of management and technology to improve an organization's energy performance' (Carbon Trust 2013). Although the terms 'carbon management' and 'greenhouse gas management' are primarily applied to industrial applications, many energy managers will need to consider non-energy GHGs, such as refrigerants, along with water conservation when dealing with commercial property.

Both energy and water management require the following elements when starting with a blank piece of paper:

- policy and management framework – development of a management system;
- survey and data analysis;
- strategy, action plan and priorities;
- monitoring and targeting;
- housekeeping and engagement; and
- procurement strategy.

Water management policy will typically include hygiene and water quality issues, such as legionellosis prevention (see below), as well as conservation, whilst energy management will impact indoor environment through the relationship between energy consumption, ventilation rates and indoor air quality, for example.

There is an abundance of standards and guidance dealing with establishing and operating energy management systems (EMSs) and carrying out energy audits. These include a suite of international standards rooted in ISO 50001:2011 Energy management systems, leading to ISO 50002, 50003, 50004, 50006 and 50015, all published in 2014, which set out detailed guidance and requirements for audit and certification bodies; implementation and improvement; and measuring, verification and benchmarking energy performance.[6] National versions of these ISO standards have been produced by the British Standards Institute (BSI) in the UK (BS EN ISO 50001) and DIN in Germany (DIN EU ISO 50001), for example, whilst between 2012 and 2014 the BSI produced a suite of EU standards for energy auditing – BS EN 16247 – in response to the EU's 2006 *Directive on Energy End-Use Efficiency and Energy Services* (predecessor to the EED). This suite comprises separate publications dealing with buildings, processes and transport. Article 8 of the EED sets out requirements for energy audits and energy management systems, enshrined in individual EU member legislation, such as the UK's Energy Savings Opportunity Scheme (ESOS) Regulations 2014 (Environment Agency 2016). ESOS is a mandatory assessment scheme for qualifying organisations administered by the Environment Agency. UK companies with either more than 250 employees, an annual turnover of more than

€50 million or a balance sheet of more than €43 million fall within the scheme's ambit. Although certification under BS EN ISO 50001, an up-to-date Display Energy Certificate (DEC) or Green Deal Assessment can form part of a compliance package, ESOS requires that at least 90 per cent of total energy consumption be assessed, including not only buildings but also transport, processes and construction activities.[7] It also applies to overseas-based companies with more than 250 employees in the UK.

In the US, there is no legislation that mandates energy- and water-saving targets for property other than that owned and operated by the federal government. Both the 2005 Energy Policy Act and the 2007 Energy Independence and Security Act (EISA) incorporate sections that require action to improve the energy efficiency of existing federal buildings, setting targets and requiring audits, reinforced by a 2006 Memorandum of Understanding called Federal Leadership in High Performance and Sustainable Buildings and the 2009 Presidential Executive Order 13514 Federal Leadership in Environmental, Energy and Economic Performance. Note the emphasis on 'leadership', further exemplified by the President's 2011 Better Buildings Initiative, subsequently rebranded as the Better Buildings Challenge (BBC) administered by the US Department of Energy,[8] which has been subsequently augmented by the 2013 Climate Action Plan (Executive Office of the President 2013). At the heart of the Better Buildings Challenge is a commitment by organisations to reduce energy consumption by 20 per cent over a ten-year period. The Climate Action Plan introduced financing of rural energy efficiency investments and a Multifamily Energy Efficiency Investment Fund, whilst also extending the BBC through a series of 'Accelerators', including a project focusing on establishing public–private partnerships to fund energy databases and facilitate benchmarking across a wide range of building types. Other projects encourage energy efficiency programmes for data centres, industrial facilities, state and local government buildings, schools, outdoor lighting and homes.

The US Environmental Protection Agency (EPA) has developed a suite of tools to support energy managers under the Energy Star umbrella, including *Guidelines for Energy Management* (EPA 2013), the Energy Star Portfolio Manager for commercial buildings and the Energy Tracking Tool and Energy Performance Indicator system for industrial sites.[9]

The policy and framework required for managing energy and water require high-level endorsement within an organisation as well as the informed engagement of occupants. Adequate resources (financial and human) must be allocated to the energy and water management action plans based on the outcome of energy audits, ongoing monitoring and feasibility studies. The key technical challenges are:

- carrying out energy and water consumption surveys/audits;
- establishing baselines and benchmarks/targets;
- identifying energy-and water-saving measures;
- implementing and managing ongoing energy and water conservation; and
- ongoing monitoring and targeting (M&T).

There is no space here to provide detailed guidance on energy and water audits; for guidance on procedures and benchmarks, see Chapters 4.1 and 4.2 of Appleby (2013); Carbon Trust (2011); CIBSE (2012); ASHRAE (2011).

Whilst auditing methodologies can be applied internationally, benchmarks are likely to vary from country to country. For example, Chapter 20 of CIBSE's 2012 *Guide F* provides energy benchmarks for a wide range of building use types for good and typical design practice in the UK, based on data drawn from a variety of sources, but mostly determined prior to the year 2000. CIBSE also publishes the benchmarks used in DECs in its Technical Memorandum TM46 (CIBSE 2008). In the US, the benchmarking BBC Accelerator referred to above is augmenting the databases being managed under the EPA's Energy Star programme, comprising its Portfolio Manager for commercial buildings and Energy Tracking Tool for industrial facilities.[10]

Benchmarking databases for water consumption are also available. Although these may be translatable between some countries, it is usually best to use those that are derived for a country where water use habits are similar: for example, per capita domestic water consumption in the US in 2006 was 575 litres/day, in Australia 493, in the UK 150 and China 86, whilst in Cambodia and Ethiopia it was around 20 (UNDP 2006). There are a number of different approaches to benchmarking available worldwide. For example, Scottish Water's 'business stream' website allows customers to enter their water consumption data into an online calculator to enable comparison with others in their sector.[11] Sydney Water provides benchmarks for typical and best-practice water consumption for a number of commercial applications.[12] The Energy Star Portfolio Manager referred to above includes benchmarks for water use as well as energy consumption, whilst a review of benchmarks from 2009 by BIO Intelligence Service and Cranfield University for the European Commission provides useful data from across the EU (BIO Intelligence Service 2012). For the UK, the government-commissioned Watermark scheme was launched in 2002,[13] and its updated and enhanced replacement, Aquamark, has recently been launched.[14]

Many of these tools and references also provide guidance on the identification and implementation of energy- and water-saving measures. These would typically be identified during an audit, whole-life costed to determine potential pay-back period and divided into 'low-hanging fruit' and longer-term objectives. Some of these measures may involve changes in behaviour, housekeeping or maintenance regimes, whilst others may require retrofit or significant refurbishment. Key to the ongoing management of energy and water use is the introduction of a suitable monitoring and targeting (M&T) regime. Targets are different from benchmarks in that they need to reflect both baseline energy and water consumption and a desire to improve efficiency with time. Targets should be set 'that will stimulate management to make improvements. These targets must be realistic and achievable, taking into account the likely savings from improvements in housekeeping, maintenance and other efficiency measures' (CIBSE 2012). For example, a reduction in energy and water consumption of 10 per cent per annum may be targeted until a predicted economic minimum is reached. Monitoring through building energy management systems (BEMs) allows trend analysis and exception reporting so that problems, such as plant being run overnight, can be quickly identified and mitigated. Where an existing building does not have a BEMS, the retrofitting of additional monitoring equipment may be an early investment decision.

16.3 Indoor environment

Of all the challenges that a facilities manager faces, according to a 2009 survey by the International Facility Management Association, the indoor environment is frequently the source of the greatest number of complaints from occupants.[15] In particular, a densely occupied or badly managed office can present a potentially stressful environment resulting in high levels of dissatisfaction and discomfort, and sometimes in illness, absenteeism and low productivity (HSE 1995). For the worst buildings, this can be manifested as a collection of symptoms that might be described as 'sick building syndrome' (SBS) with potentially disastrous long-term consequences. At the same time, millions of homes around the world fall below decency standards, or are unpleasant or unhealthy to occupy. A large proportion of these fall within the private rented sector: apartments, homes of multiple occupation (HMOs) and multi-family homes, for example.

Of course, the indoor environment comprises a large number of variables, some of which are interactive. For example, thermal comfort, which is dependent primarily upon temperature, humidity and air movement, interacts with certain aspects of indoor air quality (IAQ) – for example, when low humidity produces drying of the mucosa, resulting in inadequate flushing of airborne irritants from eyes and nasal passages. Indoor environmental quality is also impacted by the visual and aural environments which, either individually or combined with thermal and air quality factors, can significantly impact comfort, health, wellbeing and productivity. The World Green Building Council produced a landmark report in 2014, *Health, Wellbeing and Productivity in Offices*, which provides a comprehensive overview of the subject along with a framework for assessing health, wellbeing and productivity in the workplace (World GBC 2014). This report clearly demonstrates the relationship between a poor indoor environment and the reduced productivity of an office-based workforce. For example, 'being productive in the modern knowledge-based office is practically impossible when noise provides an unwanted distraction'. On the other hand, research studies have demonstrated significant productivity gains due to proximity to windows, partly due to access to daylight and partly a view to outside, 'particularly where the view offers a connection to nature'. Similar benefits accrue from a good thermal environment and IAQ. Bearing in mind that on average 90 per cent of business operating costs are from staff salaries and benefits, compared with 1 per cent from energy costs, investing in improving the indoor environment offers potentially major returns.

The facilities manager should be proactive in identifying potential causes of indoor environmental problems and instigating preventative measures. The key is to introduce a monitoring regime that includes, for example:

- taking 'snapshots' of environmental conditions with portable equipment;
- analysing data harvested from a BEMS;
- making observations during a walk-through survey; and
- obtaining occupant feedback.

Critical to all of these techniques is the quality and quantity of data. For example, the sensors connected to the BEMS must be both accurate and in locations that represent the conditions experienced by occupants. Feedback obtained from occupants must take account of all the factors that could lead to dissatisfaction with environmental conditions, such as work-related stress. No decisions should be made without evidence from multiple sources – for example, occupant complaints, supported by a walk through with simple measurements, followed if necessary by a detailed investigation.

16.4 Hygiene

In the context of facilities management, there are two main areas in which hygiene is a factor: cleaning of occupied spaces and prevention of microbial growth in building services. Although day-to-day cleaning falls within the ambit of housekeeping and soft FM (see below), the prevention and control of mould growth on and within building fabric will be considered in this section.

The common thread here is microbial contamination, resulting in either the emission of allergens or pathogens into the air or the pollution of water, which may also result in the release of airborne microorganisms, such as *Legionella*. Some allergens have their origins in larger organisms such as the excreta from house dust mites and animal dander, whilst pollens may originate from outdoors, and mould spores from indoors or out. Airborne allergens tend to be very small– typically less than 10 microns (mm) mass mean aerodynamic diameter – and hence respirable.

Whereas house dust mites tend to thrive in the proliferation of soft furnishings found in most homes, commercial buildings are more likely to suffer from hidden mould growth on cold and damp surfaces that are either poorly insulated or poorly ventilated or contain suitable substrates for microbial growth. Condensation occurs when surface temperature falls below the dew point, such as in bathrooms and kitchens, and within poorly ventilated ceiling voids (interstitial condensation). Mould can also grow within ventilation systems, particularly downstream from humidifiers and within poorly maintained cooling coils.

Some bacteria, on the other hand, thrive in warm conditions, such as those found in open cooling towers, some domestic water systems, spa baths and elsewhere. The most common of these are *Legionella*, which can multiply in static water where there is a suitable substrate (growth medium) and be dispersed into the atmosphere in an aerosol that, when inhaled by a susceptible individual, may result in that person suffering a form of pneumonia known as Legionnaires' disease. Although this is a treatable form of pneumonia, fatalities are not uncommon.

The list below, adapted from Appleby (2013), sets out the main risk factors that can contribute to the contamination of water systems with *Legionella* or a Legionnaires' disease outbreak:

- water temperatures between 20°C and 50°C;
- nutrients available for growth, such as proteins and rust;
- niches that protect *Legionella* from the penetration of heat and biocides, such as limescale, sludge and algae (see Figure 16.1);
- generation of a fine, invisible aerosol such as that produced from taps, shower heads, cooling towers and spray humidifiers;
- low water turnover, which may create conditions during which temperatures drift into the risk zone, biocides decay and sediments precipitate to form sludge;
- a water system open to the ingress of animals, insects, dirt and sunlight (direct sunlight encourages algal growth and warms water); and
- susceptible people exposed to aerosol, such as those with impaired lung function or compromised immune system.

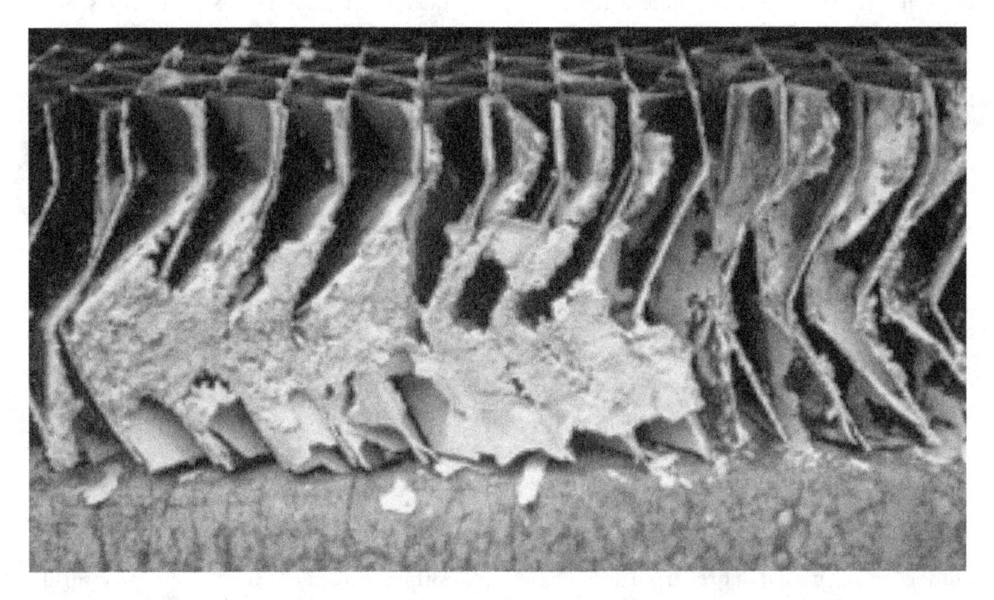

Figure 16.1 Scale can also have a major impact on cooling tower performance, here blocking air inlets to a pond

Source: www.debaltd.co.uk

Very heavily contaminated water systems have been found to contain a cocktail of fungi, bacteria, actinomycetes and protozoa which, when inhaled, has resulted in a number of different illnesses apart from Legionnaires' disease, including humidifier fever, extrinsic allergic alveolitis and asthma attacks (Sherwood Burge 1992).

Minimising the risk of exposure to these microorganisms is mandated under the 2002 Control of Substances Hazardous to Health (COSHH) Regulations in the UK, which has also required the registration of cooling towers and evaporative condensers for many years. In the US, there is no equivalent federal legislative framework, although there are numerous competing guidelines, whilst some states, counties and cities have set out their own specific requirements and guidelines.

The key features of risk management general to all types of equipment and application comprise risk assessment; written scheme; records of water treatment, temperatures and microbial quality; responsibilities and periodic condition surveys of risk items. These might include open cooling towers, evaporative condensers, water treatment equipment, humidifiers, hot water cylinders, cold water storage tanks, hot water service (hws) dead legs (pipes having flow only when taps operated), showers and other outlets, spa baths, fountains, car wash equipment, etc.

For a detailed risk assessment methodology, refer to the BSRIA *Guide to Legionellosis: Risk Assessment* (Brown and Roper 2000), which includes a pro-forma risk assessment form.

16.5 Waste management

In 1962, Rachel Carson provided a damning critique of the devastating impacts that humans were having on the planet at that time: 'The most alarming of all man's assaults upon the environment is the contamination of air, earth, rivers and seas with dangerous and even lethal materials' (Carson 1962). Although she was focusing in particular on the then widespread use of DDT and its deadly effects on wildlife, these impacts continue into modern times. For example, in 2008 around one trillion plastic bags were manufactured, contributing to some 3.5 million tonnes of non-recyclable plastic wrapping being discarded. Although much of this ends up degrading slowly in landfill, it has been estimated that between 1 and 3 percent of plastic bags have been discarded as litter, clogging drains and polluting rivers and oceans (PAG 2008). According to the World Wildlife Fund, tens of thousands of albatross chicks die every year by ingesting plastic from the waters around the remote Midway Atoll in the north Pacific (see Figure 16.2), the centre of a giant vortex or gyre that draws flotsam from across the entire Pacific Ocean.[16] However, much of this plastic is degraded and enters into the food chain as tiny particles known as neuston.

Globally, there has been a concerted effort to reduce the amount of waste produced in manufacture, packaging, retail and consumption. In 2008, the EC Waste Directive introduced the principles of the 'waste hierarchy' and 'polluter pays', transposed into UK law, for example, through the Waste (England and Wales) Regulations 2011:

- prevention
- preparing for re-use
- recycling
- other recovery (e.g., energy from waste)
- disposal.

Much waste arises from a combination of over-purchasing perishable products, inefficient material use or over-packaging. Waste streams should be divided into those that are re-usable, recyclable, disposable to landfill and hazardous. Electronic goods, for example, contain a variety of hazardous substances, which have historically ended up contaminating landfill sites. Hence in Europe, the 2002 Waste Electrical and Electronic Equipment (WEEE) Directive (recast in 2012),

Figure 16.2 Photograph of a dead Laysan albatross chick with plastic in its stomach

Source: Young, Vanderlip, Duffy, Afanasyev and Shaffer (2009)

implemented in the UK for example through the 2006 WEEE Regulations (amended in 2013), established a free Distributor Take-Back Scheme (DTS) for household WEEE, whilst commercial WEEE has to be processed for a fee by a specialist waste contractor.

16.6 Landscape and ecology

In many instances, 'landscape' may comprise no more than a strip of concrete or paving, some wrought-iron fencing and perhaps a couple of trees. Indeed, the majority of sites will not be formed of pristine biodiverse landscapes, but are likely to incorporate a mixture of car parking, trees, shrubs, flower beds and lawns, frequently designed to minimise maintenance. In these instances, a sustainable landscape plan is required that incorporates a biodiversity action plan (BAP) where appropriate. The former should include programmes and procedures for plant and soil stewardship; invasive species management; water and irrigation management; composting, fertilising and mulching practices; equipment use and maintenance.

In the UK, frameworks have been established for managing BAPs at national, local, corporate and site levels. Companies and organisations are encouraged to develop BAPs through an award scheme developed by the Wildlife Trusts known as the Biodiversity Benchmark.[17] This may form a component of reporting under an environmental management system, such as EMAS (EC 2009) or ISO 14001:2015.[18]

16.7 Soft facilities management and sustainable procurement

Soft FM includes such functions as catering, cleaning, mail room operations, security and care of indoor plants. Sustainable FM requires close attention to the procurement of sustainable products and the implementation of procedures that do not threaten the health and safety of operatives or others.

The European Commission has produced guidance and protocols for green public procurement (GPP) referring to relevant EU Ecolabels, Energy Star labels and sector-specific certification, such as Forest Stewardship Council (FSC) for timber, where appropriate.

In the EC *Buying Green* handbook, GPP is defined as procurement for a better environment using 'a process whereby public authorities seek to procure goods, services and works with a reduced environmental impact throughout their life cycle when compared to goods, services and works with the same primary function that would otherwise be procured' (EC 2016). EU Ecolabels have been developed for cleaning products, paints and varnishes, wooden furniture, imaging equipment, and computer and paper products, for example.[19]

The UK-based Waste and Resources Action Programme (WRAP) has published *Green Office: A Guide to Running a More Cost-Effective and Environmentally Sustainable Office* (WRAP 2014) that points out the link between a sustainable purchasing strategy and waste minimisation through:

- rationalising purchasing to minimise over-stocking and reduce risk of perished or damaged stock;
- specifying minimum packaging with re-used or recycled materials;
- identifying opportunities for repair, refurbishment or re-use;
- specifying products that use recycled materials and/or can be recycled and minimise environmental impact; and
- buying locally or minimising delivery miles.

16.8 Conclusions

There is good evidence that sustainability has grown in importance as a framework within which facilities managers must operate. The British Institute of Facilities Managers (BIFM) carries out annual surveys to establish how its membership is engaging with the sustainability agenda (BIFM 2015). It is interesting to note that the number of respondents reporting an increase in sustainability drivers more than doubled for some key measures between 2009 and 2015, with a countervailing increase in reported barriers to 'sustainability effectiveness'. In the 2015 survey, it was reported that only 36 per cent of responding organisations had not introduced one of the standards for environmental and/or energy performance, with 42 per cent using the international environmental management standard ISO 14001. Most respondents use key performance indicators (KPIs), with those for health and safety, energy, waste, carbon footprint and training being the most popular. Some organisations have introduced financial penalties for those who do not meet targets.

This survey indicated that although drivers such as legislation and corporate image have been increasing steadily in recent years, between 2014 and 2015 there appears to have been a negative impact on sustainability effectiveness, which the report suggests is partly due to inadequate communication within organisations from senior management. This is reflected in the lower levels of knowledge about sustainability reported among lower management and the general workforce compared with senior managers and directors.

It should be obvious from the above that FM comprises numerous different skills and disciplines, and that successfully coordinating and programming these is the key to success.

Although many organisations have established in-house FM teams, outsourcing is common and there are numerous specialist FM companies that provide a multi-disciplinary service, usually supported by teams of specialist designers and operatives, along with a supply chain of sub-contractors and suppliers. As buildings become more complex, the skills required to manage them sustainably will be at a premium and the FM role is likely to become increasingly one of a coordinator of multiple disciplines akin to project management in the construction industry.

Notes

1 www.thepotteries.org/dates/health.htm.
2 www.buildinghistory.org/regulations.shtml.
3 http://unfccc.int/kyoto_protocol/items/2830.php.
4 www.gov.uk/government/publications/climate-change-levy-main-and-reduced-rates/climate-change-levy-main-and-reduced-rates.
5 www.epa.gov/sites/production/files/2015-10/documents/608proposal.pdf?_cldee=aGVyYndvZXJwZWxAYWNocm5ld3MuY29t.
6 www.iso.org/iso/home/news_index/news_archive/news.htm?refid=Ref1515.
7 www.gov.uk/guidance/energy-savings-opportunity-scheme-esos.
8 www1.eere.energy.gov/buildings/betterbuildings/accelerators.
9 www.energystar.gov/buildings/about-us/how-can-we-help-you/benchmark-energy-use/use-energy-star-benchmarking-tools.
10 Ibid.
11 www.business-stream.co.uk/water-efficiency/water-audit-and-benchmarking/water-benchmark-calculator.
12 www.sydneywater.com.au/SW/your-business/managing-your-water-use/benchmarks-for-water-use/index.htm.
13 http://webarchive.nationalarchives.gov.uk/20140328084622/http://www.environment-agency.gov.uk/business/topics/water/34866.aspx.
14 http://adsm.com/services/grants/aquamark.
15 www.bdcnetwork.com/too-cold-and-too-hot-most-common-complaints-among-office-workers-says-ifma-study.
16 http://wwf.panda.org/what_we_do/endangered_species/albatross.
17 www.wildlifetrusts.org/biodiversitybenchmark.
18 www.iso.org/iso/iso14000.
19 http://ec.europa.eu/environment/ecolabel/the-ecolabel-scheme.html

References

Appleby, P. (2011) *Integrated Sustainable Design of Buildings*. London: Earthscan

Appleby, P. (2013) *Sustainable Retrofit and Facilities Management*. London: Routledge.

ASHRAE (2011) *Procedures for Commercial Building Energy Audits*. Atlanta, GA: American Society of Heating, Refrigerating and Air-Conditioning Engineers.

BIFM (2015) *Sustainability in Facilities Management Report 2015*. Bishops Stortford, UK: British Institute of Facilities Management. Available at: www.bifm.org.uk/bifm/knowledge/sustainabilityinfm/2015.

BIO Intelligence Service (2012) *Water Performance of Buildings*. Final Report prepared for European Commission, DG Environment. http://ec.europa.eu/environment/water/quantity/pdf/BIO_WaterPerformanceBuildings.pdf.

Blanco, G., Gerlagh, R., Suh, S., Barrett, J., de Coninck, H., Diaz Morejon, C., Mathur, R., Nakicenovic, N., Ofosu Ahenkora, A., Pan, J., Pathak, H., Rice, J., Richels, R., Smith, S., Stern, D., Toth, F. and Zhou, P. (2014) 'Drivers, trends and mitigation', in Edenhofer, O., Pichs-Madruga, R., Sokona, Y., Farahani, E., Kadner, S., Seyboth, K., Adler, A., Baum, I., Brunner, S., Eickemeier, P., Kriemann, Minx, J., B., Savolainen, J., Schlömer, S., von Stechow, C. and Zwickel, T. (eds.) *Climate Change 2014: Mitigation of Climate Change. Contribution of Working Group III to the Fifth Assessment Report of the Intergovernmental Panel on Climate Change*. Cambridge and New York: Cambridge University Press, pp. 351–412.

Brown, R. and Roper, M. (2000) *Application Guide AG29/2000: Guide to Legionellosis: Risk Assessment*. Bracknell: Building Services Research and Information Association.

BSI (2006) *BS EN 15221 Part 1 Facility Management Terms and Definitions*. London: British Standards Institute.
Carbon Trust (2011) *Energy Surveys: A Practical Guide to Identifying Energy Saving Opportunities*. London: Carbon Trust. Available at: www.carbontrust.com/resources/guides/energy-efficiency/energy-surveys.
Carbon Trust (2013) *Energy Management: A Comprehensive Guide to Controlling Energy Use*. London: Carbon Trust.
Carson, R. (1962) *Silent Spring*. Boston, MA: Houghton Mifflin.
Chandrappa, R. and Bhusan Das, D. (2012) *Solid Waste Management: Principles and Practice*. Heidelberg: Springer.
CIBSE (2008) *TM46: Energy Benchmarks*. London: Chartered Institution of Building Services Engineers.
CIBSE (2012) *Guide F: Energy Efficiency in Buildings*. London: Chartered Institution of Building Services Engineers.
EC (2009) *Regulation (EC) No 1221/2009 of the European Parliament and of the Council of 25 November 2009 on the Voluntary Participation by Organisations in a Community Eco-Management and Audit Scheme (EMAS), repealing Regulation (EC) No 761/2001 and Commission Decisions 2001/681/EC and 2006/193/EC*. Brussels: European Commission. Available at: http://ec.europa.eu/environment/emas/pdf/other/Informal%20consolidated%20version%20of%20EMAS%20Regulation%20-%20post%20amendement%20Annexes%20I,II,%20III%20.pdf.
EC (2016) *Buying Green: A Handbook on Green Public Procurement*. 3rd edition. Brussels: European Commission.
Environment Agency (2015) *Climate Change Agreements Operations Manual*. Bristol: Environment Agency. Available at: www.gov.uk/government/uploads/system/uploads/attachment_data/file/401113/LIT_7911.pdf.
Environment Agency (2016) *Complying with the Energy Saving Opportunities Scheme, Version 5*. Bristol: Environment Agency. Available at: www.gov.uk/government/uploads/system/uploads/attachment_data/file/509835/LIT_10094.pdf.
Environment Agency (2017) *Climate Change Agreements Operations Manual*. Version 7. Bristol: Environment Agency.
EPA (2013) *Energy Star Guidelines for Energy Management*. New York: Environmental Protection Agency. Available at: www.energystar.gov/sites/default/files/buildings/tools/Guidelines%20for%20Energy%20Management%206_2013.pdf.
EU (2014) Regulation No 517/2014 of the European Parliament and of the Council of 16 April 2014 on fluorinated greenhouse gases and repealing Regulation (EC) No 842/2006. *Official Journal of the European Union, Article 4*.
Executive Office of the President (2013) *The President's Climate Action Plan*. Washington, DC: Executive Office of the President. Available at: www.whitehouse.gov/sites/default/files/image/president27sclimateactionplan.pdf.
Griffin, R.D. (2016) *Principles of Air Quality Management*. Boca Raton: CRC Press.
HSE (1995) *How to Deal with Sick Building Syndrome: Guidance for Employers, Building Owners and Building Managers*. Sheffield: Health and Safety Executive. Available at: www.hse.gov.uk/pubns/books/hsg132.htm.
PAG (2008) *Plastic Bags: Responses to a Common Problem*. Tucson, AZ: PAG. Available at: www.pagnet.org/documents/committees/epac/2008/EPAC-2008-06-06-Presentation-PlasticBagProbs.pdf.
Seely, A. (2009) *Climate Change Levy*. Standard Note SN/BT/235. London: House of Commons Library. Available at: www.parliament.uk/briefing-papers/SN00235.pdf.
Sherwood Burge, P. (1992) 'Bacteria, fungi and other micro-organisms', in Leslie, G. and Lunau, F. (eds.) *Indoor Air Pollution: Problems and Priorities*. Cambridge: Cambridge University Press, pp. 29–42.
UNDP (2006) *Human Development Report*. New York: UN Development Program.
World GBC (2014) *Health, Wellbeing & Productivity in Offices*. London: World GBC. Available at: www.worldgbc.org/sites/default/files/compressed_WorldGBC_Health_Wellbeing__Productivity_Full_Report_Dbl_Med_Res_Feb_2015.pdf.
WRAP (2014) *Green Office: A Guide to Running a More Cost-Effective and Environmentally Sustainable Office*. Banbury: WRAP. Available at: www.wrap.org.uk/sites/files/wrap/WRAP_Green_Office_Guide.pdf.
Young, L.C., Vanderlip, C., Duffy, D.C., Afanasyev, V. and Shaffer, S.A. (2009) Bringing home the trash: do colony-based differences in foraging distribution lead to increased plastic ingestion in Laysan albatrosses? *PLoS ONE* 4(10): e7623.

17

Building Energy Efficiency Certificates and commercial property

The Australian experience

Clive Warren

17.0 Introduction

This chapter explores the introduction of Building Energy Efficiency Certificates (BEECs) in Australia and its impact on property owners and managers. It looks at the process that led to the introduction of BEECs and discusses the alternative strategies that could be used to achieve energy efficiency in commercial buildings. Having established the process by which BEECs were introduced in Australia, it then explores in detail what precisely constitutes a BEEC, who is impacted by the legislative requirements and what are some of the financial and management impacts that the statutory requirements impose. The final part of this chapter evaluates the impact that the introduction of BEECs has had on energy efficiency in commercial buildings and to what extent the use of statutory powers has achieved government objectives of reducing energy consumption in commercial buildings in order to meet the Australian Government's international commitments to reducing greenhouse gas (GHG) emissions. Australia is a signatory to the Kyoto Protocol and, as such, has committed to limiting emissions to 108 per cent of 1990 levels, a target that was exceeded and is now on track to a 13 per cent reduction on 2005 levels. These targets are equal to or better than US, Japan and Canada targets but much less than the EU target, which is set at 20 per cent below 1990 levels (CoA 2015).

Energy consumption in the commercial property sector offers an important opportunity for conserving resources. In this study, we evaluate the financial implications of two elements of 'sustainability' – energy efficiency and accessibility – in the market for commercial real estate. An empirical analysis of some 1,100 leasing transactions in the Netherlands over the 2005–10 period shows that buildings designated as inefficient (with an EU Energy Performance Certificate D or worse) command rental levels that are some 6.5 per cent lower as compared to energy-efficient but otherwise similar buildings (labelled A, B or C). Furthermore, this study shows that office buildings in multi-functional areas, with access to public transport and facilities, achieve rental premiums over mono-functional office districts. For policymakers, the results documented in this paper provide an indication of the effectiveness of the EU Energy Performance Certificate as a market signal in the commercial property sector. The findings documented here are also relevant for investors in European office markets, as the importance of energy efficiency and locational diversification is bound to increase following stricter environmental regulation and changing tenant preferences.

17.1 History

There has been a long tradition of seeking to manage commercial property in an efficient and cost-effective manner. One aspect of good management is the optimisation of operating costs associated with the day-to-day running of the building. A significant cost of occupation is the cost of electricity consumed in lighting, air conditioning, operation of lifts and provision of heating and hot water services, which amounts to between 20 per cent and 30 per cent of building non-statutory operating costs (PCA 2015). In addition to the obvious cost savings that can be achieved from efficient use of electricity, the production of electricity is inexorably linked to the burning of fossil fuels in order to generate electricity, and while renewable energy sources are increasing, many countries still need to rely on burning coal and gas to produce electricity. Electricity generation using fossil fuels results in significant emissions of carbon dioxide (CO_2), which is a GHG and a major contributor to global warming.

Energy consumed by the built environment, in our houses, shops, offices and other buildings, has been estimated to consume between 40 per cent and 70 per cent of global energy production and has a resulting carbon footprint significantly exceeding those of all transportation combined (McDonagh 2011; WBCSD 2009). High energy use and the consequent CO_2 emissions have become the most pressing issues of this generation, and, as a consequence, have brought considerable focus to bear on the way we use our buildings. The Intergovernmental Panel on Climate Change (IPCC) Fifth Assessment Report (AR5) (2015) concludes that climate change is unequivocal, and that human activities, particularly emissions of CO_2, are very likely to be the dominant cause. In addressing the role of the built environment in contributing to GHG emissions, the IPCC found that in 2010, the world's buildings accounted for 32 per cent of energy use and contributed 19 per cent of all GHG emissions (IPCC 2015). It also found that with business-as-usual projections, the use of energy in buildings globally could double or triple by 2050 (ECF 2014; IPCC 2015). Sustainability and global warming associated with high levels of CO_2 in the atmosphere have been the subject of many global initiatives. The catalyst for much of the global action on climate change was the 1987 Brundtland Commission Report, *Our Common Future* (WCED 1987). The report gave us a common definition of sustainability, which was endorsed by the 1992 UN Earth Summit in Rio and which culminated in the global agreement within the Rio Declaration and Agenda 21 requiring all member states to develop a national sustainability development strategy (UN 1992). Since these early sustainability agreements, governments have committed to GHG reduction targets in the Kyoto Protocol and increased use of renewable energy sources (IEA 2015).

The practical application of these stated sustainability objectives has been a particular challenge for the development industry, resulting in what has been widely reported as a 'circle of blame' leading to a lack of a business case for sustainable development (Cadman 2000; Ellison & Sayce 2007; RICS 2008). While over the years there has been much research undertaken to illustrate the potential cost savings from adopting sustainable and energy-efficient building practices (Ellison & Sayce 2007; Green Building Council of Australia 2008; Lorenz, Trück & Lützkendorf 2007; Lützkendorf & Lorenz 2005), the development industry remains largely sceptical of the cost benefits of sustainable development (Green Building Council of Australia 2008; Warren 2009). The WBCSD (2009: 1) see buildings as '[L]arge and attractive opportunities to reduce energy use at lower costs and higher returns than other sectors.' These are recognised as fundamental to achieving the International Energy Agency's (IEA) target of a 77 per cent reduction in the planet's carbon footprint against the 2050 baseline and to reach stabilised CO_2 levels called for by the IPCC.

The Australian Government, as a signatory to the UN Rio Declaration and Agenda 21, began to promote energy efficiency, initially by seeking to lead by example. In October 1990, the Australian Federal Government introduced a policy to reduce GHG emissions (ANAO 1992), which included a number of initiatives designed to improve energy efficiency in government buildings. The main features of this policy were:

- the appointment of energy managers in government departments and authorities;
- the annual reporting by government departments and authorities of energy use and energy efficiency improvement initiatives undertaken;
- the establishment of, and adherence to, high standards of energy efficiency in the construction and leasing of Commonwealth buildings;
- energy management training;
- the use of Commonwealth buildings for demonstrations of energy-efficient technologies; and
- the establishment of a database of Commonwealth energy use.

(ANAO 1996)

This initiative has had an impact on both the government and private sector by showing clear leadership and expectations of energy efficiency standards in buildings owned and leased by government, and some developers improved their designs to meet these standards.

Any government with a need to implement a policy that changes the way the free market operates is faced with a number of difficult decisions. How to implement a planned reduction in energy consumption is such a decision, and one with political implications. Generally, there are three options available to governments to implement change.

The first approach is legislation. Simply put, this is a strict obligation to comply with a standard. This route is obviously the harshest and a 'stick' approach to achieving industry compliance rather than a softer 'carrot' approach, which encourages voluntary compliance. There have been a number of strict compliance methods adopted, not least in the introduction of Section J, Building Energy Efficiency into the Building Code of Australia. Initially, the updated building code applied only to the residential sector, but application to commercial buildings came into effect in 2006 (ABCB 2010). The stringency of these provisions was increased for both residential and commercial buildings in the 2010 version of the Building Code but they are still not particularly demanding. While the development of new building standards has made a significant contribution to improved energy efficiency, it must be recognised that most buildings are designed for a useful life in the region of 60 years and that the rate of building stock replacement is in the region of 0.66 per cent to 3 per cent per annum (Eichholtz, Kok & Quigley 2010; Jowsey & Grant 2009; Surmann & Hirsch 2016). Consequently, if the implementation of energy-efficient building relies solely on the replacement of older buildings with those meeting the energy efficiency measures of current building practices, it will take in the region of 30 to 130 years to achieve total replacement. It is, therefore, self-evident that the existing stock of buildings is contributing a disproportionate amount to the total built environment GHG emissions and that greater attention should be paid to bringing these buildings to a more energy-efficient standard (Wilkinson 2013).

The second approach that can be adopted is the softer carrot approach, encouraging owners, investors and occupiers to see the value in more energy-efficient and sustainable buildings. The adoption by the federal and state governments of minimum levels of energy efficiency and sustainability ratings for space they lease is one such soft approach that has had a marked impact on the development industry. Both state and federal governments have set minimum energy efficiency standards for lease office space, which are incorporated into green lease agreements and as such become enforceable covenants (APCC 2012; Australian Greenhouse Office 2006; Wilkinson, van der Heijden & Sayce 2015). As a major occupier of office space, the public sector is an attractive tenant, and so the use of that market power is a strong incentive to meet stated energy standards (Warren & Huston 2011).

These soft approaches, along with a wider recognition of the need to build more sustainably, have led to the establishment of a plethora of sustainability assessment tools developed around the world, each of which seeks to assess a property development against a range of 'sustainability criteria'. These tools have been developed by governments and private organisations, both those established by the industry on a not-for-profit basis and those seeking to establish commercial measures of sustainability. Internationally, the tools most widely recognised include BREEAM in the UK and LEED in the US. In an Australian context, there are a range of measures commonly applied to new and existing developments. These measures range from those that focus just on building energy, such as NABERS (National Australian Built Environment Rating System) Energy, to those that seek to provide a more holistic approach, such as the Green Building Council of Australia Green Star rating tools for new and refurbished buildings. Each of the rating tools seeks to measure different criteria in order to award a star rating. The assessment criteria include the obvious energy consumption measures but also incorporate a range of other metrics from the size of car-parking spaces to the level of waste recycling.

While both the hard legislative approach of building standards and the soft encouragement of improved building practices have made significant improvements to the level of GHG emissions, much of these savings have occurred in the development of new buildings, leaving much of the existing building stock behind (BCI 2014; NSW OEH 2015). Recognising a need to bring the existing building stock up to a more sustainable level of operation, governments have sought to adopt the third approach to changing the habits of building owners and occupiers (Wilkinson 2013). This approach is a hybrid of a hard legislative approach with a softer coercion. By requiring a building owner to reveal the energy efficiency of a building through legislation, there is a minimal cost impost on the owner, and so government is able to minimise any negative backlash from the property industry. Yet through public disclosure of the building's energy, the market is encouraged to vote with its feet and seek to occupy better, more efficient buildings. The soft coercion of owners through disclosure is thus more likely to bring existing buildings up to a more sustainable level as they compete for tenants. This is particularly so where higher levels of office vacancy exist and tenants are able to be more selective as to the space they lease.

The EU, in 2002, ratified the Energy Performance of Buildings (European) Directive 2002/91/EC, which established minimum requirements for the energy performance of new and existing buildings. Article 7 of the directive required buildings with total floor areas greater than 1,000 square metres to have publicly displayed energy certificates. In the UK, the response to the EU directive was to mandate the Display Energy Certificates based on actual measured energy use in public buildings. This was considered a first test for building energy efficiency with an outcomes-focused approach (Cohen & Bordass 2015). Following on from the European initiatives, the Australian Federal Government sought public consultation on its own building certification regime for buildings in 2009. The result of this was the introduction of the Building Energy Efficiency Disclosure Act 2010 (Cth) and the establishment of the Commercial Building Disclosure (CBD) scheme. The legislative disclosure of energy use came into effect on 10 November 2010 with transitional arrangements for the first year of operation.

The CBD scheme initially applied to all commercial office buildings with a floor area of 2,000 square metres and required that any vendor of a commercial property either by sale of the freehold interest or a leasehold of greater than 2,000 square metres must obtain a BEEC before the building could be marketed. The scheme has undergone some minor revisions since its inception, and in 2014–15 underwent a review, including public consultation on the operation of the scheme. This review culminated in two significant changes to the legislation, the most important of which is the reduction of the building floor area to 1,000 square metres effective from 1 July 2017. The other change relates to the Tenancy Lighting Assessment that forms part of the BEEC, which, effective from 1 September 2016, is only required to be renewed every five years.

The objective of the scheme is to reduce GHG emissions through providing choice to purchasers of office buildings. All BEECs are recorded on a publicly available website, and in addition the legislation requires that any advertisement relating to the sale or lease of a disclosure-affected building must display the building energy star rating as contained within the BEEC. Failure to comply with these statutory requirements will lead to substantial fines for the building owner, amounting to $17,000 for the first day of infringement and $1,700 for each subsequent day on which compliance is not achieved.

The public register of BEECs, along with the inclusion of NABERS Energy star ratings in all advertising and the requirement to provide a copy of the BEEC to any prospective purchaser or lessee, ensures that any prospective purchaser will be cognisant of the energy rating of the building or tenancy in which they are interested. This will facilitate comparison between buildings and thus, it is hoped, lead to market forces encouraging poorly performing buildings to be less valued and owners to address the energy performance of their building. Thus, while not mandating that poorly performing buildings are brought up to modern performance standards, there is a substantial inducement to invest in more energy-efficient buildings.

17.2 What is a BEEC?

The Building Energy Efficiency Disclosure Act 2010 and Regulation came into full effect from November 2011. As noted above, any office building that is to be sold or leased with a floor area of greater than 2,000 square metres (1,000 square metres from 1 July 2017) must obtain a BEEC.

A BEEC is a certificate provided by the CBD administration of the federal government. Each BEEC remains valid for one year from the date of issue and must be current at the time the disclosure-affected property is transacted. The BEEC is made available online via the Building Energy Efficiency Register (see: http://cbd.gov.au/registers/find-a-rated-building). The register includes details of the building or part of the building to which the BEEC relates; a NABERS Energy star rating for the building, together with an assessment of the energy efficiency of tenancy lighting in the area of the building that is being sold or leased; and general energy efficiency guidance for commercial offices (DOEE 2016b).

The process for obtaining a BEEC requires an owner to undertake the following steps:

1 Ascertain if the building or affected tenancy area already has a BEEC. This can be achieved by undertaking a search of the publicly available register on the CBD website.
2 Engage a CBD-accredited assessor to evaluate the building and apply on the owners' behalf for a BEEC. Only accredited assessors are permitted to apply for a BEEC. Accredited assessors are registered by the Department of the Environment and Energy as CBD assessors in accordance with the Building Energy Efficiency Disclosure Act 2010. Accreditation requires completion of a NABERS training course together with a contract agreement with

the NSW Office of the Environment and Heritage as a NABERS assessor. CBD assessors must also complete training in tenancy lighting.

3 Collate all the energy data required to undertake the CBD assessment. See NABERS assessment below.
4 Undertake NABERS Energy star rating assessment.
5 Undertake Tenancy Lighting Assessment if no current valid assessment exists.
6 Apply for the BEEC based on the NABERS Energy star rating and Tenancy Lighting Assessment.

(DOEE 2016b)

Once approved, the BEEC will be added to the national register with all the building details recorded, and advertisement of the building can proceed.

If a building already has a registered BEEC, then marketing and sale of the building can proceed. However, for most buildings, a new BEEC will be required or the existing one will need to be renewed, as each BEEC expires after 12 months and must be current at the time of sale. In order to obtain a BEEC, the building owner will require both a NABERS Energy star rating and a Tenancy Lighting Assessment.

17.3 Obtaining a NABERS Energy star rating

In Australia, a voluntary rating scheme, the Australian Building Greenhouse Rating (ABGR), was developed for existing buildings, initially under the federal government's Department of the Environment and Water Resources, while at the same time the New South Wales state government was developing a similar rating tool. These tools were amalgamated into the NABERS. It is important to note that the NABERS Energy star rating scheme was not designed to rate buildings on their electricity use per se, but to evaluate them on their GHG emissions. The rating, while based on total energy input to the building, converts this figure to a GHG emission factor and adjusts this for the building's climate location and hours of use before arriving at a star rating.

NABERS Energy star ratings are available at three separate levels for any given building. The rating obtained depends on the needs of the owner seeking a rating. The three levels are:

1 *Tenancy Space Rating*, which evaluates only items under the control of the tenant. These are generally tenancy lighting and power and exclude central services provided by the lessor such as air conditioning and common area services.
2 *Base Building Rating*, which is perhaps the most useful rating in that it assesses the energy used by the building's base services including common area lighting, lifts and air conditioning. It specifically excludes the energy utilised by a tenant of the building, which is beyond the direct control of the owner. This rating is required for a BEEC.
3 *Whole Building Rating*, which is a combination of the two ratings above and measures all energy entering the building with no differentiation between who controls its use. This is perhaps the least useful measure from a building management viewpoint.

The Base Building Rating is most commonly used to obtain a BEEC. In order to obtain a Base Building Rating, a NABERS assessor has to undertake an assessment of the building's energy sources. These sources are not limited to electricity and will also include gas and oil which may be utilised for heating. The assessment will also take account of any Green Energy that is purchased by the owner. The assessment will include a rating both with and without the Green Energy allowances.

To undertake an energy assessment, the NABERS assessor will need to first establish the Rated Area of the building. The Rated Area of an office building differs from the commonly used Net Lettable Area (NLA), which is the widely accepted area used for the leasing of office space and measured in accordance with the Property Council of Australia Method of Measurement (PCA 1997). The Rated Area broadly follows the NLA but makes adjustments for certain areas within the building, including common area cafes, gyms and certain specific tenant areas including storage and computer server rooms. These adjustments in the Rated Area are designed to exclude energy uses that are atypical for an office building.

Having established the area of the building, the assessor will then require detailed energy consumption records for the 12-month period preceding the assessment. For a Whole Building Rating, this can simply be the supplier's billing information with total energy supply details. In order to undertake a Base Building Rating, it is necessary to provide electricity consumption data that is distinct from tenancy area consumption data. This necessitates the use of sub-meters for each separate area of the building. While many building management systems provide estimated energy consumption for individual plant items and for individual tenancies, these systems generally use a form of current clamp that surrounds the supply cable and provides an estimate of the quantum of electricity passing along the cable. The accuracy of these systems is not always sufficient for an assessment, and it is a requirement that sub-meters meet strict accuracy standards and undergo regular validation of their accuracy as described in the NABERS rules for collecting data (OEH 2013).

Adjustments are made to the energy consumption data to take account of the hours of operation of the building. For example, a building that is occupied 16/7 will have a greater level of base building consumption than a building that operates 9 to 5. The standard imposed relates to the total hours of occupation per week. In the case of a Base Building Rating, the number of hours is measured as the hours for which a tenant has requested that an occupied space is safe, lit and comfortable. These hours will include the core hours of operation specified within any lease agreement plus additional hours for which a tenant has requested after-hours air conditioning be provided.

The NABERS Energy star rating is then calculated based on the data collected for the building. The 12-month energy use is multiplied by a constant NABERS GHG factor. This adjusts the different energy sources to a standard factor. The energy consumed is further adjusted to take into account the hours of occupancy and the location of the building. Location adjustment is required, as there are marked climatic differences between buildings operating in tropical Darwin to those operating in temperate Melbourne. This location-based adjustment allows like buildings in specified locations to be compared with one another, but limits the ability to compare buildings in markedly different regions. The adjusted output from this calculation is referred to as a 'benchmark factor', and it is this measure that is used as the basis for comparison with other star-rated buildings in the region to arrive at a star rating for the subject property.

The NABERS Energy star rating system awards ratings from 0 to 6 stars based on benchmarks with like buildings within the region. Each star rating is assigned a descriptive standard from poor to market leading as follows:

- 6 stars: market-leading performance
- 5 stars: excellent performance
- 4 stars: good performance
- 3 stars: average performance
- 2 stars: below average performance
- 1 star: poor performance
- 0 stars: very poor performance.

It is quite evident from these star rating categories that any building owner awarded a 0- or 1-star rating is likely to have some level of motivation to improve the building's performance in order to attract new tenants or obtain a good selling price for the property.

In addition to the star rating of the building, the NABERS Energy assessment also provides other details that are included in the BEEC and that provide useful information to facilitate comparison between buildings. The annual GHG emission is reported in terms of kilograms of CO_2 equivalent per year ($kgCO_2/m^2$–e per annum) emitted from the building together with the rate of GHG emission per square metre of rated building area. The total energy consumption of the building in mega joules per annum is also included in the BEEC.

17.4 Tenancy Lighting Assessment

The second element within a BEEC is the Tenancy Lighting Assessment (TLA). This survey of the building's tenancy areas was, at the commencement of the CBD legislation, required to accompany every application for a BEEC and be updated for each application on an annual basis. Following the review of the legislation in 2014–15, from September 2016, the TLA is only required to be revised every five years.

A TLA of a building is a detailed survey of the lettable-area lighting and is described as the lighting that would reasonably be expected to remain in the office space after the current tenant vacates and undertakes the lessee's obligations to make good the tenancy (DOEE 2016a). A lighting assessment may be conducted on a whole-building basis or just in respect of defined floors depending on the owner's requirements. In the case of a sale of the freehold, a whole-building assessment is required. If, however, only a portion of the building is to be leased, then only that area needs to be assessed.

A TLA comprises two parts: an assessment of the nominal lighting power density (NLPD) expressed in watts per square metre and an assessment of the lighting control systems. The assessment must be undertaken by a registered CBD assessor and completed in accordance with the CBD Tenancy Lighting Assessment for Offices Rules (DOEE 2016a).

The lighting power density is calculated for each distinct area of the building and is expressed in watts per square metre (W/m^2). Thus it is a measure of energy consumption in watts and does not reflect the level of light provided, which is normally expressed in Lux per square metre. It does not determine the lighting level or compliance with Australian Standard AS/NZ1680:2006 on interior and workplace lighting.

The first task of the assessor is to identify the NLA to be assessed and then within that space determine the functional areas. Functional areas are defined within the rules of assessment as each individual floor and, within that, each individual or distinct tenancy.

The assessor will then collect data on all lamps within the functional area in order to determine the total power consumption, in watts, for each fitting and the aggregated amount for the functional area. This requires detailed knowledge of lighting systems and interpretation of the rules relating to differing lamp types, as detailed records and photographs of each individual lamp type need to be recorded. Most lamps will state the wattage of some part of the bulb, but in the absence of clear stated energy use, the rules for assessment provide a table of default lamp values to be used.

The aggregate level of watts for any functional area is calculated and divided by the NLA for the area to arrive at an NLPD expressed in watts per square metre. In addition, a rating of the NLPD is provided based on the levels in Table 17.1.

The second element in a TLA is the lighting control assessment. This seeks to evaluate the ability of the installed lighting control system to meet the operational requirements and needs of occupants for each functional space. Lighting control systems can range from simple

Table 17.1 Nominal lighting power density (NLPD)

NLPD	*Appearance on BEEC*
7.0 W/m^2 or less	Excellent
7.1 W/m^2 to 10.0 W/m^2	Good
10.1 W/m^2 to 15.0 W/m^2	Median
15.1 W/m^2 to 18.0 W/m^2	Poor
18.1 W/m^2 or more	Very Poor

Source: DOEE (2016a)

occupier-controlled switching, which may be direct switches, or sensor-controlled lighting that switches automatically. The latter is considered to be the highest and most efficient level of control. Alternatives are timer-based controls that can be individually controlled or switched at a supervisor level or via a building management system on a timer basis. The least favoured control mechanism is a manual-based system that does not meet the requirements of either occupancy control or timer-based control. The BEEC will report the level of control for each functional space against the criteria of Good, Moderate or Poor based on the above criteria.

The TLA is included in the BEEC and comprises a detailed list of functional areas, the NLA, the NLPD in watts per square metre, plus the descriptive rating for that area together with an assessment of the control capacity and any general comments. This detail is published on the public register of BEECs.

On completion of the NABERS Energy assessment and award of a NABERS Energy star rating, the CBD assessor submits the application for the issue of a BEEC to the CBD administrator. The certificate is then issued and the data made publicly available.

17.5 Conclusions and the impact on commercial property

The stated objective of the CBD program and statutory requirements for owners to obtain a BEEC has had a marked impact on the office property sector. Evidence suggests that the NABERS Energy star rating had a limited uptake outside of NSW prior to the introduction of mandatory disclosure, with 55 per cent of ratings coming from NSW and only 208 rated buildings in Australia (Warren 2010). Research relating to the total number of NABERS Energy star ratings issued up to 2014 shows that 1,153 individual buildings have been certified and that on comparison of the same building ratings over time, energy intensity savings occur on average over the first five re-certification periods (Gabe 2016).

The CBD program has published statistics for the first and second years of operation of the program from November 2011 to 2013. This shows that 1,081 BEECs were issued in Year Two in respect of 862 buildings representing 11.1 million square metres of office space with an average NABERS Energy star rating of 3.03 stars. More significantly, 175 buildings received more than one BEEC in Year Two and showed a 28 per cent improved energy rating, while 11 per cent declined and 61 per cent showed no change (DOI 2014). These are early indications that as more BEECs are issued, data should start to show an improvement in energy efficiency resulting from improved management and retrofitting of older buildings. While this data is limited, there is evidence of improving energy ratings in some buildings, although it is not clear that energy saving is the driver of this change. Despite increasing energy costs in Australia, building operation energy still only accounts for 20 per cent to 30 per cent of building operating costs and is only in the range of $10 to $20 per square metre per annum (PCA 2015). Thus energy represents around 1 per cent to 2 per cent of gross office rents, and consequently a saving of 10 per cent of energy consumption resulting from sustainability initiatives will only reduce operating costs by around 0.1 per cent or $1 to $2 per square metre per annum. What is driving sustainability initiatives is the investment market and tenants opting to occupy more energy-efficient buildings. The PCA/IPD Green Building Index, which measures the performance of investment-grade property across 1,418 assets in Australia, shows that buildings with 4-star NABERS Energy ratings or above have much lower vacancy rates compared to other similar-quality buildings and that total return on investment from those assets is higher by between 0.1 per cent and 0.9 per cent per annum (MSCI/IPD 2016). The combined effect of the mandatory requirement to obtain a BEEC together with cost savings and improved building performance is that buildings managed in a more sustainable and energy-efficient way attract and retain tenants and provide building owners and investors with a greater return on investment while also reducing the environmental impact of the buildings.

In a global context, the sustainability criteria of office buildings in Australia resulting from the combined initiatives of government with mandatory schemes such as the use of BEECs and voluntary schemes such as Green Star place the region as one of the most sustainable in the world. The annual GRESB survey of institutional property investors around the world has consistently ranked sustainability in Australia as well above the world average and leading in the Asia Pacific region. The most recent GRESB data shows that Australia has further increased sustainability measures by 13 per cent on the previous year and has an overall score of 25 per cent above the global average (GRESB 2015).

References

ABCB (2010), *Information Handbook: BCA Section J – Assessment and Verification of an Alternative Solution*, Australian Building Codes Board, Canberra.

ANAO (1992), *Energy Management of Commonwealth Buildings*, Australian National Audit Office, Canberra.

——— (1996), *Energy Management of Commonwealth Buildings*, Australian National Audit Office, Canberra.

APCC (2012), *National Green Leasing Policy*, Australasian Procurement and Construction Council, Canberra.

Australian Greenhouse Office (2006), *Energy Efficiency in Government Operations (EEGO) Policy*, Commonwealth of Australia, Department of the Environment and Water Resources, Canberra.

BCI (2014), *Green Building Market Report: Australia/New Zealand*, BCI Media Group, Sydney.

Cadman, D. (2000), 'The vicious circle of blame', cited in: Keeping, M. (2000), *What about demand? Do Investors Want 'Sustainable Buildings'?* RICS Research Foundation, London.

CoA (2015), *Setting Australia's Post-2020 Target for Reducing Greenhouse Gas Emissions*, final report of the UNFCCC Taskforce, Commonwealth of Australia, UNFCCC Taskforce at the Department of the Prime Minister and Cabinet, Canberra.

Cohen, R. & Bordass, B. (2015), 'Mandating transparency about building energy performance in use', *Building Research & Information*, vol. 43, no. 4, pp. 534–52.

DOEE (2016a), *CBD Tenancy Lighting Assessment for Offices Rules Version 3*, Commonwealth of Australia, Department of Environment and Heritage, Canberra.

——— (2016b), *Guidance Note: Disclosure Affected Buildings*, Commonwealth of Australia, Department of Environment and Energy, Canberra.

DOI (2014), *CBD Second Year of Mandatory Disclosure: Statistical Overview*, Commonwealth of Australia, Department of Industry, Canberra.

ECF (2014), *Climate: Everyone's Business – Climate Change: Implications for Buildings*, European Climate Foundation, Cambridge.

Eichholtz, P., Kok, N. & Quigley, J. M. (2010), 'Doing well by doing good? Green office buildings', *American Economic Review*, vol. 100, no. 5, pp. 2492–509.

Ellison, L. & Sayce, S. (2007), 'Assessing sustainability in the existing commercial property stock: establishing sustainability criteria relevant for the commercial property investment sector', *Property Management*, vol. 25, no. 3, pp. 287–304.

Gabe, J. (2016), 'Successful greenhouse gas mitigation in existing Australian office buildings', *Building Research & Information*, vol. 44, no. 2, pp. 160–74.

Green Building Council of Australia (2008), *The Dollars and Sense of Green Buildings 2008*, GBCA, Sydney.

GRESB (2015), *2015 GRESB Report: Australia/NZ Snapshot*, GRESB, Amsterdam.

IEA (2015), *Energy Matters: How COP21 Can Shift the Energy Sector into a Low-Carbon Path that Supports Economic Growth and Energy Access*, International Energy Agency, Cedex, France.

IPCC (2015), *Climate Change 2014: Synthesis Report*, Intergovenmental Panel on Climate Change, Geneva.

Jowsey, E. & Grant, J. (2009), 'Greening the existing housing stock', paper presented to the Pacific Rim Real Estate Society Conference, Sydney, 18–21 January.

Lorenz, D., Trück, S. & Lützkendorf, T. (2007), 'Exploring the relationship between the sustainability of construction and market value: theoretical basics and initial empirical results from the residential property sector', *Property Management*, vol. 25, no. 2, pp. 119–49.

Lützkendorf, T. & Lorenz, D. (2005), 'Sustainable property investment: valuing sustainable buildings through property performance assessment', *Building Research & Information*, vol. 33, no. 3, pp. 212–34.

McDonagh, J. (2011), 'Electricity use trends in New Zealand office buildings, 1990–2008', *Property Management*, vol. 29, no. 2, pp. 160–80.

MSCI/IPD (2016), 'Australian property investment seminar', *Proceedings of MSCI/IPD Quarterly Update*, Brisbane, August.

NSW OEH (2015), *NABERS Annual Report 2013–2014*, NSW Government, Office of Environment and Heritage, Sydney.

OEH (2013), *NABERS Energy and Water for Offices: Rules for Collecting and Using Data*, NSW Government, Office of Environment and Heritage, Sydney.

PCA (1997), *Method of Measurement for Lettable Area*, Property Council of Australia (PCA), Sydney.

——— (2015), *Benchmarks of Operating Expenses: Office & Retail*, Property Council of Australia (PCA), Sydney.

RICS (2008), *Breaking the Vicious Circle of Blame: Making the Business Case for Sustainable Buildings*, FiBRE, Royal Institution of Chartered Surveyors, London.

Surmann, M. & Hirsch, J. (2016), 'Energy efficiency: behavioural effects of occupants and the role of refurbishment for European office buildings', *Pacific Rim Property Research Journal*, vol. 22, no. 1, pp. 77–100.

UN (1992), *Agenda 21*, United Nations Conference on Environment & Development, Rio de Janeiro, Brazil, 3–14 June. https://sustainabledevelopment.un.org/content/documents/Agenda21.pdf.

Warren, C. M. J. (2009) 'Who needs a Green Star', *Proceedings of Pacific Rim Real Estate Society – 15th Annual Conference, Sydney*, 18–21 January.

——— (2010), 'Measures of environmentally sustainable development and their effect on property asset value: an Australian perspective', *Property Management*, vol. 28, no. 2, pp. 68–79.

Warren, C. M. J. & Huston, S. (2011), 'Promoting energy efficiency in public sector commercial buildings in Australia', *Proceedings of COBRA 2011*, Salford, 12–13 September.

WBCSD (2009), *Energy Efficiency in Buildings: Transforming the Market*, World Business Council for Sustainable Development, London.

WCED (1987), *Our Common Future*, World Commission on Environment and Development – Brundtland Commission, Geneva.

Wilkinson, S. J. (2013), 'Are sustainable building retrofits delivering sustainable outcomes?', *Pacific Rim Property Research Journal*, vol. 19, no. 2, pp. 211–22.

Wilkinson, S. J., van der Heijden, J. & Sayce, S. L. (2015), 'Hybrid governance instruments for built environment sustainability and resilience: a comparative perspective', *Proceedings of COBRA – AUBEA*, Sydney, 8–10 July.

18

Workplace ecology

Craig Langston and Abdallah Al-khawaja

18.0 Introduction

A 'workplace' is defined as a shared location where people collaborate in organisational settings. It consists of physical space, such as an office building, shop or factory, complete with supporting infrastructure and business technologies. It is distinct from a place where people do work, such as home or school, and where they are either self-employed or unemployed and not part of an organisational structure for the purpose of business activities. Workplaces act as eco-systems that are important to business goals and ultimate success. They need to be conducive to gaining and retaining valuable knowledge workers, enabling them to perform at their best in a supportive and well-designed environment (Vischer, 2005). Workplace ecology, therefore, is when organisation, space and technology are in harmony to support human endeavour.

It has been recently said that "Americans are increasingly unhappy with their jobs and work environments" (Tam, 2013). This phenomenon may apply more broadly. The aim in this chapter is to explore a new method for measuring workplace ecology on the understanding that a workplace is a human eco-system supported by organisation, space and technology infrastructure. Through such an approach, a better understanding of workplace performance can be established.

18.1 Underpinning literature

Post-occupancy evaluation (POE) has typically focused on the physical characteristics of buildings, including their technical compliance with design documentation and regulatory codes. In the case of green buildings, this is logically extended to encompass environmental performance and impact (Agha-Hossein, El-Jouzi, Elmualim, Ellis & Williams, 2013). The physical issues take prominence over the human issues (Akimoto, Tanabe, Yanai & Sasaki, 2010; Altomonte & Schiavon, 2013), so it is not surprising that how well a building functions from the perspective of its occupants and business managers is often overlooked. POE is also more likely to occur in new or updated buildings than older buildings.

The Royal Institute of British Architects Research Steering Group (RIBA, 1991: p.191) defined POE as "a systematic study of building in use to provide architects with information about the performance of their designs and building owners and users with guidelines to achieve the best out of what they already have". The basic assertion is that POE is the process of evaluating buildings in a systematic and rigorous manner after they have been built and occupied for some time (Preiser, Rabinowitz & White, 1988). This process includes "any and all activities that originate out of an interest in learning how a building performs once it is built, including if and how well it has met expectations" (Vischer, 2001: p.23).

In an attempt to integrate environmental impact and job satisfaction, research from the disciplines of business and environmental auditing has been used to help construct a new POE conceptual framework suitable for indoor office workplaces. In particular, the Environmental Legitimacy, Accountability and Proactivity (ELAP) model (Alrazi, De Villiers & Van Staden, 2010) has been employed as inspiration for the proposed model. The components of their model are reviewed below in the context of their underpinning literature.

18.1.1 Legitimacy

Legitimacy has been defined as "a generalised perception or assumption that the actions of an entity are desirable, proper, or appropriate within some socially constructed system of norms, values, beliefs, and definitions" (Suchman, 1995: p.574). According to Brønn and Vidaver-Cohen (2009), meeting social expectations is becoming necessary for firms to maintain legitimacy in the public eye, and the changing global business climate may now require firms to invest in a social programme to maintain legitimacy within their organisational fields.

On the other hand, in exploratory research conducted with managers from over 500 Norwegian companies to examine corporate motives for engaging in social initiatives, Brønn and Vidaver-Cohen found that social responsibility can be legitimatised in three main ways: image building, altruism and profitability. They also found that the most important considerations related to the legitimacy motives are 'improve our image', 'be recognised for moral leadership' and 'serve long-term company interests'. The researchers also recognised that managers need to believe that they are responding to these forces significantly by caring for their firm's image, encouraging goodwill among stakeholders and enhancing the reputation of the industry to which the firm belongs. Bertels and Peloza (2006) and Vidaver-Cohen (2007) provided further explanation about legitimacy and reputation enhancement.

Owen (2008) found that social and environmental accounting studies employed a legitimacy theory lens. Mobus (2005) studied employing legitimacy theory as an explanatory tool. He argued that the theory remains immature and using it to make specific predictions is hard. POE is a technique that has developed over the last 50 years to legitimise design via an audit of a building and, more recently, the opinions of its occupants (Bortree, 2009).

POE has particular advantages for facilities management (Preiser, 1995; Hadjri & Crozier, 2009). Many sources recommend that an essential shift in the style of building procurement and practice, mainly inside the client/developer and design communities, is necessary to truly acknowledge the idea of POE and ensure that its benefits are realised (Green & Moss, 1998; Zimmerman & Martin, 2001; Bordass & Leaman, 2005; Hadjri & Crozier, 2009).

18.1.2 Accountability

Accountability in management is one of the most elusive concepts. Gray, Owen and Adams (1996) found that accountability leads to two sorts of responsibility: responsibility to report and responsibility to act. This means that firms should not only focus on environmental responsibility by protecting the natural environment or minimising negative impacts on the environment (Shafer, 2006), but should also report any efforts undertaken when considering the community (Marr & Schiuma, 2003).

Environmental accountability emerged from two different concepts: environmental reporting and environmental performance. Hence, in this case, environmental accountability is the extent to which an entity acts sensibly and ethically towards the natural environment and reports on its ecological performance. Accountability theory is focused more on stakeholder issues than on cleaning up poor business behaviour (Deegan, 2006; Delmas & Blass, 2010; Gray et al., 1996).

In the late twentieth century, public awareness of the impacts of both rising populations and technology increased (Halal, 2015). Western nations in particular started to focus their efforts on decreasing their ecological impact; further, buildings have been recognised as major contributors to the world's energy usage, landfill waste and diminishing green space (IFMA Foundation, 2010). Building rating schemes have emerged as a means of guiding the design and operation of more environmentally friendly buildings in an increasing number of countries, although the total number of rated buildings is still relatively insignificant.

The most popular green rating systems include the Building Research Establishment's Environmental Assessment Method (BREEAM) and Leadership in Energy and Environmental Design (LEED). The BREEAM system is perceived as being elastic to local regulations but strict in areas where local regulations are not valid (Steemers & Manchanda, 2010). BREEAM is one of the major green rating schemes in the world, and there is a requirement for the assessor to be involved in all stages of the process for new construction (Julien, 2009). Green rating schemes are designed to bring accountability to the process of producing high-environmental-performance buildings, including the renovation of existing buildings in recent times.

Firms also need to be accountable for their management of people. This can be described as the process of planning, putting in order, orienting and managing the procurement, development, compensation and maintenance of human resources (Flippo, 1984; Guest, 2011). Job satisfaction is more likely to be a function of the human resource policies and types of managerial relationships that exist in the workplace than the physical characteristics of the office environment, yet productivity is potentially affected by both satisfaction with and comfort of the physical environment.

It is quite normal to support the argument that in an upgraded or new work environment, workers tend to be happier and so they perform better (Frontczak et al., 2012; Schwede, Davies & Purdey, 2008). With a change or upgrade, people have been found to become more focused. Many real-life examples have been observed in which the performance of workers has increased due to positive changes in environmental aspects. For example, an 8 per cent increase in the performance of employees at a US post office was observed upon the improvement in lighting and acoustics (Browning, 1997). In another study, when indoor ambience was improved, a 10 per cent increase in performance followed (Roelofsen, 2002). More recent studies continue to demonstrate that sustainability initiatives make good economic sense (e.g. Eichholtz, Kok & Quigley, 2009). However, behaviour can be influenced by the fact that people know that they are being observed – a phenomenon called the 'Hawthorne Effect'.

Furthermore, a Canadian study by Veitch and Newsham (2000) found that a valuable improvement in mood, room appraisal, environmental satisfaction and self-assessed productivity was observed from the introduction of lighting dimming controls. Kruk (1989) gave much importance to the furniture present in the office, as he claimed that a comfortable well-designed chair increased the performance of employees by 27 per cent, while well-designed office furniture increased it by 15.4 per cent. If the workplace is full of innovation and creativity, then it is reported to attract and retain more workers who possess creativity (Haynes & Price, 2004). It is often postulated that there is a direct and strong relationship between employee environment (i.e. comfort) and employee satisfaction (e.g. Carlopio, 1996; Obasan Kehinde, 2011). Modern office environments include functional design for higher levels of workplace engagement, collaboration and problem-solving, strategies like hot-desking and hoteling to use existing space more efficiently, and support for offsite work underpinned by mobile technology and high-speed data access.

18.1.3 Proactivity

González-Benito & González-Benito (2006) classified environmental proactivity as improving ecological performance through the voluntary realisation of practices and initiatives. They translated the concept of environmental proactivity into three main components: planning and organisational practices such as environmental management systems, operational practices and communicational practices.

Measurement issues are becoming gradually more significant, which explains why the attention paid to a firm's environmental performance by government regulators, shareholders and the general public has increased, particularly in the developed world. González-Benito & González-Benito also revealed that while a firm's participation in unpaid conservation programmes may be interpreted as a proactive environmental stance, making political donations may suggest a desire to avoid or mitigate non-compliance. So aspects of performance, the managerial system and stakeholder relations are analogous to environmental proactivity. In a similar vein, Kolk and Mauser (2002) indicated that environmental operational indicators and environmental management indicators symbolise environmental proactivity.

Environmental management systems (EMSs) have been defined as "formal systems and databases that integrate procedures and processes for the training of personnel, monitoring, summarising and reporting of specialised environmental performance information to internal and external stakeholders of the firm" (Melnyk, Sroufe & Calantone, 2003: p.332). However, the main purpose for using an EMS is to build up, employ, control, organise and observe company environmental activities to accomplish two goals. The first goal is compliance, which means "reaching and maintaining the minimal legal and regulatory standards for acceptable pollution levels for the purpose of avoiding sanctions" (Sayre, 1996: p.332). Compliance is beneficial for a company: for instance, a firm may have increased costs if they are fined for breaching environmental requirements. The second goal is waste reduction, which goes beyond compliance and focuses a firm's behaviour on the dramatic reduction of negative environmental impact (Sayre, 1996; Govindan, Khodaverdi & Jafarian, 2013). González-Benito and González-Benito (2006) found that companies should consider developing environmental policies and goals with clear long-term environmental plans, knowledge of their environmental responsibilities, training programmes and performance measurement systems and evaluations.

An EMS can be defined as a collective internal effort to analyse, evaluate, enforce and derive policies (Coglianese & Nash, 2001). Training the workforce regarding ecological topics, building up ecological execution indicators and objectives and enforcing contract-based environmental laws and internal ecological audits are all a part of an EMS (Netherwood, 1998; Hillary, 2010). Unlike regulations that impose external constraints on firms, an EMS arises from within a firm and consists of a voluntary self-regulatory structure (Coglianese & Nash, 2001; Thakore, Lowe & Nicholls, 2012).

While a number of companies have adopted EMSs, some have gone even further and had their systems certified for the international standard ISO–14001 (Jiang & Bansal 2003; Curkovic & Sroufe, 2011; Thakore et al., 2012). ISO–14001 is "a set of management processes that requires firms to identify, measure, and control their environmental impacts" (Bansal & Hunter, 2003: p.290). ISO–14001 is not a performance standard; rather, it is a process-based standard, and indicates that a firm has implemented a management system that documents its pollution features and impacts, and classifies a pollution anticipation process.

Bansal & Hunter (2003) found six steps that need to be followed in order to act in accordance with the ISO–14001 standard: (1) expand environmental policy; (2) classify the firm's activities, products and services that interrelate with the environment; (3) identify legislative/regulatory necessities; (4) identify the firm's preference with regards to setting the objectives and targets for reducing environmental impacts; (5) adjust the organisational structure to meet those objectives, such as conveying responsibility, training, conversing and documenting; and (6) verify and correct the EMS. These represent communication practices that must be observed to fully comply with the standard. ISO–14001 compliance helps to move an organisation forward, but this momentum needs to be maintained.

18.2 Towards a conceptual framework

The notions of proactivity, accountability and legitimacy are applied to firms. Alrazi et al. (2010) found that firms should ensure a sensible level of stakeholder satisfaction, which is why they put environmental legitimacy and stakeholder satisfaction in the same group. This involves attention to two main aspects of environmental accountability, namely environmental reporting and environmental performance. These are achieved through having more robust accounting and environmental management systems and increasing stakeholder communication and engagement.

It is proposed in this chapter that the health of an office workplace eco-system can be measured at the level of an individual occupant and combined into a star rating to compare different organisational settings over time. A conceptual framework inspired by ELAP is constructed and described by Al-khawaja (2015) to define workplace ecology as the combination of organisational satisfaction, spatial comfort and worker productivity, each based on a scale of −10 to +10. His work is summarised in Figure 18.1. Organisation, space and technology are usually the three biggest cost centres for any business activity (Purdey, 2010).

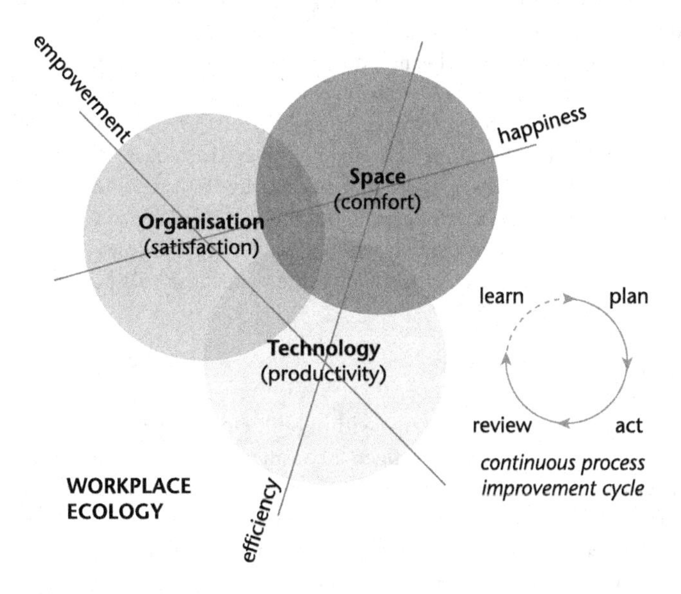

Figure 18.1 Workplace ecology conceptual framework

Source: Al-khawaja (2015)

18.2.1 Organisation

'Organisation' refers to the management of the workplace and its human resources (Bakiev, 2012). Organisational ecology (OE) is the study of life within an organisational setting. It is a conceptual and empirical approach in the field of social sciences that is an important part of organisational studies. Organisational demography and population ecology of organisations are also terms used to refer to OE. In this field of study, information from economics, biology and sociology are also used along with statistical analysis so that the conditions suitable for the emergence, growth and downfall of organisations can be determined (Robbins & Judge, 2013).

People are the biggest resource of most organisations (Schwede et al., 2008; Purdey, 2010), and therefore it is important that effective management of this resource takes place. OE includes aspects as diverse as line management responsibilities, promotion, performance review, conflict resolution, remuneration, conditions of employment, organisational values and communication, and is underpinned by a substantial body of knowledge. Workers generally need to be engaged with the organisation and be effective in performing their role. The relationship between a worker and his or her 'boss' is critical to maintaining a harmonious environment. Dissatisfaction can lead to a range of problems, not only for the worker but for the organisation as a whole. OE is a vital part of workplace ecology (Abbott, Green & Keohane, 2013).

18.2.2 Space

'Space' refers to the physical environment designed to accommodate the organisation and its people. Organisations that are seeking a competitive advantage in the present-day market need to make optimal use of their space as this incurs the highest expenses for an organisation after labour (Purdey, 2010). It has been asserted that companies are seeking a decrease in occupancy costs, and for this they carefully evaluate the way they utilise space. Organisations have faced costly inefficiencies in space planning and building utilisation due to misunderstandings about the way accommodation impacts business activity over time (Dao, Langella & Carbo, 2011; Delgado-Ceballos, Montiel & Antolin-Lopez, 2014; Isa, Hin & Yunus, 2011). However, one strategy to reduce costs is to improve the environmental performance of spaces through better design. Examples include a greater reliance on natural light, natural ventilation and more flexible fit-out that adopts open-plan layouts and activity-centred workspaces. Of course, these strategies are not always appropriate.

It is widely reported that such spaces have positive effects on occupant comfort (Leaman & Bordass, 1999; Bluyssen, Janssen, Van den Brink & De Kluizenaar, 2011; de Young, 2013; Deuble & de Dear, 2012). However, this is not necessarily the case, and skilled consultants are required to deliver high-performance 'green' buildings that work well (Bordass, Morrell & Ballentine, 2011). There is a body of knowledge that suggests green buildings can reduce absenteeism and be healthier environments for people than those that have a heavy reliance on mechanical systems (Bluyssen, 2010).

18.2.3 Technology

'Technology' refers to the tools and systems within built environments to improve productivity and work flow. Over the years, there has been a considerable increase in the use of technology in workplaces. Organisations utilise technology in various ways to save money, time and generally to refine processes. Automation, for example, enables time saving with respect to the collection, processing and distribution of information and even a decreased margin of error that can occur due to manual entry. Computerisation has improved the processing speed of work tasks, enabling information to be more readily shared. Personal computing and mobile devices, in particular, have revolutionised the modern office (Lai, 2010).

Quick and economical communication is another reason for the use of technology. Irrespective of their location and at an affordable cost, workers are able to communicate in real time through the internet and with mobile phones and other handheld devices. Documents can be edited or presentations shared between out-of-area workers by meeting online in internet-based conference rooms or with wireless networks. Technology has now become even more mobile, enabling work processes to be undertaken outside the organisation's facilities, enabling higher levels of productivity to be achieved. Many workers are now free to work anywhere, anytime, and still stay connected to their colleagues and customers (Purdey, 2010; Ramaseshan, Rabbanee & Hui, 2013). This, however, may have implications for work–life balance.

18.2.4 Continuous process improvement

Continuous process improvement can enhance the quality and efficiency of the workplace over time. It is a cyclical process and involves the following four stages (see Figure 18.2):

1. *Plan*: Data is gathered for the identification and definition of issues or problems that need to be addressed.
2. *Act*: This activates the process and involves the full-scale implementation of the plan.
3. *Review*: This involves reflection on the implemented plan to understand the actual performance level.
4. *Learn*: Often forgotten, this final step ensures that we learn from our actions and correctly influence new plans.

18.2.5 Measurement

Using a philosophy of continuous process improvement, each of the base attributes would need to be measured regularly, perhaps annually, to create an index that describes the health of the workplace eco-system. Ideas for improvement are developed, implemented and evaluated to gain understanding of the nature of the eco-system over time. Peter Drucker's mantra is that

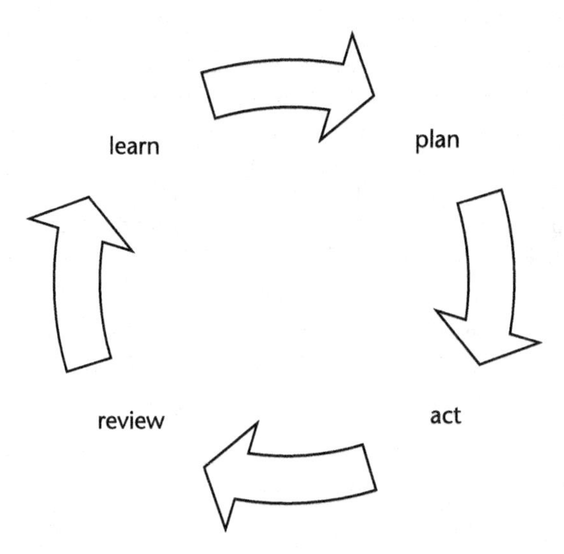

Figure 18.2 The continuous process improvement cycle

Source: Adapted from Kolb & Fry (1975)

Table 18.1 Workplace ecology star rating

Workplace ecology	*Star rating*
Above 8	★★★★★
Above 6 and <= 8	★★★★
Above 4 and <= 6	★★★
Above 2 and <= 4	★★
0–2	★
Below 0	

Source: Al-khawaja (2015)

"what gets measured gets improved" (Shore, 2014), so it is critical to have a robust method available. The approach adopted to translate this into an easy-to-interpret star rating is illustrated in Table 18.1.

It is hypothesised that there is a relationship between comfort and satisfaction (noted in the framework as 'happiness'), between comfort and productivity (noted as 'efficiency') and between satisfaction and productivity (noted as 'empowerment'). It is expected that all relationships are positively correlated (i.e. increases in comfort lead to increases in satisfaction) in healthy workplace eco-systems (Preiser & Vischer, 2005).

It is likely that variances in opinion will occur according to the job role and responsibilities of workers, and a number of other demographics including age, gender, education level, salary and length of time employed by the organisation (Guerin, Brigham, Kim, Choi & Scott, 2012). These demographics are considered along with specific occupant opinion on job complexity to determine the level of stress that knowledge workers (including managers) and/or support staff (hereafter called 'workers') may be under in performing their duties. In a healthy workplace, worker stress should not be excessive.

18.2.6 Post-occupancy evaluation

A survey instrument has been developed by Al-khawaja (2015) to collect personal opinions from workers regarding their job (work tasks) and their environment (workspace). Table 18.2 outlines the scoring mechanism adopted for each question.

Data about workers' jobs was derived from ten questions pertaining to job satisfaction and ten questions pertaining to job complexity. Data about workplace environment were derived from ten questions concerning comfort and a further ten questions concerning productivity. The survey instrument is a form of POE and is described in Tables 18.3–18.7.

Table 18.2 Workplace ecology scoring mechanism

Personal opinion	*Score*
Strongly agree	+2
Agree	+1
No opinion	0
Disagree	–1
Strongly disagree	–2

Source: Al-khawaja (2015)

Table 18.3 Demographic data

ID	*Questions on demographics*
1a	Age: less than 25 years old, 25–40 years old, 41–55 years old, more than 55 years old
1b	Gender: male, female
2a	Current role: contractor/casual, employee, line manager, senior manager
2b	Employed in organisation: less than 1 year, 1–10 years, 11–20 years, more than 20 years
3a	Primary workspace type: closed-cell office, open-plan office, shared space, activity-based
3b	Time worked in this space: less than 1 year, more than 1 year
3c	Location of workspace: high-performance green building, heritage-listed building, other office building, non-office building
4a	Formal education: diploma/certificate, bachelor's degree, master's/doctorate, none
4b	Main motivation for study: advancement, mandatory, promotion, not applicable/other
5a	Annual salary (f/t equiv. $USD gross): less than $50,000, $50,001–$100,000, $100,001–$150,000, more than $150,000
5b	Dominant work ethic: hard-working, reliable, creative flair, team player

Source: Al-khawaja (2015)

Table 18.4 Job satisfaction data

ID	*Questions on satisfaction*
6	I love my job
7	I get on well with those in higher positions
8	I have a good relationship with my immediate work colleagues
9	I receive generous rewards for my work
10	I feel appreciated when I do good work
11	Change in the workplace is generally handled openly
12	I can easily speak with my supervisor/manager when I need to
13	I feel engaged with the organisation and its mission
14	Bullying and harassment in my workplace do not exist
15	I have opportunities for advancement in the organisation

Source: Al-khawaja, 2015

Table 18.5 Job complexity data

ID	*Questions on job complexity*
16	My work tasks are usually undertaken collaboratively
17	I regularly have to work extra hours
18	I am an influential member of a dynamic team
19	I often need to solve complex problems myself
20	Effective time management is a critical attribute in my position
21	I am responsible for the work of others in the organisation
22	I am a person others frequently come to for help
23	I spend a lot of time in formal and informal meetings
24	I am allowed to work at home or offsite at my discretion
25	I enjoy a job that is challenging albeit stressful at times

Source: Al-khawaja (2015)

Table 18.6 Spatial comfort data

ID	*Questions on comfort*
26	Office air temperature is normally conducive to my work tasks
27	Indoor air quality/ventilation is excellent
28	My workspace has a good combination of natural and artificial light
29	I have a clear view of what is going on outside the building
30	Office noise disturbance is minimal
31	I have control over my personal comfort settings
32	I have reasonable visual privacy when working
33	There are no workspace issues that impact negatively on my health
34	My office furniture is comfortable and adjustable
35	I have appropriate work and storage space

Source: Al-khawaja (2015)

Table 18.7 Worker productivity data

ID	*Questions on productivity*
36	I have access to the necessary IT services to fulfil my role
37	All equipment I use in the office is provided by the organisation
38	I can access the data I need wherever it might be
39	My work is automatically backed up every day
40	I use IT extensively to communicate and stay well informed
41	Prompt IT support is available if I have a problem
42	Data compatibility and transfer among office staff is straightforward
43	If I need specialist IT equipment for a particular task, it is provided
44	I consider myself to be an effective user of technology
45	Use of IT enables me to be more organised and professional

Source: Al-khawaja (2015)

Table 18.8 Question weighting scores

Personal relevance	*Score*
Very important	5
Slightly important	4
Neutral	3
Slightly unimportant	2
Not important	1

Source: Al-khawaja (2015)

Occupant opinion needs to be considered in the light of the relevance of each issue to the individual respondent. Obviously not all questions are of equal importance. Occupants are asked to rate each of the opinion questions using a simple scale of 1 to 5, as shown in Table 18.8.

Opinion and relevance scores are multiplied together for each question. For example, when asked to respond to question Q6 'I love my job', an opinion of 'strongly agree' with relevance of 'very important' would compute a weighted score for that question of +10. At the other extreme, an opinion of 'strongly disagree' with relevance of 'very important' would compute a weighted score for that question of −10. Having no opinion, regardless of importance, computes a score of zero.

18.3 Case study

To illustrate the process of measuring and interpreting workplace ecology, Bond University's Sustainable Development Building (Figure 18.3) was used as a small pilot study. The building is an award-winning 6-star Green Star rated educational building constructed on the Gold Coast, Australia, in 2010. This building is primarily office space for academic and support staff and comprises many significant sustainability features that include information board viewing stations (1–13).

Occupants of the building comprise a combination of academic staff and administrative staff, and their opinions take into account non-office spaces including teaching rooms, computer laboratories and auxiliary spaces. The majority of the floor area is office usage, and the building formed part of a study of green offices in Australia by Armitage, Murugan and Kato (2011). A total population of 35 staff were surveyed, and 24 responses were collected, representing 68.57 per cent. The survey was undertaken in 2012. Figure 18.4 summarises the results.

18.3.1 Happiness

In this case study, there is a weak to moderate correlation between comfort (the independent variable) and satisfaction (the dependent variable). Comfort explains 28.73 per cent of the

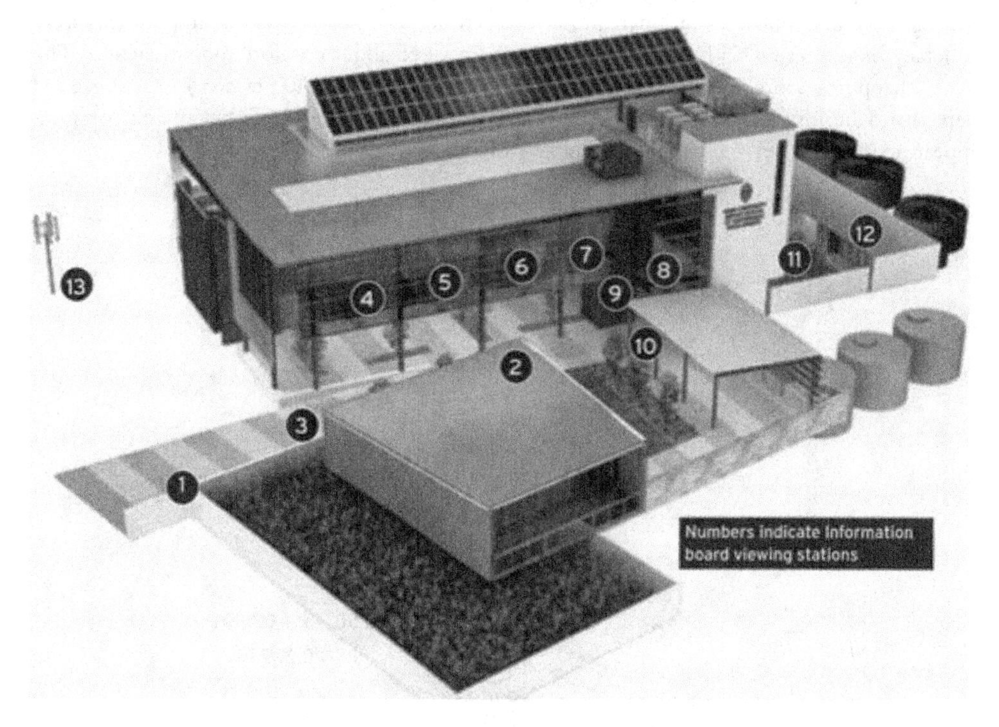

Figure 18.3 Bond University's Sustainable Development Building

Source: Image courtesy of Bond University

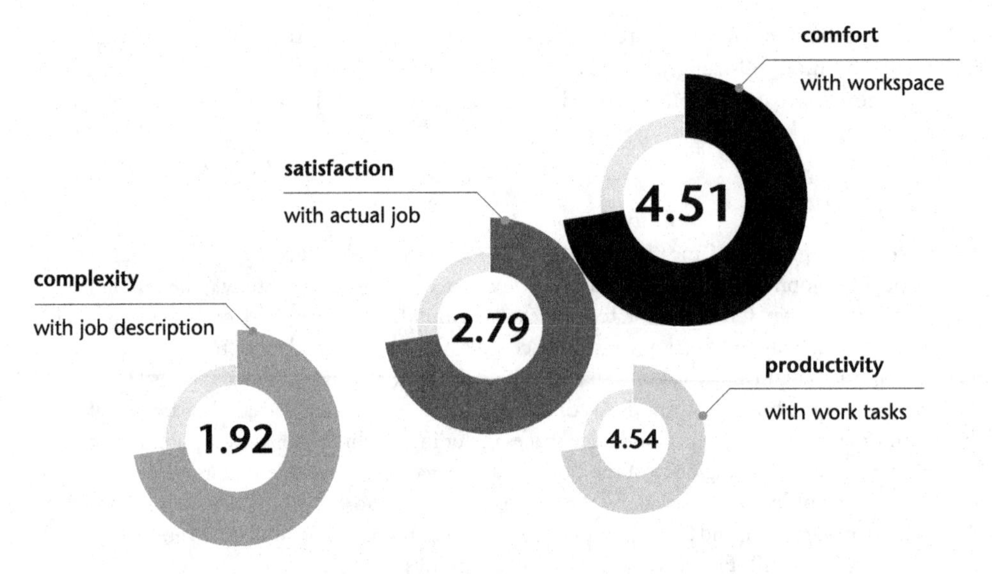

Figure 18.4 Case study findings

Source: Al-khawaja (2015)

Note: Size of circle has no significance here.

regression equation (y = 0.5012x + 0.5301), and it is clear that this relationship is a positive one (i.e. increases in comfort lead to increases in satisfaction). The data shows that respondents were generally in agreement with the questions that were asked about comfort and satisfaction. The overall happiness score was 3.68 on a scale of −10 to +10 (i.e. 68.40 per cent). It is suggested here that a healthy workplace should have a concentration of positive respondent opinion appearing in the upper-right-hand quadrant of the chart (Figure 18.5).

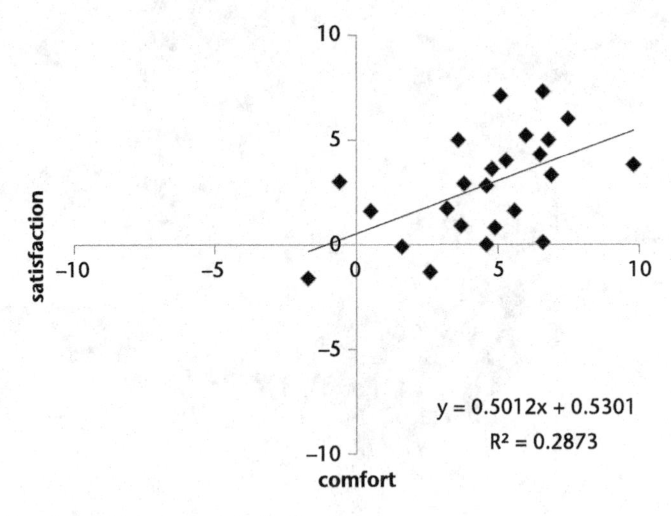

Figure 18.5 Happiness (comfort v satisfaction)

Source: Al-khawaja (2015)

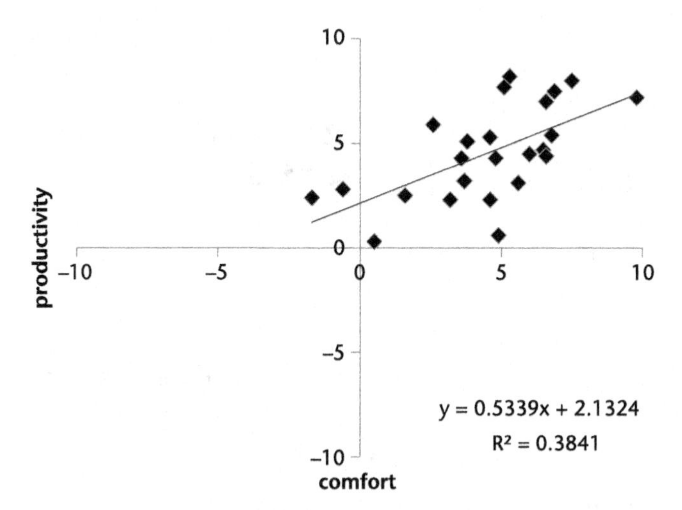

Figure 18.6 Efficiency (comfort v productivity)
Source: Al-khawaja (2015)

18.3.2 Efficiency

A moderate correlation between comfort (the independent variable) and productivity (the dependent variable) is also found. Comfort explains 38.41 per cent of the regression equation ($y = 0.5339x + 2.1324$), and it is clear that this relationship is a positive one (i.e. increases in comfort lead to increases in productivity). The data shows that respondents were generally in agreement with the questions that were asked about comfort and productivity. The overall efficiency score was 4.55 on a scale of −10 to +10 (i.e. 72.75 per cent). A healthy workplace should also have a concentration of opinion in the upper-right-hand quadrant of the chart (Figure 18.6).

18.3.3 Empowerment

Finally, there is a moderate correlation between satisfaction (the independent variable) and productivity (the dependent variable). Satisfaction here explains 31.38 per cent of the regression equation ($y = 0.5160x + 3.1011$), and it is once again clear that this relationship is a positive one (i.e. increases in satisfaction lead to increases in productivity). The data shows that respondents were generally in agreement with the questions that were asked about satisfaction and productivity. The overall empowerment score was 3.67 on a scale of −10 to +10 (i.e. 68.35 per cent). As before, a healthy workplace should have a concentration of opinion in the upper-right-hand quadrant of the chart (Figure 18.7).

18.3.4 Other relationships

Position expectation (PE) is a control variable developed in this research to test if job complexity scores are aligned to the demographics for each respondent. Age, rank, seniority, salary and qualification data are used to compute PE and provide some objectivity about how complex a respondent's job should be. If job complexity is considerably higher than PE, it might suggest an environment where people are stressed.

In this case study, the mean score for PE was 4.46 on a scale of 0 to +10. The overall job complexity score was 1.92 on a scale of −10 to +10 (i.e. 59.60 per cent). This means that

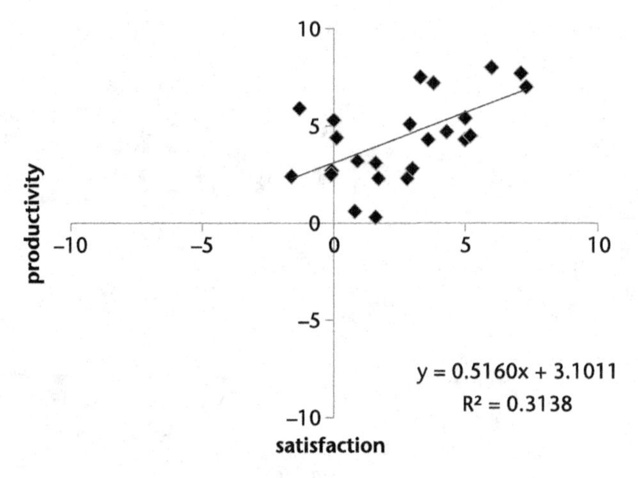

Figure 18.7 Empowerment (satisfaction v productivity)
Source: Al-khawaja (2015)

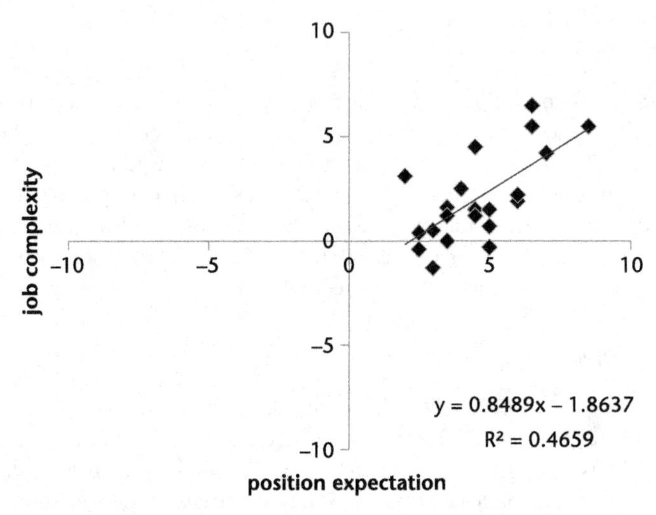

Figure 18.8 Position expectation v job complexity
Source: Al-khawaja (2015)

organisational expectation and job complexity were reasonably balanced, and therefore staff were probably not suffering undue stress to perform at a level higher than their position might dictate. PE explains 46.59 per cent of the regression equation ($y = 0.8489x - 1.8637$), as shown in Figure 18.8.

The rating for this workplace was computed as 2.50 stars (standard deviation = 1.10; coefficient of variation = 44.13 per cent). Realised performance is illustrated by the mean response to satisfaction, comfort and productivity questions, and is called workplace ecology index (WEI). The value for WEI in this case study was 3.97 on a scale of −10 to +10 (i.e. 69.85 per cent). Expected performance is illustrated by the mean PE score, less the mean response to job complexity questions, and is called workplace performance index (WPI). The value for WPI

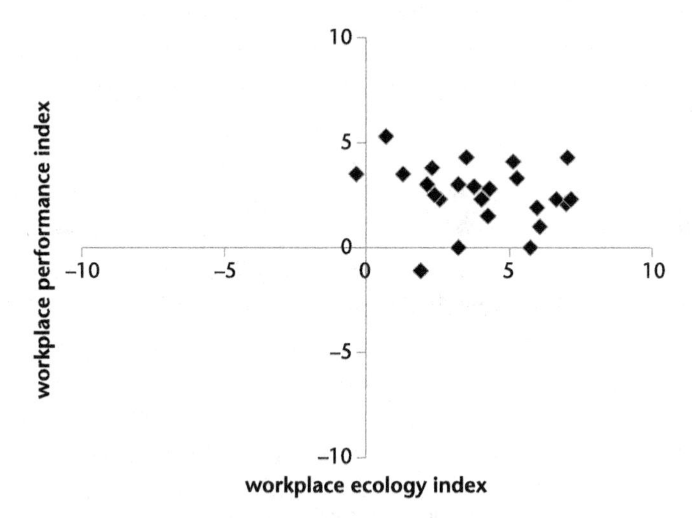

Figure 18.9 Workplace ecology index v workplace performance index

Source: Al-khawaja (2015)

in this case study was 2.54 on a scale of −10 to +10 (i.e. 62.70 per cent). A notional target of 75 per cent of respondents falling within the upper-right-hand quadrant is considered by the researchers to reflect a well-balanced workplace. In this case study, the actual value was 83.33 per cent (Figure 18.9).

18.4 Further studies

In the opinion of the researchers, healthy workplaces should ideally have a rating of at least 3 stars. This translates to mean scores for satisfaction, comfort and productivity around 4 or higher (the pilot study exceeded this threshold for comfort and productivity). In addition, at least 75 per cent of occupants should demonstrate that their WEI and WPI scores are both positive. This will imply that realised and expected performances are somewhat in balance and that occupants are not unduly stressed.

Four more studies have been undertaken since the pilot, involving a total of 315 occupant opinions, as shown in Table 18.9 and Figure 18.10. Interestingly, all four buildings failed to meet any of these benchmarks. The point of these studies is to demonstrate the framework on a range of different building typologies. Additional case studies are needed to demonstrate the practical range that modern workplaces can achieve.

Table 18.9 Field testing

Case study	*Responses*	*Star rating*
6-star Green Star building	76 (33.04%)	1.17
Low-quality non-green building	32 (50.00%)	1.59
High-quality non-green building	29 (65.91%)	1.59
Medium-quality non-green building	178 (35.60%)	1.35
All four studies (combined)	315 (37.59%)	1.36

Source: Al-khawaja (2015)

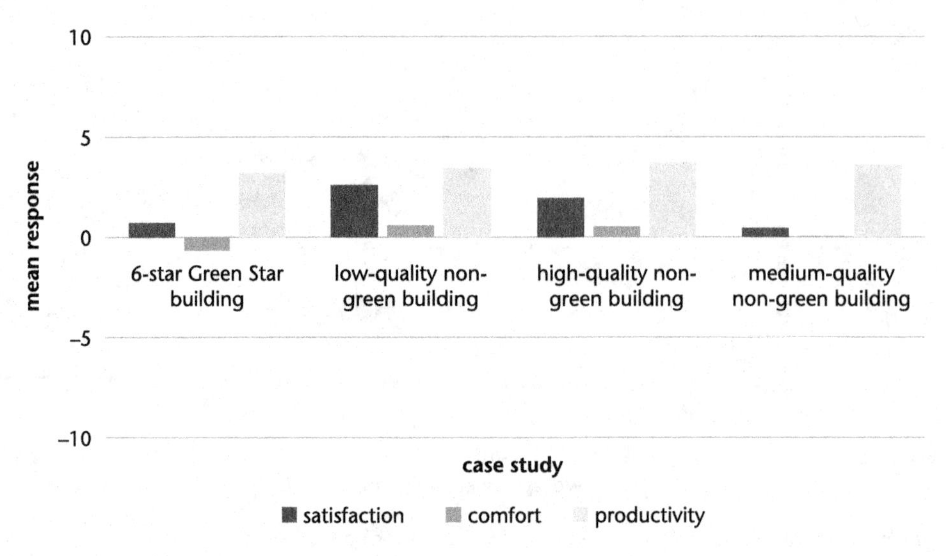

Figure 18.10 Summary of satisfaction, comfort and productivity scores

Source: Al-khawaja (2015)

No inference is being drawn that green buildings perform more poorly than traditional buildings, but it can be concluded that just because a building has a high environmental rating does not mean that a healthy workplace eco-system will automatically occur. Every workplace needs to be assessed on a case-by-case basis, and the framework outlined here is merely a method to evaluate occupant opinion.

18.5 Conclusions

Many more case studies in a range of different settings are needed to provide more insight into the relationships between job satisfaction, spatial comfort and worker productivity. The work completed so far does suggest, however, that results will be difficult to generalise. It is expected that each workplace has unique attributes that affect the performance of its occupants, and that these attributes change over time. They therefore need to be measured regularly and continuously improved. An annual occupant survey would be desirable, and its analysis could be largely automated.

Workplace ecology is an emerging field of study for creating or adapting office environments where people are more likely to be happy, efficient and empowered in their work activities. It is now possible to ensure that continuous process improvement for workplaces is able to occur using a systematic and rigorous procedure such as the one described herein. Healthy workplaces require occupants to be satisfied with their job, comfortable in their space, productive with supplied technology and regularly evaluated. This applies equally to new and existing or adapted office environments.

References

Abbott, K. W., Green, J. F. & Keohane, R. O. 2013. Organisational ecology and organisational strategies in world politics. Harvard Project on Climate Agreements, discussion.

Agha-Hossein, M. M., El-Jouzi, S., Elmualim, A. A., Ellis, J. & Williams, M. 2013. Post-occupancy studies of an office environment: energy performance and occupants' satisfaction. *Building and Environment*, 69, 121–130.

Akimoto, T., Tanabe, S. I., Yanai, T. & Sasaki, M. 2010. Thermal comfort and productivity: evaluation of workplace environment in a task conditioned office. *Building and Environment*, 45(1), 45–50.

Al-khawaja, A. 2015. Environmental auditing: modelling office workplace ecology. PhD thesis, Bond University.

Alrazi, B., De Villiers, C. & van Staden, C. 2010. A framework for the integration of Environmental Legitimacy, Accountability and Proactivity (ELAP). 6th Asia Pacific Interdisciplinary Research in Accounting (APIRA) Conference 2010, Sydney, New South Wales, Australia.

Altomonte, S. & Schiavon, S. 2013. Occupant satisfaction in LEED and non-LEED certified buildings. *Building and Environment*, 68, 66–76.

Armitage, L., Murugan, A. & Kato, H. 2011. Green offices in Australia: a user perception survey. *Journal of Corporate Real Estate*, 13(3), 169–180.

Bakiev, E. 2012. *Organisational ecology*. Zirve University. Available from http://slideplayer.com/slide/9145396.

Bansal, P. & Hunter, T. 2003. Strategic explanations for the early adoption of ISO 14001. *Journal of Business Ethics*, 46(3), 289–299.

Bertels, S. & Peloza, J. 2006. Running to stand still: managing CSR reputation in an era of ratcheting expectations. CCGRM Working Paper.

Bluyssen, P. M. 2010. Towards new methods and ways to create healthy and comfortable buildings. *Building and Environment*, 45(4), 808–818.

Bluyssen, P. M., Janssen, S., van den Brink, L. H. & de Kluizenaar, Y. 2011. Assessment of wellbeing in an indoor office environment. *Building and Environment*, 46(12), 2632–2640.

Bordass, B. & Leaman, A. 2005. Making feedback and post-occupancy evaluation routine 1: a portfolio of feedback techniques. *Building Research & Information*, 33(4), 347–352.

Bordass, B., Morrell, P. & Ballentine, I. 2011. The Usable Buildings Trust and new professionalism. ACE Annual Conference, London, 25 May.

Bortree, D. S. 2009. The impact of green initiatives on environmental legitimacy and admiration of the organisation. *Public Relations Review*, 35(2), 133–135.

Brønn, P. S. & Vidaver-Cohen, D. 2009. Corporate motives for social initiative: legitimacy, sustainability, or the bottom line? *Journal of Business Ethics*, 87, 91–109.

Browning, W. D. 1997. Boosting productivity with IEQ improvements. *Building Design and Construction*, 38(4), 50–52.

Carlopio, J. R. 1996. Construct validity of a Physical Work Environment Satisfaction Questionnaire. *Journal of Occupational Health Psychology*, 1(3), 330.

Coglianese, C. & Nash, J. 2001. Environmental management systems and the new policy agenda, in C. Coglianese & J. Nash (eds), *Regulating from the inside: can environmental management systems achieve policy goals*? Washington, DC: Resources for the Future, 1–25.

Curkovic, S. & Sroufe, R. 2011. Using ISO 14001 to promote a sustainable supply chain strategy. *Business Strategy and the Environment*, 20(2), 71–93.

Dao, V., Langella, I. & Carbo, J. 2011. From green to sustainability: information technology and an integrated sustainability framework. *The Journal of Strategic Information Systems*, 20(1), 63–79.

de Young, R. 2013. Environmental psychology overview, in A. Hergatt Huffman & S. R. Klein (eds), *Green organisations: driving change with IO psychology*. London: Routledge, 19–45.

Deegan, C. 2006. Legitimacy theory, in Z. Hoque (ed.), *Methodological issues in accounting research: theories, methods and issues*. London: Spiramus Press, 161–181.

Delgado-Ceballos, J., Montiel, I. & Antolin-Lopez, R. 2014. What falls under the corporate sustainability umbrella? Definitions and measures. 25th Annual Conference of the International Association for Business and Society (IABS).

Delmas, M. & Blass, V. D. 2010. Measuring corporate environmental performance: the trade-offs of sustainability ratings. *Business Strategy and the Environment*, 19(4), 245–260.

Deuble, M. P. & de Dear, R. J. 2012. Green occupants for green buildings: the missing link? *Building and Environment*, 56, 21–27.

Eichholtz, P., Kok, N. & Quigley, J. 2009. *Doing well by doing good: an analysis of the financial performance of green office buildings in the USA*. London: RICS Research.

Flippo, E. B. 1984. *Personnel management*. New York: McGraw-Hill.

Frontczak, M., Schiavon, S., Goins, J., Arens, E., Zhang, H. & Wargocki, P. 2012. Quantitative relationships between occupant satisfaction and satisfaction aspects of indoor environmental quality and building design. *Indoor Air*, 22(2), 119–131.

González-Benito, J. & González-Benito, Ó. 2006. A review of determinant factors of environmental proactivity. *Business Strategy and the Environment*, 15(2), 87–102.

Govindan, K., Khodaverdi, R. & Jafarian, A. 2013. A fuzzy multi-criteria approach for measuring sustainability performance of a supplier based on triple bottom line approach. *Journal of Cleaner Production*, 47, 345–354.

Gray, R., Owen, D. & Adams, C. 1996. *Accounting and accountability: changes and challenges in corporate social and environmental reporting*. London: Prentice Hall.

Green, S. D. & Moss, G. W. 1998. Value management and post-occupancy evaluation: closing the loop. *Facilities*, 16(1-2), 34–39.

Guerin, D. A., Brigham, J. K., Kim, H. Y., Choi, S. & Scott, A. 2012. Post-occupancy evaluation of employees' work performance and satisfaction as related to sustainable design criteria and workstation type. *Journal of Green Building*, 7(4), 85–99.

Guest, D. E. 2011. Human resource management and performance: still searching for some answers. *Human Resource Management Journal*, 21(1), 3–13.

Hadjri, K. & Crozier, C. 2009. Post-occupancy evaluation: purpose, benefits and barriers. *Facilities*, 27(1-2), 21–33.

Halal, W. 2015. Business strategy for the technology revolution: competing at the edge of creative destruction. *Journal of the Knowledge Economy*, 6(1), 31–47.

Haynes, B. & Price, I. 2004. Quantifying the complex adaptive workplace. *Facilities*, 22(1-2), 8–18.

Hillary, R. 2010. Environmental auditing: concepts, methods and developments. *International Journal of Auditing*, 2(1), 71–85.

IFMA Foundation, 2010. Sustainability how-to guide series. Available from https://foundation.ifma.org/research/sustainability-how-to-guide-series.

Isa, F. M., Hin, C. W. & Yunus, J. M. 2011. Change management initiatives and job satisfaction among salespersons in Malaysian direct selling industry. *Workforce*, 1(7), 106–121.

Jiang, R. J. & Bansal, P. 2003. Seeing the need for ISO 14001. *Journal of Management Studies*, 40(4), 1047–1067.

Julien, A. 2009. Assessing the assessor: BREEAM vs LEED. *Sustain Magazine*, 9(6), 30–33.

Kolb. D. A. & Fry, R. E. 1975. Toward an applied theory of experiential learning, in C. L. Cooper (ed.), *Theories of group processes*, London: John Wiley.

Kolk, A. & Mauser, A. 2002. The evolution of environmental management: from stage models to performance evaluation. *Business Strategy and the Environment*, 11(1), 14–31.

Kruk, L. B. 1989. Why consider seating last? *The Office*, 109(6), 45–50.

Lai, T.-K. 2010. How does information and communication technology affect workplace organisation? Available from https://editorialexpress.com/cgi-bin/conference/download.cgi?db_name=CEA2010&paper_id=672.

Leaman, A. & Bordass, B. 1999. Productivity in buildings: the 'killer' variables. *Building Research & Information*, 27(1), 4–19.

Marr, B. & Schiuma, G. 2003. Business performance measurement: past, present and future. *Management Decision*, 41(8), 680–687.

Melnyk, S. A., Sroufe, R. P. & Calantone, R. 2003. Assessing the impact of environmental management systems on corporate and environmental performance. *Journal of Operations Management*, 21(3), 329–351.

Mobus, J. L. 2005. Mandatory environmental disclosures in a legitimacy theory context. *Accounting, Auditing & Accountability Journal*, 18(4), 492–517.

Netherwood, A. 1998. Environmental management systems, in R. Welford (ed.), *Corporate environmental management 1*. London: Earthscan, 35–58.

Obasan Kehinde, A. 2011. Impact of job satisfaction on absenteeism: a correlative study. *European Journal of Humanities and Social Sciences*, 1(1), 25–49.

Owen, D. 2008. Chronicles of wasted time?: A personal reflection on the current state of, and future prospects for, social and environmental accounting research. *Accounting, Auditing & Accountability Journal*, 21(2), 240–267.

Preiser, W. F. E. 1995. Post-occupancy evaluation: how to make buildings work better. *Facilities*, 13(11), 19–28.

Preiser, W. F. & Vischer, J. C. 2005. The evolution of building performance evaluation: an introduction, in W. F. Preiser & J. C. Vischer (eds), *Assessing building performance*. London: Elsevier, 3–13.

Preiser, W. F. E., Rabinowitz, H. Z. & White, E. T. 1988. *Post-occupancy evaluation*. New York: Van Nostrand Reinhold.

Purdey, B. 2010. Conditions under which the performance of teams is compromised: the role of workspace density in triggering the collapse of workgroups in commercial office settings. PhD thesis, University of Sydney.

Ramaseshan, B., Rabbanee, F. K. & Hui, L. T. H. 2013. Effects of customer equity drivers on customer loyalty in B2B context. *Journal of Business & Industrial Marketing*, 28(4), 335–346.

RIBA 1991. *Architectural knowledge: the idea of a profession*. London: E & FN Spon.

Robbins, S. P. & Judge, T. A. 2013. *Organisational behaviour*. London: Pearson Higher Education.

Roelofsen, P. 2002. The impact of office environments on employee performance: the design of the workplace as a strategy for productivity enhancement. *Journal of Facilities Management*, 1(3), 247–264.

Sayre, D. 1996. *Inside ISO 14000: the competitive advantage of environmental management*. Delray Beach, FL: St. Lucie Press.

Schwede, D. A., Davies, H. & Purdey, B. 2008. Occupant satisfaction with workplace design in new and old environments. *Facilities*, 26(7/8), 273–288.

Shafer, W. E. 2006. Social paradigms and attitudes toward environmental accountability. *Journal of Business Ethics*, 65, 121–147.

Shore, J. 2014. These 10 Peter Drucker quotes may change your world. *Entrepreneur*, 16 September. Available from www.entrepreneur.com/article/237484.

Steemers, K. & Manchanda, S. 2010. Energy efficient design and occupant well-being: case studies in the UK and India. *Building and Environment*, 45(2), 270–278.

Suchman, M. C. 1995. Managing legitimacy: strategic and institutional approaches. *The Academy of Management Review*, 20(3), 571–610.

Tam, M. 2013. A happy worker is a productive worker. *Huffington Post*, 31 July. Available from www.huffingtonpost.com/marilyn-tam/how-to-be-happy-at-work_b_3648000.html.

Thakore, R., Lowe, C. & Nicholls, T. 2012. Financial impact of certified ISO 14001 Environment Management Systems in UK and Ireland. *ICSDEC Proceedings*, Fort Worth, November, 894–902.

Veitch, J. A. & Newsham, G. R. 2000. Exercised control, lighting choices, and energy use: an office simulation experiment. *Journal of Environmental Psychology*, 20(3), 219–237.

Vidaver-Cohen, D. 2007. Industry legitimacy and organisational reputation: a model of reciprocal processes. Paper presented at Faculty Research Colloquium, Department of Management and International Business, Florida International University, 16 March.

Vischer, J. 2001. *Post-occupancy evaluation: a multifaceted tool for building improvement: learning from our buildings*. Washington, DC: National Academy Press.

Vischer, J.C. 2005. *Space meets status: designing workplace performance*. New York: Routledge.

Zimmerman, A. & Martin, M. 2001. Post-occupancy evaluation: benefits and barriers. *Building Research & Information*, 29(2), 168–174.

19

Creating a green index based on tenant demand for sustainable office buildings and features

Spenser J. Robinson and Robert A. Simons

19.0 Introduction

An oft-overlooked aspect of the sustainability spectrum in commercial real estate revolves around specific tenant preferences. While this discussion focuses primarily on US institutional grade-core-type office stock, the lessons and theory apply broadly to international and other commercial property types. Tenants have been shown to assume higher rental costs for green certified buildings (Devine & Kok, 2015; Fuerst & McAllister, 2011; Reichardt, Fuerst, Rottke, & Zietz, 2012; Robinson & McAllister, 2015). Paying these premiums makes intuitive sense if tenants perceive benefits in human talent retention, increased productivity, and/or corporate social responsibility (CSR) advantages. A number of studies support human talent retention (Heerwagen, 2000; Kats, Alevantis, Berman, Mills, & Perlman, 2003). As many as 94 percent of workers in the US have stated preferences for working in environmentally sound and efficient buildings (Reiss & Costello, 2007). CSR arguments as motivation for office tenants to use green buildings are well established (Eichholtz, Kok, & Quigley, 2009). Productivity benefits are less empirically established but growing in anecdotal evidence and widespread acceptance (Allen et al., 2015; Miller, Pogue, Gough, & Davis, 2009). Potential rental premiums may also exist due to savings on utilities, but this is contingent on lease structure (Janda, Bright, Patrick, Wilkinson, & Dixon, 2016).

While the reasons why tenants may pay for sustainable buildings features are evident, little discussion or research exists on specifically what tenants are paying for. Much of the research revolves around green labels such as Leadership in Energy and Environmental Design (LEED), Energy Star, or BREEAM. This chapter will discuss and highlight growing knowledge about green building attribute preferences rather than the prescribed bundles inherent in green labels.

19.1 Background on our research process

In 2012, CBRE Inc. announced the initiation of the Real Green Research Challenge whereby they offered their resources, financial and corporate, to further knowledge about sustainable real estate. One of five research grants was awarded to a team co-led by the chapter authors.

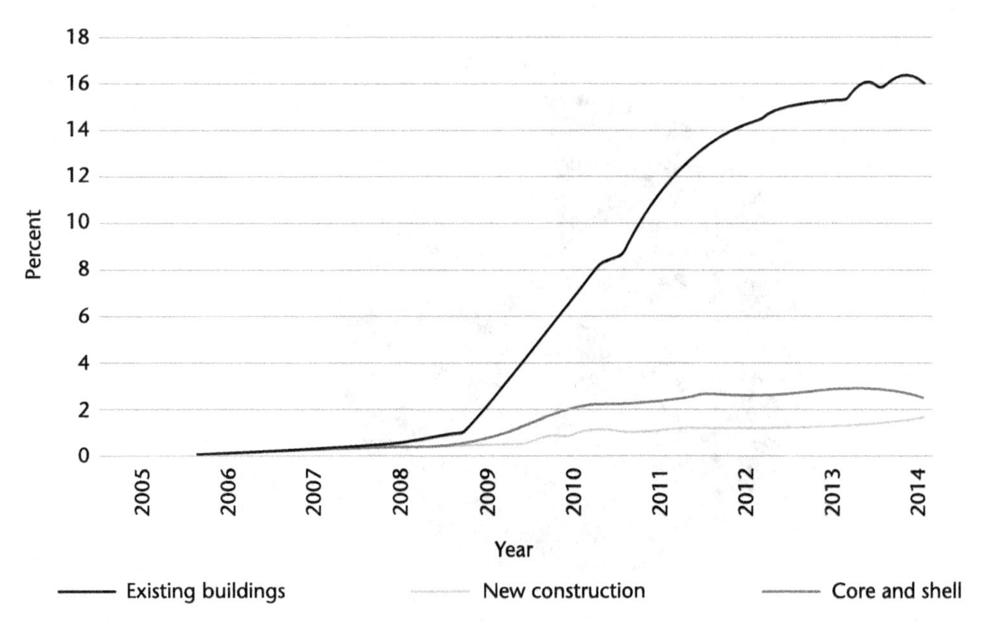

Figure 19.1 LEED adoption for existing building, new constructions, and core and shell programs

Source: Kok et al. (2015)

Our key strategic outcome was to rigorously determine which sustainability characteristics and certifications are actually valued by occupants and develop that into an index. Another sponsored study, by Kok, Holtermans, and Pogue (2015), found that the number of certifications annually has plateaued, as shown for LEED in Figure 19.1. The same report further revealed that despite slowing growth, approximately 87 percent of US office building stock—over 30,000 buildings —remains without LEED certification. This is demonstrated in Figure 19.2.

Existing eco-labels have advanced the operation and environmental soundness of buildings in the US and internationally. However, it appears that a large number of office buildings have not attained or cannot attain these labels. Yet, they may contain other green attributes that have value in the marketplace. Perhaps a different way of looking at eco-labeling is now called for. The intended outcome of the process and research described below is an eco-labeling system that can supplement existing ones for buildings that structurally are unable to qualify for labels such as LEED or whose owners choose not to bear the expense associated with it. We seek to "unpack" the bundle of attributes embedded in the LEED and Energy Star brands, and examine green attributes, by looking at which green features are valued by the market, especially tenants. The question then becomes more granular: what features will tenants pay for or which features add the most value?

19.2 First steps: Focus groups

The first step in understanding tenant preferences was simply to generate a potential list of features by asking tenants what features they value. This was accomplished by convening a series of focus groups in four major US markets, plus a pilot in a fifth. We approached the focus groups with a completely clean slate. No prior list of attributes or labels was provided. Questions were designed

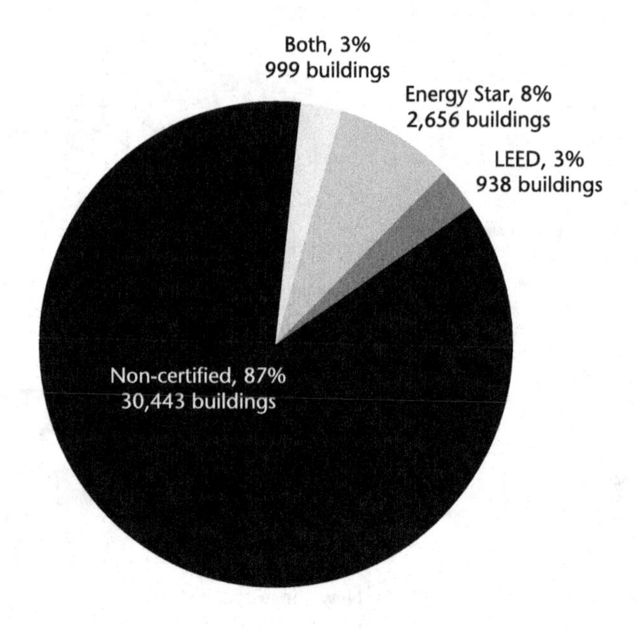

Figure 19.2 Eco-label adoption in the US for Energy Star and LEED

Source: Kok et al. (2015)

specifically to solicit brainstorming-style input and to generate lists of attributes from scratch. A pilot focus group was first held in Detroit, Michigan. The purpose was to ensure some common language, discuss the proper mix of tenant representatives, tenant brokers, property managers, and other related stakeholders for a broad tenant-based perspective. Then, a series of national focus groups in Chicago, Illinois; Denver, Colorado; Washington, D.C.; and the San Francisco Bay area in California were held. Participants were specifically directed to represent tenants as a whole and not their own personal opinions. This direction was generally followed, and in some cases participants expressed their feedback for the focus group and stated that their personal preference may differ but that the tenant community they represent would be best served by their targeted input. The focus groups helped form a list of important attributes for further research and discussion; detailed results are available in Simons, Robinson, and Lee (2014). The results were intended to be nationally representative, and focus group locations were chosen to help ensure that. More detailed strengths and weaknesses of the research are discussed in the paper itself.

Table 19.1 reproduces the various green office building attributes identified by Simons et al. (2014) as most important. The count (frequency) column identifies how many of the participants included the attribute in their top five most important sustainable building attributes. The maximum possible count (frequency) is 48, and a higher number means that the attribute was more often expressed by the participants as important. The average ranking column is on a five-point Likert scale ranging from 1 (most important) to 5 (least important), so a lower number means that the attribute is ranked as more important.

We can see that access to natural light is most often mentioned, showing that it is almost ubiquitously valued. However, its median ranking of 2.5 shows that it is rarely the top valued attribute. Lease structure that rewards green behavior is much less frequently mentioned in the focus groups but shows the highest mean rating when it is. The vast majority of tenant

Table 19.1 Focus group results for sustainable building attributes

Green office building attributes	*Count (frequency)*	*Ave. ranking*
Access to natural light in work space	42	2.5
Public transportation (within five-minute walk)	31	2.5
Indoor air quality	36	3.0
Temperature control/comfort	31	3.1
Efficient lighting system	26	3.5
Efficient electrical and gas use for heating and cooling	30	2.7
Open-space layout	11	2.8
Lease structure rewards green behavior	10	1.6
Shower on-site	10	4.0
Water conservation	8	3.0
LEED designation	8	3.4
Recycling provided on-site	12	3.8
Outdoor amenities	5	3.6
Green cleaning	6	3.8
Floor plate size/depth	3	3.0
New/fresh/young/cool	2	2.0
Green roof	2	4.0
Column spacing	1	3.0
Bike racks	1	4.0

Source: Simons et al. (2014)

representatives were well aware of whether their lease structure encourages green behavior or if they had no financial incentives (e.g. a Full Service Gross Lease). The high ranking suggests that tenant awareness of the behavioral influences based on the lease structure is high and perhaps growing.

Interestingly, most of the highly rated features are related to productivity. Features such as natural light, public transportation, indoor air quality, and localized temperature control generally concern employee well-being more so than resource conservation. Tenants were asked to frame the issues around green buildings in the triple bottom line (Elkington, 1997), where Planet, People, and Profit (PPP) distinctions arise. The goal of this discussion was to better understand the motivation behind tenant preferences. Applying this PPP framework to green buildings, resource conservation associated with better and more efficient buildings clearly relates to the planet. Productivity-related issues would be associated with people. Profit relates to financial profits. However, multiple focus groups asked us to distinguish between tenant profit and landlord profit for this section. The results show that the most highly ranked attributes appear strongly related to the People category. Participants ranked relevance on a 0–2 scale, with 0 as not relevant and 2 as most relevant. The four highest ranked attributes were natural light (1.9), public transportation (2.0), indoor air quality (1.8), and localized temperature control (2.0). This clearly shows a tenant preference for comfort-related features that may be associated with improved productivity.

We also used the later focus groups to field test early versions of a survey instrument, to be discussed below. Now, having discovered the 18 most desirable green features, we moved toward collecting data from both the tenants (their stated demand via surveys) and from their corresponding office building leases.

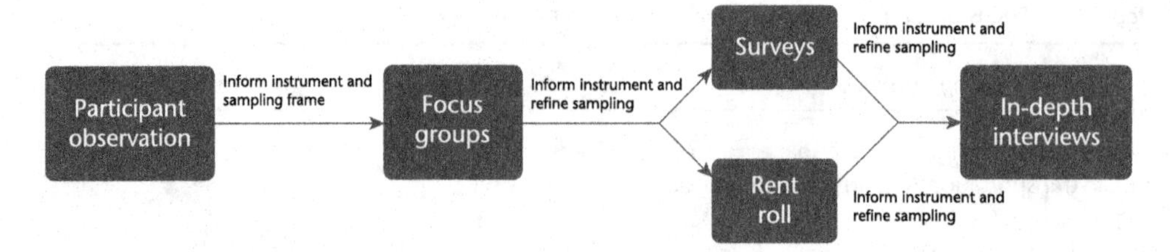

Figure 19.3 Data collection process: an exploratory sequential design
Source: Christensen et al. (2016)

19.3 Process for development of the green office building index

At this point, tenants and owners had expressed a need for something to supplement LEED and other eco-labels. The results of the focus groups defined the need for a tenant-based system. In order to ensure a rigorous, peer-reviewed process to incorporate both quantitative and qualitative input, the research team custom developed a product development and construct validation (PDCV) methodology (Christensen, Robinson, & Simons, 2016). This system follows the mixed-methods data collection and refinement process shown in Figure 19.3.

Historically, most meaningful market-driven indices tend to primarily or exclusively use quantitative data. Using qualitative feedback loops, such as interviews and focus groups, enhances the richness and applicability of a product. Few formal processes exist that real estate, finance, and urban economics researchers can use to formally merge qualitative and quantitative data. The PDCV methodology helps ensure fidelity of data during collection, and enhances the rigor and science behind a market-driven product with comprehensive stakeholder input.

The process includes ten stages of market-driven PDCV. The process is designed to integrate multiple stakeholder feedback to create an instrument that meets the needs of the market incorporating a 360-degree feedback loop. The ten phases are outlined in Table 19.2. Adopting the numbering system from this exhibit, we recognize the focus groups as Phases 1–2, and the following tenant surveys as Phases 3–6 (Part 1). The tenant survey is discussed next.

19.4 Tenant survey (tenant willingness to pay)

The focus groups were a foundation for an online survey of over 3,000 tenants in 19 Metropolitan Statistical Areas (MSAs) producing a response rate of nearly 25 percent. The survey was also rigorously tested for internal and external validity. Over half of the respondents were office managers, known to have been involved in lease discussions, and decision makers. Part of the rigorous testing was to ensure that no statistical differences were evident between the responses of leaders and their representatives; full results of those tests and other survey results are reported in Robinson, Simons, Lee, and Kern (2016).

Using the foundation of the focus groups, the survey asked respondents to rank their top nine of the 18 green building attributes. Results from the broader survey are similar to those of the focus groups. Table 19.3 shows how many respondents indicated that an attribute was in their top nine "most important" attributes of the 18 green features presented. The N column is the number of respondents who included the attribute in their top nine with a corresponding percentage of total possible. The Score column weights the number of times represented by the ordinal importance ranking (1–9) of the attribute. A variable that is ranked 1 by all respondents would be 100. The Score column jointly measures the frequency and importance of the rankings.

Table 19.2 Ten stages of market-driven product development and construct validation (PDCV)

Phase number and title	*Details*
Phase 1: Identify market need and develop common language	Through literature review and industry analysis, identify market need for proposed product (e.g. green index). Ensure commonality of language and uniform definition of terms among researchers, key informants, and market participants.
Phase 2: Solicit key customer input	Through focus groups, gain qualitative insight and quantitative data. Use to develop terminology for survey and identify primary attributes or foci of the market that need to be studied.
Phase 3: Design initial quantitative feedback instrument	Using information from initial phases, design initial stakeholder feedback instrument (e.g. survey) with careful attention to terminology. Include some open-ended items to provide qualitative feedback on instrument.
Phase 4: Pilot test initial feedback instrument	Pilot test feedback instrument with expert respondent group to test for content- and construct-validity. Additional data may also be gathered (e.g. focus group) to gain qualitative feedback to enhance instrument development. Small sample size.
Phase 5: Final design and field test of revised survey	Having received quantitative and qualitative data on initial feedback instrument, revise feedback instrument according to feedback. Sample size should be large enough to conduct statistically valid analysis of results using appropriate econometric tools. Qualitative feedback should be used to ensure construct- and content-validity.
Phase 6: Part 1: Validate feedback: Quantitative analysis phase	Analyze quantitative data to assess content-, criterion-, and construct-related validity. Appropriate econometric tools should be used to analyze survey responses.
Phase 6: Part 2: Collect additional data to demonstrate market relevance of product	Where possible, collect additional external data to demonstrate face-validity (i.e. market preference) and conduct relevant econometric analysis of data.
Phase 7: Part 1: Validate feedback: Qualitative analysis phase	Analyze qualitative data to address one or more of Greene, Caracelli, and Graham's (1989) purposes for mixing methods.
Phase 7: Part 2: Collect additional data to demonstrate market relevance of product	Where possible, incorporate additional input from key stakeholders who may adopt the product. This may be optimally done through a series of individual interviews.
Phase 8: Validate initial product design: Mixed-methods phase: Qualitative dominant crossover analyses	Conduct one of crossover analysis strategies to further address Greene et al. (1989). In a mixed-method sequential design, this phase may also involve collection of additional qualitative data, which should be integrated with previous data as part of an integrated data reduction process. Goal at this phase is to identify product design and market relevance.
Phase 9: Validate revised product design: Mixed-methods phase: Quantitative dominant crossover analyses	Builds on previous phase but emphasis is on quantitative dominant crossover analyses. In a mixed-method sequential design, this phase may also involve collection of additional quantitative data, which should be integrated with previous data as part of integrated data reduction process. Goal at this phase is to further validate product design and market relevance.

(continued)

Table 19.2 Ten stages of market-driven product development and construct validation (PDCV) *(continued)*

Phase number and title	*Details*
Phase 10: Pilot launch of product and evaluation of the market-driven product development process and product	Analysis of product using findings in Phases 6–9 to further refine product. Evaluate product via launch of a pilot program to assess key benefits and weaknesses of the product through industry launch. Based on evaluation data, product developer may need to return to earlier phases; like the IDCV process, the PDCV framework is also an iterative, cyclical process.

Source: Christensen et al. (2016)

Table 19.3 Difference between "preferred attribute" and "perceived as present" of green building features

Green building feature	*N*	*% Total (N=708)*	*Score*
Indoor air quality	659	93%	67.8
Access to natural light	627	89%	67.1
Recycling on-site	583	82%	39.2
Energy-efficient lighting	542	77%	40.0
Efficient electrical and gas use for heating and cooling	540	76%	43.7
Walking access to services and restaurants	526	74%	37.9
Comfortable temperature control system	518	73%	47.9
Public transportation nearby	425	60%	33.0
Fitness facility on-site	387	55%	25.9
Lease structure	304	43%	18.9
Green cleaning products used on-site	277	39%	15.0
Water conservation	274	39%	12.7
Energy Star designation	254	36%	14.2
LEED designation	211	30%	10.4
Shower on-site	181	26%	9.7
Bike racks on-site	111	16%	5.9
Electric car charging station	54	8%	2.4
Green roof	53	7%	2.1

Source: Robinson et al. (2016)

Indoor air quality and access to natural light rank as by far the most important and frequently mentioned attributes. Recycling, while frequently included in the top nine, demonstrated consistently lower importance. Comfortable temperature control systems, while less frequently mentioned, showed high importance levels when they were.

With an overview of tenant preferences, established developers, owners, and Tier 1 suppliers have particular interest in what tenants are willing to pay for each of these features. When describing the results of this study, we often use the cocktail party story. Developers usually want to do the right thing but without it costing them anything. They want to be able to go out to an event and tell their friends about how they helped the planet today: "Look at this beautiful green building we just built!" Most just want to do it in a financially profitable way. For example, if developers and investors know that increasing the natural light in a building earns them on average 1.33 percent more rent, then they can earn a par return by increasing window coverage. If tenants have expressed a willingness to pay an extra 1.06 percent for land close to public transit,

Table 19.4 Tenant willingness to pay for green building features (N=708)

Green building feature	*Average WTP (%)*	*% of respondents (≥ 2% of WTP)*
Access to natural light	1.33%	243 (34%)
Indoor air quality	1.29%	250 (35%)
Comfortable temperature control system	1.27%	187 (26%)
Lease structure	1.17%	111 (16%)
Efficient electrical and gas use for heating and cooling*	1.09%	175 (25%)
Walking access to services and restaurants	1.06%	158 (22%)
Public transportation nearby	1.06%	140 (20%)
Energy-efficient lighting*	1.03%	179 (25%)
Fitness facility on-site	0.98%	107 (15%)
Water conservation*	0.97%	79 (11%)
LEED designation	0.82%	47 (7%)
Shower on-site	0.78%	40 (6%)
Green roof	0.70%	11 (2%)
Recycling on-site	0.65%	110 (16%)
Energy Star designation	0.63%	53 (7%)
Bike racks on-site	0.54%	18 (3%)
Green cleaning products used on-site	0.42%	28 (4%)
Electric car charging station	0.41%	10 (1%)

Source: Robinson et al. (2016)

Note:* The assumption of 2 percent annual building operation savings is provided for the "Efficient electrical and gas use for heating and cooling," "Energy-efficient lighting," and "Water conservation" questions, based on the tenant's lease structure.

that can be factored in as well. The survey results report willingness to pay for each of the attributes, shown in detail in Table 19.4. The table shows the stated willingness to pay (WTP) of each respondent for their top nine "most important" sustainable building amenities. Respondents were asked how much more they would be willing to pay for a building with that amenity over one without. Respondents were given information regarding the current size of their lease, current market rates for leases, the dollar costs associated with percentage increases, and their current lease structure. The second column is the overall average WTP for the respondents. The third column is the number of responses indicating a 2 percent or greater WTP for that amenity.

The WTP column is a key take-away from this analysis. Admittedly, stated WTP does not always correlate 100 percent with revealed preferences in the market. However, preferences are fast evolving in the field of sustainability, and this survey represents the most cutting-edge data available, dated from summer 2014. Also, as a caveat, there are no outside sample frames of reference for statistical significance tests associated with these results.

The survey suggested that regional differences exist among preferences. For example, water conservation is less of a concern in the water-rich Great Lakes regions of the US. Increased HVAC systems efficiency appears to be more of a concern in hot climates that require greater air-conditioning loads. In cities with high density like Chicago, walkability is almost taken for granted, while in cities like Houston it is considered an important amenity. Walkability scores are based generally on access to goods, services, and public transit. Electric charging stations are more important on the West Coast where there is a higher penetration of electric cars. Although the survey suggested these regional differences, anecdotal reports from sustainability executives of large holding companies reveal little difference in their state-to-state treatment of sustainability issues. Still, tenant preferences are a fast-evolving area. In the summer of 2014, while California suffered a severe drought, water conservation came to the forefront of people's minds. However, since the drought passed, that apparent priority may have ebbed.

The next section discusses analysis of rent rolls, whose strength is revealed in price points in the market. However, their weakness is that they are the average preferences of leases signed in the last three to ten years, while the tenant survey only represents 2014 preferences.

19.5 Rent roll analysis (revealed willingness to pay)

Whereas the tenant surveys cover the stated WTP for green attributes, this section presents revealed preferences on the rent roll/supply side. Robinson, Simons, and Lee (forthcoming) reported on the effect of the presence of most of the 18 green office building features on office rents. The database includes 2,246 tenants in 197 buildings in 19 major US markets, provided by CBRE. The sample includes a reasonable distribution of LEED, Energy Star only, and non-certified buildings. The sample is primarily composed of and highly representative of institutional-grade real estate.

The unit of analysis for this section is lease rates for office tenants. Each model is a multivariate mixed-effects regression with a dependent variable of "lnrent" (the natural log of annual rent per square foot), including appropriate random effects for metropolitan area and building and fixed effects for lease level differences such as size, age, and so on; the controls are shown in Table 19.5. The green building attributes were superimposed onto the rent rolls through a large-scale survey of building managers. For a detailed review of how each variable was measured, see Robinson et al. (forthcoming).

Three sets of models are shown in Table 19.5. The models were constructed to test first the standalone impact of green eco-labels, then the effect of the bundle of attributes independent of green labels, and finally they were analyzed jointly. The first model showed that rent premiums for LEED and Energy Star are 18 percent and 4 percent, respectively. These levels are consistent with prior research and lend validity to the sample and testing methodology. The second model run omits the brands but has all remaining green features. Many variables are statistically significant, showing green premiums that are directionally appropriate (except bike rack, which has a counterintuitive sign).

Finally, the third, and arguably most complete, model includes all green variables. Results show that adding the full complement of features reduces the significance of the LEED brand variable by almost half (from 0.182 to 0.092), and the Energy Star variable becomes statistically insignificant when the other underlying green variables are considered. Eight of the green variables are statistically significant in this model, including access to natural light, public transportation nearby, walking access to services and restaurants, efficient electrical and gas use for heating and cooling, water conservation, and LEED designation.

As a final step, we introduced an additional set of models designed to estimate the standalone impact of each variable. Since it is not feasible to measure the univariate (or single) impact of a variable without appropriate geographic and building controls, these models are colloquially called semi-univariate. Detailed results are found in the source article but are summarized along with the results in Table 19.6.

Table 19.6 jointly looks at the semi-univariate and multivariate model results and applies a weight to each green variable. Since the multivariate results are considered stronger estimates of market impact, they are weighted more heavily. The variables are then ranked highest to lowest, considering statistical significance and parameter size, so they can be brought out of this analysis and merged with the tenant stated preference models.

The seven top ranking variables consistently show statistical significance, and their weighted parameter estimates (in natural log form) are used in full. The percentage influence on rent

Table 19.5 Multivariate mixed-effect regressions

Variable	*Model 1*	*Model 2*	*Model 3*
Energy Star	0.039*		–0.023
LEED	0.182***		0.092***
Access to services/CBD		0.040***	0.034***
Access to transit		0.072***	0.055***
Bike racks		–0.074***	–0.076***
Indoor air quality		0.004	0.000
Electric car charging station		0.082***	0.075***
Electricity efficiency		–0.004	–0.006
Fitness on-site		0.020	0.011
HVAC relative		0.065***	0.057***
Access to natural light		0.048***	0.070***
Recycling on-site		0.099***	0.054
Shower on-site		–0.014	–0.025
Water conservation		0.096***	0.075***
Intercept	4.564***	3.457***	3.940***
lnbldg_sf	–0.133***	–0.056***	–0.096***
Lnsf	0.002	0.005	0.003
Freeway access		0.085***	0.063***
Lease term	–0.001	–0.001	–0.001
Tenant floor level	0.000	0.001	0.001*
Renovated lobby	0.047**	0.057***	0.058***
Retail	0.123***	0.124***	0.117***
Annual occupancy rate	0.003***	0.003***	0.003***
Floors	0.008***	0.002***	0.004***
Lnage	0.054**	–0.060	–0.006
Lnage^2	–0.017	0.006	–0.0002
CBSA random effects	*Included*	*Included*	*Included*
AIC	524	480	449
BIC	494	428	393
Model N	1,845	1,841	1,841

Source: Robinson et al. (forthcoming)

Note: *, **, and *** represent statistical significance at 90 percent, 95 percent, and 99 percent levels, respectively.

factors of the variables ranked 8–14 is viewed as less certain since not all model specifications were statistically significant. The final factor (bike racks) had counter-intuitive results, and was largely discarded. The weighting assumptions are contained in Table 19.6's footnotes. Thus, the weighted premium column is the take-away from this section.

As a bridge to implementing this index, we next combined the tenant survey data with the rent roll data, and vetted it with 34 institutional property owners and other sustainability innovators. This is the focus of the next section.

19.6 Institutional owner input and feedback

The authors conducted separate structured interviews with 34 senior-level real estate portfolio managers and sustainability influencers through our own contacts and assistance from Dave Pogue of CBRE. We got their input about their strategic decision making regarding sustainability, their

Table 19.6 Rank order of relative effects on lease values

Green building feature	*Semi-univariate*	*Multivariate (Model 3)*	*Weighted premium*	*Overall rank*
LEED designation	0.136***	0.092***	0.103	1
Water conservation	0.134***	0.075***	0.090	2
Access to natural light	0.075***	0.070***	0.071	3
Efficient electrical and gas use for heating and cooling	0.094***	0.057***	0.066	4
Electric car charging station	0.033*	0.075***	0.065	5
Public transportation nearby	0.084***	0.055***	0.062	6
Walking access to services and restaurants	0.051***	0.034***	0.038	7
Recycling on-site	0.089***	0.054	0.022	8
Shower on-site	0.087***	–0.025	0.022	9
Fitness facility on-site	0.081***	0.011	0.020	10
Energy Star designation	0.065***	–0.023	0.016	11
Comfortable temperature control system	0.055**	n/a	0.014	12
Indoor air quality	0.045***	0.000	0.011	13
Energy-efficient lighting	0.020**	–0.006	0.005	14
Bike racks on-site	–0.025	–0.076***	0.000	15
Lease structure				

Source: Robinson et al. (forthcoming)

Note: Parameter estimates are weighted 25 percent from semi-univariate results and 75 percent from multivariate results as shown in Robinson et al. (forthcoming). Statistically insignificant and negative variables are considered zero value for the weighted premium. *, **, and *** represent statistical significance at 90 percent, 95 percent, and 99 percent levels, respectively.

operational and purchase processes, and the potential usefulness of a green office building index. Detailed results are included in a working paper (Christensen, Robinson, and Simons, 2017). The interview process was designed to achieve the strategic objectives outlined in Table 19.7.

The typical interviewee was an institutional investor/manager who controlled between 50 to 100 US office buildings. Most respondents reported internal use of some green practices; they additionally stated that buildings' green status (LEED or Energy Star) could affect acquisition

Table 19.7 Strategic objectives of interview process

Strategic objectives	*Management process*	*Purchase and renovation considerations*	*Interpretation of tenant feedback*	*Utility of a new index*
Decision drivers	Decision criteria	Impact on purchase decision	Tenant survey results impressions*	Define need for new index tool
Value creation	Data collection strategy	Impact on renovation decision	Rent roll regression results impressions**	Assess market response
Regionalization	Strategic implementation	Economic value added by green building features	Owner perspectives	Measure price elasticity

Source: Christensen et al. (2017)

Note: *Survey feedback on stated willingness to pay for building attributes from Robinson et al. (2016). **Regression output on revealed preferences for building attributes from Robinson et al. (forthcoming).

strategy. Some managers liked the idea of adding value by moving a building's status up the green totem pole. Many managers already reported to voluntary reporting initiatives such as GRESB (n.d.), and almost all monitored buildings for energy efficiency. They were less positive about tenant attitudes toward green: the consensus was that only the biggest tenants seemed to value it highly enough for it to be a necessary condition of leasing space. This perception is somewhat counter to the results of the survey but may have to do with willingness to pay for green labels. Perhaps smaller tenants, who do not benefit as much from CSR reporting, are more concerned with features than the current set of labels.

The primary decision driver for sustainable buildings for these institutional owners relies on some type of cost–benefit analysis. A key strategic objective that influences the decision process for both purchasing and retrofit of assets for virtually all interviewees was the creation of financial value. Less regionalization than expected was found, but participants did adjust for specific market needs.

The decision making of the majority of institutions, but by no means all, tended to be data driven. They tended to, at a minimum, track energy usage and often water and waste as well. Their data collection strategies often focused on "cumbersome" reporting requirements such as GRESB or LEED. Strategic focus on sustainability differed from firm to firm: whereas some had national policies and procedures in place, others tended to focus on a project-by-project process.

Many participants preferred to purchase buildings without eco-labels, seeing it as a potential opportunity to add value. Low-to-no-cost renovations, especially for those with national policies, were usually the first to be made. Other renovations tended to have paybacks under three years.[1] The economic value added by green building features was viewed as the other operational savings associated with "green" retrofits or, in a market that offers it, hope of a rent increase through eco-labeling.

Part of the process of the interview was to preview working results for the tenant survey and rent roll analyses described above. What participants saw was an early and simplified version of what is described in earlier sections. They generally thought that the results were consistent with their expectations. However, transit was frequently thought of as "undervalued" by the tenants. The participants found it interesting that indoor air quality was so important in the survey, but not so much in the rent roll results. The general consensus was that this may signal an opportunity to tell tenants about air quality.

Finally, over 80 percent of the managers indicated that they would be likely to use an index based on tenant demand, if it had a reasonable price for both the initial audit and ongoing annual charges. They stated that they would prefer the index to be transparent, have third-party verification, be presented on a 1–100 scale, and build on existing sustainability data, rather than forcing managers to "reinvent the wheel" with another onerous set of data demands.

19.7 An early form of the green office building index

Having defined the process and outlined the key elements of a new green office building index, this section builds on the foundation described above and reports on the calibration of the matrix. As of this writing, the work is in beta form (Simons, Robinson, & Lee, 2017) and has advanced from that presented at conference[2]; although the ultimate index may differ from that presented here, the foundations for it are all now published or accepted for publication. The index is intended to provide a market-driven green rating for US office buildings that complements existing systems like LEED to encompass a broader market spectrum of buildings. The matrix data are drawn from "supply-side" data from hedonic analysis of rent rolls and "demand-side" data from tenant surveys for individual green office building features.[3] The

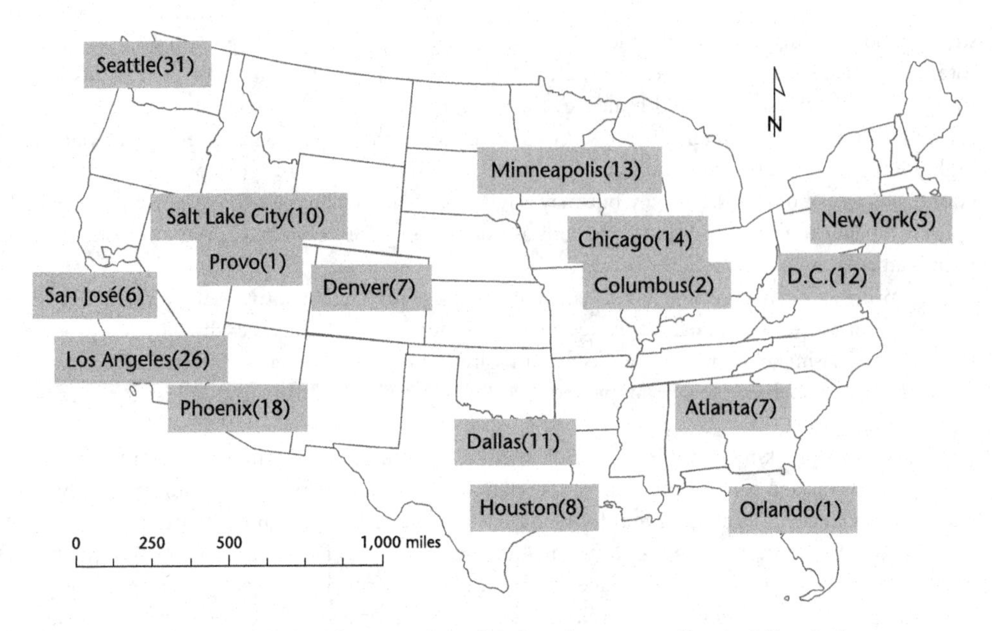

Figure 19.4 Map of office buildings used to develop the green office building index

Source: Simons et al. (2017)

matrix also includes qualitative input from institutional industry leaders on how it would be useful to them in practice.

This section details the creation and first steps toward implementation of the scoring system. A variety of proposed models, including some that separate out LEED and non-LEED buildings, are analyzed, discussed, and optimized, and tested on a captive sample of 197 office buildings, shown in Figure 19.4.

Note that at this point the unit of observation moves to buildings, not tenants or leases. This index can be likened to a credit score like FICO. Credit scores are generally generated based on the statistical analysis of a person's credit report, and credit bureaus maintain a record of a person's financial activities. In the same context, this new green building scoring system represents an assessment of the sustainability of an office building, and the likelihood that the building's owners and tenants will benefit from its sustainable attributes.

The next step is to experiment with different supply and demand weights. A total of 197 office buildings with complete data are used. Various models are applied to every building, which are then compared. The first four models combine data from all office buildings. Since LEED status is also a market driver, a series of LEED subgroups is created for one set of models. Consequently, the last three models aggregate the weights and scores as described for LEED buildings, for a total of ten models. Each of these index configurations gives a score for each building:

1 pure demand preference
2 50 percent (tenant) demand/50 percent (rent roll) supply
3 75 percent demand/25 percent supply
4 75 percent demand/25 percent supply weighted by preference
5 preference based, LEED and non-LEED buildings separately (5a and 5b)

6 LEED and non-LEED buildings separately, 75 percent demand/25 percent supply (6a and 6b)
7 LEED and non-LEED buildings separately, 75 percent demand/25 percent supply weighted by preference (7a and 7b).

The index generates raw scores with different ranges dependent on the maximum of the model being tested. Raw scores are then scaled on a 1–100 basis as a percentage of the maximum possible. After this step, the highest building scores are usually in the mid-to-upper 90s, with lows in the single digits, although some models generate highs only in the 80s. For most model presentations, the lowest scores are truncated at 30 (about 5 percent of the buildings fell into this category). The reason for truncations is practical more than anything else; the introduction of a baseline provides for easier adoption of a portfolio without "blacklisting" a building. The following tables show the output format for each model. Figures 19.5, 19.6 and 19.7 compare model results in graphic form. Observations are made on fit, and the number of possible models is winnowed down to the most parsimonious and consistent ones, for further use. Most interestingly, a reasonable model shows that about one-third of the non-LEED buildings score higher on this scoring system than the lower-scoring LEED building, and this is the primary market for this index.

The list of green weighting factors, winnowed down primarily from the tenant demand factors set forth in Table 19.4 and from the rent rolls weighted premium in Table 19.6, is now presented below in Table 19.8. Some models also use weights considering the ordinal preferences from Table 19.3. Thus, all ten models employ these factors, and have varying weights, as explained below.

To visualize better the inner workings of this index, we have taken one model (combined, weighted 75 percent tenant demand and 25 percent rent roll supply, and shown in Table 19.9)

Table 19.8 Factor estimates used in all ten models

Green building feature	*Survey (demand)*	*Rent rolls (supply)*	*Preference-based weight*
Energy Star designation	0.63%	1.60%	2.89%
LEED designation	0.82%	10.30%	2.12%
Indoor air quality	1.29%	1.10%	16.84%
Walking access to services and restaurants	1.06%	3.80%	7.71%
Access to natural light	1.33%	7.10%	13.65%
Water conservation*	0.97%	9.00%	2.58%
Fitness facility on-site	0.98%	2.00%	5.27%
Shower on-site	0.78%	2.20%	1.97%
Bike racks on-site	0.54%	0.00%	1.20%
Electric car charging station	0.41%	6.50%	0.49%
Recycling on-site	0.65%	2.20%	7.97%
Efficient electrical and gas use for heating and cooling*	1.09%	6.60%	8.89%
Energy-efficient lighting*	1.03%	0.50%	8.14%
Public transportation nearby	1.06%	6.20%	6.71%
Lease structure	1.17%	N/A	3.84%
Comfortable temperature control system	1.27%	1.13%	9.74%

Source: Simons et al. (2017)

Note:* The respondents were asked to value the rent impacts when the benefits of these features were set at 2 percent.

Table 19.9 Sample data set: top five, middle five, and bottom five buildings by score (Model 3)

	Top five buildings and associated metro					*Middle five buildings and associated metro*					*Bottom five buildings and associated metro*				
Green building feature	*San Francisco*	*Atlanta*	*Minneapolis*	*San Diego*	*San Francisco*	*Seattle*	*Houston*	*Chicago*	*Los Angeles*	*Dallas*	*Chicago*	*Chicago*	*Minneapolis*	*Chicago*	*Orlando*
ESTAR	100%	100%	100%	100%	100%	0%	0%	0%	100%	100%	0%	0%	0%	0%	0%
LEED	100%	100%	100%	100%	100%	0%	0%	0%	0%	0%	0%	0%	0%	0%	0%
Clean air	100%	100%	100%	100%	100%	100%	100%	33%	100%	33%	33%	33%	67%	33%	67%
Walkability	100%	59%	84%	93%	29%	95%	82%	30%	38%	75%	31%	19%	6%	40%	36%
Natural light	100%	100%	100%	75%	100%	100%	100%	100%	100%	75%	0%	100%	100%	100%	0%
Water conservation	100%	100%	100%	100%	100%	100%	100%	100%	100%	100%	100%	0%	0%	0%	0%
Fitness on-site	100%	100%	100%	0%	100%	0%	100%	0%	0%	0%	0%	0%	0%	0%	0%
Shower on-site	100%	100%	100%	100%	100%	0%	0%	0%	0%	0%	0%	0%	0%	0%	0%
Bike racks	100%	100%	100%	100%	100%	0%	100%	100%	100%	100%	0%	0%	0%	100%	0%
Elec. car station	100%	100%	0%	100%	100%	0%	0%	0%	100%	100%	0%	0%	0%	0%	0%
Recycling	100%	100%	100%	100%	100%	100%	100%	100%	100%	100%	100%	100%	100%	100%	0%
HVAC efficiency	100%	75%	100%	100%	75%	50%	50%	50%	50%	50%	25%	50%	50%	25%	25%
Electric efficiency	33%	100%	67%	100%	100%	33%	33%	100%	33%	33%	67%	33%	0%	67%	0%
Transit	100%	100%	100%	100%	0%	100%	0%	100%	0%	0%	0%	0%	0%	0%	0%
NNN	100%	0%	100%	0%	0%	0%	100%	100%	0%	100%	100%	100%	100%	0%	0%
Index (0–100)	93%	92%	86%	84%	81%	51%	51%	51%	51%	51%	27%	26%	26%	25%	8%

Source: Adapted from Simons et al. (2017)

Note: NNN = Triple net lease where the tenant pays most express.

for 15 of the 197 buildings in our data set. The top five buildings set are all high-scoring LEED buildings, the middle five are non-LEED but have a lot of green features, and the bottom five lack most or all green features. All the factors in this table are summed to produce a total index score. The top group's scores (normalized) are in the 90s and high 80s; the middle group's scores are in the 50s, and the lowest group has the default 30 score (indicating a score of less than 30).

Another consideration is the relative importance of individual green features across index models. This is shown in Figure 19.5. Some variables, like bike racks, have little or no weight across all specifications. Some variables, like indoor air quality, exhibit high importance over all model types. Some variables, like recycling, appear as zero in models where they are bundled with the LEED designation. The other variables show moderate variance but again demonstrate consistency in the models.

A quick look indicates that the most influential green features are access to natural light and LEED designation, followed by public transport nearby and efficient electrical and gas use for heating and cooling. Also influential are fitness facility on-site and indoor air quality. LEED impacts on the score are high in the subset of models with LEED-only buildings. In the absence of LEED designation, other green features – notably access to natural light, indoor air quality, fitness facility on-site, efficient electrical and gas use for heating and cooling, public transportation nearby, and energy-efficient lighting – go up in importance. Some variables – like Energy Star designation, electric car charging station, and bike racks on-site – never seem to have much of an impact on index score.

Moving to the index results themselves, the graphs in Figure 19.6 show the normalized combined results for all seven all-building index configurations, on a 1–100 scale, with lower-scoring buildings at the bottom left, and higher-ranking ones at the top right. In general, the index models appear to track similarly, indicating that the mix of the underlying model components (i.e. supply and demand weights) are not that influential on index scores. With the exception of Model 7, which shows a noticeable upturn when non-LEED buildings give way to LEED structures around Observation 140 (middle right of the line graph), the model lines are mostly parallel. This supports the idea that value may be added by separate modeling of LEED and non-LEED buildings.

The next and final set of indices break out LEED and non-LEED buildings separately. Figure 19.7 compares the building score between LEED and non-LEED buildings, modeled separately. The data set includes 58 buildings with LEED designation. The final model configuration, including preference weights and a 75 percent/25 percent demand/supply split, is shown separating LEED and non-LEED buildings. As expected, LEED buildings, indicated by the uppermost line, generally scored high in the models.

There are a total of 139 non-LEED buildings in the data set. The non-LEED buildings indicated by the black line have scores greater than or equal to the minimum score of LEED buildings. In the model shown, nearly 40 percent of the non-LEED buildings fall into this category. These buildings could clearly benefit from being able to advertise (to tenants and portfolio managers) that they are already in fact modestly green. These top-scoring non-LEED buildings likely have Energy Star certification. The bottom-scoring non-LEED buildings likely are not proximate to public transit, do not have Energy Star certification, and are potentially inferior in other key areas.

19.8 Conclusions

This chapter reports on the research into tenant demand, providing the backbone for the creation of a new green office building rating matrix in the US. It attempts to provide a market-driven

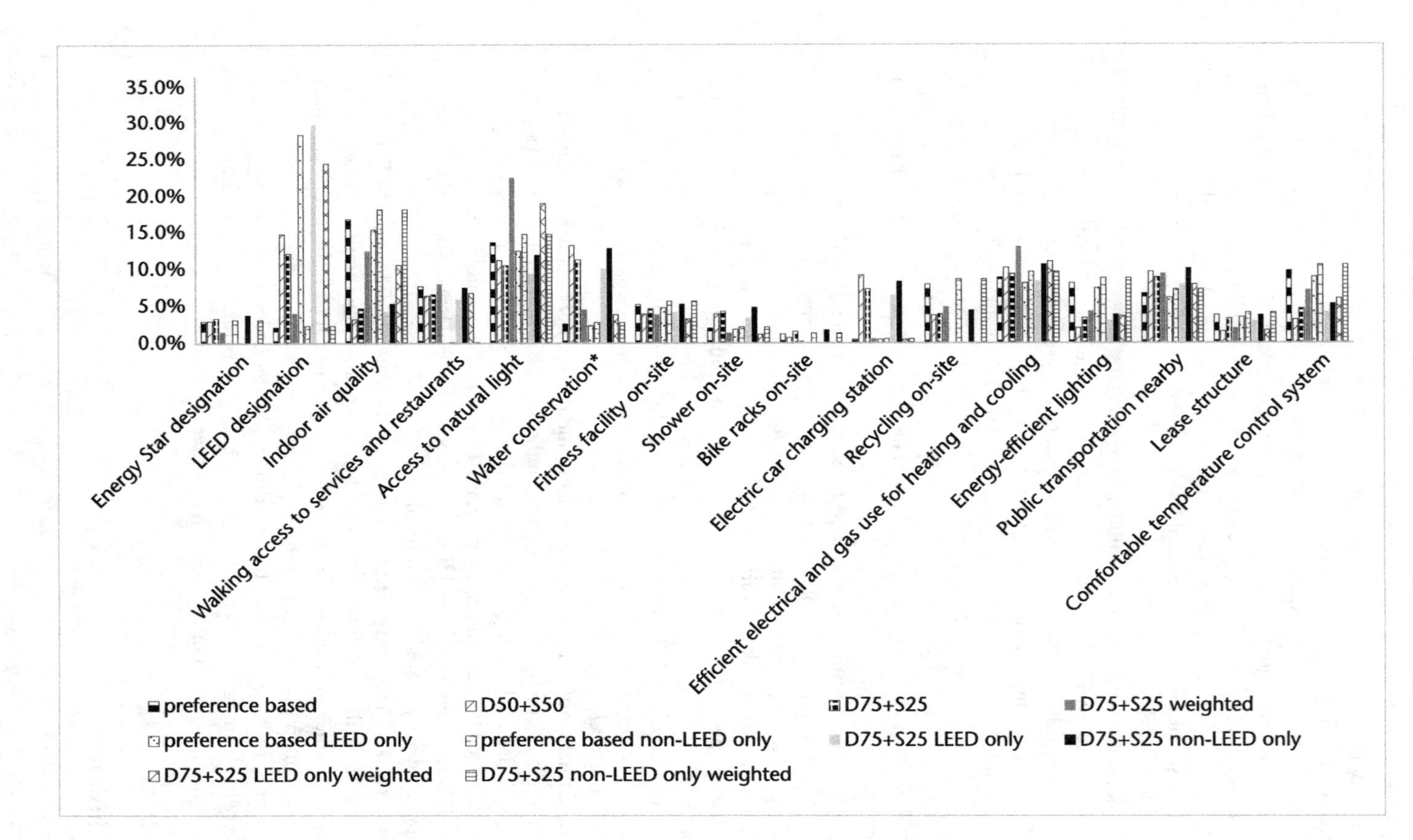

Figure 19.5 Contribution of green factors to overall index score

Source: Adapted from Simons et al. (2017)

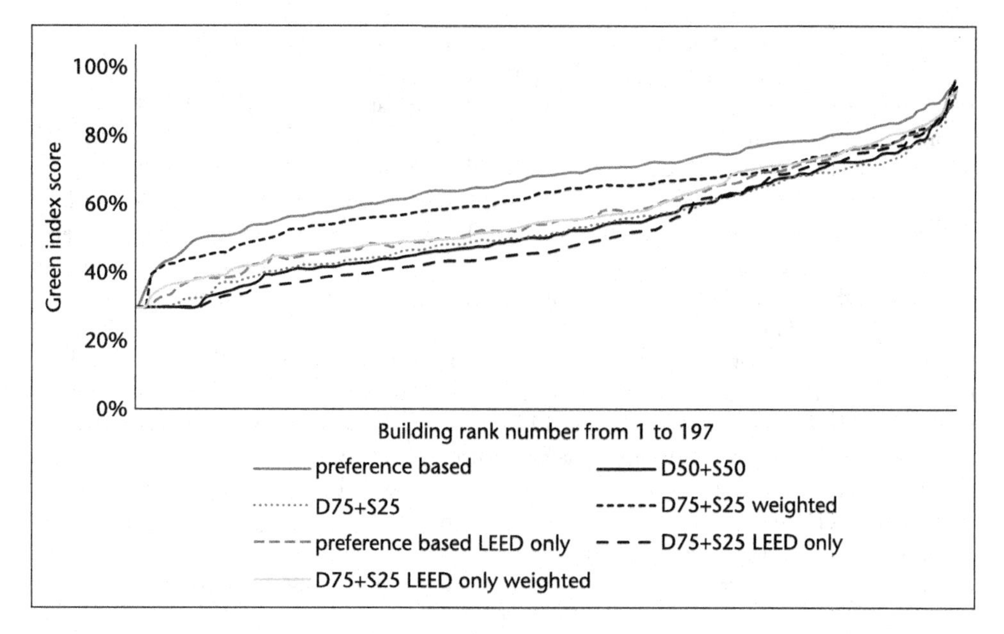

Figure 19.6 Normalized graphs of each index

Source: Simons et al. (2017)

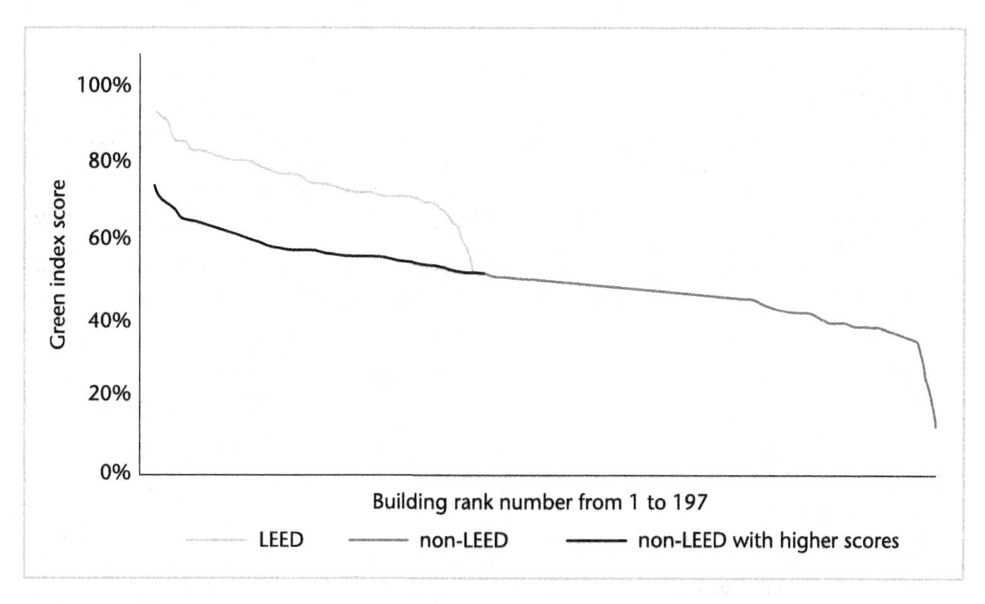

Figure 19.7 Comparison between LEED and non-LEED building index

Source: Simons et al. (2017)

Note: Non-LEED buildings with higher scores indicate those buildings with scores greater than or equal to the minimum score of LEED-designated buildings.

green rating for US office buildings that targets buildings below the level of LEED as well as to dissect feature and design preferences. The index data are drawn from demand-side data from tenant surveys for individual green office building features and from supply-side data through hedonic analysis of rent rolls. The matrix also includes qualitative input from institutional industry leaders on how the matrix could be useful to them. This chapter details the creation and first steps toward implementation of the scoring system. A variety of proposed models, including some that separate out LEED and non-LEED buildings, are analyzed, discussed, and optimized, and tested on a sample of 197 office buildings. A reasonable model shows that about one-third of the non-LEED buildings score higher on this scoring system than the lowest-scoring LEED building.

The results show that a substantial number of office buildings that offer green amenities currently desired by tenants are unable to signal the presence of those features through the current eco-labels. These high-scoring non-LEED office buildings could benefit from being able to advertise to tenants and portfolio managers that they are already modestly green. In other words, this analysis reveals that there are a substantial number of buildings that are not currently certified by LEED but that, according to this green office building scoring system, possess green practices or features and have a higher ranking score than lower-to-medium-scored LEED buildings. It provides the rationale for developing a new green building index that can be applied to any office building, not just limited to high-performance buildings.

Notes

1 Interestingly, many of these highly sophisticated organizations tended to rely on payback rather than Net Present Value or Internal Rate of Return calculations preferred by the financial community.
2 2016 American Real Estate Society Conference, Denver, Colorado.
3 The authors recognize that these are not supply and demand indicators in a traditional economic sense but have been frequently used for ease of exposition to the practitioner stakeholders.

References

Allen, J. G., MacNaughton, P., Satish, U., Santanam, S., Vallarino, J., & Spengler, J. D. (2015). *Associations of Cognitive Function Scores with Carbon Dioxide, Ventilation, and Volatile Organic Compound Exposures in Office Workers: A Controlled Exposure Study of Green and Conventional Office Environments.* Harvard TH Chan School of Public Health's Center for Health and the Global Environment, SUNY Upstate Medical University, and Syracuse University.

Christensen, P., Robinson, S., & Simons, R. (2016). The Application of Mixed Methods: Using a Crossover Analysis Strategy for Product Development in Real Estate. *Journal of Real Estate Literature, 24*(2), 429–451.

Christensen, P., Robinson, S., & Simons, R. (2017). Institutional Investor Motivation, Process, and Expectations for Sustainable Building Investment. Working paper.

Devine, A., & Kok, N. (2015). Green Certification and Building Performance: Implications for Tangibles and Intangibles. *The Journal of Portfolio Management, Special Edition: Real Estate*, 151–163.

Eichholtz, P., Kok, N., & Quigley, J. M. (2009). Why Do Companies Rent Green? Real Property and Corporate Social Responsibility. Program on Housing and Urban Policy Working Paper No.W09–004.

Elkington, J. (1997). *Cannibals with Forks: The Triple Bottom Line of 21st Century Business.* Oxford: Capstone.

Fuerst, F., & McAllister, P. (2011). Green Noise or Green Value? Measuring the Effects of Environmental Certification on Office Values. *Real Estate Economics, 39*(1), 45–69.

Greene, J., Caracelli, V., & Graham, W. (1989). Toward a Conceptual Framework for Mixed-Method Evaluation Designs. *Educational Evaluation and Policy Analysis, 11*(3), 255–274.

GRESB (n.d.). Accessed on 4 October 2017 at: https://gresb.com.

Heerwagen, J. (2000). Green Buildings, Organizational Success, and Occupant Productivity. *Building Research & Information, 28*(5), 353–367.

Janda, K., Bright, S., Patrick, J., Wilkinson, S., & Dixon, T. (2016). The Evolution of Green Leases: Towards Inter-Organizational Environmental Governance. *Building Research & Information, 44*(5-6), 660–674.

Kats, G., Alevantis, L., Berman, A., Mills, E., & Perlman, J. (2003). *The Costs and Financial Benefits of Green Building*. Washington, DC: Capital E.

Kok, N., Holtermans, R., & Pogue, D. (2015). *National Green Building Adoption Index*. Los Angeles, CA: CBRE.

Miller, N., Pogue, D., Gough, Q., & Davis, S. (2009). Green Buildings and Productivity. *The Journal of Sustainable Real Estate, 1*(1), 65–89.

Reichardt, A., Fuerst, F., Rottke, N., & Zietz, J. (2012). Sustainable Building Certification and the Rent Premium: A Panel Data Approach. *Journal of Real Estate Research, 34*(1), 99–126.

Reiss, D., & Costello, M. (2007). Location Alone Can't Retain Today's Workforce. *Financial Executive, 23*(7), 50–54.

Robinson, S., & McAllister, P. (2015). Heterogeneous Price Premiums in Sustainable Real Estate? An Investigation of the Relation between Value and Price Premiums. *Journal of Sustainable Real Estate*, 7(1), 1–20.

Robinson, S., Simons, R., & Lee, E. (forthcoming). Which Green Office Building Features Do Tenants Pay For? A Study of Observed Rental Effects. *Journal of Real Estate Research*.

Robinson, S., Simons, R., Lee, E., & Kern, A. (2016). Demand for Green Buildings: Office Tenants' Stated Willingness-to-Pay for Green Features. *Journal of Real Estate Research, 38*(3), 423–452.

Simons, R. A., Robinson, S., & Lee, E. (2014). Green Office Buildings: A Qualitative Exploration of Green Office Building Attributes. *The Journal of Sustainable Real Estate, 6*(2), 211–232.

Simons, R., Robinson, S., & Lee, E. (2017). A Market-Driven Green Office Building Index. Working paper.

Part 4
Redevelopment and adaptation

20

Multi-stakeholder partnership promotes sustainable housing supply chain

Connie Susilawati

20.0 Introduction

Since 2010, all new and renovated detached or attached residential buildings in Australia need to adopt the national six-star energy rating standard under the National Construction Code (NCC) (Wong, Susilawati, Miller and Mardiasmo, 2015a). The star rating correlates to the amount of energy required to maintain a comfortable temperature in homes all year round. For example, a six-star rated home in Brisbane requires 43 MJ/m^2.annum. A six-star-rated home indicates a good, but not outstanding, thermal performance (NatHERS, 2016). The six-star house will require less energy for artificial heating and cooling to be made comfortable than older houses and this will reduce ongoing electricity costs (Building Codes Queensland, 2014). In addition, multi-unit residential buildings are required to achieve a minimum five-star energy rating. All three levels of government in Australia (federal, state and local) have developed a comprehensive guideline to the design of five-star and six-star energy equivalent housing (Department of the Environment and Energy, 2016; Department of Housing and Public Works, 2016; Brisbane City Council, 2016), though these guidelines need continuous updating because of advancements in technology, building materials and construction methods, as well as environmental changes due to climate change.

The New South Wales government has used the Building Sustainability Index (BASIX) since 1 July 2004 to incorporate the NCC standard into the development application process (New South Wales Government, 2016). The Nationwide House Energy Rating Scheme (NatHERS) is administered by the Commonwealth Government, Department of Industry, Innovation and Science, and is used by local governments (Commonwealth Government, 2016). The NatHERS provides homes with an energy efficiency rating out of ten, based on their design or their potential heating and cooling energy use. A high-rated (ten-star) house allows occupants to feel comfortable without artificial cooling or heating (NatHERS 2016). Energy-efficient houses use a passive design that takes advantage of the climate to maintain a comfortable temperature range in the home, thereby reducing the need for auxiliary heating or cooling (Australian Government, 2016; Your Home, 2016). NatHERS divides Australia into 69 climate zones (the country has a diverse climate) to model the heating and cooling requirements for an Australian home and to show the maximum energy loads for each star rating.

Well-designed houses may increase the energy efficiency and performance of a house; however, this information is neither measured nor provided to stakeholders. This is a problem because it creates a knowledge gap between the design and operation of sustainable housing. The focus of this chapter is to propose the incorporation of building information into the existing property database prior to housing transactions, so it will bridge the knowledge gap in the sustainable housing supply chain. The existing property databases used in Australia are CoreLogic RP Data and PriceFinder, both owned by private companies that collect data from multiple sources. CoreLogic RP Data collects data from government sources, major real estate franchises, website portals, developers and advertisements in newspapers (CoreLogic RP Data, 2016). PriceFinder collects property data from multiple sources including property ownership, phone numbers, property zoning, title information and over 30 years of sales history and real-time auction results (PriceFinder, 2016). The information distribution flow requires the active involvement of all stakeholders, not just in the development process, but the post-occupancy period as well. Finally, well-informed buyers are thought to be likely to promote and create demand for sustainable housing, not just for new homes, but also for existing homes. The hypothesis is that, other things being equal, when all new houses meet minimum energy efficiency requirements, sustainable houses will become the norm and houses without sustainable features will be discounted in value.

20.1 Energy-efficient housing: What does it mean for consumers?

An energy-efficient house, in theory, will need less artificial heating and cooling, so occupiers will pay less for energy without compromising their thermal comfort. The savings made from energy-efficient housing may be offset by 10–30 per cent because of the rebound effect as a result of more disposable income (Saunders, 2000; Swan, Wetherill and Abbott, 2010). The most energy-efficient home uses more than theoretical consumption (Majcen, Itard and Visscher, 2013).

In Victoria, six-star homes are projected to use 24 per cent less energy through heating and cooling compared to five-star homes (Victorian Building Authority, 2016). The figures for Australian household expenditure in 2009 and 2010 (ABS 6530, 2011) show that energy cost is only 3 per cent of their weekly household bills. The impact of reduced energy bills is significant for low-income households as the bills arrive every three months (around AUD468 based on AUD39 per week in 2012 (ABS 4670, 2013)).

Passive energy-efficient design will produce an energy-efficient house, which includes orientation, cross ventilation, insulation and shading. The energy-efficient requirements relate to the building envelope, lighting, hot water systems and heating and cooling devices. In addition, energy-efficient design involves using renewable energy such as solar photovoltaic (PV) system. The water-efficient requirements include the appliances and rainwater-harvesting tanks.

Designers and contractors provide evidence that the house plans meet the NCC energy efficiency requirements. The NCC-required documents are building plans and specifications, certification of star-rated energy from NatHERS or BASIX and certification of air movement, building sealing, glazing, hot water system insulation and insulation of services and lighting (Harrington and Miller, 2015). Although the documents are required by legislation, not all are submitted and kept in the local government database. The energy efficiency assessment is based on simulated data from the building plan, not on the actual energy performance of the building post-construction. Martinaitis, Zavadskas, Motuziene and Vilutiene (2015) stated that an energy performance assessment discrepancy of 14–21 per cent between design and performance is caused by human behaviour and the preference of occupants.

Germany's sustainable building quality scheme, which is applicable to all new and refurbished buildings (offices, retail and residential), is categorised into six quality labels: ecological, economical, social-cultural and functional, technical, process and location (Bock, Linner and Hartmann, 2010; Lützkendorf and Lorenz, 2011). In the ARC linkage research project led by Queensland University of Technology, the German features have been adapted to fit Australian conditions and ten features are used for five categories: spatial planning, occupant health and safety, occupant comfort, operation and services and building durability (Miller, Stenton, Worsley and Wuersching, 2014; Wong, Susilawati, Miller and Mardiasmo, 2015b). The physical category is spatial planning, which relates to the physical aspect of the house, such as house and room size, site area, coverage, land use, number of bedrooms/bathrooms, layout, ceiling height and access to modes of transport. The features in spatial planning, such as site area and number of bathrooms, are recorded in the real estate database, such as CoreLogic RP Data (Wong, Susilawati, Miller and Mardiasmo, 2016b).

A pilot study was conducted on a small sample of repeated sales of a suburb in Townsville, Queensland, Australia, which has a limited number of houses for sale advertising sustainable features (Wong, Susilawati, Miller and Mardiasmo, 2016a). Sustainable features were manually extracted from the advertisements recorded in the CoreLogic RP Data property transaction database (Wong et al., 2016a). Real estate agents may advertise the existence of visible features, such as solar panels and water tanks, as sustainable features, but a lack of information on the size and capacity will not make the information very useful to compare for potential buyers.

Two comparative analyses, listing periods and house sizes, were conducted for the existence of sustainable features and their impact on sales price. The sales prior to 2009 have shown that the listing time is longer for houses with sustainable features. Potential buyers have a lower awareness of sustainable features prior to the introduction of the mandatory sustainability declaration on 1 January 2010 (Queensland Government, 2012) and the Queensland Government solar bonus scheme (Department of Energy and Water Supply, 2015). The median size of houses with sustainable features is now higher and the median house price is 10 per cent higher than that of properties without sustainable-related features (Wong et al., 2016a). A smaller-sized house is targeted at low-to-medium income buyers, for whom price is a very important purchasing decision. The market perception is that sustainable housing is more expensive because of the additional capital expenditure required to install solar panels, insulation and water tanks, therefore increasing the house price (Susilawati and Miller, 2013). Furthermore, cost saving on future energy bills is not considered sufficiently significant by buyers to compensate for the higher price of an energy-efficient house (Eves and Kippes, 2010; Hurst, 2012).

Although the existing property databases (CoreLogic RP Data and PriceFinder) include multiple sources of data, they do not include information collected by local governments during the building approval process. Local governments capture some of the sustainable housing feature requirements during the development approval and building approval process. However, the information is not available in a searchable format, only as scanned copies of applications/ proposals and supporting images.

20.2 Mismatched problems of energy-efficient house design and its performance

The evaluation of whether or not a building meets the minimum six-star energy equivalence of building standard requirements is conducted at the design stage. The energy efficiency performance of a house or building is checked using building simulation software, such as BERS

Pro, AccuRate or FirstRate5, at the design stage, as required by BASIX and NatHERS. NCC (2016) states that five main categories of energy efficiency requirements relate to the heating and cooling loads:

1 building fabric that impacts on thermal performance;
2 external glazing;
3 building sealing;
4 air movement; and
5 energy efficiency of main building services.

Although almost all local governments use either private certifiers or local government certifiers to conduct onsite inspections, not every building and every stage are checked (Harrington and Miller, 2015). In addition, the information is not easily accessible by the subsequent owners.

In Queensland, each new house is required to be inspected by a building certifier at least four building stages (Queensland Department of Housing and Public Works, 2016). The first stage for single detached housing is the foundation and excavation stage, before the footings are poured. The second stage is the slab stage, before the concrete is poured. The third stage is the frame stage, before the cladding is installed and before the wall cavities are filled. The final stage is the comprehensive site works, at which point drainage, fire safety, and energy and water efficiency are assessed. The energy efficiency requirements, as per building development approval, are energy-efficient lighting and hot water supply systems.

Almost all the local governments in Australia require a building certifier to check that the building has the energy efficiency requirements that have been proposed in the building permit application. However, the final inspection conducted by a building certifier will not guarantee that the building will have the same performance as the simulated outcomes at the design stage; for example, the quality of building sealing, which impacts on air movement in the house, may not be as high in reality as at the design stage. The simulation at the design stage uses the assumption that the contractor will use specific building materials and a specific construction process.

Harrington and Miller (2015) introduced an Electronic Building Passport to be used to keep the building information in one database that is searchable by a local government as well as the owner of the building. Ideally, any changes made at any stage of the building process need to be updated in the same portal (Miller and Lützkendorf, 2016). The same building information portal will have the 'building plan' from the planning stage, 'as-built' information after the completion of the construction stage and 'as-occupied' information from the post-occupancy stage.

The building information management (BIM) may store this information, if available, but many house builders do not use BIM (Fabris, 2010). Figure 20.1 shows how the building information flows from the designer to the end users. This will provide the occupier of the building with a 'user manual' for the building. It is important for the user or occupier to conduct an as-built audit of performance measurement and the energy-monitoring system.

Potential homebuyers will need to conduct building and pest inspections to check the as-built building performance. Building and pest inspections are usually conducted by one company prior to finalising the property purchase. The purpose of the inspection is mainly soundness of the building (structural risk) and any other fault that may increase the risk of high repair and maintenance costs in the short term. The inspection is based on the condition, without checking whether the design has impacted performance, as the inspector will not have access to the design information. There is no penalty if the house does not meet the energy efficiency requirements. The house does not have to be energy efficient; as long as it is safe and structurally sound, the

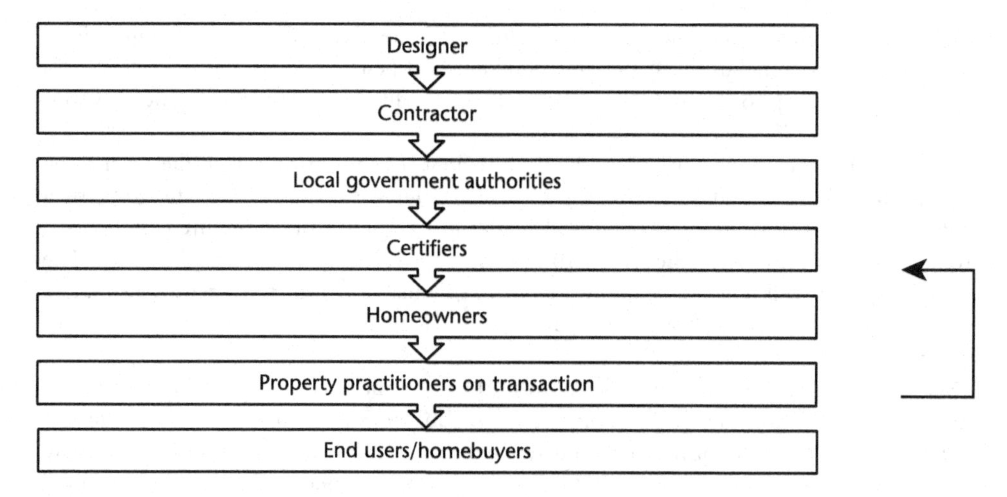

Figure 20.1 Building information flow from the designer to the end users

homeowner does not have to refurbish it to improve energy efficiency. In the exception of an auction, potential homebuyers will have an opportunity to have a building inspection done by an assigned professional inspector prior to making their purchasing decision. There is no requirement to do a full survey when changing occupation, only an entry and exit condition checklist conducted by the occupiers or real estate agent. In addition, lenders do not need to include energy efficiency requirements in the home loan evaluation.

20.3 How can building information prior to the housing transaction process be established?

One of the important stages shown in Figure 20.1 is the housing transaction process that will continue, using the loop of end users/homebuyers who will be the owners for the next housing transaction sale. The building information flow during the transaction process requires facilitation by property practitioners. Figure 20.2 shows the stakeholders involved during the housing transaction process; the seller will provide information required by the buyer and/or the real estate agent, valuer, mortgager/financier, building and pest inspector and insurer. The real estate agents mainly work for the seller to promote and process a sale contract. They have interaction

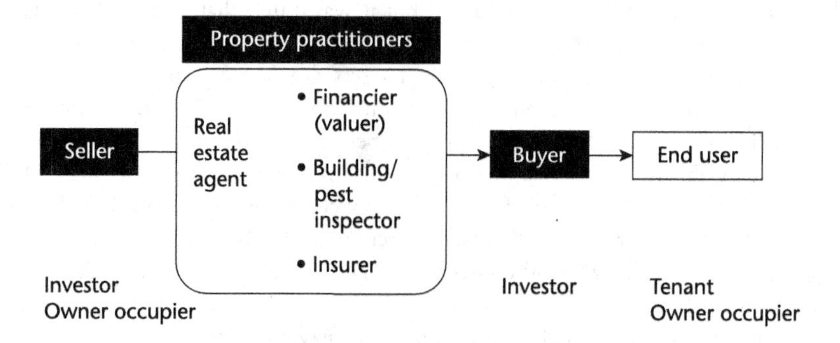

Figure 20.2 Building information flow during the housing transaction process

with both buyers and sellers. The other stakeholders who are involved in housing transactions, who will relate to homebuyers, are insurers, building and pest inspectors and financiers, who may engage a valuer prior to approving a home loan. A valuer is liable to the financier who is the client.

The property practitioners in Figure 20.2 can request the information from the seller to be provided to the buyer. In addition, the buyer can demand information that is related to thermal comfort and their health and safety. The availability and accessibility of the required information throughout the process, from the owner to the end user, is key to its successful transfer. The information flows will increase client awareness of sustainable housing. Currently, real estate agents receive little evidence of growing interest in sustainable features among their clients (Wong et al., 2016b).

Mandatory Energy Performance Certificates (EPCs) are included in houses for sale and for rent across Europe, from design to implementation (Buildings Performance Institute Europe, 2010, 2014). The EPC is information included to enhance decision-making criteria in real estate transactions, which can be a "market tool to create demand for energy efficiency in buildings" (Buildings Performance Institute Europe, 2014).

The Australian Capital Territory has introduced a mandatory disclosure scheme of energy efficiency rating for new housing since 1999, which is measured by an accredited energy rating assessor. The Queensland State Government introduced it on 1 January 2010, but it was abolished on 1 July 2012 (Queensland Government, 2012) to reduce "real estate red tape" (Queensland Parliamentary, 2012).

The Queensland mandatory sustainability declaration is a checklist signed by the seller about the sustainability features relating to energy, water and access/safety of a home for sale. In the first year of introduction, the mandatory sustainability declaration has raised awareness in homeowners of the importance of sustainability features (Bryant and Eves, 2011). However, the majority of the real estate agents in Queensland surveyed stated that sustainability had not become a purchasing decision criterion for homebuyers in Queensland. There was anecdotal evidence that home sales that had a sustainability declaration completed achieved an increased sale price of up to 3 percent on average (PBC, 2016a).

When homebuyers are aware of the benefits of sustainable features, they will either discount non-sustainable housing or pay more for sustainable housing. Some researchers have claimed that the homebuyers pay price premiums for high-rated houses (in the Netherlands, Brounen and Kok, 2011; in Germany, Cajias and Piazolo, 2013; and in Ireland, Hyland, Lyons and Lyons, 2013). Bruegge, Carrion-Flores and Pope (2016) find that homeowners are willing to pay a premium for new Energy Star homes in Florida, but there is no difference of resale price between an Energy Star home and a traditional home. The builders promote the energy efficiency label when they sell brand new Energy Star homes, but it was found that homeowners do not do so when reselling their Energy Star homes.

The introduction of energy efficiency rating systems should have an effect on the level of house prices, based on the theory of buyer perception (Hurst, 2012). Fuerst, McAllister, Nanda and Wyatt (2016) stated that there are differences between investors (landlords) and owner-occupiers in their willingness to pay a premium for the potential of energy savings. Occupiers will be concerned with thermal comfort and the energy cost, which does not directly impact an investor. An owner-occupier may wish to wish to renovate a house to increase its performance, but Bruegge et al. (2016) show that the energy efficiency of a home will have little impact on future resale value.

Some researchers have endorsed the importance of having a 'user manual' (an Electronic Building Passport) to improve occupants' understanding of their home's energy efficiency and

Table 20.1 Main barriers to pilot project for Electronic Building Passports.

Categories	*Barriers*
Lack of regulatory enforcement	• Lack of regulation • Lack of pull policies • Weak enforcement fails to ensure that the performance of a house will be consistent with the simulated performance at the design stage
Lack of consumer awareness	• Poor consumer understanding of energy-efficient homes • Poor consumer understanding of the regulatory system
Poor quality of information collection and storage	• Information collected using a non-systematic approach without clear accountability and responsibility for information collection and storage • Privacy concerns • Low demand for good documentation processes and information

Source: Adapted from Harrington and Miller (2015)

performance (Harrington and Miller, 2015). However, an Electronic Building Passport is not readily available during the housing transfer process because currently it is not available in the housing transaction database (Wong et al., 2016a). Although the potential buyers may not understand the advantage of having sustainable features in their future homes, providing a 'user manual' in housing transactions will increase consumer awareness and allow buyers to make better-informed purchasing decisions.

Energy-efficient housing may not have a significant impact on energy consumption if the occupants do not maximise sustainable characteristics (Hurst, 2012). Almost no buyers and occupiers will have as-built documents or a 'user manual' for the effective operation and maintenance of the building. In comparison, when consumers buy any appliances or tools, they receive a user manual and warranty notice.

The user manual for a building is called an Electronic Building Passport, or building files in Finland, the Netherlands, Germany, the UK and France (Harrington and Miller, 2015). The Electronic Building Passport pilot project encountered a lack of central storage of building files in Australia, and it will reduce compliance costs via electronic document lodgement and storage (Harrington and Miller, 2015). The process will facilitate audits, compliance, consistency and certainty of regulatory requirements, voluntary disclosure and documentation accountability. The information accessibility will increase industry and consumer awareness, which will drive value chains and produce more quality, safe and comfortable buildings.

The main information barriers to a pilot project for the introduction of Electronic Building Passports in Australia (Harrington and Miller, 2015) include the lack of regulatory enforcement, the lack of consumer awareness, poor quality of information collected and storage (Table 20.1). Using an online portal will assist central storage to store all information during housing transactions.

20.4 Multi-stakeholder partnership on information distribution to promote sustainable housing

Currently, designers, contractors, local governments and building certifiers are collecting and storing building information for their own purposes. Local governments demand many documents during the development approval and building approval process, the information mainly being provided by designers and contractors. The information is available to the public during the approval process. All the development approval and building approval certificates are based on the design or plan. Government officers or private building certifiers check the plan, but not every detail. Moreover, there is no building performance check after the building is completed to compare with simulation results. Although local governments and certifiers collect building information, homeowners do not have access to complete information on their own house. Limited building plan information is required to be handed over on sale, especially during the resale process.

Real estate agents will select the most important features to promote their houses for sale or for rent (Hurst and Wilkinson, 2015). Although they will select housing features that are unique and add value to the house, real estate agents generally do not include sustainable housing features among these. Real estate agents and financiers are concerned about the compliance of the building, such as having current smoke alarms, rather than the sustainable features that improve consumer comfort. Wong et al. (2016b) conclude that potential buyers are not interested and not many demand information on the sustainable features.

When potential buyers sign a contract for purchasing agreement, they can add two conditions: subject to financial approval and building and pest inspection. Financiers are more interested in potential buyers' ability to pay back the loan from their equity and income to guarantee their repayments. Not all financiers will ask valuers to do an appraisal of the house. The desktop valuation that the financier has used will only use the available information in real estate databases, such as CoreLogic RP Data or PriceFinder. If a valuer is engaged by financiers, he or she will use the same data plus a physical site inspection, which will capture only observable items. The actual energy efficiency performance of the house will not be included in the evaluation. Potential buyers will need to buy building insurance as part of the requirement by the financiers. Insurers will only be concerned with features that increase the risk of house fires or other disasters, not the thermal comfort of occupants.

Similar to the valuer, the building inspector at the time of the sales transaction will conduct a visual inspection of the minimum work to ensure the structural integrity of the building. Some building and pest inspectors will use additional equipment such as moisture meters, thermal imaging cameras, acoustic emissions devices and Termatrac to produce comprehensive reports to evaluate the property's structural integrity and identify replacement and repair requirements (PBC, 2016b). Termatrac is used to detect the presence of termites using termite detection radar, a thermal sensor with a laser guide and a moisture sensor (Termatrac, 2016). A moisture meter is used to check humidity in bathroom and kitchen areas. The certified infrared thermal imaging cameras produce an image of heat patterns of anomalies beyond visual inspection, such as below the surface of a floor, behind walls and above the ceiling.

There is no requirement to check the energy efficiency of the building and there is no exit clause for cancelling an agreement because the house does not have sustainable features, especially when buying an existing house (rather than a new home). Some building inspectors can provide additional services by conducting sustainability inspections, indoor air quality testing and energy audits to show missing insulation, energy loss calculation and air infiltration (PBC, 2016b). The thermal images will help to show missing insulation and any leaking on the ceilings, below carpets or behind the walls.

The most significant motivators for people considering building a new energy-efficient house are financial incentives, both the short-term rewards of government subsidies and the long-term financial benefits from cost savings on future energy bills (Raisbeck and Wardlaw, 2009). Raisbeck and Wardlaw (2009) conducted a survey in 2008 of randomly selected people at a Melbourne Home Show who were considering building their new home in the next 12 months. The research showed that younger people were more interested in the government subsidies compared to the lifestyle and health benefits. Nearly 90 per cent of people under 30 years old were primarily motivated by government subsidies, while more than 80 per cent of people over 40 years old were primarily motivated by lifestyle and health benefits.

New homebuyers generally have limited knowledge of sustainability opportunities in new homes and it is perceived as a lack of awareness and interest in sustainability initiatives by property professionals (Warren-Myers and Heywood, 2016). Adding an extra process to the existing data collection will minimise the risk and cost of promoting sustainable features. The local government submit information collected during the approval stage in the Electronic Building Passport. This information on sustainability features will be available to anyone. All sustainable feature aspects are recorded in the database, will be available to the potential homebuyers and can be included in the price determination of the house.

Figure 20.3 lists all stakeholders involved in using and promoting sustainable features in the housing supply chain. The larger the font size used for a particular stakeholder, the more frequently that stakeholder was mentioned for their involvement in promoting sustainable housing. Government intervention has been proven to speed up the process of the promotion of sustainable housing. A combination of the introduction of various enablers, such as changes to regulations, building standards and market stimulation incentives programmes, helps to make the economic advantages evident, makes an exemplar of public sustainable building and, increasingly, promotes the acceptance of social and environmental corporate responsibility.

The key stakeholders are the owner/seller/tenant and the real estate agent (Figure 20.3). Designers, developers and contractors can help to implement sustainable housing if the owner of the building is required to do so. In a housing transaction, the seller and the investor have important roles in passing on building information to the new owner. Financiers, insurers and valuers do not actively look for information on sustainable features. The RenoValue project is developing a training toolkit for property valuers across the EU on how to factor sustainability into the valuation process (RenoValue, 2016). RenoValue integrate sustainability into four valuation methodologies (comparable sales method, investment method, replacement cost approach and discounted cash flow). Other stakeholders still have a minimum role in promoting sustainable housing.

The main challenge of low sustainable housing supply is not having obvious additional value with a higher cost. In Ireland, Hyland, Lyons and Lyons (2013) show that energy efficiency has

Figure 20.3 Stakeholders involved in the sustainable housing supply chain

Note: Realestate = real estate agent

a strong positive effect on sales prices of residential property and not as strong a positive effect on rental prices. In addition, the same positive association of energy performance and sales price is found in England; however, the association is stronger for attached dwellings than for detached and semi-detached dwellings (Fuerst, McAllister, Nanda and Wyatt, 2015). Qualitative descriptions of the direct and indirect benefits of sustainable buildings are illustrated using building or process quality and effects for the developer/owner/investor/landlord, the user/tenant, society and the environment (Lützkendorf and Lorenz, 2010).

Governments can be enablers by providing incentives for renewable energy generation investment, and can also develop policies and guidelines to support the implementation of sustainable housing. A minimum standard is integrated in six-star energy requirements. For existing buildings, when the government abolished the solar bonus scheme, there was limited incentive for a homeowner to install PV on his or her roof (Department of Energy and Water Supply, 2015). In addition, more protection is needed for the investor to be able to gain benefit from solar energy generation investment. The feed-in tariff agreement will be broken when occupiers change. If investors keep the electricity connection under their names to retain the benefit of solar generation, the electricity billing process can get complicated.

Finally, multi-stakeholder collaboration and involvement in information distribution will enhance the promotion of sustainable housing features and increase the demand for sustainable housing. The local government can initiate the information distribution by providing homeowners with access to their house's plans that were submitted for building permit/approval. The new homeowners can update the initial information when they conduct a building inspection, which will have information on the snapshot of the building's performance during the transaction process, as well as a continuous performance measurement during occupancy periods. Any upgrades to the building will be recorded to continue updating the building's information. The availability and accessibility of up-to-date information on sustainable features on housing will break the vicious circle of blame, and may increase the value of energy-efficient houses or decrease the value of non-energy-efficient houses.

20.5 Conclusions

Home occupiers, even of new energy-efficient houses, are not well informed about how to effectively operate and maintain their houses. Home occupiers may use heating and cooling more cost effectively if they know the building material and system installed in their home. Regular maintenance will retain the house's comfort and value. Sellers provide neither information on a building's sustainable features prior to a house transaction, nor a 'user manual' (Electronic Building Passport) after the transfer of title to the buyers. Information on new and renovated housing design is available and kept by local governments, designers, contractors and certifiers, but is not accessible for potential buyers during their decision-making process. Buyer awareness of potential value will help to drive value chains and promote sustainable housing as a desirable and more valuable asset.

This pilot project on the Electronic Building Passport facilitates the accessibility of complete and accurate building information in a useful/understandable format at low cost (Harrington and Miller, 2015). Providing building information in one Electronic Building Passport database will improve documentation accountability, reduce compliance cost and facilitate audits, not just as design-stage certification, but also as-built and as-occupied information. Providing access to available and reliable information is recommended as the cost-effective way to gather building information. This information will enrich the existing house sales transaction information that is used by real estate agents, valuers and home loan institutions to validate the market value of houses for sale. It will increase consumer awareness of the impact of sustainable features on house price.

Governments need to facilitate the distribution of building information across multiple stakeholders to bridge the knowledge gap in the whole life cycle of the sustainable housing supply chain. The information collected for compliance purposes is currently often seen as a barrier rather than an opportunity to improve occupiers' comfort, reduce future energy bills and increase property value. Furthermore, there is a lack of incentives for investors, developers and homeowners to add expensive sustainable features, such as solar PV, as there will be very limited financial benefits either for the landlord or tenants. The existence of historical data on the building will allow effective renovation and additional investment in existing houses. Wilkinson, Sayce and Christensen (2015) suggested how to add to the value of property by incorporating sustainable features. If many buyers demand sustainable housing, the price for houses without sustainable features will be discounted, as buyers need to incur additional costs to improve their comfort and the quality of their housing.

Acknowledgements

The Electronic Building Passport pilot project was funded by the South Australia Department of State Development as part of a larger National Energy Efficiency Building Project and conducted in collaboration with pitt&sherry and Queensland University of Technology.

The ARC Linkage Project entitled "From innovators to mainstream market: a toolkit for transforming Australian housing and maximising sustainability outcomes for stakeholders" was funded by the Australian Research Council, Queensland University of Technology, Karlsruhe Institute of Technology, Bondor, Electriq Pty Ltd, Finlay Homes Pty Ltd, Townsville City Council and PRDNationwide.

References

ABS 6530 (2011). *Household Expenditure Survey*, Australia.

ABS 4670 (2013). *Household Energy Consumption Survey*, Australia.

Australian Government (2016). Your energy savings: Understand heating and cooling. Retrieved from: http://yourenergysavings.gov.au/energy/heating-cooling/understand-heating-cooling.

Bock, T., Linner, T. and Hartmann, K. (2010). Sustainable/'green' buildings in Germany: Status quo, advantages and future strategies. *Building*, 54(12), 14–19.

Brisbane City Council (2016). Building sustainable homes. Retrieved from www.brisbane.qld.gov.au/planning-building/do-i-need-approval/residential-projects/dwellings/new-house/building-sustainable-homes.

Brounen, D. and Kok, N. (2011). On the economics of energy labels in the housing market. *Journal of Environmental Economics and Management*, 62(2), 166–179.

Bruegge, C., Carrion-Flores, C. and Pope, J.C. (2016). Does the housing market value energy efficient homes? Evidence from the Energy Star program. *Regional Science and Urban Economics*, 57(2), 63–76.

Bryant, L. and Eves, C. (2011). Sustainability and mandatory disclosure in Queensland: An assessment of the impact on home buyer patterns. In: *17th Pacific Rim Real Estate Society Conference*, 16–19 January, Gold Coast, Australia.

Building Codes Queensland (2014). *6-Star Energy Efficiency Rating for New Houses and Townhouses*. Retrieved from www.hpw.qld.gov.au/SiteCollectionDocuments/SixStarEnergyEfficiencyRating FactSheet.pdf.

Buildings Performance Institute Europe (2010). *Energy Performance Certificates Across Europe: From Design to Implementation*. Retrieved from: http://bpie.eu/wp-content/uploads/2015/10/BPIE_EPC_report_2010.pdf.

Buildings Performance Institute Europe (2014). *Energy Performance Certificates Across the EU: A Mapping of National Approaches*. Retrieved from: http://bpie.eu/wp-content/uploads/2015/10/Energy-Performance-Certificates-EPC-across-the-EU.-A-mapping-of-national-approaches-2014.pdf.

Cajias, M. and Piazolo, D. (2013). Green performs better: Energy efficiency and financial return on buildings. *Journal of Corporate Real Estate*, 15(1), 53–72.

Commonwealth Government (2016). NatHERS (Nationwide House Energy Rating Scheme). Retrieved from: www.nathers.gov.au.

CoreLogic RP Data (2016). CoreLogic RP Data. Retrieved from: www.corelogic.com/au.

Department of Energy and Water Supply (2015). Solar Bonus Scheme 44c feed-in tariff. Retrieved from: www.dews.qld.gov.au/electricity/solar/installing/benefits/solar-bonus-scheme.

Department of the Environment and Energy (2016). Energy information. Retrieved from: www.environment.gov.au/energy/information.

Department of Housing and Public Works (2016). Smart and sustainable housing. Retrieved from: www.hpw.qld.gov.au/construction/Sustainability/SmartSustainableHomes/Pages/SmartSustainable Housing Tips.aspx.

Eves, C. and Kippes, S. (2010). Public awareness of 'green' and 'energy efficient' residential property: An empirical survey based on data from New Zealand. *Property Management*, 28(3), 193–208.

Fabris, P. (2010). BIM for home builders. *Professional Builders*, 1 March. Retrieved from: www.probuilder.com/bim-home-builders.

Fuerst, F., McAllister, P., Nanda, A. and Wyatt, P. (2015). Does energy efficiency matter to home-buyers? An investigation of EPC ratings and transaction prices in England. *Energy Economics*, 48, 145–156.

Fuerst, F., McAllister, P., Nanda, A. and Wyatt, P. (2016). Energy performance ratings and house prices in Wales: An empirical study. *Energy Policy*, 92, 20–33.

Harrington, P. and Miller, W.F. (2015). *Electronic Building Passport – Final Report: Project 2 National Energy Efficient Building Project Phase 2*. State of South Australia, Adelaide, S.A. Retrieved from: www.sa.gov.au/__data/assets/pdf_file/0004/192928/NEEBP-phase-2-project-2-electronic-building-passport-final-report.pdf

Hurst, N. (2012). Energy efficiency rating systems for housing: An Australian perspective. *International Journal of Housing Markets and Analysis*, 5(4), 361–376.

Hurst, N. and Wilkinson, S. (2015). Housing and energy efficiency: What do real estate agent advertisements tell us? *In RICS COBRA 2015*, Sydney, 8–10 July.

Hyland, M., Lyons, R.C. and Lyons, S. (2013). The value of domestic building energy efficiency: Evidence from Ireland. *Energy Economics*, 40(C), 943–952.

Lützkendorf, T. and Lorenz, D. (2010). Socially responsible property investment: Background, trends and consequences. In: Newell, G. and Sieracki, K. (Eds.), *Global Trends in Real Estate Finance*, Chichester and Armes, IO: Wiley-Blackwell, 194–237.

Lützkendorf, T. and Lorenz, D. (2011). Capturing sustainability-related information for property valuation. *Building Research & Information*, 39(3), 256–273.

Majcen, D., Itard, I. and Visscher, H.J. (2013). Theoretical vs. actual energy consumption of labelled dwellings in the Netherlands. *Energy Policy*, 54, 125–136.

Martinaitis, V., Zavadskas, E.K., Motuziene, V. and Vilutiene, T. (2015). Importance of occupancy information when simulating energy demand of energy efficient house: A case study. *Energy and Buildings*, 101, 64–75.

Miller, W. F. and Lützkendorf, T. (2016). Capturing sustainable housing characteristics through Electronic Building Files: The Australian experience. In: *Sustainable Built Environment Conference 2016, Hamburg: Strategies, Stakeholders, Success Factors*, 7–11 March; Conference Proceedings.

Miller, W.F., Stenton, J., Worsley, H. and Wuersching, T. (2014). *Strategies and Solutions for Housing Sustainability: Building Information Files and Performance Certificates*. Brisbane: Queensland University of Technology.

NatHERS (2016). Star rating scale overview. Retrieved from: www.nathers.gov.au/owners-and-builders/star-rating-scale-overview.

NCC (2016). *National Construction Code: Volume 2*. Building Code of Australia: Class 1 and Class 10 Building.

New South Wales Government (2016). BASIX (Building Sustainability Index). Retrieved from: www.basix.nsw.gov.au/iframe.

PriceFinder (2016). PriceFinder. Retrieved from: www.pricefinder.com.au/realestate.

Prime Building and Pest Inspection Consultants (PBC) (2016a). Sustainability reports. Retrieved from: www.pbcinspections.com.au/page.aspx?id=85&tSustainability%20Reports.

Prime Building and Pest Inspection Consultants (PBC) (2016b). What a PBC inspector inspects. Retrieved from: www.pbcinspections.com.au/page.aspx?id=4&tWhat%20a%20PBC%20Inspector%20Inspects.

Queensland Department of Housing and Public Works (2016). Inspection timing. Retrieved from: www.hpw.qld.gov.au/construction/BuildingPlumbing/Building/BuildingApprovals/Pages/InspectionTiming.aspx.

Queensland Government (2012). *Treasury (Cost of Living) and Other Legislation Amendment Act 2012*. Retrieved from: www.legislation.qld.gov.au/Bills/54PDF/2012/TreasCostLivB12.pdf.

Queensland Parliamentary (2012). *Treasury (Cost of Living) and Other Legislation Amendment Bill 2012 (Qld): Repeal of Sustainability Declarations Provisions*. Retrieved from: www.parliament.qld.gov.au/documents/explore/ResearchPublications/ResearchBriefs/2012/RBR201207.pdf.

Raisbeck, P. and Wardlaw, S. (2009) Considering client-driven sustainability in residential housing. *International Journal of Housing Market and Analysis*, 2(4), 318–333.

RenoValue (2016). RenoValue. Retrieved from: http://renovalue.eu.

Saunders, H. (2000). A view from the macro-side: Rebound, backfire and Khazzoom-Brookes. *Energy Policy*, 28(6–7), 439–449.

Susilawati, C. and Miller, W.F. (2013). Sustainable and affordable housing: A myth or reality.? In: Kajewski, S. L., Manley, K., and Hampson, K. D. (Eds.), *Proceedings of the 19th CIB World Building Congress*, Queensland University of Technology, Brisbane Convention & Exhibition Centre, Brisbane, Queensland, 1–14.

Swan, W., Wetherill, M. and Abbott, C. (2010). *Research Report 5: A Review of the UK Domestic Energy System*. Salford: SCRI.

Termatrac (2016). Termatrac T3i. Retrieved from: www.termatrac.com/products/termatrac-t3i.

Victorian Building Authority (2016). Six Star Standard. Retrieved from: www.vba.vic.gov.au/consumers/6-star-standard.

Warren-Myers, G. and Heywood, C. (2016) Investigating demand-side stakeholders' ability to mainstream sustainability in residential property. *Pacific Rim Property Research Journal*, 22(1): 59–76.

Wilkinson, S.J., Sayce, S.L. and Christensen, P.H. (2015). *Developing Property Sustainably*. London: Routledge.

Wong, S.Y, Susilawati, C., Miller, W. and Mardiasmo, A. (2015a). Enhancing information about sustainability features for sustainable housing delivery. In: Koskinen, K.T., Kortelainen, H., Aaltonen, J., Uusitalo, T., Komonen, K., Mathew, J., et al. (Eds.), *Proceedings of the 10th World Congress on Engineering Asset Management* (WCEAM 2015), Series: Lecture Notes in Mechanical Engineering, 407–414. Cham: Springer.

Wong, S.Y., Susilawati, C., Miller, W. and Mardiasmo, A. (2015b). A comparison of international and Australian rating tools for sustainability elements of residential property. In: *RICS COBRA 2015*, Sydney, 8–10 July.

Wong, S.Y., Susilawati, C., Miller, W. and Mardiasmo, A. (2016a). Assessing the impact of sustainability-related features on residential property price. In: *22nd Annual Pacific Rim Real Estate Society Conference*, Sunshine Coast, Queensland, 17–20 January.

Wong, S.Y., Susilawati, C., Miller, W. and Mardiasmo, A. (2016b). Understanding Australian real estate agent perspectives in promoting sustainability features in the residential property market. In: *7th International Conference Energy and Environment of Residential Buildings Conference*, Brisbane, Queensland, 20–24 November.

Your Home (2016). Passive design. Retrieved from: www.yourhome.gov.au/passive-design.

21

Scaling up commercial property energy retrofitting

What needs to be done?

Tim Dixon

21.0 Introduction

Often, the research focus in commercial property (which includes retail, offices and industrial space) has been on 'new build' as the growth in 'green' and 'sustainable' buildings has taken root (Dixon, Ennis-Reynolds, Roberts and Sims, 2009; Leishman, Orr and Pellegrini-Masini, 2012). However, there is an increasing concern that the rate of progress in tackling energy, water and waste inefficiencies in existing commercial property stock is too slow (Carbon Trust, 2010, WEF, 2011). This is a challenging issue for actors and policy makers because, for example, in the UK it is estimated that by 2050 some 75–85 per cent of today's buildings will still be standing, with 40 per cent built prior to 1985 (when Part L of the Building Regulations was first introduced (BBP, 2010)), and 60 per cent built prior to 2010 (Mackenzie , Pout, Shorrock, Matthews and Henderson, 2010). The importance of existing stock is also highlighted when it is appreciated that the addition to the building stock in the UK is very slow, with less than 1–2 per cent each year being 'new build' (Dixon, Eames, Hunt and Lannon, 2014a).

Retrofitting (or literally to 'provide (something) with a component or accessory not fitted during manufacture' (*Oxford English Dictionary*[1]) is therefore seen as a powerful way of tackling and reducing energy emissions in a sector that globally produces substantive greenhouse gas (GHG) emissions. Retrofitting measures can therefore be taken to include energy efficiency projects, but can also sometimes include water and waste measures and accelerated replacement of existing services and plant (BBP, 2010). This chapter discusses the meaning of retrofit in relation to commercial property in more detail.

Commercial property (which includes retail, offices and industrial space) plays a vital role in the UK and other international economies. With a total of 680 million square metres of floor space, commercial property is worth about £787 billion (13 per cent of the value of all buildings in the UK), with retail (at £340 billion) the largest sub-sector within this overall figure. The sector is also diverse and complex, with more than half of commercial property being rented (57 per cent) rather than owned, compared with 37 per cent of housing being rented (PIA, 2015).

The sector also has a substantial environmental impact. Commercial property produces 10 per cent of the UK's GHG emissions and consumes 7 per cent of UK energy, and there is increasing concern from the independent body the Committee on Climate Change and other commentators that the rate of progress in tacking energy inefficiency in existing commercial stock is too slow (CCC, 2013).

For example, it is estimated that UK business is overlooking a potential cost saving of £1.6 billion through under-investment in energy efficiency, with the UK's commercial retrofit market potential estimated at £9.7 billion (or US$16 billion) (Westminster Sustainable Business Forum/Carbon Connect, 2013). The market potential for commercial property retrofit is significant and also growing elsewhere; the World Economic Forum (2011) quotes data suggesting that in the US the market potential of retrofit is $16 billion per annum (with a potential energy saving of US $41 billion per annum). A key challenge for the sector, then, is how existing commercial building stock can be retrofitted in an integrated way and across building, portfolio and city levels. Drawing on recent research and consultation findings from the UK, the US and elsewhere, this chapter focuses primarily on energy retrofitting. The chapter also:

- identifies key characteristics and conceptualisations of UK commercial property retrofit;
- defines what is meant by commercial property retrofit;
- identifies and compares the key drivers and barriers facing the UK commercial retrofit sector; and
- concludes with a discussion of the changes in policy and practice that are needed in the UK and elsewhere to speed up the rate of progress in the sector.

21.1 UK commercial property retrofitting: Characteristics and conceptualisations

The UK commercial property sector's characteristics are important in understanding the context of retrofit. First, there is a relatively high level of tenanted property in the UK commercial property sector (57 per cent) (PIA, 2015). This is because many businesses have become increasingly reluctant to commit the capital and management time required for owner occupation, and owner-occupiers took advantage of high prices in the mid-2000s to participate in 'sale and leaseback' deals (PIA, 2015). This, therefore, places the issue of the 'split incentive' centre stage. The split (or 'misaligned') incentive refers to transactions where the benefits do not accrue to the person who pays for the transaction. In the context of building-related energy, it refers to the situation where the building owner pays for energy retrofits and efficiency upgrades, but cannot recover savings from reduced energy use that accrue to the tenant (because, for example, the tenant is responsible for the payment of energy bills) (Economidou, 2014).

Second, we also know that the sector is complex. For example, the Carbon Trust report *Building the Future* (2010) highlighted the complexity of the sector in terms of its diversity, building types and range of stakeholders, and the report also spoke about the 'conservatism' of the sector and its risk-averse nature.

The Climate Change Act 2008 is the overarching statute governing the UK's GHG emissions targets. The Act makes it the duty of the Secretary of State to ensure that by 2050 the UK's GHG emissions are at least 80 per cent lower than the 1990 baseline. In the commercial property sector, there have been a variety of legislative measures that have sought, directly or indirectly, to encourage retrofit projects focusing on energy efficiency, although some of these measures have turned out to be relatively short lived (Dixon, 2014). For example, the mandatory Carbon Reduction Commitment Energy Efficiency Scheme (CRC-EES), which was first introduced under enabling powers in Part 3 of the Climate Change Act 2008, aims to drive energy efficiency improvements in large commercial and public-sector organisations by requiring participants to purchase and surrender allowances corresponding with their annual carbon dioxide emissions from their energy use. However, following criticism over its bureaucracy from some members of the business community, and its perceived failure to drive genuine behavioural change, the UK government simplified the scheme by reducing the fuel types covered and abolishing the scheme's league table in 2012. More recently, the UK government has announced plans to abolish the scheme after the 2018–2019 compliance year, and, in its place, increase the Climate Change Levy (a tax on energy delivered by non-domestic users, which is designed to incentivise energy efficiency and reduce carbon emissions) (DECC, 2016). In the UK (following the introduction of the EU's Energy Performance of Buildings Directive in 2006), there is also a requirement to provide Energy Performance Certificates (EPCs) on the sale, rent or construction of buildings other than dwellings with a floor area greater than 50 square metres from 6 April 2008 (similar schemes operate in other countries – for example, Australia (Building Energy Efficient Certificates)). However, only 39 per cent of leased commercial premises in the UK have provided EPCs as required, with more than 18 per cent receiving the lowest rating of F and G, and only 8 per cent of EPCs rated at B or above (CCC, 2013). Using a database of 400,000 properties, research for the Green Construction Board (Mactavish, Woolham and Sayce, 2014) also found that some 19 per cent of buildings are likely to fall below the minimum standard, with a further 17 per cent E-rated, which would be directly affected either if the threshold is raised after 2018 or if the assessment process were to be reviewed.

In the commercial property sector, there have been a number of conceptualisations developed to explain individual organisational behaviour in relation to commercial property retrofit, rather than understand the sector *as a whole* (Cooremans, 2012; Pellegrini-Masini and Leishman, 2011). As has been argued in the domestic retrofit sector (Bergman, Whitmarsh and Kohler 2008; Horne, Maller and Dalton, 2014), there is, however, a logical rationale for taking a socio-technical sectoral view of retrofitting in the commercial property sector, rather than a more deterministic view (Geels, 2002; Axon, Bright, Dixon, Janda and Kolokotroni, 2012; Janda, 2014). In this way, we can potentially build a better understanding of how we might shape and influence the future trajectory of retrofitting in the commercial property sector as a whole.

The overall complexity of the commercial property sector has produced challenges when it comes to developing conceptual frameworks, particularly at a sectoral level. There have, nonetheless, been a number of frameworks that have attempted to provide insights into how we should analyse energy efficiency decision making (as part of wider retrofitting) at an individual firm level. For example, at an organisational level, Pellegrini-Masini and Leishman (2011) used an investment decision-making framework based on cost–benefit analysis to understand why some companies adopted energy efficiency policies in offices whilst others did not. Similarly, Cooreman's (2012) study of companies undertaking energy efficiency measures placed a strong emphasis on understanding investment decision making as a systematic process within an organisation, by positioning behaviour centre stage and recognising socio-cultural factors, regulation and policy impact, and material and market domain influences. These ideas are also taken further by Axon et al.'s (2012) 'Communities of Practice' framework, which focuses on looking across the scales of building, company and its overall portfolio; this research has been taken further through the development of links to multi-level research, including transition theory (Janda, 2014; Janda and Moezzi, 2014).

However, if we are to understand sector change, we need to understand not only temporal change, but also the landscape of policy and regulations that may or may not influence changes in technological innovation in commercial property retrofitting. This was at the heart of the arguments made by DECC (2012) in suggesting a model that engenders a much closer understanding of socio-technical dimensions. The Multi-Level Perspective (MLP) offers a conceptualisation that incorporates a socio-technical focus, and has been applied within the domestic retrofit arena at a sectoral level (Bergman et al., 2008; Horne et al., 2014; Gibbs and O'Neill, 2014). A key feature of the MLP is its focus on co-evolution (or a framing of the issue in socio-technical terms), and so understanding the diverse interactions that exist between industry, technology, markets, policy, culture and society is important in this perspective (Geels, 2002; Smith, 2007). In the MLP, the underlying premise is that transitions (or movements between current state and future pathways to sustainability) are best seen as non-linear processes that result from interactions and interplay between multiple developments at three levels: 'niches' (the focus for radical innovations), socio-technical 'regimes' (the focus for established practices and associated rules) and the concept of an exogenous 'landscape' of policy and regulation (Geels, 2011).

Drawing on the MLP, recent research has characterised the different actor groups in the UK commercial property 'regime' (Figure 21.1) (Dixon, Britnell and Watson, 2014b). At the heart of the regime lies the 'producer network' comprising those who 'produce' (or supply) buildings that are to be retrofitted. These include investors/developers, owners and occupiers and their advisors. Commercial buildings are also owned and occupied (or owner-occupied), and so the 'user group' represents these actors. Furthermore, the financial network provides finance in the sector to both 'producers' and 'users' (because retrofit may be owner driven or occupier driven). 'Influencer' groups such as Building Research Establishment (BRE) and the Better Buildings Partnership (BBP) also play an important role in shaping practice in retrofit by helping provide learning spaces and advice through toolkits and guidance. Similarly, technology suppliers provide retrofit technologies to undertake the retrofit work and energy companies may also provide allied retrofit products and services. Finally, public authorities play a potentially important role through

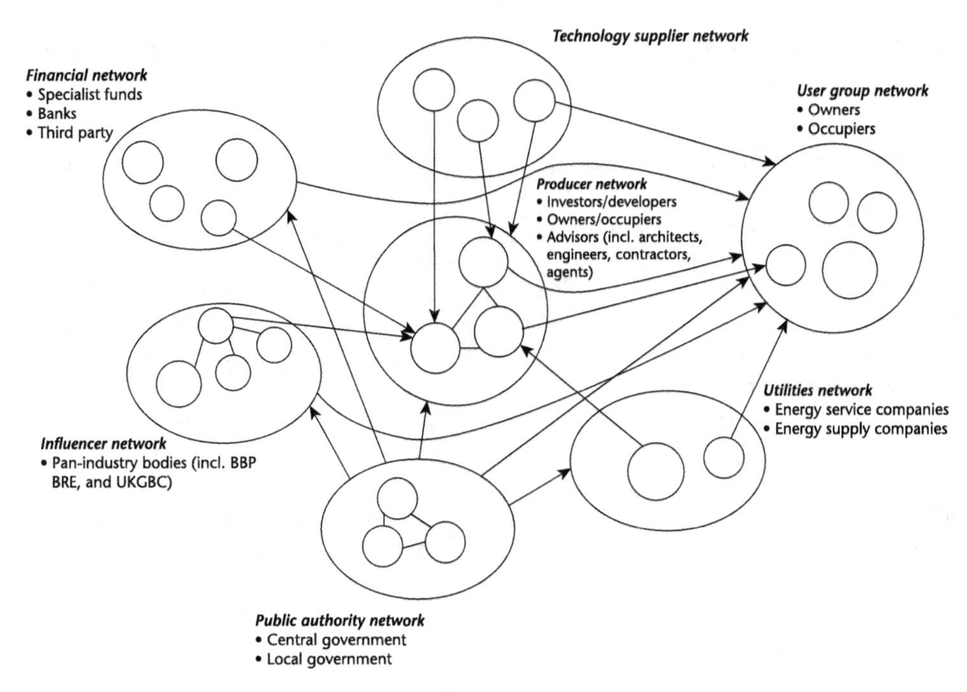

Figure 21.1 Conceptualisation of the UK commercial property retrofit regime

Source: Dixon et al. (2014b)

Note: Owners and occupiers feature in both producer and user groups because they may provide retrofitted buildings or use them, depending on context.

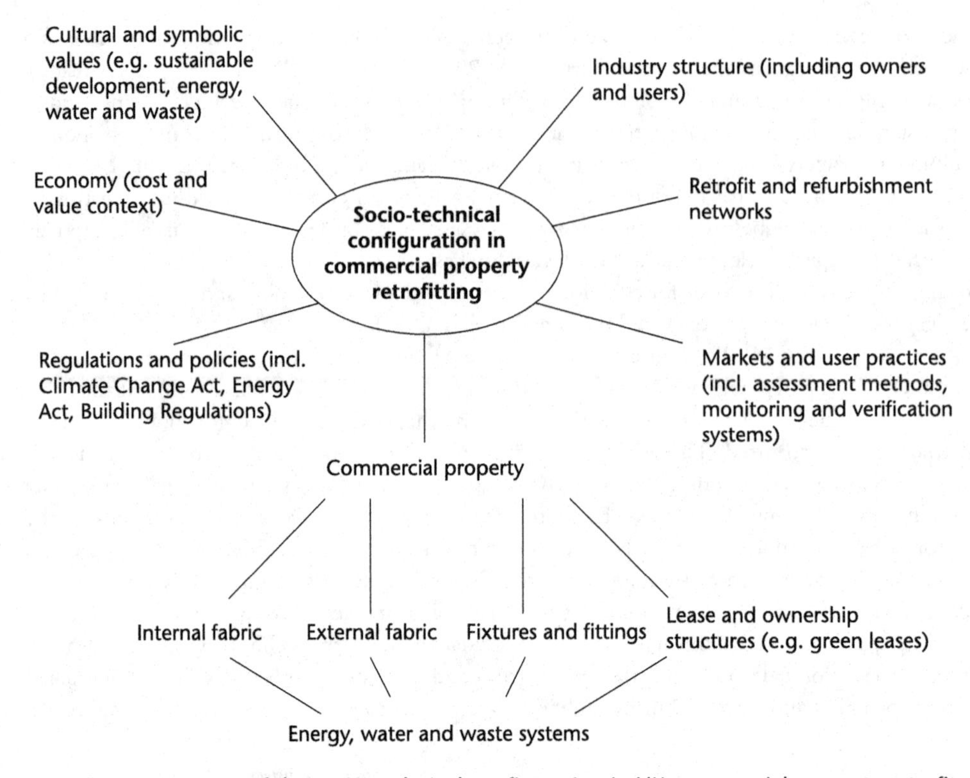

Figure 21.2 Overview of the socio-technical configuration in UK commercial property retrofit

Source: Dixon et al. (2014b)

influencing national and local policy and practice, and through financial and legal interventions such as Building Regulations and the Climate Change Levy.

The commercial property retrofit 'regime' can be viewed, therefore, as comprising a complex array of stakeholders who interact in a variety of ways when a retrofit project is undertaken (Figure 21.1). In this sense, the term 'regime' can be used to describe 'relatively stable but nevertheless dynamic configurations of buildings and infrastructures, networks of actors and institutions, technologies, policies and regulations, social norms, practices and shared expectations' (Eames, Dixon, May and Hunt, 2013: 509).

Based on this, we can also postulate how technological innovation in the sector needs to be seen in the context of a 'socio-technical' framework that recognises that simply overcoming 'barriers' in itself may well not result in a desirable outcome (Dixon et al., 2014b). Rather, using an MLP perspective, we can potentially see the existing regime within a landscape of regulations and policies, user practices, and existing norms and values, as shown in Figure 21.2. Understanding this is vital to developing relevant policy and practice for change in the sector.

21.2 Defining retrofit

Retrofitting takes on an important significance in the context of commercial property. In the academic literature, there has been much debate over the meaning of 'retrofit' and its distinction, if any, from 'refurbishment' or 'renovation'. In a literal sense, to retrofit can be defined as 'to provide (something) with a component or accessory not fitted during manufacture; to add (a component or feature) to something that did not have it when first constructed' (*Oxford English Dictionary*). In other words, the term, which originated in the US in the late 1940s and early 1950s, is essentially a blend of the words 'retroactive' (applying or referring to the past) and 'fit' (to equip).

Within the context of the built environment, the term 'retrofit' has been used to imply substantive physical changes to a building or buildings (for example, mitigation activities to improve energy efficiency), and is often linked to the concept of 'adaptation' (i.e., intervention to adjust, reuse or upgrade a building to suit new conditions or requirements) (Douglas, 2006; Wilkinson, 2012). Confusingly, the term has also been used interchangeably with other terms such as 'refurbishment', 'conversion', 'renovation' and 'refit' (see Mansfield (2002) and Wilkinson (2012) for a discussion of this point at a property level). However, at a city level it can be argued that the term 'retrofit' is distinguishable from these terms because the defining characteristics of urban retrofitting are: (1) its comprehensive nature and large scale; (2) its integrated nature, requiring a high degree of private–public partnership arrangements; (3) the sustainable nature of its funding; and (4), a clearly defined set of goals and metrics for monitoring (Institute for Sustainable Communities/Living Cities, 2009; Dixon and Eames, 2013).

Based on 37 in-depth interviews with key players, recent UK research funded through EPSRC[2] (Dixon et al., 2014b) found that in many instances a distinction was indeed made between 'retrofit', where a building(s) could be refitted with relatively 'light-touch' energy efficiency measures, for example, whilst a tenant was still in occupation, and 'refurbishment', which entails a much 'deeper' level of refit with changes to the internal and external fabric of the building, with the latter frequently occurring at lease renewal. However, in other cases 'refurbishment' was used rather than 'retrofit'.

First, the research found that there needs to be a much clearer consensus over what the term 'retrofit' means (Table 21.1), as this is hampering progress because of a lack of common language and understanding. For example, although the RICS provides guidance on sustainability and valuation, the current edition of the guide does not define 'retrofit' and 'refurbishment' explicitly (RICS, 2013a).

Second, there is also a conflation of terminology around 'retrofit' and 'renovation'. At a European level, the main policy lever relating to the reduction of energy consumption in buildings is the Energy Performance of Buildings Directive (EPBD, 2002/91/EC), which was recast in 2010 (EPBD Recast 2010/31/EU).[3] Under this Directive, member states must establish

Table 21.1 Suggested definitions for 'retrofit' and 'refurbishment'

Commercial property retrofit	*Commercial property refurbishment (or renovation)*
The process of making planned interventions in a building to install or replace elements or systems which are designed to improve energy and/or water and waste performance.	The cyclical process of improving a building above and beyond its initial condition in order to increase asset value. The focus is on systemic upgrading and renewal of building elements, finishes and mechanical services, with a potential impact on energy and/or water and waste efficiencies.
Typical characteristics	*Typical characteristics*
Non-intrusive whole-system upgrades or new elements added to existing systems.	Major alterations to fabric and/or services at a systemic, whole-building level.
Carried out during lease or during ownership.	Carried out on lease renewal (or lease end) or on a cyclical basis in owner-occupied property.

Source: Dixon et al. (2014b)

and apply minimum energy performance requirements for both new and existing buildings, ensure the certification of building energy performance and ensure the regular inspection of boilers and air-conditioning systems in buildings. The Directive also requires member states to ensure that by 2020 all non-domestic buildings are 'nearly zero-energy buildings', and is perhaps one of the most ambitious programmes for the renovation of existing buildings. For example, the EPBD defines a 'major renovation' as one where the total cost of the renovation is more than 25 per cent of the value of the building (excluding land value), or where more than 25 per cent of the building envelope undergoes renovation (RICS, 2013b).

This raises the issue of what is meant by 'renovation' or 'retrofit'. Certainly, the term 'renovation' tends to be used in Europe, whereas the term 'retrofit' is perhaps more common in the US and Australasia, although the latter term is also used in the UK (Dixon and Eames, 2013). Both terms would apply (at the building level, at least) to an improvement to the building envelope and/or a building's mechanical systems. However, if as the Global Buildings Performance Network (GBPN) (2013) suggests, we adopt the term 'DR' (to represent 'deep renovation' or 'deep refurbishment'[4]), then that also implies a substantial improvement in the energy performance of a building. Indeed, the most recent EU guidance on the subject suggests that 'deep renovation' means an improvement in energy performance in a building of at least 80 per cent (European Parliament, 2012).

Global evidence suggests that it is possible to achieve a significant (50–70 per cent) reduction in the energy use of existing buildings with 'deep' measures (Ürge-Vorsatz et al., 2012). A 'standard' renovation, however, in the majority of cases, is likely to achieve only 20–30 per cent energy savings, which would be insufficient to meet the long-term carbon emissions target of 50 per cent by 2050 in the buildings sector (GBPN, 2013). In commercial buildings, DR includes envelope upgrades, replacement and reconfiguration of HVAC and heating/cooling systems, better control systems and lighting improvements, although much will depend on the type of building under consideration.

Retrofit measures can encompass not only energy, but also water and waste as well. For example, light-touch retrofit, such as energy-efficient lighting and controls, building services and management systems and controls, can reduce energy costs by up to 30–40 per cent per annum, but recycling water and waste (for example, in shopping centres) can also have significant and positive sustainability and cost impacts (Dixon et al., 2014b). Moreover, changes to structure, such as green roofs, can also be seen as 'retrofit' activities. Nonetheless, recent research from both the UK (Dixon et al., 2014b) and the US (Institute for Building Efficiency, 2013) suggests that the following retrofit measures are the most common in the commercial property sector:

- energy-efficient lighting and controls;
- management systems and controls; and
- building services.

21.3 Drivers for commercial property retrofit

Recent research found that the most important drivers in UK commercial property retrofit relate to policy, economic factors (for example, rising energy costs) and marketing/reputation (Dixon et al., 2014b). Cost-related factors, such as energy cost savings, improved asset value and energy security, were also found to be key drivers in the Americas and Europe (including the UK) in a recent global survey (Institute for Building Efficiency, 2013).

In the UK, the Carbon Reduction Commitment (CRC) Energy Efficiency Scheme (which is due to be abolished in 2019) was seen in this research as being important in driving change in organisations. There was a strong feeling that retrofit was landlord driven, particularly in relation to larger and 'deeper' projects, and in these instances there was a strong interrelationship with cost and with a desire to reposition the asset in the property portfolio. In this context, a number of interviewees spoke about the distinction that exists between the drivers for owners and those for occupiers. For owners, the drivers often relate to what can be described as an energy-related risk factor associated with premature obsolescence, and a potential depreciation of assets from a future 'lettability' point of view. Owners are increasingly realising that higher energy performance standards are an essential part of marketing a property, and can be an enabler for commanding potentially higher rents. Moreover, the emergence of green leases and green building management groups, although relatively sporadic in the UK, has also helped bring owners and occupiers together around a common agenda (Bright and Dixie, 2013; Dixon et al., 2014b).

Other research (Dixon et al., 2014b; Westminster Sustainable Business Forum/Carbon Connect, (2013) also highlights the important potential role of the Energy Act 2011, which from April 2018 will make it unlawful to let residential or commercial properties with an EPC rating of F or G (currently 18 per cent of the total stock) (Mactavish et al., 2014). This raises the broader question of which policies are seen as important in driving change in the sector. A report by the Institute for Building Efficiency (2013), for example, found that tax credits and incentives or rebates for energy efficiency measures and low-interest financing were the most favoured by respondents globally, although there were variations between countries (for example, according to the report, China, Australia and France favoured stricter building codes).

21.4 Barriers to commercial property retrofit

In the UK, the most significant barriers to retrofit relate to economic factors (overall cost and value impact), organisational issues (where retrofit projects may be competing for funds) and lease structures (Dixon et al., 2014b). The economic barriers relate to the fact that energy use is often only a small proportion of business costs, although this is increasing as energy prices rise. Aside from continued problems over the split-incentive issue (in tenanted property, the landlord is responsible for the building but the tenant is responsible for energy costs), the required payback periods in leased premises are also shorter and often limited to a maximum of five years (and normally two to three years), which restricts the type of retrofit measures that can be adopted. This is partly driven by perceptions of 'risky' technology requiring longer paybacks, but also declining lease lengths in the sector (for example, the average length of a new lease in 2015 is 6.8 years, compared to 9.6 years in 1999 (PIA, 2015)).

However, the significance of organisational barriers should not be underestimated. For some commentators, the term 'barriers' carries the sense that in some way if these were removed then energy efficiency would automatically act as a precursor to 'rational' behaviour in the marketplace. But this ignores the organisational context for decisions, the interrelationship between the barriers themselves and the fact that they should best be seen in the context of the wider legislative landscape and how companies arrive at investment decisions (DECC, 2012). For example, often leadership is lacking at the executive level when it comes to retrofit projects, which may also be competing for core business funds from new construction or even bigger capital projects (Dixon et al., 2014b).

These findings are borne out in the Institute for Building Efficiency (2013) study, which found that lack of funding and internal financial criteria were the most important barriers to energy efficiency investments. Perhaps significantly, more respondents in the US and Canada suggested lack of capital as their top barrier, compared with Europe and other countries.

21.5 Funding vehicles for commercial property retrofit

Retrofit financing is crucial to success, and lack of funding, particularly in the small and medium-sized enterprise (SME) sector, is hampering progress in the UK and elsewhere (Dixon et al., 2014b). With the demise of the 'Green Deal' (a 'pay-as-you-save' scheme), which in any event never really took off in the commercial sector, the majority of UK retrofit projects are paid for through self-financing or service charge arrangements, although there have been recent examples of the increasing take-up of energy performance contracting arrangements where retrofitting is financed through projected future energy savings. In summary, besides self-financing, other financing options for retrofit include:

- *Service charge*: where a landlord can claim the costs of retrofit back through the 'hard services' part of the service charge payable by a tenant, although this is dependent on the wording used.
- *Energy performance contracting (EnPC)*: where retrofitting is financed through projected future energy savings. Typically, an energy supply company (ESCO) provides customised engineering, installation and maintenance with the guarantee of reduced energy consumption as a result of their work.
- *Managed Energy Service Agreements*: where the contractor takes over responsibility for the energy bill and manages the relationship with the utility provider(s). The building owner then pays the contractor the historical energy bills corrected for weather and other factors (or what they would have paid) (e.g., SciEnergy).
- *Investment funds*: where specialist funders provide capital for retrofit. An example here is the Green Investment Bank's underwriting of the partnership between Sustainable Development Capital and BRE. Other examples include Equitix and Low Carbon Workplace.

Perhaps best classified now as a wasted opportunity, the UK Green Investment Bank (UKGIB) became operational in October 2012, with £3 billion in UK taxpayer capital dedicated to its mission of 'accelerating the UK's transition to a more green economy, and creating an enduring institution, operating independently of government' (UKGIB, 2016). The bank backed 21 green projects and committed over £700 million, mobilising a further £2 billion in private finance, but struggled to gain traction as its powers were limited. On 3 March 2016, the UK government launched the process to move the Green Investment Bank into the private sector and details were set out in statements from the UK government and the bank. The transaction is expected to involve both the sale of existing shares owned by the UK government and the commitment of additional capital for the bank by new investors (UKGIB, 2016).

Learning from international experience, in the US Property Assessed Clean Energy (PACE) finance is being used at a municipal level to scale up commercial property retrofit. In 2013, there were some 16 commercial PACE programmes, which work through the property owner who receives financing support from a local government or approved financial institution and the investment is repaid by an assessment (or equivalent) and added to the owner's property tax bill for a period of up to 20 years (Managan and Klimovich, 2013). Indeed, a more radical interpretation of local tax assessment has been mooted recently by the British Retail Consortium (2014) through one of its four options for reforming the business rates system in the UK. This would be a fundamental reform that would shift the basis for taxing property by abolishing rates and replacing them with a new energy tax based on usage.

In fact, cities have become the focus for much of the discussion and debate about how to scale up energy efficiency and retrofit measures. A report from the C40 Cities Climate Leadership

Group/Tokyo Metropolitan Government (Takagi et al., 2015) showed that a number of cities around the world were developing their own codes for new buildings and major retrofits that were more stringent than national or state codes. Examples in this group include Melbourne's 1200 Buildings Program (where success has been limited), New York's Greener Greater Buildings Plan (a mandatory benchmarking scheme) and Philadelphia's Building Energy Benchmarking Ordinance. Similarly, in some US cities, and also Tokyo, mandatory reporting and retro-commissioning is becoming more common (for example, in Singapore and Hong Kong). Tokyo itself is a unique example of how a city can introduce a mandatory cap and trade programme, with an emissions target, focused on buildings.

21.6 Improving transparency in actual energy performance

A recent report by the Committee on Climate Change (2013) in the UK revealed the lack of progress in UK commercial property retrofit: of the 427,814 EPCs that had been issued by mid-June 2013, more than 18 per cent received the lowest rating of F and G, and only 8 per cent of EPCs were rated at B or above. This showed little improvement since 2012, and in fact only 39 per cent of leased commercial premises have so far provided EPCs as required. Research (JLL, 2012) has also shown that in a comparative study of two EPC-rated government buildings in London, the building with the worse EPC out-performed the higher-scored building in terms of its actual energy performance by 66 per cent.

Research findings also suggest that in the UK mandatory Display Energy Certificates (DECs), which measure the actual energy consumed within a building rather than the theoretical performance, would drive change. This is not surprising since many in the UK commercial sector argue strongly for the introduction of DECs, but the UK government continues to resist calls to do so. The current position may well be influenced by contrary arguments, for example, that DECs were designed originally for public buildings with very different occupancy characteristics from the commercial sector, that increasing intensity and efficiency of use may worsen DEC performance and that existing data is based on a limited number of case studies (Westminster Sustainable Business Forum/Carbon Connect, 2013).

This suggests that the UK needs much better data on energy performance in commercial property because it could inform more successful practice. In the EPSRC Retrofit 2050 research (Dixon et al., 2014b), there were several references to Australia, where there was considered to have been substantial success with the National Australian Built Environment Rating System (NABERS) scheme (based on actual energy performance). This was seen to be not only driving retrofit in commercial offices, but also embedded in the marketplace and influencing renting decisions. This has been largely driven by the Australian government as an occupier through a stipulation that they would only occupy buildings with a minimum energy rating of three stars, which has forced landlords to improve stock and management practices, albeit in fairly localised areas. In response, other corporate occupiers in Australia have started to follow the government's lead and have mandated minimum energy ratings for their own premises. Over half of all Australian commercial office buildings are covered by the scheme and, on average, have increased energy efficiency by 12 per cent. Importantly, the market is now rewarding investment in energy-efficient design: buildings with high ratings have higher asset values and lower vacancies (NABERS, 2014; Snoxall, 2013).

Clearly, there are important differences between the UK and Australian contexts, but the lessons learned from the EPSRC Retrofit research suggest that mandatory actual energy performance certification could help drive change in the UK sector.

21.7 Consistency in measurement, verification and assessment standards

The EPSRC Retrofit 2050 research (Dixon et al., 2014b) found that payback periods and contracting arrangements can also hinder innovation. Difficulties with the process of validating and approving technologies were also identified as being a problem. Other respondents (who were senior decision makers in the commercial property sector) in the research spoke about how the selection and procurement of technologies can be limited by existing technology lists that permit tax breaks through enhanced capital allowances. A broad set of reforms is needed in the UK, and the EPSRC Retrofit 2050 research (Dixon et al., 2014b) makes the following recommendations:

- More consistency in assessment of retrofit through an agreed and confirmed use of RICS Ska and BREEAM Refurbishment and Fit-Out. RICS Ska is an environmental assessment tool for sustainable fit-outs and BREEAM Refurbishment and Fit-Out is used for major non-domestic refurbishment and fit-out projects.
- The introduction of post-occupancy evaluation and 'soft landings' approaches (which allow building occupiers and managers the opportunity to become involved with the design and construction process to help ensure improved operational performance and better working environments) in the commercial property sector to verify that retrofit designs do deliver in practice and to adjust misalignment of design.
- Improved metering, monitoring and verification to measure delivery and impact of retrofit on energy performance. For example, there is a growing recognition of the International Performance Measurement and Verification Protocol (IPMVP®), which defines standard terms and sets out best practice for quantifying the results of energy efficiency investments and/or investment in energy and water efficiency, demand management and renewable energy projects.

21.8 Conclusions

In what is essentially a complex, diverse and conservative sector, rolling out retrofit at scale in the UK is challenging. For one thing, commercial property investors and developers tend to see retrofitting through the lens of individual buildings and portfolios rather than at the city level. This, combined with the diversity of commercial stock (in terms of age, type and leasing arrangements) and its geographical spread, can lead to discontinuities between key stakeholders in the sector and retrofit projects across wider urban areas. Currently, there is a major independent study underway into whether the UK government's energy and carbon policies are having the desired impact in the property sector (APGEBE, 2013).

Achieving a consensus on what we mean by 'retrofit' is essential (not least to ensure a common language and understanding feeds into RICS guidance). For large-scale commercial property retrofit to succeed at all scales in the UK, there also needs to be urgent action in both policy and practice, drawing on lessons from overseas where relevant and appropriate. This is based on four key principles (Dixon et al., 2014; Westminster Sustainable Business Forum/ Carbon Connect, 2013):

- *Financing is crucial to success.* There should be further strengthening of the financial incentives around retrofitting, which could then be used to offer financial support at a city level to retrofit projects and/or to SMEs. Radical options such as tying the business rates system in commercial property to energy performance could also be examined in detail.

- *Actual energy performance should be transparent.* DECs should be mandatory in the sector, perhaps incentivised through business rates and stamp duty reductions for more energy-efficient properties. Other suggestions include increasing financial penalties for those failing to fulfil both EPC and DEC requirements.
- *Better integrated leadership at the city level is needed.* In the UK, local authorities have a role to play in helping drive the retrofit agenda, but they face funding constraints. Local Economic Partnerships in England and the wider business community both have a key role to play through partnerships and innovative financing models, and the move towards greater city powers through the Cities and Local Government Devolution Bill (2015–2016) may give UK cities the impetus required to take more independent actions on energy efficiency and retrofitting, as is happening in other cities globally. 'Sticky' infrastructure projects,[5] such as district heating schemes supported by improved incentives, could also provide further opportunities for city-wide retrofit to attract commercial property stakeholders.
- *Consistency in standards is needed at a number of levels.* There needs to be a clearer consistency in commercial retrofit assessment standards around BREEAM, RICS Ska and other related standards. An approved products and suppliers list is also needed for commercial property retrofit, with more transparent performance-in-use data, and better support for emerging technologies, so that companies have more certainty over technology choice. There should also be more consistency in monitoring and verification standards, perhaps based around the IPMVP. As the Westminster Sustainable Business Forum/Carbon Connect research (2013) suggests, this could also be underpinned by a comprehensive database of UK commercial buildings, which could create a performance benchmark and help foster competition.

Acknowledgements

The author would like to thank Emerald for their permission to allow some of the material from this chapter to be based on a previously published paper by the author (Dixon, 2014).

Further details on EPSRC Retrofit 2050 can be found at: www.retrofit2050.org.uk.

Notes

1 See www.oed.com.

2 EPSRC Retrofit 2050 programme (www.retrofit2050.org.uk).

3 The recast EPBD requires that from 2019 onwards 'all the new buildings occupied and owned by public authorities are nearly zero-energy buildings' (nZEBs) and by the end of 2020 'all new buildings are nearly zero-energy buildings'. Whilst the focus is on new buildings, member states must also create a national plan for increasing the number of nZEBs, which also includes existing buildings. These plans must include: (1) an nZEB definition; (2) intermediate targets for the energy performance of new buildings by 2015; and (3) details of how nZEBs (new and existing) will be promoted.

4 The term 'deep renovation' is also important to recognise because there has been an increasing focus on renovation roadmaps in member states. Article 4 of the EU's recent Energy Efficiency Directive (EED) (2012/27/EU) requires that by 30 April 2014 long-term national strategies to stimulate cost-effective deep renovations of buildings should be in place (Policy Partners, 2013).

5 These are projects that can increase the cohesion and ability to bind in stakeholders as they provide benefits for a wide range of commercial and non-commercial interests.

References

All Party Group for Excellence in the Built Environment (APGEBE) (2013) *Re-Energising the Green Agenda.* Report from the Commission of Inquiry into Sustainable Construction and the Green Deal. House of Commons, London.

Axon, C. J., Bright, S. J., Dixon, T. J., Janda, K. B. and Kolokotroni, M. (2012) 'Building communities: reducing energy use in tenanted commercial property', *Building Research & Information*, 40(4). 461–472.

Bergman, N., Whitmarsh, L. and Kohler, J. (2008) *Transition to Sustainable Development in the UK Housing Sector: From Case Study to Model Implementation.* Tyndall Centre for Climate Change Research. Working Paper 120.

Better Buildings Partnership (BBP) (2010) *Low Carbon Retrofit Toolkit: A Roadmap to Success.* BBP, London.

Bright, S. and Dixie, H. (2013) 'Evidence of green leases in England and Wales', *International Journal of Law in the Built Environment*, 6(1–2). 1–17.

British Retail Consortium (2014) *Business Rates: The Road to Reform.* British Retail Consortium, London.

Carbon Trust. (2010) *Building the Future.* Carbon Trust, London.

Committee on Climate Change (CCC) (2013) *Meeting Carbon Budgets: 2013 Progress Report to Parliament.* CCC, London.

Cooremans, C. (2012) 'Investment in energy efficiency: do the characteristics of investments matter?' *Energy Efficiency*, 5(4). 497–518.

DECC (2012) *What Are the Factors Influencing Energy Behaviours and Decision-Making in the Non-Domestic Sector? A Rapid Evidence Review.* DECC, London.

DECC (2016) *2010 to 2015 Government Policy: Energy Demand Reduction in Industry, Business and the Public Sector.* Accessed May 2016: www.gov.uk/government/publications/2010-to-2015-government-policy-energy-demand-reduction-in-industry-business-and-the-public-sector/2010-to-2015-government-policy-energy-demand-reduction-in-industry-business-and-the-public-sector#appendix-3-crc-energy-efficiency-scheme.

Dixon, T. (2014) 'What does "retrofit" mean, and how can we scale up action in the UK sector?' *Journal of Property Investment & Finance*, 32(4). 443–452.

Dixon, T. and Eames, M. (2013) 'Scaling up: the challenges of urban retrofit', *Building Research & Information*, 41(5). 499–503.

Dixon, T., Britnell, J. and Watson, G. B. (2014b) *'City-Wide' or 'City-Blind?' An Analysis of Emergent Retrofit Practices in the UK Commercial Property Sector.* Project Report. EPSRC Retrofit 2050, Cardiff.

Dixon, T., Eames, M., Hunt, M. and Lannon, S. (eds) (2014a) *Urban Retrofitting for Sustainability: Mapping the Transition to 2050.* Routledge, London.

Dixon, T., Ennis-Reynolds, G., Roberts, C. and Sims, S. (2009) 'Is there a demand for sustainable offices? An analysis of UK business occupier moves (2006–2008)', *Journal of Property Research*, 26(1). 61–85.

Douglas, J. (2006) *Building Retrofit.* Butterworth Heinemann, Oxford.

Eames, M., Dixon, T., May, T. and Hunt, M. (2013) 'City futures: exploring urban retrofit and sustainable transitions', *Building Research & Information*, 41(5). 504–516.

Economidou, M. (2014) *Overcoming the Split Incentive Barrier in the Building Sector: Workshop Report.* JRC Science and Policy Reports. European Commission, Luxembourg.

European Parliament (2012) *Report on the Proposal for a Directive of the European Parliament and of the Council on Energy Efficiency and Repealing Directives 2004/8/EC and 2006/32/EC.* European Parliament, Brussels.

Geels, F. (2002) 'Technological transitions as evolutionary reconfiguration processes: a multi-level perspective and a case study', *Research Policy*, 31. 1257–1274.

Geels, F. (2011) 'The multi-level perspective on sustainability transitions: responses to seven criticisms', *Environmental Innovation and Societal Transitions*, 1(1). 24–40.

Gibbs, D and O'Neill, K. (2014) 'Rethinking socio-technical transitions and green entrepreneurship: the potential for transformative change in the green building sector', *Environment and Planning A*, 46. 1088–1107.

Global Buildings Performance Network (GBPN): (2013) *What is a Deep Renovation Definition? Technical Report.* GBPN, Paris.

Horne, R., Maller, C. and Dalton, T. (2014) 'Low-carbon, water-efficient house retrofits: an emergent niche?', *Building Research & Information*, 42(4). 539–548.

Institute for Building Efficiency (2013) *Energy Efficiency Indicator Survey.* Institute for Building Efficiency, Washington.

Institute for Sustainable Communities/Living Cities (2009) *Scaling Up Building Energy Retrofitting in US Cities.* Institute for Sustainable Communities/Living Cities, Vermont.

Janda, K. (2014) 'Building communities and social potential: between and beyond organisations and individuals in commercial properties', *Energy Policy*, 67. 48–55.

Janda, K., and Moezzi, M. (2014) 'Broadening the energy savings potential of people: from technology and behaviour to citizen science and social potential', *Proceedings of the American Council for an Energy Efficient Economy Summer Study on Energy Efficiency in Buildings*, 17–22 August 2014, Asilomar, CA, 7. 133–146.

JLL (2012) *A Tale of Two Buildings: Are EPCs a True Indicator of Energy Efficiency?* JLL, London.

Leishman, C., Orr, A. and Pellegrini-Masini, G. (2012) 'The impact of carbon emission reducing design features on office occupiers' choice of premises', *Urban Studies*, 49(11). 2419–2437.

Mackenzie, F., Pout, C., Shorrock, L., Matthews, A. and Henderson, J. (2010) *Energy Efficiency in New and Existing Buildings: Comparative Costs and CO_2 Savings*. BRE, Watford.

Mactavish, A., Woolham, C. and Sayce, S. (2014) 'Sustainability: minimum energy performance standards for commercial buildings', *Building*, 26 June. Accessed May 2016: www.building.co.uk/sustainability-minimum-energy-performance-standards-for-commercial-buildings/5069388.article.

Managan, K. and Klimovich, K. (2013) *Setting the PACE: Financing Commercial Retrofits*. Institute for Building Efficiency, Washington.

Mansfield, J. R. (2002) 'What's in a name? Complexities in the definition of "refurbishment"', *Property Management*, 20(1). 23–30.

NABERS (2014) *National Australian Built Environment Rating System*. Accessed February 2014: www.nabers.gov.au/public/WebPages/Home.aspx.

Pellegrini-Masini, G. and Leishman, C. (2011) 'The role of corporate reputation and employees' values in the uptake of energy efficiency in office buildings', *Energy Policy*, 39(9). 5409–5419.

Policy Partners (2013) *Renovation Roadmaps for Buildings*. Policy Partners, London.

Property Industry Alliance (PIA) (2015) *Property Data Report, 2015*. PIA, London.

RICS (2013a) *Sustainability and Commercial Property Valuation*. RICS, London.

RICS (2013b) *Sustainable Construction: Realising the Opportunities for Built Environment Professionals*. RICS, London.

Smith, A. (2007) 'Translating sustainabilities between green niches and socio-technical regimes', *Technology Analysis and Strategic Management*, 19(4): 427–450.

Snoxall, J. (2013) 'Race for the prize', *Building*, 16 January. Accessed November 2017: www.britishland.com/~/media/Files/B/British-Land-V4/documents/Race_for_the_prize.pdf.

Takagi, T., Sprigings Z., Nishida, Y., Graham, P., Horie, R., Lawrence, S. and Miclea, C.P. (2015). *Urban Efficiency: A Global Survey of Building Energy Efficiency Policies in Cities*. Tokyo Metropolitan Government and C40 Cities, Tokyo.

UK Green Investment Bank (UKGIB) (2016) UK Green Investment Bank Website. Accessed May 2016: www.greeninvestment bank.com.

Ürge-Vorsatz, D., Eyre, N., Graham, P., Harvey, D., Hertwich, E., Jiang, Y., Kornevall, C., Majumdar, M., McMahon, J. E., Mirasgedis, S., Murakami, S. and Novikova, A. (2012) 'Energy end-use: building', in International Institute for Applied Systems Analysis (IIASA), *Global Energy Assessment: Toward a Sustainable Future*. Cambridge University Press, Cambridge and New York, and the International Institute for Applied Systems Analysis, Laxenburg. 649–760. Accessed February 2014: www.iiasa.ac.at/web/home/research/Flagship-Projects/Global-Energy-Assessment/Chapte10.en.html.

Westminster Sustainable Business Forum/Carbon Connect (2013) *Building Energy Efficiency: Reducing Energy Demand in the Commercial Sector*. Westminster Sustainable Business Forum/Carbon Connect, London.

Wilkinson, S. (2012) 'Analysing sustainable retrofit potential in premium office buildings', *Structural Survey*, 30(5). 398–410.

World Economic Forum (WEF) (2011) *A Profitable and Resource Efficient Future: Catalysing Retrofit Finance and Investing in Real Estate*. WEF, Geneva.

22

Sustainable building conversion and issues relating to durability

Hilde Remøy and Peter de Jong

22.0 Introduction

On 12 December 2015, the Paris Agreement set a new global target to reduce greenhouse gas emissions (GGE) as part of the approach to keep global warming well below 2 °C. In order to realise these Paris targets, the building industry has to adjust on many fronts. The built environment, including the upstream emissions from heat and electricity, is responsible for approximately 45 per cent of the total global GGE. Additionally, the built environment uses approximately 40 per cent of the materials produced globally (DGBC, 2013). Whereas the Paris Agreement focuses solely on energy use, it has become evident that material and resource (including land) use also has a big influence on sustainability. Also, focusing on energy use has some negative side effects: new buildings seemingly are more sustainable than existing buildings, which leads to new developments without taking account of existing buildings and their embodied energy, leading to vacancy and obsolescence of older real estate, and to more land-take. Notwithstanding these developments, newly constructed buildings comprise only 1–2 percent of the total stock annually (Soeter, Koppels, & de Jong, 2009). Even though these newly constructed buildings should include sustainable features in the design and operation phase, Kelly (cited in Wilkinson, Remøy, & Langston, 2014) estimated that, according to the current rates of replacement, 87 percent of the built environment that will exist in 2050 has already been built. Based on these calculations, the Inter Governmental Panel on Climate Change (IPCC) concluded that, due to the slow turnover of stock, the largest portion of carbon savings by 2030 could be achieved by retrofitting existing buildings and replacing energy-using equipment. In order to achieve significant cuts in the GGE, the upgrade of existing buildings is considered to be a more climate-friendly and immediate strategy than the production of new energy-efficient buildings (Conejos, Langston, & Smith, 2014).

Although several studies and publications have shed light on adaptive reuse as a sustainable real estate strategy, the results are not significantly implemented in practice.[1] Some possible reasons for this lack of action could be the ease of new development compared to adaptive reuse, risks related to adaptive reuse of existing real estate, and the focus on 'new is better' that prevails in the 'throw-away' culture of contemporary society. This chapter aims to shed light on the sustainability of adaptive reuse by

- offering insights about obsolescence as opposed to new developments;
- presenting results about the extent to which adaptive reuse is seen as sustainable development; and
- linking quantification of sustainability to adaptive reuse.

22.1 Office vacancy: The example of the Netherlands

In the top 15 most expensive cities to live (Cordell & Wolff, 2016), it probably makes sense financially to demolish less well-performing buildings in order to replace them with new developments with higher density and efficiency. Obsolete office buildings are demolished to give way to brand new, efficient offices or residential buildings. In the Dutch market, dealing with the high level of obsolete office buildings demands another approach. According to a report of the EIB, the Dutch Economic Institute for the Built Environment (Arnoldussen, Van Zwet, Koning, & Menkveld, 2016), the current office stock has a gross floor area (GFA) of almost 85 million m^2, based on Key Registers Cadastre[2] of 2014. More than 15 per cent of the stock[3] is vacant (DTZ Zadelhoff Research, 2016), of which a third is assumed to be frictionally vacant, a third cyclically vacant, and a third marginal and outdated (Arnoldussen et al., 2016).[4] The latter part is optimistic, especially where it is used to define the boundaries of the impact of the proposed additional regulations. Moreover, office users do not consider outdated office space as potential office accommodation because it does not meet their general requirements of location or building characteristic. Only properties that meet these general requirements influence the perceived supply of office space. Structurally, or long-term, vacant offices are rarely taken up, whereas new buildings are regularly added to the market.

Additionally, in a post-global financial crisis (GFC) period, hidden vacancy, as described by Remøy, Koppels, and Lokhorst (2013), is becoming perceptible as office users are gaining the

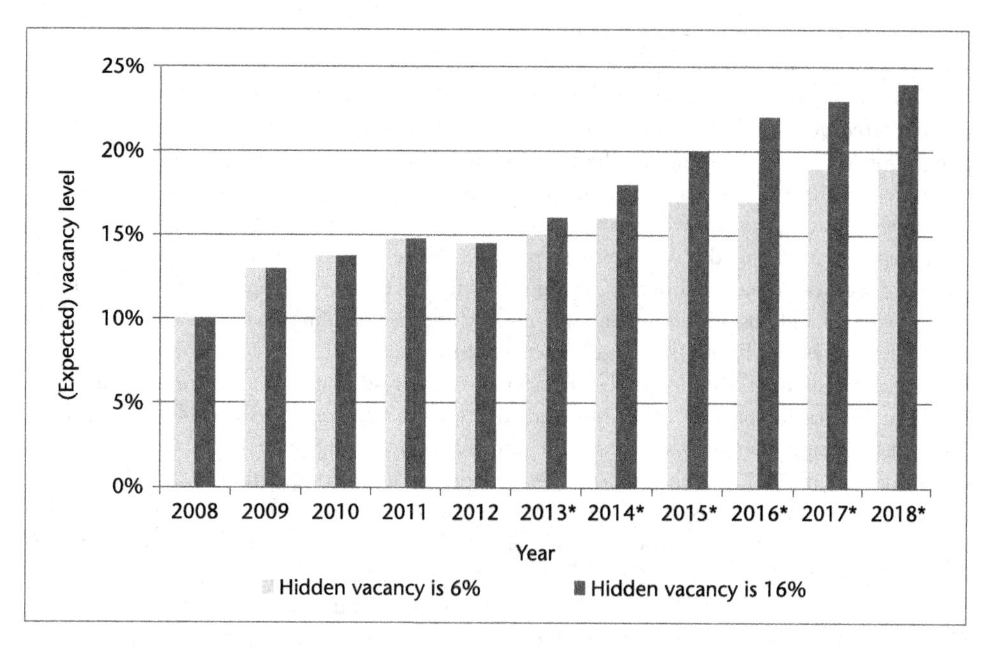

Figure 22.1 Expected future vacancy as a percentage of total stock

Source: Remøy et al. (2013)

Note: * = prognosis

ability to optimise their accommodation needs (Figure 22.1). The current office market in the Netherlands is a replacement market. The persistent vacancy originates from oversupply developed in the late 1990s and early 2000s before the GFC. The oversupply was caused by overproduction, due to overly optimistic assessments of demand for office use (direct return on investment) and high expectancies of value development of real estate as compared to other sectors (indirect return on investment). However, new technology and innovations have led to new ways of working, including flexible working, desk sharing, and working from home or a different location. Following this, there is a lower perceived status to having a private office than previously. Finally, this leads to overall lower demand for office space, and office organisations moving to smaller premises, often in more centrally located areas.

22.2 Lifespan and the life cycle cost approach

Probably one of the most challenging tasks for sustainable real estate development is establishing a life cycle cost (LCC) approach throughout the development cycle. Ultimately, LCC is more relevant to investors and clients than the current focus on investment costs. Although mandatory core competencies are set (Fisher, Coll, Pelly, & Percy, 2008) and awareness is rising, there is an observed lack of knowledge and expertise (Dixon et al., 2008). The 'vicious circle of blame' was described in 2000 by Cadman (as cited in RICS Europe, 2008). Cadman suggested that investors, developers, occupiers, and contractors/designers sequentially blamed each other for the failure to adopt sustainability in building practices. In an effort to give the circle a positive connotation, RICS Europe followed up with a statement that 'not going green' may eventually lead to a building becoming obsolete, as over time actors will prefer sustainable buildings to the non-sustainable or less sustainable buildings (RICS Europe, 2008). However, the authors believe that as current low levels of sustainable buildings in the total market attest, the circle of blame is not broken yet, and the virtuous circle of sustainability has not yet been achieved. The building industry is focused on new production, and in the real estate sector, new developments are getting more attention than the maintenance and upgrading of the existing stock, even where this is not supported by the division of the economic value between new-build and existing buildings. The performance of a building for the first user remains the primary driver for development. In research on the measurements of flexibility to increase the lifespan of buildings, it became clear that there are still investors who prefer to ignore the consideration of a second function for a building at the start of the development process. 'If the project is not a guaranteed success, we won't start' is a typical quote representing this view (Remøy, de Jong, & Schenk, 2011). In the Netherlands and possibly elsewhere, the number of office buildings that are vacant after a few years of use, or even without being used, contradict this view.

Adaptable buildings are designed with a view to later adaptations, and can accommodate several functional lifespans in order to avoid functional obsolescence. Adaptable buildings are designed for longer lifespans, and hence are sustainable by definition (Remøy, 2010; Remøy, de Jong, & Schenk, 2011), as they involve less material use in the long run. Duffy (1990) was one of the first to describe robust buildings with a high capacity to adapt to the changing requirements of their users. He defined buildings as systems with several subsystems or layers. His way of defining adaptability was based on experience with refurbishments of office buildings and which layers may be altered in order to renew the functional lifespan of the building without influencing its technical functioning. The distinction between different layers was related to the dissimilar lifespans of the layers. In Europe and North America, calculating with an estimated building life of 50–70 years is common, although some parts of the building have far shorter lifespans, and other parts, or layers, like the load-bearing structure, could have a much longer lifespan (Nutt, 1988).

Duffy's approach was adapted by Brand (1994) and Leupen (2006), who both described the building as a system of layers. A high independence of the layers makes adaptations possible. According to Leupen (2006), buildings that consist of several layers are sustainable and are more likely to be adapted than buildings where the different layers are dependent on each other.

Still, the 'business case' for sustainable development is not fully convincing. The calculation of LCC and life cycle assessment (LCA) requires the use of an estimated lifespan for the buildings under consideration. As LCC and LCA are projections of future impacts and cost, the lifespan used in their calculation is also a prediction. Derived concepts of lifespan, like economic, functional, technical, or fiscal lifespan, are mainly focused on a defined shorter period of the total lifespan for a specific building, without detailed considerations for its total lifespan. The predicted lifespans adopted are short, typically between 30 to 50 years. In many cases, buildings are replaced in rezoning schemes after shorter periods to accommodate increased density and real estate value. Paradoxically, demolition and new build is chosen to improve the energy performance, without taking any account of the material and embodied energy wasted as a result. The problem of using short lifespans is that the opportunities and values of longer lifespans are missing from the calculations, with the apparent risk of a self-fulfilling prophecy that buildings are regarded as obsolete after reaching the targeted lifespan.

LCC was defined in 2008 (NEN-ISO, 2008) to help to improve decision-making and evaluation processes at relevant stages of any project. The discounted cost[5] consists of/distinguishes between acquisition costs,[6] capital costs,[7] operational costs,[8] disposal costs,[9] end-of-life costs,[10] and external costs.[11] With this definition, LCC becomes an all-inclusive approach. Environmental costs and end-of-life costs are addressed, enabling the incorporation of circular economy principles in the scope of LCC: to quantify the LCC for input into a decision-making or evaluation process and include inputs from other evaluations (e.g., environmental assessment, design assessment, safety assessment, functionality assessment, regulatory compliance assessment). However, in 2008 it was seen as necessary to give room to the industry to exclude both environmental costs and end-of-life costs. It is probable that keeping environmental costs and end-of-life costs out of the equation was necessary in order for the market to adopt the measure; however, this exclusion prevented further development towards an integrated approach to decision making and policy making.

22.3 National regulations and evidence

In a letter to the parliament (Blok, 2016), the Dutch Minister of Housing and Spatial Planning revealed plans to enforce a minimum energy label for Dutch commercial real estate. An energy label C or higher would be required for public and private offices larger than 100 m^2 before 2023. The EIB calculated the cumulative reduction of energy use by this measurement at 8.6 petajoules (PJ).[12] The proposed regulation follows existing Dutch building regulations, which require energy-saving measures in buildings that have a lifespan of five years or longer. Both regulations are necessary to fulfil national ambitions to meet international agreements, including the Paris Agreement of 2015 (UN, 2015). As such, the new requirements can be seen as an example of national regulation based on international agreements. This is likely to happen in many countries, though it is uncertain when and with what ambitions.

In reaction to this regulation, Arnoldussen et al. (2016) acknowledged an increase in vacancy due to this measurement following market adoption of the new requirements. A number of Dutch offices with lower energy performance will be taken out of the supply[13] basically at the cost of owners who are not able to finance the cost of improvements. The renewed focus on the energy agenda could easily cause a surge in new buildings being brought onto the market, again causing a replacement market where office users move to new buildings, leaving older buildings behind (Remøy & Wilkinson, 2012). A first example of such behaviour was an outcome of the RICS survey on the engagement of their members with the sustainability agenda (Dixon et al., 2008): members ranked energy use much higher than waste management and natural resources management. As market parties ranging from users to developers and investors do not regard lifespan to be as important as energy use, there seems to be a need for policy on this topic; otherwise, new legislation on energy saving could end in an increased use of natural resources.

A study on the Prague office market by Dvořáková (2016) confirmed the findings from the 2008 RICS study (Dixon et al., 2008). Using a Delphi approach, the importance of predefined sustainability factors for the developer, investor, and tenant were studied. Semi-structured interviews were held to gain a more thorough understanding of the current state of the Prague market, its development and stakeholders' overall awareness of sustainability, as well as which sustainability factors were perceived more as barriers and which were perceived as drivers of sustainable development.

When comparing the three types of stakeholders, some similarities were noteworthy. For example, both developers and investors saw the return on investment and selling price as the most important factors for sustainable development. This indicates the prevailing financial focus of the developers and investors in the market. Developers invest in sustainable solutions when they expect it to lead to a premium on selling price or a decrease in the time on the market, influencing their return on investment. In the case of the investor, this notion is supported by other factors seen as very important and strongly connected to one another, such as the asset value, exit yield, and occupancy or rent level. This confirms findings from more developed markets as well (Pivo & Fisher, 2010; Nappi-Choulet & Decamps, 2013). The knowledge of sustainability is connected with the interest in sustainability, and was found unimportant by all stakeholders. The perceived unimportance of knowledge and interest in sustainability is linked to one of the biggest barriers to sustainability, namely unsatisfactory education about sustainability and the reluctance of market stakeholders to immerse themselves fully in the topic. From the interviews conducted by Dvořáková, it seems that for most stakeholders in the field, sustainability is limited to earning points in the environmental certification systems; they are not making much effort to understanding the problem in a more systematic way. Over–reliance on rating tools, where a slavish adoption of tools can be seen to drive design and result in attainment of targets to give a market edge and premium to value, is also found in other markets. Earlier studies have also pointed out that the stakeholders here may have missed the point of sustainability (Wilkinson & Remøy, 2015a).

The opinions of stakeholders regarding the increase in design and construction costs for top-level certified buildings varied from an estimate of 2 per cent to 15 per cent (Dvořáková, 2016). These findings confirm other studies, proving that it is very hard to establish certainty about cost increase due to the immaturity of the field, the project phase in which developers decide to certify a building, the original design quality of a building before certification, and so on. However, the cost increase is clearly linked to the targeted level of certification, as low levels of certification may be achieved by small improvements and 'easy' credits (not requiring changes in the design), whereas the top levels require stronger devotion, higher financial inputs, and early decision making. Another factor that was often mentioned as hindering not only sustainable development but development of buildings in Prague in general is the complex regulatory environment and long permit procedures in cases of big developments (Dvořáková, 2016). The study further showed that investors in the Prague market demand certified office buildings, hoping for lower running costs and attractiveness of the buildings for A-class tenants. Some respondents saw positive impacts of the certification on the building's management, the way that the building performs in the operating period, and the way that it is commissioned.

In Dvořáková's interviews held with tenants, the emphasis on the rent level indirectly shows that the market is not yet fully prepared to pay a premium for buildings with increased environmental quality. However, as stated by some respondents, the market shows a trend for paying more attention to the wellbeing of employees and using sustainability as a means of attracting and keeping a skilled workforce. The main decision-making factors were still location, amenities in the neighbourhood, and architecture; however, some companies (usually big multinationals) are pushed by their mother organisation to value the building's energy certificate as well, and thereby enforce their corporate social responsibility and company image. The difference in motivation of the company management and employees was pointed out during the interviews, although the trend is moving towards paying more attention to the needs of employees. This trend of emphasising the wellbeing and satisfaction of the occupiers is also triggered by the certification organisations, as, for example, the new 2016 version of BREEAM values the category Health & Wellbeing higher than the previous version.

As markets are usually driven by tenants or demand, the tenant could be the stakeholder to break the circle of blame. However, a more plausible scenario is probably that if an accelerated development of sustainable offices is aimed for, the push has to come from all the involved actors simultaneously, influencing one another in the decision-making process through adjusting demand and supply.

22.4 Quantification of sustainability

Current rating tools are focused on new construction, recurrent building operations, and maintenance. Worldwide, more than 600 tools are used to measure or evaluate the social, environmental, and economic dimensions of sustainability (Conejos et al., 2014). The most used rating tools are LEED, Green Star, and BREEAM (Damwijk, 2015). These three tools provide a broad assessment of the environmental impact of a building. Each tool is based on a checklist, and points are awarded based on answers to the questions in the checklist (Jansz, 2012). The tools are used to rate a building, and hence to market the building according to its sustainability label. The tools have been developed in different geographical, climatic, and cultural contexts, and hence focus more or less on different aspects. As an example, Green Star was developed in Australia and focuses more on water usage and less on energy usage than LEED, developed in the US, and BREEAM, developed in Northern Europe (Jansz, 2012). Next to these large, holistic tools, a range of smaller instruments exist that focus on one single aspect of sustainability. For example, energy is seen as an important aspect to sustainability in the Dutch context, and so the Dutch government developed the Energy Performance Coefficient (EPC) in 1995, followed by the Energy Label in 2008. Whereas the EPC is a tool to certify construction plans, the Energy Label was developed as a design tool, and gives recommendations for how to improve the energy label of a design (Jansz, 2012).

An international comparison of rating tools (Reed, Bilos, Wilkinson, & Schulte, 2009) showed some of the challenges of measuring sustainability. The assessment criteria used in the tools are different. Comparing 15 assessment criteria, the study showed that all criteria were used in several tools, but no tool used all criteria. The assessment methods and weighting of the criteria are also different, hence buildings that get the top rating by one tool score average with another tool. Moreover, the differences between tools are unclear. A common basis and agreement about criteria, assessment methods, and weighting of criteria is needed.

The building's lifespan plays a minor role in current methodologies, and the standard lifespan applied in the models is not supported by hard data. In 2017, BREEAM is the only rating tool that has an application to assess buildings in use. None of the existing rating tools appraises adaptive reuse performance (Conejos et al., 2014). Some authors (de Jonge, 2005; Langston, 2011) call for the use of LCA models. However, commonly used LCA models apply a standard lifespan expectation, which means that the specific building's qualities or shortcomings are not taken into account. Haapio and Virtaniemi (2008) emphasise that the effect of the assumed lifespan on the results of environmental assessments should be thoroughly analysed. The first explorations of estimating the lifespan of buildings and including this estimated service life (ESL) in an environmental assessment showed a profound influence (Jansz, 2012; Van den Dobbelsteen & Van der Linden, 2005). To be able to fully rate the sustainability of new, adapted, and existing buildings, LCA tools are needed that fully incorporate the lifespan of buildings, including the material use and energy use. A rating tool should include durability, sustainability, and adaptability.

22.5 Decision making

Many assessment methods can be used to assess and measure sustainability in the built environment. In the Netherlands and most European countries, BREEAM is the most commonly used method. BREEAM is a checklist-based assessment method. Depending on which aspects are applicable to a certain project, a sustainability score is determined. BREEAM-NL includes a weighting for the different aspects, while in LEED and Green Star all aspects are weighted equally to calculate the final sustainability score (Damwijk, 2015). Even though the checklist system in these methods is rather simple to understand, the approval of aspects from the checklist is difficult and time consuming and needs to be conducted by an approved assessor. The sustainability attributes that could be embraced by the market in conversion adaptations are illustrated in Table 22.1. The table shows the sustainability attributes of BREEAM and Green Star, and shows that although the categories are different, most attributes measured are similar.

Table 22.1 Sustainability characteristics in adaptive reuse

Characteristics	*Green Star*	*BREEAM*
Management	Green Star accredited Professional commissioning and tuning Adaptation and resilience Building information Commitment to performance Metering and monitoring Construction Environmental management Operational waste	Performance assurance Building site and surroundings Environmental impact of building site User manual Consultation Safety Knowledge transfer Maintenance/serviceability Life cycle costing analysis
IEQ in Green Star, Health and Wellbeing in BREEAM	Indoor air quality Acoustic comfort Lighting comfort Visual comfort Indoor pollutants Thermal comfort	Daylight admittance View/vista Daylight control High-frequency lighting Indoor and outdoor artificial lighting Lighting control Purge ventilation Internal air quality Volatile organic compounds Thermal comfort Temperature control Acoustics Private outdoor space Accessibility

	Sustainability criteria Green Star	*Sustainability criteria BREEAM*
Energy	Greenhouse gas emissions Peak electricity demand reduction	Energy efficiency Sub-metering energy use Energy-efficient outdoor lighting Renewable energy sources Energy-efficient cooling/freezer space Energy-efficient elevators Energy-efficient escalators and ribbons Guaranteed thermal quality facade
Transport	Sustainable transport	Availability of public transportation Distance to facilities Alternative transportation Pedestrian and bike safety Traffic plan and parking policy Traffic information point
Water	Potable water	Water usage Water meter Leak detection of the main water supply Self-closing water supply sanitary functions Water recycling Irrigation system Car-cleaning service
Materials	Life cycle impacts Responsible building materials Sustainable products Construction and demolition waste	Building materials Sustainable products Robust design Building flexibility
Waste		Waste management on building site Use of recycled material Storage for reusable material Compost Interior waste management
Land use and ecology	Ecological value Sustainable sites Urban heat island effect Stormwater Light pollution Microbial control Refrigerant impacts	Reuse of land Polluted land Plants and animals on building site Plants and animals as users of the site Long-term sustainable co-use with plants and animals Efficient land use
Pollution		Global warming potential (GWP) of refrigerants for climate control Prevent leakage of refrigerants for climate control GWP of refrigerants for cool and freeze storage Heating-related NO_x emissions Minimize light pollution Noise pollution
Innovation	Innovative technology or process Market transformation Improving on Green Star benchmarks Innovation challenge Global sustainability	

Source: Wilkinson & Remøy (2015b)

The technological aspects considered in sustainable conversion are a combination of the physical building aspects, using fewer materials and resources, and less transport energy and pollution during construction. Embodied energy is considered and structure and fabric is retained where possible. In addition, reduction of operational energy and water consumption belongs to the technological aspects. Societal aspects include increased amenities and wellbeing for residents and building users where possible. Environmental assessment tools such as Green Star in Australia or BREEAM in the Netherlands, the UK, and elsewhere could be adopted to evaluate the reuse of existing buildings. However, where excessive amounts of deleterious materials, such as asbestos, exist or sick building syndrome is prevalent, adaptive reuse might not be desirable or viable.

The potential of office to residential conversion is well recorded (Geraedts & Van der Voordt, 2007; Remøy, 2010; Remøy & Van der Voordt, 2014; Wilkinson et al., 2014) and determined by market, location, and building (functional, technical, financial, aesthetic, and legal) characteristics. Adding sustainability assessment to existing adaptive reuse assessment methods would improve decision-making support for adaptive reuse and sustainability.

22.6 Conclusions

The Brundtland Commission (1987) defines sustainable development as 'development that meets the needs of the present without compromising the ability of future generations to meet their own needs'. This definition can be broadened to discuss the expected lifespan of a system or parts or aspects in the system. From this proposition, sustainability goals can be defined and achieved, nowadays resulting in climate emission restrictions and aims for reduction. While during the 1990s sustainable development was mostly understood as development without growth, this train of thought has been replaced by ideas of cradle-to-cradle developments that consider recycling or upcycling second-hand building materials (McDonough & Braungart, 2002) and extended lifespan (de Jonge, 1990) to reduce waste production and energy use in construction.

Extending building lifespan is a means to improve the sustainability of the built environment by increasing the durability of buildings (de Jonge, 1990). From this perspective, change is seen as a constant, while the implications of the changes are seen as uncertain (Leupen, 2006). The building is seen as a frame and possible functions or activities as infill. A slightly different approach focuses on buildings that are robust and flexible in use but specific in appearance, and that can accommodate several programmes. By developing new buildings with a sturdy structure and without a functional programme, a proposal is made for buildings that can become dear to their users and last for at least 100 years.

In many models, 50 years is applied as lifespan for commercial buildings, for which no rationale can be determined. This is a very short average lifespan for buildings, as most survive far longer. The aim should preferably be 200 years in order to fulfil the demand of quality (urban, architectural), sustainability (embodied energy), and circular economy (reduction of waste). During this longer lifespan, maintenance and operation costs, including refurbishment and periodic renovations, will surpass initial cost and demand a new approach to assessing the costs and benefits of the development and use of real estate.

Notes

1 Dynamis (2016) constructed an index for offices with a high/reasonable/low take-up potential with the supply–demand ratio. Being the largest group with a ratio of 8:1, the mid-section reveals that many adaptive reuse opportunities are still missed.

2 https://bagviewer.kadaster.nl/lvbag/bag-viewer.

3 The stock used by DTZ, defined as office and industrial property market, is a segment, 49.5 million m², of the stock EIB/BAG is referring to.
4 Frictional vacancy is the result of reconstruction (transformation/refurbishment), cyclical vacancy refers to the gap to enable transactions and moving, and marginal vacancy results from obsolescence.
5 Resulting cost when the real cost is discounted by the real discount rate or when the nominal cost is discounted by the nominal discount rate.
6 All costs included in acquiring an asset by purchase/lease or construction procurement route, excluding costs during the occupation and use or end-of-life phases of the life cycle of the constructed asset.
7 Initial construction costs and costs of initial adaptation where these are treated as capital expenditure.
8 Costs incurred in running and managing the facility or built environment, including administration support services.
9 Costs associated with the disposal of the asset at the end of its life cycle, including taking account of any asset transfer obligations.
10 Net cost or fee for disposing of an asset at the end of its service life or interest period, including costs resulting from decommissioning, deconstruction, and demolition of a building, and recycling, making environmentally safe, recovering and disposing of components and materials, and transport and regulatory costs.
11 Costs associated with an asset that are not necessarily reflected in the transaction costs between provider and consumer and that, collectively, are referred to as externalities.
12 A commitment to label A would deliver a reduction of 10.9 PJ. Total energy consumption in the Netherlands was 1,792 PJ in 2015 (CBS StatLine, 2017) – a reduction of 0.6 per cent.
13 In the search for key attributes for adaptation (Wilkinson & Remøy, 2011) the energy performance, as part of the Property Council of Australia grade, seems to be one of the most important attributes.

References

Arnoldussen, J., van Zwet, R., Koning, M., & Menkveld, M. (2016). *Verplicht energielabel voor kantoren*. EIB, Amsterdam. Retrieved from: www.eib.nl/pdf/verplicht_energielabel_voor_%20kantoren.pdf.

Blok, S. (2016). *Kamerbrief over energiebesparing gebouwde omgeving*. Den Haag.

Brand, S. (1994). *How Buildings Learn: What Happens After They're Built*. New York, Viking.

Brundtland, G. H. (1987). *Our Common Future*. Retrieved from: www.worldinbalance.net/agreements/1987-brundtland.php.

CBS StatLine (2017). Energy balance sheet: Key figures. Retrieved from: http://statline.cbs.nl/Statweb/publication/?DM=SLEN&PA=37281eng&D1=a&D2=0-1,4,7-12&D3=153&LA=EN&HDR=G2,G1&STB=T&VW=T.

Conejos, S., Langston, C., & Smith, J. (2014). Designing for better building adaptability: A comparison of adaptSTAR and ARP models. *Habitat International*, *41*, 85–91.

Cordell, K., & Wolff, S. (2016). *The Routledge Handbook of Ethnic Conflict*. London, Routledge.

Damwijk, R. M. (2015). *Comparing Adaptation and Demolition & New Build for Office Buildings in the Newly Developed ADNB Indicator*. MSc, Delft University of Technology, Delft.

de Jonge, H. (1990). The philosophy and practice of maintenance and modernisation. Paper presented at the *Property Maintenance Management and Modernisation*, Singapore, 7–9 March 1990.

de Jonge, T. (2005). *Cost Effectiveness of Sustainable Housing Investments*. Delft, DUP Science.

DGBC. (2013). *DGBC Handleiding voor BREEAM-NL Assessments*. Rotterdam, Dutch Green Building Council.

Dixon, T., Colantonio, A., Shiers, D., Reed, R., Wilkinson, S., & Gallimore, P. (2008). A green profession? A global survey of RICS members and their engagement with the sustainability agenda. *Journal of Property Investment & Finance*, *26*(6), 460–481.

DTZ Zadelhoff Research. (2016). *The Netherlands, A National Picture: Fact Sheets Office and Industrial Property Market*. Retrieved from: www.dtz.nl/en/market-information.

Duffy, F. (1990). Measuring building performance. *Facilities*, *8*(5), 17–20.

Dvořáková, K. (2016). *Sustainability Drivers and Barriers: Mapping the Motives for Sustainable Office Development in Prague*. MSc, Delft University of Technology, Delft.

Dynamis (2016). Sprekende *Cijfers Kantorenmarkten*. Utrecht, Dynamis B.V.

Fisher, R., Coll, L., Pelly, L., & Percy, J. (2008). Surveying sustainability: A short guide for the property professional. *Appraisal Journal*, *76*(1), 15–22.

Geraedts, R. P., & van der Voordt, D. J. M. (2007). A tool to measure opportunities and risks of converting empty offices into dwellings. In: *ENHR Proceedings of International Conference on Sustainable Urban Areas*, Rotterdam.

Haapio, A., & Viitaniemi, P. (2008). A critical review of building environmental assessment tools. *Environmental Impact Assessment Review, 28*(7), 469–482.

Jansz, S. (2012). *The Effect of the Estimated Service Life on the Sustainability of Vacancy Strategies.* (MSc), Delft University of Technology, Delft.

Langston, C. (2011). Estimating the useful life of buildings. Paper presented at the *Australian University Building Educators Association*, Gold Coast.

Leupen, B. (2006). *Frame and Generic Space.* Rotterdam, 010 Publishers.

McDonough, W., & Braungart, M. (2002). *Cradle to Cradle: Remaking the Way We Make Things.* New York, North Point Press.

Nappi-Choulet, I., & Decamps, A. (2013). Can sustainability enhance business district attractiveness? A survey of corporate property decisions in France. *Urban Studies, 50*(16), 3283–3304.

NEN-ISO. (2008). Buildings and constructed assets – Service-life planning – Part 5: Life-cycle costing (ISO 15686-5:2008,IDT).

Nutt, B. (1988). The strategic design of buildings. *Long Range Planning, 21*(4), 130–140.

Pivo, G., & Fisher, J. D. (2010). Income, value and returns in socially responsible office properties. *Journal of Real Estate Research, 32*(3), 243–270.

Reed, R., Bilos, A., Wilkinson, S., & Schulte, K. W. (2009). International comparison of sustainable rating tools. *Journal of Sustainable Real Estate, 1*(1), 1–22.

Remøy, H. (2010). *Out of Office: A Study on the Cause of Office Vacancy and Transformation as a Means to Cope and Prevent.* Amsterdam, IOS Press.

Remøy, H. & van der Voordt, D. J. M. (2014). Adaptive reuse of office buildings: Opportunities and risks of conversion into housing. *Building Research & Information*, 42(3), 381–390.

Remøy, H., de Jong, P., & Schenk, W. (2011). Adaptable office buildings. *Property Management, 29*(5), 11. Retrieved from: www.emeraldinsight.com/journals.htm?articleid=1955890&ini=aob.

Remøy, H., Koppels, P., & Lokhorst, J. (2013). Verborgen leegstand. Paper presented at the *Proceedings of the 20th European Real Estate Society 20th Annual Conference*, Vienna, 3–6 July 2013; authors' version.

Remøy, H., & Wilkinson, S. J. (2012). Office building conversion and sustainable conversion: A comparative study. *Property Management*, 30(3), 218–231.

RICS Europe. (2008). *Breaking the Vicious Circle of Blame: Making the Business Case for Sustainable Buildings.* Findings in Built and Rural Environments. Brussels, RICS Europe.

Soeter, J. P., Koppels, P., & de Jong, P. (2009). Market interdependencies between real estate, investment, development and construction. In: Ruddock, L. (Ed.), *Economics for the Modern Built Environment.* London, Taylor & Francis, 229–248.

United Nations (UN). (2015). *Paris Agreement.* Retrieved from: https://treaties.un.org/Pages/ViewDetails.aspx?src=IND&mtdsg_no=XXVII-7-d&chapter=27&clang=_en.

van den Dobbelsteen, A., & van der Linden, A. (2005). Managing the time factor in sustainability: A model for the impact of the building lifespan on environmental performance. In: Yang, J., Brandon, P., & Sidwell, A. (Eds.), *Smart & Sustainable Built Environments.* Oxford, Blackwell, 223–233.

Wilkinson, S. J., & Remøy, H. (2011). Sustainability and within use office building adaptations: A comparison of Dutch and Australian practices. In: *PRRES 2011: Proceedings of the 17th Pacific Rim Real Estate Society Annual Conference.*

Wilkinson, S. J., & Remøy, H. (2015a). Sustainability and office to residential conversions in Sydney. In: *ZEMCH 2015 Conference Proceedings*, Lecce, 22–25 September 2015.

Wilkinson, S. J., & Remøy, H. (2015b). Building resilience in urban settlements through conversion adaptation. In: *RICS Cobra Aubea 2015 Conference Proceedings*, Sydney, 8–10 July 2015.

Wilkinson, S. J., Remøy, H., & Langston, C. (2014). *Sustainable Building Adaptations.* Oxford, Wiley-Blackwell.

23 Sustainable urban redevelopment in the Netherlands

Erwin Heurkens

23.0 Introduction

Spatial planning and real estate development in the Netherlands is based on a strong tradition of utilising land in an efficient way and creating a well-structured built environment. By making use of a comprehensive integrated approach to planning (Dühr, Colomb, and Nadin, 2010), decisions on building new spatial infrastructures have always been based on a very coordinated way of working between layers of government bodies. On a more operational level, the Dutch have become known for their cooperative way of developing cities' urban areas with public and private actors, named integrated urban area development (Bruil, Hobma, Peek, and Wigmans, 2004). This development approach is characterised by mixed-use real estate and infrastructure development aimed at delivering places of high spatial and design quality. However, some authors argue that the coordinative planning doctrine is diminishing in the Netherlands (Roodbol-Mekkes, Van der Valk, and Korthals Altes, 2012) because of socio-economic, political, and financial reasons that point to the reduced influence of government institutions in spatial decision making and urban development in general.

Especially since the start of the global financial crash (GFC) in 2008, it has become apparent that the Dutch municipal active land development policies impose too much financial risk on local authorities, and this has created the need for alternative development strategies (Van der Krabben and Heurkens, 2015). These development strategies have put the private sector (i.e., developers and investors) in the lead in developing urban areas, with a more facilitative role for municipalities (Heurkens, 2012; Heurkens, 2013; Heurkens and Hobma, 2014; Heurkens, Adams, and Hobma, 2015). Others (Buitelaar et al., 2012) indicate that local communities and entrepreneurs are more likely to play an important role in urban development. Within such a changing cooperative context for urban development, there are debates about how to develop Dutch cities, urban areas, and real estate in a more sustainable manner, as they are vulnerable to climate change and resource scarcity.

This chapter will shed light on the ways that the Dutch conceive, organise, and practice sustainable urban redevelopment. We do so first by explaining some indicators and concepts of sustainable urban and real estate redevelopment applicable to Dutch practice. This is followed by sections on development strategies and partnership models that are in place in the Netherlands to create sustainable urban areas. To illustrate the various approaches to realising sustainable urban redevelopment – that is, economically viable, socially responsible, and environmentally friendly places (Heurkens, 2016) – two contemporary contrasting case studies in Rotterdam and Amsterdam are discussed. Finally, we conclude with some major implications of these findings for Dutch and international practices of sustainable urban and real estate redevelopment.

23.1 Introduction to Dutch sustainable urban real estate development

The Dutch Green Building Council (DGBC) issues BREEAM certificates for sustainable buildings similar to other countries around the globe. As a network organisation, it works together with participants in multiple projects on making the built environment sustainable (DGBC, 2016). DGBC uses various assessment tools such as BREAAM-NL New-Build, BREAAM-NL In-Use, BREEAM-NL Demolition and Deconstruction, and BREEAM-NL Urban Areas to make distinctions between buildings, materials, and urban development. Over the last decade, DGBC has issued numerous BREEAM certificates for the first three buildings and materials categories (BREEAM-NL, 2016a). However, a quick glance at the BREEAM-NL Urban Areas illustrates that there are just five certified projects registered (BREAAM-NL, 2016b). Moreover, these urban development projects are mainly industrial or office parks, and not inner-city brownfield mixed-use redevelopment projects, which are more complex in nature and also quite numerous in Dutch practice. Based on the assumption that certifying such urban development projects is preferable for benchmarking and creating sustainable neighbourhoods, the question remains: to what extent do the Dutch deliver sustainable urban redevelopments?

Looking beyond such certifications, different perceptions exist in Dutch urban area development practice about what is conceived as sustainable urban redevelopment. According to Puylaert and Werksma (2011), sustainable urban development links spatial quality to aspects of people, planet, and profit, which adds value for all stakeholders involved and society as a whole, now and in the future. Puylaert and Werksma (2011) further argue that sustainable urban development can be achieved by focusing spatial interventions on several aspects: soil, water, urban green, nature and landscape, energy, mobility and transport, health and safety, heritage and identity, transformation and redevelopment, economic vitality, process and programmatic flexibility, and social vitality. Despite the fact that such aspects are recognised by public and private stakeholders as important to creating sustainable urban areas, development projects in practice often include just a few aspects of sustainable urban areas, and therefore can hardly be earmarked as sustainable.

Steen (2016: 5) argues that the 'underlying problem in [Dutch] practice is that there is uncertainty in the field of urban area development on how to develop sustainable mixed-use urban areas, both in terms of product (what to develop) and process (how to develop it)'. Buskens (2016) adds that these multiple conceptions make it harder to identify whether development projects are truly sustainable. Based on a survey of Dutch urban redevelopment professionals, interviews and literature reviews, Buskens (2016) concludes that there seem to be several obstacles present in Dutch practice that limit the delivery of sustainable urban redevelopment projects:

- Sustainability is mainly driven by municipalities.
- The primary focus is on environmental aspects of sustainability.
- Sustainability discussion focuses on real estate building levels.
- There is a lack of incentives for developers to commit to sustainable urban development.
- The focus on sustainability is relatively new for developers.
- Approaches to sustainability are reactive instead of proactive.

These findings imply that institutional conditions for realising sustainable urban redevelopment seem far from optimal (Heurkens, 2016). Despite the reputation of Dutch urban development practice as being able to realise comprehensive high-quality urban areas and real estate, Buskens and Heurkens (2016) argue that the development industry and municipalities miss opportunities to make sustainable urban redevelopment common practice and a focal point of spatial decisions and organisational commitment. Haak and Heurkens (2015) have also indicated that the Dutch building sector is amongst the least innovative in the country, and relatively slowly changes its ways of working and products into more sustainable variants. The following sections elaborate on two broad development approaches and subsequent development strategies that can be used to develop urban areas in the Netherlands in a more sustainable manner.

23.2 Dutch urban development approaches

Dutch urban redevelopment practice has witnessed some changes in terms of prevailing development approaches over the last two decades. Buitelaar et al. (2014) have introduced a useful categorisation of urban development approaches currently present in Dutch practice (Table 23.1), which are elaborated on hereinafter. These are:

- integrated urban development, and
- organic urban development.

As early as the beginning of the millennium, Bruil et al. (2004) indicated that a so-called Dutch form of 'urban area development' came into existence, which is known for its integrated comprehensive approach towards urban and real estate development. This, among other things, involves linking public and private interests, joining up spatial issues across different scales, financing infrastructure through real estate development revenues, and linking various professional disciplines. The integrated development approach has been used extensively as a basis for decisions on greenfield development, as well as brownfield redevelopment projects with a mixed-use function emphasis. According to Franzen, Hobma, De Jonge, and Wigmans (2011), this integrated development approach can be considered as quite complex, which brings forward the need for professional public and private actors to organise and manage urban redevelopment processes in a merely 'top-down' manner. Due to its complex and comprehensive nature, this integrated urban development strategy came under pressure during the GFC from 2008 onwards. This event caused actors to be more risk averse in real estate investment and development, and to down-scale development activity and to re-schedule development phasing to realistic market uptake estimates.

Table 23.1 Development approaches in the Netherlands

	Integrated urban development	*Organic urban development*
Approach	At once	Gradually
Scale of development	Large	Small
Type of management	Project management	Process management
Plan type	Blueprint	Strategic
Type of developer	Large developers	Small developers and individuals
Role of local authority	Active and risk prone	Facilitative
Development and management	Sequential	Mixed

Source: Adapted from Buitelaar et al. (2014)

More recently, Buitelaar et al. (2014) have argued that integrated urban development is gradually being replaced by a more 'bottom-up' organic urban development approach. Organic urban development is more gradually phased over time and involves starting a variety of separate real estate developments wherever and whenever there is demand for them. In addition, the scale of development is rather small with real estate redevelopment on a plot-by-plot basis rather than large-scale land development. Moreover, the organic approach stresses the importance of process management over project management. Also, the role of plans seems to be more strategic and flexible than blueprint variants in integrated urban development to cope with changing needs. Furthermore, large professional real estate developers are not necessarily involved in organic urban redevelopment, which is delivered by smaller local developers, private entrepreneurs, and property owners. Additionally, whereas local authorities were very actively steering integrated urban development, they are more facilitative to private and community initiatives in organic development. And finally, urban and real estate development and management occur in a more mixed manner rather than sequentially, which means that redevelopment is favoured.

Despite these notable differences, Buitelaar, Galle, and Sorel (2014) argue that mixed strategies with elements from both top-down integrated and bottom-up organic urban development occur in Dutch practice (see also Robles-Duran, 2011). It also seems that new ways of developing urban areas and real estate are highly dependent on existing institutional conditions (Heurkens, 2016). Established values, behaviours, systems, and rules are hard to change and offer limited ground and opportunities for completely allowing new approaches to become common practice. Although the term 'organic urban development' seems ambiguous, it nevertheless does occur in practice and represents an alternative way of approaching urban redevelopment. To a certain degree, one could argue that organic urban development is just a collection of (smaller-scale) real estate developments taking place within the context of a more largely defined urban project, and that the focus on smaller-scale real estate development and investment is caused by stagnating market demand and a more risk-averse attitude of the development industry towards big schemes. Already we see that with the rising demand for housing in various parts of the Randstad, signs point towards a revival of the integrated urban development approach (Buitelaar et al., 2014).

23.3 Private-sector-led urban development strategies

Nonetheless, fundamental changes, such as a more facilitative role for municipalities and a more manageable size of developments, seem to have gained ground in Dutch urban development practice. These fundamental socio-economic changes and development approaches have resulted in an increasing number of private-sector-led urban development projects (Heurkens, 2012; Heurkens and Hobma, 2014) and the privatisation of Dutch planning powers (Hobma and Heurkens, 2015). These projects and development strategies were used increasingly in the pre-GFC 2000s period and are currently experiencing a comeback in the post-GFC period. As a result of the more risk-prone, less active land development role of local authorities, urban redevelopment projects in the Netherlands, to an increasing extent, see private actors taking the lead in delivering urban and real estate projects (Heurkens, 2013). As such, Van der Krabben and Heurkens (2015) indicate that roughly two types of development strategies have come to represent the existing two mainstream development approaches indicated earlier:

1 private-sector-led urban development concessions, and
2 private-sector-led incremental piecemeal development.

In theory, the concession model encompasses the previously mentioned integrated urban redevelopment strategy, while piecemeal development embraces the organic way of developing urban areas. Both of these development strategies move away from public-led or public–private-led urban development practices common for decades in the Netherlands (Heurkens, 2012). Both embrace the facilitative role of local planning authorities that is increasingly concerned with enabling market initiatives, and planners who increasingly operate as essential market actors themselves (Adams and Tiesdell, 2010; Heurkens et al., 2015). The leading private-sector actors can be 'traditional' real estate developers, investors, or owners, or non-traditional real estate industry actors such as corporations, entrepreneurs, and communities. To understand the major differences between both strategies, some features are described below.

23.3.1 Private-sector-led urban development concessions

In organisational-legal terms, a private-sector-led urban development concession is:

> A contract form with clear preconditioned agreements between public and private parties, in which a conscious choice from public parties has been made to transfer risks, revenues, and responsibilities for plan development, land preparation, land and real estate development and possible operation of the entire development plan towards private parties, within a previously defined public brief [or tender] in which the objective is to create an effective task division and a clear separation of public and private responsibilities (Gijzen, 2009: 19).

In essence, the concession is a contractual agreement between public and private partners under private law. The concession to develop the land is given to a private entity once a public procurement/tender, formulated by a municipality, has been awarded to the private entity, often based on a development competition. The initiative for a concession partnership, in most cases, lies with the municipality that formulates various objectives related to the urban development project and provides market actors with assessment criteria and other procedures in a public brief or tender. Private actors are required to design a development plan and provide economic-financial feasibility studies to back up their bidding for the land. At the same time, municipalities use their public law mandate, such as land use plans, to regulate the land for development and give planning permission once private actors are awarded a concession that meets the requirements stated in the public tender. In Dutch concessions, the management or operation of public space is a task mostly performed by the municipality, as the development industry is inexperienced with this, and local authorities consider the management of the public realm a core responsibility.

Despite its formal contractual nature and a strict public–private role division, various Dutch case studies have shown that concessions allow for and require informal public–private interaction (Gijzen, 2009; Heurkens and Peek, 2010; Heurkens, 2012; Heurkens and Hobma, 2014). In other words, there is room for negotiation between municipalities and developers about the development conditions, and often some programmatic flexibility about the development plan. Furthermore, the formal nature of the public tenders can provide fruitful ground for a clear formulation of public objectives concerning sustainable urban development. These objectives are then to be met by the private actors who have to come up with their own specific, sometimes innovative, solutions. Thus, in brief, the private-sector-led urban development concessions could be an effective formalised partnership arrangement to deliver sustainable urban redevelopment.

23.3.2 Private-sector-led incremental piecemeal development

The second development strategy that appears in the Netherlands is private-sector-led incremental piecemeal development. This model is very much a representation of the recent organic development approach. In this model, the municipality develops a broad vision for the redevelopment of a certain location and 'invites' the private sector to come up with plans that fit in with the broad vision for the location (Peek and Van Remmen, 2012; Buitelaar et al., 2012). 'The private sector initiatives may concern small developments situated in the (re)development location and do not have to cover the whole location' (Van der Krabben and Heurkens, 2015: 73). This is in line with the risk-prone behaviours of both public and private actors, and the often limited financial liquidity and urban and real estate development knowledge of the organisations involved in this strategy. For instance, such private actors may involve local entrepreneurs, property owners, collective groups of homebuilders, architectural offices, and even energy or technology companies. They may initiate redevelopment in the first place, or they may wish to contribute to (part of) an urban development vision initiated by the municipality. Moreover, such private initiatives often favour incorporating some sort of sustainability aspect in the development strategy, such as circularity principles or energy efficiency measures, with a strong focus on local opportunities and benefits.

As with every new way of working, the efficient introduction of this incremental piecemeal development strategy – considering the Dutch public-led planning doctrine – requires both a change of attitude by public and private actors, and increased flexibility in planning procedures (Van der Krabben and Heurkens, 2015). For instance, effective private–private partnerships between energy companies and collective groups of homebuilders need to be constructed that represent the direct relationship between the actors without public interference. Moreover, local authorities search for ways to build effective public–private partnerships that are often tailor-made and less generic than development concessions can be. Therefore, as of yet, no panacea for organisational and legal arrangements exists that represents the formal and informal relationships between public and private actors in organic urban development. Moreover, it remains uncertain whether, for instance, infrastructure can be financed in this strategy through some sort of value capturing. Nevertheless, private-sector-led incremental piecemeal developments are an increasingly popular way of redeveloping urban areas and real estate.

23.3.3 Dutch private-sector-led development strategies

Table 23.2 illustrates the main characteristics of the two private-sector-led development strategies. The most prominent question now is: to what extent do these Dutch urban development strategies and their particular public and private partners allow for the delivery of sustainable urban places? In terms of sustainability, the traditional integrated urban development approach mainly focused on delivering places with an attractive spatial quality and design (Franzen et al., 2011). This involved making trade-offs between user value, future value, and

Table 23.2 Private-sector-led development strategies in the Netherlands

	Private-sector-led urban development concessions	*Private-sector-led incremental piecemeal development*
Development scale focus	Urban area	Real estate
Private organisations	Developers, development consortium, investors	Small developers, architects, homebuilders
Legal agreement/entity	Concession	Private realisation
Planning law/rules	Tenders, requirements	Guidelines, visions
Financial value capturing	Developer contributions	Uncertain
Public–private relations	Formal	Informal

experience value of a development. However, in Dutch urban development practice before the 2008 GFC, there was hardly any attention paid to sustainability aspects such as climate proofing, energy neutrality, resilience, adaptation, circularity, and the like. Such aspects have gained more ground over the last five years in both integrated and organic urban development approaches, and have found their way into both private-sector-led urban development strategies. The following sections illustrate how sustainability aspects are incorporated into a development concession project in Rotterdam and an incremental piecemeal urban redevelopment project in Amsterdam.

23.4 Case study: Rotterdam Rijnhaven

Rotterdam is the second-largest city in the Netherlands with about 630,000 inhabitants; it has the biggest port in Europe, is an important economic area in the country, and is recognised as a city with inspiring contemporary architecture. As port activities in the last decades have shifted outside the city boundaries towards the sea (Frantzeskaki, Wittmayer, and Loorbach, 2014), Rotterdam has created several strategies to redevelop its industrialised waterfront locations into mixed-use urban areas (Daamen, 2010). Since the 1990s, integrated urban area development approaches functioned as the focal point of developments like the Boompjes, Kop van Feijenoord, Kop van Zuid, and Katendrecht, and some of these areas are still in progress. In this process, the Municipality of Rotterdam (in the role of city planner) and the Port Authority (in the role of major landowner) founded a separate organisation, the Stadshavens (City Ports) Project Office, in the 2000s to envision the future direction of and oversee urban developments of the City Ports area.

According to Ernst, De Graag-Van Dinther, Peek, and Loorbach (2016: 2993), 'the City Ports development program is closely related to the city's programs for sustainable development, CO_2 reduction and climate adaptation. Its objectives are to connect a stronger economy with an attractive city by combining inner-city waterfront development with broadening the "mainport" and making it more sustainable.' Various partnerships and planning policies concerning the City Ports regeneration process are in place (Frantzeskaki et al., 2014). The Clean Tech Delta and Rotterdam Climate Initiative are the most notable partnership arrangements for implementing sustainability agendas in the city. They mainly function on strategic and tactical governance levels (Loorbach, 2010), and are valuable for institutional transitions, policy making, networking, and learning. However, we are mainly interested in the role of operational partnerships that deliver concrete sustainable urban redevelopment projects.

As such, one of the most appealing recent concrete development initiatives by the municipality is the realisation of a 'floating' development in the former harbour water basin Rijnhaven (Figure 23.1). 'The rationale [behind building on water] is that increasing water levels (river, groundwater) will make innovative resilient living arrangements and settlements necessary. Floating urbanisation is conceptualised and envisaged as the adaptation option for Rotterdam as a deltaic city to climate change pressures' (Frantzeskaki et al., 2014: 411) by basically combining water management with urban regeneration. This Rijnhaven project is a private-sector-led urban development concession area, which is located adjacent to the dense mixed-use Kop van Zuid Willeminapier area and more residential Katendrecht area on the southern banks of the River Maas. Ernst et al. (2016) argue that after an organised market consultation by the municipality in 2012, the scope had shifted from a floating development to an urban development (on water).

This led to the decision to tender the development to the market in 2013, which involved a bid book *Rijnhaven: Metropolitan Delta Innovation* (Stadshavens Rotterdam, 2013) and public procurement directory (Gemeente Rotterdam, 2013) for the Rijnhaven concession. According

Figure 23.1 The Rijnhaven harbour water basin, with the Kop van Zuid development in the background and the floating sustainability pavilion on the right

to Ernst et al. (2016), the ambitions of the development had been broadened to incorporate new municipal policy objectives, including delta metropolitan innovation, quality of life improvement, shaping the Rotterdam Waterfront, and continuous creation of added value. Procurement rules asked for a creative and flexible development strategy and for specification of public and private roles. Moreover, future private concession holders would carry responsibility for all development and plan costs, hold the concession in management for 30 years, and transfer the land back to the municipality without causing costs for the municipality.

This ambitious concession for a sustainable urban redevelopment carried out by private consortia was based on procurement experiences with another urban development in Rotterdam called Hart van Zuid. Ernst et al. (2016: 2995) set out that the procurement 'offered a 30-year concession to design, build, finance, maintain and operate the area [and] a competitive dialogue between municipality and consortia of private parties'. Also, a committee of global experts in sustainability, transitions, and urban planning had to assess the extent to which the private plans lived up to the municipal ambitions. Moreover, during the process the municipality organised innovation markets creating private meeting points for bidders and other market parties. Despite these facilitative activities by the municipality, the stringent set of requirements and high ambitions caused a drop-out of interested market parties. According to Ernst et al. (2016), in 2015, after two years of dialogue between bidders and municipal officials, the municipality of Rotterdam concluded that neither of the two remaining (out of seven initially interested) consortia of private parties had submitted a proposal that met the ambitions and prerequisites (Stadshavens Rotterdam, 2015). The municipality is currently reconsidering the way forward with the area's development.

De Zeeuw (2015) argues that the failure of the Rijnhaven project could have been expected. The main reasons for this were the over-ambitious requirements in terms of sustainability, which included the development of social educational programmes for the inhabitants of adjacent neighbourhoods and innovative solutions for floating development. Furthermore, the winning consortium had to pay €3 million to the municipality upfront as a compensation fee for municipal labour on the project, while the 30 year concession period already involved some financial risks for the private consortia, certainly in this type of development. When looking at the lessons from previous generations of concessions (Heurkens, 2012), one might conclude that the public–private partnership involved both building informal relationships and establishing a formal procurement relationship between market actors and the municipality, which combined a facilitative role with regulatory tasks. However, what becomes clear from this case is that the high sustainability ambitions combined with the precarious financial viability of such a business case was asking too much from the development industry, at least for the time being. Moreover, according to Ginter (2013), overcoming this barrier would also involve the emergence of new institutions in Rotterdam that are more supportive of sustainable urban development practices.

23.5 Case study: Amsterdam Buiksloterham

Amsterdam is the largest city and capital of the Netherlands; with about 840,000 inhabitants, it is the most globally oriented economic area in the country, and is recognised as magnet for young talent, international companies, and tourism. The population is growing at a steady rate and the City of Amsterdam has the ambition to build 50,000 new dwellings by 2025 (Grim, 2016). While its city centre is UNESCO listed and its famous water canal structure and densely built-up area do not allow for a significant contribution to reaching the municipal housing target, the city has turned its eye towards the various remaining former industrial (waterfront) sites alongside the River IJ, mainly on Amsterdam's north bank. This is a continuation of municipal spatial policies targeted at redeveloping waterfronts and piers into mixed-use urban areas. Over the last two decades, similar to Rotterdam, the municipality has already redeveloped waterfront locations such as Java, KNSM and Borneo islands, Zeeburg, and Ijburg, and other sites such as Cruquius, Houthavens, and Overhoeks are currently under construction.

At the same time, the municipality has formulated a structural vision for creating a strong economy and a sustainable city (Gemeente Amsterdam, 2011). In addition, various more specific policies and visions exist that embrace the ambition of becoming a smart city (Amsterdam Smart City, 2016) and a circular city (City of Amsterdam, 2015). Grim (2016) argues that the ambition of building 50,000 homes and creating a smart and circular city through urban and real estate development might be in conflict with each other, and opts for learning from existing initiatives and projects as examples for developing in a smart and circular manner. The most prominent recent example of a circular urban redevelopment in Amsterdam is Buiksloterham on the northern bank of the River IJ (Figure 23.2). This former industrial area was home to a Fokker aeroplane factory, a Shell oil laboratory, a number of shipbuilding firms, and other manufacturing (Reimerink, 2016). As a lot of companies either ceased trading or left over time, the polluted site resulted in an opportunity for redevelopment.

Buiksloterham can be considered a private-sector-led incremental piecemeal development. This incremental approach proved to be the only viable way forward during the 2008 GFC, which coincided with the start of the redevelopment. Before 2008, the municipality had initially tendered the redevelopment of four locations as office developments, dictating high sustainability demands, but developers backed out of the project due to the GFC. As a result of these circumstances, in Buiksloterham the city leaders eventually opted for a more bottom-up organic approach. 'They changed the zoning to allow for a mix of uses, and they created a relatively hands-off path to allow Buiksloterham to slowly fill in with residences and offices on whatever land was safe to inhabit' (Grim, 2016, author's translation).

Figure 23.2 Circular collectively commissioned housing in Buiksloterham

In 2010, the municipality started a tender for a ten-year lease of a land parcel called De Ceuvel, backed by the idea to put the waterfront location to temporary use until the market picked up, and the desire for creative approaches to sustainable urbanism. The winning idea from a group of young entrepreneurs focused on redeveloping the polluted site with retrofitted houseboats pulled up onto land and connected by wooden walkways and with special plants to clean the soil within ten years. It also encompasses a waterfront café, shared workspaces, an organic restaurant, and various sustainable technologies. As a result of this project, 'Buiksloterham has evolved into a creative hub for the so-called "circular economy" attracting devotees of the idea that renewable power, rainwater harvesting, recycling and other techniques can allow an urban neighbourhood to handle all its own energy, water and food needs without creating waste' (Grim, 2016, author's translation).

Plot by plot, the rest of the Buiksloterham's development is progressing, with individual and collective homebuilders, creative designers and architects, energy and water companies, and more traditional real estate developers and housing associations active in redeveloping the area with housing. 'In 2011, the municipality decided to sell off a small number of housing lots to attract people who wanted to build their own homes using sustainable building practices such as recycled materials and generating their own electricity' (Grim, 2016, author's translation). By doing so, Buiksloterham could contribute to the municipal housing development and sustainability ambitions. As the circular economy narrative spread, parties other than homebuilders and creative people began to show interest in the area, such as developers, investors, public utility companies, and researchers. For instance, housing association De Alliantie, real estate developer Hurks, and real estate development investor Amvest are currently developing several housing projects in the area.

In March 2015, about 20 public and private organisations – both traditional and non-traditional real estate parties such as energy and water management companies and citizens – signed the so-called *Manifest Circulair Buiksloterham*. With the manifesto, the parties expressed their aim to strengthen a collective ambition of making Buiksloterham a test case of circular urban redevelopment through Living Labs, and a catalyst for broader transition in Amsterdam. Several formal and informal private–private and public–private partnerships have come into existence in Buiksloterham (De Ridder, 2014) related to various initiatives and projects (Buiksloterham, 2016), which makes this incremental development a complex governance challenge. The municipality's facilitative role particularly allowed the area to flourish organically from the grassroots. However, now that developers are moving in and market demand for housing is high, some active parties in Buiksloterham fear that 'the enthusiasm for cutting-edge sustainability practices will wane' (Grim, 2016, author's translation).

Therefore, the regulatory role of the municipality for sustainable urban development remains important. Steen (2016: 195) argues that

> the sustainability-oriented tenders and selection procedures for PC [private commissioning] and CPC [collective private commissioning] in Buiksloterham prove that by including high requirements to sustainable performance in the selection procedures, highly sustainable development results can be achieved. ... It must be taken into account that the development within the set requirements stays feasible for the developer, which can be ensured by lower land- or leasehold prices, subsidies, or helping investments in for example basis infrastructure.

In fact, the Buiksloterham case nowadays can be considered a combination of a bottom-up private-sector-led incremental piecemeal development strategy (individual plot development) and a private-sector-led urban development concession strategy (mixed-use housing developments). In other words, both development strategies co-exist in the area, albeit executed by different actors and partnerships. Thereby, the chances increase that each development strategy incorporates aspects from the other. This in turn might positively influence the institutionalisation of sustainable urban development principles in both planning systems and development practices (Buitelaar, Galle, and Sorel, 2011).

23.6 Conclusions

This chapter has illustrated the main approaches and strategies for sustainable urban redevelopment in the Netherlands. Although the practice of developing sustainable real estate is becoming more common (as evidenced by the increasing number of BREEAM-certified office buildings), sustainable urban places seem more difficult to realise. There is simply no consensus in Dutch urban development practice about what sustainable urban development is and how it can be achieved. Moreover, in urban areas, many sustainability issues can be taken into account, whether they are economically, socially, and environmentally focused, or more specifically targeted at smart, circular, energy-neutral, climate-adaptive principles and objectives. Nonetheless, both the integrated and organic approaches currently co-existing in Dutch urban development seem to offer the potential means to deliver sustainable urban development. These overarching approaches have resulted in both top-down, private-sector-led urban development concession and bottom-up, private-sector-led incremental piecemeal development strategies applied to urban projects.

Examples of these contrasting strategies in Rotterdam and Amsterdam illustrate that Dutch urban development practice is incorporating multiple sustainability aspects into urban redevelopment projects with varying degrees of success. What can be learned from the Rijnhaven case is that municipal ambitious and a risk-prone tender for a sustainable floating urban development proved to be unviable for private consortia. Buiksloterham, in Amsterdam, illustrates that a circular urban redevelopment can be achieved by building various alliances between public and private agencies. In essence, both cases indicate that formal, legal public–private arrangements on the one hand, and intensive, informal public–private interactions on the other hand, are necessary to define what sustainable urban development means for a particular area and how it can best be achieved. Also, it has become clear that neither development strategy is preferable or superior for achieving sustainable urban areas. Ultimately, when actor attitudes change and experience grows, established institutions in Dutch practice might prove to be more receptive to sustainable urban redevelopment in the future. Hence, other countries and practices each have to discover their own effective approaches and strategies to realise sustainable urban redevelopment.

References

Adams, D., & Tiesdell, S. 2010, 'Planners as market actors: rethinking state–market relations in land and property', *Planning Theory and Practice*, vol. 11, no. 2, 187–207.

Amsterdam Smart City 2016, *About ASC*, Amsterdam Smart City, viewed 25 May 2016, http://amsterdamsmartcity.com/about-asc.

BREEAM-NL 2016a, *BREEAM-NL*, Dutch Green Building Council, viewed 25 May 2016, www.breeam.nl.

BREEAM-NL 2016b, *Gecertificeerde Projecten*, Dutch Green Building Council, viewed 25 May 2016, www.breeam.nl/projecten/projecten-gecertificeerd?type=gebied#type=gebied.

Bruil, A. W., Hobma, F. A. M., Peek, G. J., & Wigmans, G. (eds) 2004, *Integrale Gebiedsontwikkeling: Het Stationsgebied's-Hertogenbosch*, SUN, Amsterdam.

Buiksloterham 2016, *Projecten*, Buiksloterham, viewed 25 May 2016, http://buiksloterham.nl/web/lijst/projecten.vm?reset=true.

Buitelaar, E., Feenstra, S., Galle, M., Lekkerkerker, J., Sorel, N., & Tennekes, J. 2012, *Vormgeven aan de Spontane Stad: Belemmeringen en Kansen voor Organische Stedelijke Herontwikkeling*, Planbureau voor de Leefomgeving & Urhahn Urban Design, Den Haag & Amsterdam.

Buitelaar, E., Galle, M., & Sorel, N. 2011, 'Plan-led systems in development-led practices: an empirical analysis into the (lack of) institutionalisation of planning law', *Environment and Planning A*, vol. 43, 928–941.

Buitelaar, E., Galle, M., & Sorel, N. 2014, 'The public planning of private planning: an analysis of controlled spontaneity in Netherlands', in D. A. Andersson & S. Moroni (eds), *Cities and Private Planning: Property Rights, Entrepreneurship and Transaction Costs*, Edward Elgar, Cheltenham, 248–268.

Buskens, B. 2016, 'De duurzame ontwikkelaar: hoe en waarom projectontwikkelaars zich kunnen committeren aan duurzame gebiedsontwikkeling', MSc thesis, Delft University of Technology.

Buskens, B., & Heurkens, E. W. T. M. 2016, 'De duurzame private gebiedsontwikkelaar', *Real Estate Research Quarterly*, vol. 15, no. 3, 38–46.

City of Amsterdam 2015, *Towards the Amsterdam Circular Economy*, City of Amsterdam, Amsterdam.

Daamen, T. A. 2010, *Strategy as Force: Towards Effective Strategies for Urban Development Projects – The Case of Rotterdam City Ports*, IOS Press, Amsterdam.

De Ridder, E. 2014, 'Buiksloterham in transition: developing tools to support processes of urban transition', MSc thesis, Delft University of Technology.

De Zeeuw, F. 2015, 'Aanbesteding Rijnhaven valt in het water', *Praktijkleerstoel Gebiedsontwikkeling*, viewed 25 May 2016, www.gebiedsontwikkeling.nu/artikelen/aanbesteding-rijnhaven-valt-in-het-water.

DGBC 2016, *Projectoverzicht*, Dutch Green Building Council, viewed 25 May 2016, www.dgbc.nl/projectoverzicht.

Dühr, S., Colomb, C., & Nadin, V. (eds) 2010, *European Spatial Planning and Territorial Cooperation*, Routledge, London.

Ernst, L., De Graag-van Dinther R. E., Peek, G. J., & Loorbach, D. A. 2016, 'Sustainable urban transformation and sustainability transitions: a conceptual framework and case study', *Journal of Cleaner Production*, vol. 112, 2188–2199.

Frantzeskaki, N., Wittmayer, J., & Loorbach, D. A. 2014, 'The role of partnerships in "realising" urban sustainability in Rotterdam's City Ports Area, the Netherlands', *Journal of Cleaner Production*, vol. 65, 406–417.

Franzen, A., Hobma, F. A. M., De Jonge, H., & Wigmans, G. (eds) 2011, *Management of Urban Development Processes in the Netherlands: Governance, Design, Feasibility*, Techne Press, Amsterdam.

Gemeente Amsterdam 2011, *Structuurvisie Amsterdam 2040: Economisch Sterk en Duurzaam*, Gemeente Amsterdam, Amsterdam.
Gemeente Rotterdam 2013, *Aanbestedingsleidraad Deel 1, 1-506-12. Gebiedsontwikkeling Rijnhaven, Concessie*, College van Burgemeester en Wethouders Rotterdam, Rotterdam.
Gijzen, M. H. M. 2009, 'Zonder loslaten geen concessie: inzicht in de recente toepassing van deze publiek-private samenwerkingsvorm in de Nederlandse gebiedsontwikkelingspraktijk met "evidence-based" verbetervoorstellen', MCD thesis, Erasmus University Rotterdam & Delft University of Technology.
Ginter, D. 2013, 'Vermogen tot verandering: een onderzoek naar de nieuwe werkwijze in stedelijke ontwikkeling', MCD thesis, Erasmus University Rotterdam & Delft University of Technology.
Grim, S. 2016, 'Grote bouwopgave vs. circulaire ambitie', *Steden in transitie*, viewed 25 May 2016, https://stedenintransitie.nl/stadbericht/grote-bouwopgave-vs-circulaire-ambitie.
Haak, M., & Heurkens, E. W. T. M. 2015, 'Innovatie bij vastgoedontwikkelaars: typologieën en strategieën', *Real Estate Research Quarterly*, vol. 14, no. 2, 48–54.
Heurkens, E. W. T. M. 2012, *Private Sector-Led Urban Development Projects: Management, Partnerships and Effects in the Netherlands and the UK*, Architecture and the Built Environment, Delft.
Heurkens, E. W. T. M. 2013, 'Een nieuwe rolverdeling: privaat "in the lead", publiek faciliteert', *VHV Bulletin*, vol. 40, no. 3, 15–16.
Heurkens, E. W. T. M. 2016, 'Institutional conditions for sustainable private sector-led urban development projects: a conceptual model', in ZEBAU – Centre for Energy, Construction and the Environment (eds), *Proceedings of the International Conference on Sustainable Built Environment: Strategies – Stakeholders – Success Factors (SBE16)*, Hamburg, 726–735.
Heurkens, E. W. T. M., & Hobma, F. A. M. 2014, 'Private sector-led urban development projects: comparative insights from planning practices in the Netherlands and the UK', *Planning Practice and Research*, vol. 21, no. 4, 350–369.
Heurkens, E. W. T. M., & Peek, B. 2010, 'Effecten van de toepassing van het concessiemodel bij gebiedsontwikkeling', *Real Estate Magazine*, vol. 71, 42–45.
Heurkens, E. W. T. M., Adams, D., & Hobma, F. A. M. 2015, 'Planners as market actors: the role of local planning authorities in the UK's urban regeneration practice', *Town Planning Review*, vol. 86, no. 6, 625–650.
Hobma, F. A. M., & Heurkens, E. W. T. M. 2015, 'Netherlands', in S Mitschang (ed.) *Privatisation of Planning Powers and Urban Infrastructure*, Peter Lang Verlag, Frankfurt am Main.
Loorbach, D. 2010, 'Transition management for sustainable development: a prescriptive, complexity-based governance framework', *Governance*, vol. 23, 161–183.
Peek, G. J., & van Remmen, Y. 2012, *Investeren in Gebiedsontwikkeling Nieuwe Stijl; Handreiking voor Samenwerking en Verdienmodellen*, Ministerie I&M, Den Haag.
Puylaert, H., & Werksma, H. 2011, *Duurzame Gebiedsontwikkeling: Doe de Tienkamp!*, Praktijkleerstoel Gebiedsontwikkeling TU Delft & H2Ruimte, Delft.
Reimerink, L. 2016, 'How Amsterdam turned a polluted industrial site into its most interesting neighborhood', *City Scope*, viewed 25 May 2016, http://citiscope.org/story/2016/how-amsterdam-turned-polluted-industrial-site-its-most-interesting-neighborhood.
Robles-Duran, M. 2011, 'Prelude to a brand new urban world', *Archis Volume, Privatize!*, vol. 30, no. 4, 54–57.
Roodbol-Mekkes, P. H., van der Valk, A. J. J., & Korthals Altes, W. K. 2012, 'The Netherlands spatial planning doctrine in disarray in the 21st century', *Environment and Planning A*, vol. 44, 377–395.
Stadshavens Rotterdam 2013, *Rijnhaven: Metropolitan Delta Innovation*, Stadshavens Rotterdam, Rotterdam.
Stadshavens Rotterdam 2015, *Ontwikkeling Rijnhaven Heroverwogen*, Stadshavens Rotterdam, viewed 25 May 2016, http://stadshavensrotterdam.nl/area_page/ontwikkeling-rijnhaven-heroverwogen.
Steen, K. 2016, 'Developing sustainable urban areas: recommendations on urban form and development based on theory and top-down & bottom-up planning examples in Overhoeks and Buiksloterham', MSc thesis, Delft University of Technology.
van der Krabben, E., & Heurkens, E. W. T. M. 2015, 'Netherlands: a search for alternative public–private development strategies from neighbouring countries', in G. Squires & E. W. T. M. Heurkens (eds) *International Approaches to Real Estate Development*, Routledge, London, 66–81.

24

Smart growth and real estate development in Saudi Arabia

Abdullah Alhamoudi and Peter Lee

24.0 Introduction: Ten smart growth principles

Smart growth and its principles have attracted much attention in real estate and planning circles. In addition to theory, both practice and policy have seen the emergence of smart growth via governmental documents, planning policies, and regulations – with much of smart growth's inception emerging from sustainable development concerns in the touchstone Brundtland Report (Edwards and Haines, 2007; Grant, 2009). Furthermore, smart growth has been a rallying point to normatively promote 'various policies and planning practices that create more compact and multimodal communities, in contrast to sprawl, which results in more dispersed and automobile-dependent development' (Litman, 2015).

There are a number of reasons that led to the appearance of this smart growth approach and its principles, particularly by Western countries that have been facing 'sprawling' development. The sprawl pattern of development is defined by a complex pattern of land use, transportation, and social and economic development (Frumkin, 2002). This pattern of development leads to several economic, environmental, and social issues, particularly those that have a significant concern for planners and property decision makers in addition to the community. Examples of these aspects include increasing dependency on cars, the rising cost of infrastructure, increasing levels of pollution, housing affordability issues, and the ongoing transfer of agricultural and natural lands for residential and commercial uses (Litman, 2015; Geller, 2003). The following sections will summarise (see Table 24.1) and explain the principles of smart growth (U.S. Environmental Protection Agency, 2011) in addition to providing a number of benefits of these principles.

24.0.1 Encourage mixed land uses

One of the main principles of smart growth is to encourage mixed land use for residential, commercial, and recreational purposes. These land usages ought to be located proximate to each other in order to achieve better sustainable outcomes (Dempsey, 2010). Proponents of this principle often argue that adopting mixed land use will result in a number of social, economic, and environmental benefits (Litman, 2015; Jenks and Burgess, 2000). One benefit is that smart growth will offer several transportation options that are much cheaper for the community, such as buses and trains, instead of depending on private cars. Another social benefit is a more secure

Table 24.1 Smart growth principles

1. Encourage mixed land uses
2. Take advantage of compact building design
3. Create a range of housing opportunities and choices
4. Create walkable neighbourhoods
5. Foster distinctive, attractive communities with a strong sense of place
6. Preserve open space, farmland, natural beauty, and critical environmental areas
7. Strengthen and direct development towards existing communities
8. Provide a variety of transportation choices
9. Make development decisions predictable, fair, and cost effective
10. Encourage community and stakeholder collaboration in development decisions

and safe community for local residents, in addition to an increased number of activities located near to open spaces, green areas, or other local facilities (Jacobs, 1961; Litman, 2015). Further benefits of adopting a mixed land use principle is that it will increase social interaction in the local community, particularly by providing the opportunity for local residents to meet each other within these local areas and places (Bramley and Power, 2009; Jenks and Burgess, 2000; Litman, 2015).

24.0.2 Take advantage of compact building design

The second principle of smart growth is to support and encourage residential density, mostly by advocating for buildings to be developed vertically instead of horizontally (Jenks et al., 1996). Proponents of this type of development believe that it will result in many sustainability benefits. For instance, accommodating more people on the same land footprint size will offer and preserve more open spaces and green areas for the local community (Jenks et al., 1996; Litman, 2015). Offering more and better public transportation choices can reduce air pollution from traffic congestion and save money for the local community. This type of development and building is preferred to sprawl development, as it increases the density and demand for such modes of transport.

24.0.3 Create a range of housing opportunities and choices

Creating a range of housing opportunities and choices is the third principle of smart growth. For instance, smart growth can provide more housing choices and more affordable housing, and make a neighbourhood more diverse, by allowing people of different backgrounds, incomes and ages to be housed in the same place (Talen, 2008; Alexander and Tomalty, 2002; Litman, 2015). There are a number of restrictions that influence residents' choice of place and type of housing. For example, some people might choose an apartment or choose to live in a particular area due to its local facilities and activities, noting that others might feel that this apartment option is a trade-off for some other aspects such as affordability or accessibility (Easthope and Judd, 2010). Thus, smart growth will offer more housing choices that meet different groups' preferences and needs, such as types of apartments, villas, and detached, semi-detached, or terraced houses (Talen, 2008; Alexander and Tomalty, 2002; Litman, 2015). Another benefit of smart growth is that it offers more affordable housing, in addition to financial savings for the community, by using the current infrastructure system instead of building a new infrastructure system (Alexander and Tomalty, 2002; Litman, 2015).

24.0.4 Create walkable neighbourhoods

Creating walkable neighbourhoods is the fourth principle of smart growth. Walkability can be defined as 'the extent to which walking is readily available to the consumer as a safe, connected, accessible and pleasant activity' (Fitzsimons D'Arcy, 2013). Several researchers claim that adopting this principle will result in a number of benefits for the local community. One example of these benefits is that it will encourage local residents to walk as a green mode of transport for daily trips such as getting to school or work, which is almost impossible in a sprawl development where most residents are dependent on private cars. Another benefit is increasing social interaction among neighbours, allowing them to do more physical activities that lead to better health (Jacobs, 1961; Litman, 2015).

24.0.5 Foster distinctive, attractive communities with a strong sense of place

The fifth principle of smart growth is to foster distinctive and attractive communities with a strong sense of place. This sense of place can create a community and environment that reflects the culture and values of local residents. Furthermore, development patterns may use several environmental landmarks to enable a definition of place. Place may also be enhanced by providing communities with high-quality elements such as building design, as well as landscape and neighbourhood design, that reflect local residents' interests and concerns. Improving place will result in a number of benefits for the community, such as increasing its cohesion and enhancing the sense of community.

24.0.6 Preserve open space, farmland, natural beauty, and critical environmental areas

Another principle of smart growth is to preserve open space, farmland, natural beauty, and critical environmental areas. This principle has a number of touted social benefits in that 'city parks make inner-city neighbourhoods more liveable; they offer recreational opportunities for at-risk youth, low-income children, and low-income families; and they provide places in low-income neighbourhoods where people can experience a sense of community' (Sherer, 2006). Some other social benefits of open spaces and parks are reducing crime and offering different age groups in the community a safe and secure place to relax and play, thus increasing their sense of community.

24.0.7 Strengthen and direct development towards existing communities

The seventh principle of smart growth is to encourage the development of existing communities via infill development and brownfield development projects, particularly as these development projects have become a critical part of the smart growth programs in many countries, such as the US and the UK (McConnell and Wiley, 2012; Williams and Dair, 2007; Adams and Watkins, 2002). It is argued that adopting this principle will result in many sustainability benefits, such as less development in rural areas, reducing traffic congestion, offering more affordable houses with good accessibility to the main facilities and activities, improving the quality of life, and making the community more active (Steinacker, 2003; McConnell and Wiley, 2012; Litman, 2015).

24.0.8 Provide a variety of transportation choices

Another important principle of smart growth is to provide more transportation options for the community. Offering such transportation options will have many benefits, including reducing traffic congestion due to dependence on private cars. Another benefit of widening the variety of transportation choices is that it will encourage people to use more sustainable modes of transportation. In addition, offering different groups within the community (according to age, income, and physical ability) more transport options will more effectively meet their varied and extensive needs and preferences – hence another social benefit from this principle (Litman, 2015).

24.0.9 Make development decisions predictable, fair, and cost-effective

Another smart growth principle is that development decisions should be made predictable, fair, and cost effective. Decisions by the public sector to involve the private sector can help, especially when the private sector can leverage finance for smart growth development projects involving large sums of money. Decisions for different levels of government could be those that provide facilities and financial incentives (notably profit) to the private sector (Morris, 2009). There are several tools that can help government to offer facilities to the private sector, such as facilitating the approval process of a project by offering a consultation about the process, or giving priority to the processing of the application (Morris, 2009). Another facility, for example, is allowing developers to raise the density (see Principle 2 above) of the building, which will enable a more economically viable project (Morris, 2009).

24.0.10 Encourage community and stakeholder collaboration in development decisions

Encouraging the community and stakeholders to participate in the development decision-making process is the final principle of smart growth. Different communities have different needs and preferences; for example, some communities prefer to have more housing choices, while other groups might prefer to have better access to local facilities and amenities as offered in a mixed land use development, while still other groups might seek to have more transport options. Thus, it is best to communicate with local residents and workers as they know the needs and preferences of their local community. Moreover, inducing local community and stakeholder participation in the development decision-making process will result in several benefits, such as having more inclusive opinions and suggestions about particular issues from the main users of the facility. This inclusion can then offer better solutions by enabling the participation of stakeholders (Bass, Dalal-Clayton, and Pretty, 1995). This participation will allow the community a certain level of transparency regarding the government's decisions and efforts. Participation will also have a number of benefits that can enhance social equity and community stability (Bass et al., 1995).

24.1 Smart growth principles in Saudi Arabian real estate development

The Kingdom of Saudi Arabia can be seen to be trying to integrate smart growth principles when dealing with sustainability of real estate development. For instance, the Ministry of Municipal and Rural Affairs (MOMRA) has issued a statute that allows for increasing the density in some areas in Riyadh city (the capital city of Saudi Arabia) if it meets certain conditions relating to the width of the fronting street and plot size (Alskait, 2011). This has led to several social issues such as traffic congestion, thereby increasing the demand for local services and facilities. There may also be a resulting poor-quality social fabric in neighbourhoods as more people move and reside within them. A high proportion of new renters can lead to negative social impacts, as it may reduce the sense of local community and belonging (Alskait, 2011). Moreover, highly populated neighbourhoods could lead to a decline in the deeper personal and social interactions among residents who visit the local mosque regularly (Alskait, 2011).

Moreover, although one of the main principles of smart growth is to provide more and smaller housing units (e.g., apartments), this might not necessarily mean providing more housing choices. For instance, location and housing type matter, and offering to locate people in particular areas or in particular housing types that are not preferable to them would not mean increasing the housing choice (Ohlin, 2003; Easthope and Judd, 2010). In Saudi Arabia, the majority of residents prefer a detached house in a low-density area due to religious and cultural reasons of privacy. One of the most important cultural and religious aspects in Saudi society (or Islamic society generally, for that matter) is privacy, which could be classified as a critical requirement for all real estate development projects (Eben Saleh, 1997; Shojaee and Paeezeh, 2015; Aina, Al-Naser, and Garba, 2013). Abu-Gazzah (1995) states that this has led to 'the concept of privacy becoming a subject of growing concern for people, architects, urban designers, landscape architects and social scientists involved in development projects in Saudi Arabia'.

Moreover, privacy within the Saudi household is related to the separation of gender facilities, such as bedrooms and bathrooms, the entrance, and facilities for residents and guests. In addition, most housing in Saudi Arabia has a walled boundary around the house (between three and five metres high) that offers residents full privacy from public activities outside. In addition, at most social events in Saudi Arabia, male and female members and their facilities and activities are often separated. In housing apartments, it is a common occurrence that residents redesign or rearrange their units to meet their needs for privacy. Furthermore, although most of the housing units in Saudi Arabia have gardens or open spaces, the attractiveness of these facilities is very limited compared to those of housing in Western cities (Eben Saleh, 1997; Al-Hemaidi, 2001; Aina et al., 2013). While Montgomery (1986) found examples of unsuccessful high-rise apartment projects (i.e., projects that were vacant for several years up to the mid-1980s), the rapid increase in land and housing prices, as well as the difficulty of owning a detached house (or villa), has changed the situation somewhat. Many people, especially young people, in Saudi Arabia have decided to move into the apartment market (GIZ International Services, 2009; Knowledge Corporation, 2013).

Another real estate development issue when using smart growth principles in Saudi Arabia is housing affordability, especially given increasing land and housing prices. Housing affordability is due to a number of factors, such as the limited supply of land (Alexander and Tomalty, 2002; Litman, 2015). Furthermore, the real estate development sector in Saudi Arabia is also facing the challenge of enabling lower-middle-class residents to own affordable housing. The percentage of housing ownership in Saudi Arabia is very low at around 30 per cent, whereas the global average is around 70 per cent. In the three biggest regions of Riyadh, Jeddah, and Dammam, this percentage is also very low: 33 per cent, 34 per cent, and 40 per cent, respectively. This homeownership proportion is low compared to the high salary of residents in the Saudi Arabia, and most employers give a housing bonus (Bagaeen, 2015; Abdul Salam, Elsegaey, Khraif, and Al-Mutairi, 2014; Knowledge Corporation, 2013). Thus, there is a gap between housing prices and the ability of young families and foreign residents in Saudi Arabia to pay back their loans. The majority of large real estate development companies are offering houses that are over 1 million SR, whereas the demand range is between 500,000 SR to 750,000 SR (Abou-Korin and Al-Shihri, 2015; Knowledge Corporation, 2013). Although it might be argued that this issue of housing affordability has resulted from the rapid expansion of urban areas and the sprawling development pattern, there are also a number of housing projects that have been constructed based on some of the smart growth principles. Such projects are still unaffordable for residents, especially for young families and low–medium-income residents (Abou-Korin and Al-Shihri, 2015).

24.2 Barriers to real estate development using smart growth principles in Saudi Arabia

Real estate development issues occur in Saudi Arabia due to a number of barriers. These barriers are often related to the planning and development system, and institutions whose members influence the ability and efficiency of these institutions. Moreover, these barriers are related directly either to some of the smart growth principles or to the planning and real estate development system in general. The following section will discuss these barriers using examples from the Saudi Arabian case.

One of the main barriers to real estate development is the lack of urban planners and professionals at different levels of government in Saudi Arabia. For those planners and professionals available, urban planning does not always take into account the specific Saudi Arabian cultural and social constraints. Moreover, there is no clear pattern or path of planning and development that could be followed, particularly with cities and populations that have been expanding and growing rapidly. This has arguably resulted in an urban pattern that cannot address several basic needs of the local residents and community (Alatni, Sibley, and Minuchin, 2012; Al-But'hie and Eben Saleh, 2002; Abou-Korin and Al-Shihri, 2015; Alskait, 2011; Madbouly, 2009; Kiet, 2011).

In addition to the shortage of professional planners and architects in Saudi municipalities, the ability of the planning departments in these municipalities is also limited and weak. For example, Abou-Korin and Al-Shihri (2015) show that the ability of the Dammam Urban Planning Department responsible for the planning in the Dammam Metropolitan Area (one of the most important areas in Saudi Arabia) was limited and weak and that their effort was insufficient. In addition, the employees' capacities and experiences in these departments were also insufficient. This department was responsible for 19 important tasks in the Dammam Metropolitan Area, but it had only 15 planners and architects, and the majority of those were new graduates with limited experiences, distributed into four big departments.

Another critical barrier to real estate development using smart growth principles in Saudi Arabia is related to public participation in the planning and real estate development process and decision making. A number of researchers find that there is a demonstrable failure in Saudi Arabia to allow the community to play a central role in the planning of real estate developments and investments (Alskait, 2011; Al-Shihri, 2013; United Nations Development Programme, 2010). Public participation in the planning and real estate development process (as well as decision making) in Saudi Arabia happens at three levels: (1) the national level, through the Consultative Council; (2) the regional level, through Regional Councils; and (3) the local level, through Municipalities Councils. Alskait (2011) believes that Municipalities Councils are not very effective and their efforts and powers are weak and limited. Thus, it is important that local residents should be allowed to become participants in the planning and real estate development processes, especially if participation will result in better development outcomes, as well as meeting local residents' needs and preferences in aspects such as privacy and housing design. Participation will also widen understanding among the general public as to the long-term and short-term benefits of sustainability and its smart growth principles (Susilawati and Al Surf, 2011).

There is also a lack of effort at the different levels of the Saudi government in terms of encouraging and supporting cooperation and participation between members of planning and development institutions in academia. Such effort remains limited. This lack of effort by government to integrate is argued by Abubakar (2013), who shows that universities are playing a critical but unintegrated role in the planning and development process. Many universities in the

Middle East have restructured their courses to include consideration of urban real estate development issues and challenges to meet sustainable goals, but remain largely unintegrated with government at present.

Moreover, there are governmental policies and regulations that need to be improved in order to address wider issues, such as housing affordability and finance. For example, mortgage laws are still new and under development, and thus weak. The percentage of housing loans in Saudi Arabia is only 2 per cent, which could be seen as very low compared to other countries in the Gulf Cooperation Council (GCC), such as Kuwait and the United Arab Emirates with 15 per cent and 15.5 per cent, respectively, as well as other Middle Eastern countries, such as Lebanon, Jordan, and Morocco with 7.2 per cent, 7.9 per cent, and 15.9 per cent, respectively (Knowledge Corporation, 2013). In 1974, the Saudi government established an organisation called the Real Estate Development Fund (REDF) with the aim of offering housing loans (300,000 SR, which was raised three years ago to 500,000 SR) to Saudis at zero interest. The demand for housing loans from this organisation is very high as this could be seen as the only effective and accepted option for Saudis in terms of housing loans. Up to the end of 2014, there were around 600,000 applications still pending, and one of the main challenges that this organisation now faces is the difficulty of collecting debts due to a lack of strict regulations regarding defaulters paying their yearly capital instalments (Bagaeen, 2015; Knowledge Corporation, 2013; GIZ International Services, 2009).

24.3 The direction of travel for Saudi Arabian smart growth in real estate development

In order to deal with smart growth principles and barriers in real estate development, the Saudi government has implemented a number of solutions that indicate the broad direction of travel. Implemented solutions include an increased involvement of the private sector in the planning and real estate development process in Saudi Arabia, a trend that started with the Fifth National Development Plan and the Ninth National Development Plan, and it is clear that the government is supporting increased private-sector involvement (United Nations Development Programme, 2010; Ministry of Economy and Planning, 2009). For example, the government, with the cooperation of the private sector, plans to build four new growth centres away from the three biggest cities of Riyadh, Jeddah, and Dammam, all of which will accommodate around 4.5 million people. In doing so, the government is encouraging the integration of smart growth principles, such as compact patterns of development, to deal with issues such as housing shortages and affordability. The government is also incentivising the local private sector and foreign real estate companies to invest in the real estate sector, while also facilitating requirements and the approval process (Saudi Arabian General Investment Authority, 2008; Bagaeen, 2015; Abdul Salam et al., 2014; Knowledge Corporation, 2013).

Another solution that has been implemented in Saudi Arabia with the expectation of having a positive impact on the real estate development process is the establishment of 'urban observatories'. These urban observatories have been established to support and improve the government's and planning organisations' decision making in the future, as well as monitoring the urban and real estate development process. Examples of these urban observatories can be found in Jeddah, Madinah, and the Dammam metropolitan area, and they will make use of a number of development indicators that have been prepared by the United Nations (Madbouly, 2009).

In addition, there are several planning policies that have been amended or established to play positive roles in real estate development issues. For example, in 2015 the Saudi government approved the land tax law on undeveloped land inside the urban boundary, and it is predicted that this law will lead to positive impacts, especially for housing affordability (Abou-Korin and Al-Shihri, 2015). Moreover, the Real Estate Advisory Council has recently been established to make the process more effective. According to the Saudi Arabian General Investment Authority (2008), the Council aims to 'establish a platform for real estate and construction stakeholders to align their initiatives, advocate for reform, and create meaningful change'. In doing so, the Council will offer a platform for real estate stakeholders to increase their cooperation in order to solve real estate development issues.

To demonstrate the direction of travel for smart growth in real estate development, there have been a number of recommendations that could result in better social outcomes. One example is that planning organisations such as the Ministry of Municipal and Rural Affairs, the Ministry of Economy and Planning, and the REDF in Saudi Arabia should work together with the private sector. Cooperation between them is essential, especially in the decision-making process (Al-But'hie and Eben Saleh, 2002; Al-Shihri, 2013). Moreover, one of the main policies of the Ninth National Development Plan, as mentioned by the Ministry of Economy and Planning (2009), is to 'upgrade performance of municipalities, enabling them to invest in job creating projects, and strengthening cooperation with the private sector in implementation of development projects'.

Smart growth solutions appear to take into account Western experiences and expertise, such as technology, planning practices, and urban approaches, in tandem with the cultural, social, physical, and traditional aspects of Saudi society. Jubail and Yanbu cities could be seen as successful examples in Saudi Arabia from a planning and real estate perspective. One of the main reasons for this success is that unlike other Saudi cities, Jubail and Yanbu implemented foreign planning approaches and principles while taking into account the social, cultural, and physical aspects of Saudi society. This may open up for debate the path dependency of city development as well as other reasons for success beyond what is considered a 'Western model'. For example, the cities of Jubail and Yanbu belong to a special organisation called the Royal Commission for Jubail and Yanbu (RCJY) for their planning, management, and maintenance, and thus do not follow the Ministry of Municipal and Rural Affairs. The planning system of this Royal Commission is used in most of the critical projects in Saudi Arabia, such as hospitals, airports, and universities, and this organisation has gained several awards from international organisations such as the United Nations (Al-But'hie and Eben Saleh, 2002; Costa and Noble, 1986).

Another direction of travel is in the improvement of management between all levels of government and other stakeholders. This improvement is partly related to educating urban planners and those within governmental organisations about local and foreign developers' needs (Daghistani, 1993). This is also to educate the general public about the benefits of sustainable development and its approaches such as smart growth (Susilawati and Al Surf, 2011). There is also the need to improve and increase the role of the public in the real estate development and planning process, especially in the early stages (Al-Shihri, 2013). Moreover, as mentioned, the Saudi government recently supported the REDF by investing around 40 billion SR in order to offer housing loans to Saudis at zero interest rates. As mentioned, this could be seen as the only effective and accepted option for Saudis in terms of housing loans. It is also seen as important that the government increases its efforts to achieve better mortgage law outcomes with the cooperation of financial organisations and the private sector.

Although there are several issues raised due to the use of the smart growth principles in Saudi Arabia, the direction of travel can create some broader trade-offs that need to be managed. For instance, there may be a misallocated solution if there is excessive focus on one or a few principles, while not taking into account other important principles. An example of this over-focus could be looking only at encouraging the compact urban form, to the detriment of offering more public transport options, or increasing public participation in the planning and development process. Thus, the direction of travel for smart growth principles in real estate development is a notion that to achieve more sustainable outcomes, it is important to take a holistic view (Litman, 2015; Stewart, Sirr, and Kelly, 2006; Jacobs, 1961).

24.4 Conclusions

As a result of several urban changes such as rapid population growth and the expansion of urban areas, a pattern of urban sprawl development has been prevalent. This sprawl has seen outcomes such as increasing dependency on cars, the rising cost of infrastructure, increasing levels of pollution, housing unaffordability, and an ongoing transfer of agricultural and natural lands for residential and commercial uses. To deal with these development problems, several governments around the world have moved to implement smart growth principles. Furthermore, in Saudi Arabia, there has been a move for government to implement smart growth in order to achieve more sustainable real estate development outcomes.

We have introduced the smart growth approach, summarised and explained via ten underlying principles. Further discussion has shown the integration of smart growth principles when dealing with real estate development in Saudi Arabia. Illustration was made of a number of social issues using several examples from Saudi Arabia, including: a lack of community participation in the planning process; traffic congestion and poor quality of services and facilities; a reduced sense of local community and belonging; housing options that are not preferable to the community; housing unaffordability; and cultural issues such as privacy. Barriers to smart growth in real estate development were related either to the implementation of the smart growth principles, or to the planning and development systems. Most of these reasons are related to professionals engaged in the planning and real estate development process who are unfamiliar with the social and cultural characteristics of local residents. Barriers to public participation in the planning and real estate development process, as well as a lack of cooperation between planning institutions and organisations, are also present.

Solutions that have been implemented or recommended indicate the direction of travel for smart growth and real estate development. The direction of travel for smart growth to solve some of the real estate development challenges includes: increasing the level of cooperation between government and the private sector; establishing a number of urban observatories; and, in Saudi Arabia, establishing the Real Estate Advisory Council and the Ministry of Housing. Additionally, there needs to be greater cooperation between planning organisations during the real estate development process; greater policy transfer opportunities and continual professional development practice at an international level; the creation of the conditions for greater sensitivity towards and incorporation of local cultural and social characteristics of residents; the improvement of the management between government and real estate stakeholders; and the education of urban planners in the governance of smart growth that incorporates the needs and aspirations of local residents. The government may also increase efforts in relation to offering better financial options for residents by refining more sophisticated mortgage law. In summary, it is the holistic use of the smart growth principles that will encourage real estate development to be sustainable in Saudi Arabia and throughout the rest of the world.

References

Abdul Salam, A., Elsegaey, I., Khraif, R., & Al-Mutairi, A. (2014). Population distribution and household conditions in Saudi Arabia: Reflections from the 2010 census. *SpringerPlus*, 3, 1–13.

Abou-Korin, A. (2011). Impacts of rapid urbanisation in the Arab World: The case of Dammam Metropolitan Area, Saudi Arabia. *5th International Conference and Workshop on Built Environment in Developing Countries*, ICBEDC 2011.

Abou-Korin, A. A., & Al-Shihri, F. S. (2015). Rapid urbanization and sustainability in Saudi Arabia: The case of Dammam Metropolitan Area. *Journal of Sustainable Development*, 8, 52–65.

Abubakar, I. (2013). Role of higher institutions of learning in promoting smart growth in developing countries: University of Dammam as a case study. *Smart Growth: Organizations, Cities and Communities, 8th International Forum on Knowledge Asset Dynamics*. Zagreb.

Abu-Gazzeh, T. (1995). Privacy as the basis of architectural planning in the Islamic culture of Saudi Arabia. *Architecture and Behaviour*, 11, 93–112.

Adams, D., & Watkins, C. (2002). *Greenfields, Brownfields and Housing Development*. Oxford: Blackwell Science Ltd.

Aina, Y. A., Al-Naser, A., and Garba, S. B. (2013). Towards an integrative theory approach to sustainable urban design in Saudi Arabia: The value of geodesign. In: Özyavuz, M. (ed.) *Advances in Landscape Architecture*. InTechOpen, 531–550.

Alatni, B., Sibley, M., & Minuchin, L. (2012). Evaluating the impact of the internationalisation of urban planning on Saudi Arabian cities. *WIT Transactions on Ecology and the Environment*, 155, 291–301.

Al-But'hie, I. M., & Eben Saleh, M. A. (2002). Urban and industrial development planning as an approach for Saudi Arabia: The case study of Jubail and Yanbu. *Habitat International*, 26, 1–20.

Alexander, D., & Tomalty, R. (2002). Smart growth and sustainable development: Challenges, solutions and policy directions. *The International Journal of Justice and Sustainability*, 7, 397–409.

Al-Hemaidi, W. (2001). The metamorphosis of the urban fabric in Arab-Muslim City: Riyadh, Saudi Arabia. *Journal of Housing and the Built Environment*, 16, 179–201.

Al-Rushaid, W. (2010). *Strengthening of National Capacities for National Development Strategies and Their Management: An Evaluation of UNDP's Contribution – Country Study: Saudi Arabia*. UNDP: New York. Available at: http://web.undp.org/evaluation/documents/thematic/cd/Saudi-Arabia.pdf.

Al-Shihri, F. (2013). Principles of sustainable development and their application in urban planning in Saudi Arabia. *Journal of Engineering Sciences*, 41, 1703–1727.

Alskait, K. (2011). Impacts of increasing building density on urban roads: The case of Riyadh. *The 3rd World Planning Schools Congress*. Perth.

Bagaeen, S. (2015). Saudi Arabia, Bahrain, United Arab Emirates, and Qatar: Middle Eastern complexity and contradiction. In: Heurkens, E. & Squires, G. (eds.) *International Approaches to Real Estate Development*. New York: Routledge, 101–122.

Bass, S., Dalal-Clayton, D. B., & Pretty, J. N. (1995). *Participation in Strategies for Sustainable Development*. London: Environmental Planning Group, International Institute for Environment and Development.

Bramley, G., & Power, S. (2009). Urban form and social sustainability: The role of density and housing type. *Environment and Planning B*, 36, 30–48.

Costa, F. J., & Noble, A. G. (1986). Planning Arabic towns. *Geographical Review*, 76, 160–172.

Daghistani, A.-M. I. 1993. *A Case Study in Planning Implementation: Jeddah, Saudi Arabia*. University of Newcastle-upon-Tyne, Department of Town and Country Planning.

Dempsey, N. (2010). The future of the compact city. *Built Environment*, 36, 116–121.

Easthope, H., & Judd, S. (2010). *Living Well in Greater Density*. Sydney: Shelter NSW.

Eben Saleh, M. (1997). Privacy and communal socialization: The role of space in the security of traditional and contemporary neighborhoods in Saudi Arabia. *Habitat International*, 21, 167–184.

Edwards, M., & Haines, A. (2007). Evaluating smart growth: Implications for small communities. *Journal of Planning Education and Research*, 27, 49–64.

Fitzsimons D'Arcy, L. (2013). A multidisciplinary examination of walkability: Its concept, measurement and applicability. Doctoral dissertation, Dublin City University.

Frumkin, H. (2002). Urban sprawl and public health. *Public Health Reports*, 117(3), 201–217.

Geller, A. L. (2003). Smart growth: A prescription for livable cities. *American Journal of Public Health*, 93, 1410–1415.

GIZ International Services (2009). *Urban Planning and Land Market and Construction of the Housing Sector in Saudi Arabia*. GIZ International Services.

Grant, J. (2009). Theory and practice in planning the suburbs: Challenges to implementing new urbanism, smart growth, and sustainability principles. *Planning Theory & Practice*, 10, 11–33.
Jacobs, J. (1961). *The Death and Life of Great American Cities*. New York: Random House.
Jenks, M., & Burgess, R. (2000). *Compact Cities: Sustainable Urban Forms for Developing Countries,* London and New York: E. & F. N. Spon.
Jenks, M., Burton, E., & Williams, K. (1996). *The Compact City: A Sustainable Urban Form?* London: New York: E. & F. N. Spon.
Kiet, A. (2011). Arab culture and urban form. *Focus*, 8, Article 10.
Knowledge Corporation (2013). *Housing the Growing Population of the Kingdom of Saudi Arabia*. Knowledge Corporation.
Litman, T. (2015). *Evaluating Criticism of Smart Growth*. Victoria, BC: Victoria Transport Policy Institute. Available at: www.vtpi.org/sgcritics.pdf.
Madbouly, M. (2009). Revisiting urban planning in the Middle East North Africa region. *The Global Report on Human Settlements (2009),* GRHS and UN Habitat, Kenya: Nairobi.
McConnell, V., & Wiley, K. (2010). Infill development: Perspectives and evidence from economics and planning. *Resources for the Future*, 10, 10–13.
Ministry of Economy and Planning (2009). *The Ninth Development Plan (2010–2014).* Ministry of Economy and Planning: Riyadh.
Montgomery, S. (1986). Planning & urban change in Saudi Arabia. *Planning Outlook*, 29, 74–79.
Morris, M. (2009). *Smart Codes: Model Land-Development Regulations*, American Planning Association.
Ohlin, J. (2003). *A Suburb Too Far? Urban Consolidation in Sydney*. Sydney: NSW Parliamentary Library Research Service.
Saudi Arabian General Investment Authority (2008). *The Real Estate Sector in Saudi Arabia*. Saudi Arabian General Investment Authority: Riyadh.
Sherer, P. (2006). The benefits of parks: Why America needs more city parks and open space. *The Trust for Public Land*. White Paper.
Shojaee, F., & Paeezeh, M. (2015). Islamic city and urbanism, an obvious example of sustainable architecture and city. *Cumhuriyet Science Journal*, 36, 231–237.
Steinacker, A. (2003). Infill development and affordable housing patterns from 1996 to 2000. *Urban Affairs Review*, 38, 492–509.
Stewart, D., Sirr, L., & Kelly, R. (2006). Smart growth: A buffer zone between decentrist and centrist theory? *International Journal of Sustainable Development and Planning*, 1, 1–13.
Susilawati, C., & Al Surf, M. (2011). Challenges facing sustainable housing in Saudi Arabia: A current study showing the level of public awareness. *17th Pacific Rim Real Estate Society Conference, Bond University, Gold Coast*. Queensland University of Technology.
Talen, E. (2008). New urbanism, social equity, and the challenge of post-Katrina rebuilding in Mississippi. *Journal of Planning Education and Research*, 27, 277–293.
U.S. Environmental Protection Agency (2011). *Smart Growth: A Guide to Developing and Implementing Greenhouse Gas Reduction Programs*. Local Government Climate and Energy Strategy Series. EPA: Washington, DC.
Williams, K., & Dair, C. (2007). A framework for assessing the sustainability of brownfield developments. *Journal of Environmental Planning and Management*, 50, 23–40.

25

Applying sustainability in practice

An example of new urban development in Sweden

Agnieszka Zalejska-Jonsson

25.0 Introduction

The definition of sustainability and sustainable development has been debated for many years now (Cook & Golton, 1994; Wilkinson, 2016). The discussion is intensifying and new perspectives on the terms are being brought to our attention. One of the problems has been the vagueness of the concept (Wilkinson, 2012). The fact that defining the term comes with this potential for misunderstanding escalates the risk of miscommunication, misinterpretation of goals and settling for easily achievable objectives (Kimmet, 2009). Every stakeholder may perceive sustainability differently, and therefore pursue different goals and assign different ranks to objectives, and even act to satisfy his/her own goals first. Wilkinson (2013: 261) has stated that sustainability shares the characteristics of what Söderbaum (1956) called 'an essentially contested concept'; that is, 'it means all things to all men'.

In this chapter, the problems that occur when applying sustainability in practice are highlighted. The aim is to discuss the conflicts that arise between environmental, social and economic sustainability goals when considering the perspectives of various stakeholders. A case study – an urban development in Sweden, Hammarby Sjöstad, which is an old former industrial area that has been redeveloped into a contemporary urban housing development with the ambition of achieving high levels of sustainability – is used to explore how sustainability is defined and understood in practice. In this example, attention is drawn to the risks of a lack of understanding of the various sustainability dimensions, and the consequences of disharmony in the perceptions of the different stakeholders.

The chapter presents just a fraction of the discussion on the sustainability terminology, the main interpretations and translations of the term. Different views on sustainability are described in the literature review (Section 25.1) and potential applications are discussed. Weak and strong sustainability as applied to the built environment is described in Section 25.2. The experiences and lessons learned from the case study of sustainable urban development (Section 25.3) illustrates the challenges related to the definition of sustainability, and the application of the term in practice is evaluated in Section 25.4.

25.1 Sustainability and sustainable development

The contemporary conversation about sustainability often refers to the UN report *Our Common Future*, in which sustainable development is defined as 'development that meets the needs of the present without compromising the ability of the future generations to meet their own needs' (Brundtland Commission, 1987: 41). It is a broad definition, open to wide interpretation in numerous contexts and timeframes, and in different locations. There has been a tendency to use the two concepts 'sustainability' and 'sustainable development' interchangeably. However, the concepts are not synonymous; in fact, many researchers have argued the significance of distinctions between those terms (Robinson, 2004).

There is a broad literature discussing different perspectives on and analysing the relationship between sustainability, environmentalism, climate change and sustainable development. It is not intended to engage in a conceptual analysis, but to draw attention to the difference between the terms and to clarify our understanding of the concepts as used in this chapter.

The literature on sustainability and sustainable development has attempted to clarify the concepts for some time. Researchers have analysed the meaning of the terms, ranging from a semantic (Glavič and Lukman, 2007), to a philosophical and a historical perspective (Mitcham, 1995; Mebratu, 1998; Robinson, 2004; Hector, Christensen and Petrie, 2014). Hector et al. (2014) performed a conceptual analysis based on the proposition that 'sustainability' aims at a dynamic-equilibrium 'end-state' and that 'sustainable development' is the pathway to reach that goal. Based on those assumptions, the authors concluded that the fundamental difference between the two terms is the valuation of nature, and the ways in which those values can be fulfilled (Hector et al., 2014).

Robinson (2004) suggested that the term 'sustainable development' relates mainly to technical fixes, the approach that aims at reducing the environmental impacts of economic activity, hoping that those actions will bring improvements in social equality. Robinson (2004) also indicated that sustainable development might be understood as growth, and this affects the meaning of the term, shifting the meaning to continuing economic growth. Mitcham (1995) also called attention to the duality of the term 'sustainable development', asking what society is actually aiming to sustain and how it relates to the need or wish for development. He pointed out that the ideal of sustainable development aims at bridging potentially conflicting claims, limits to growth and the need for development (Mitcham, 1995).

Robinson (2004: 370) wrote that 'sustainability' focuses on 'the ability of humans to continue to live within environmental constraints', and by those objectives 'sustainability' has a different meaning from 'sustainable development'. 'Sustainability', as described by Robinson (2004), is a broad concept, which aims at and actively works towards improvements in efficiency and social and environmental impact (technical fixes) but also individual responsibility, and a change in values and attitudes. The author emphasises that what truly defines sustainability is the transdisciplinary thinking that focuses on the connection between various fields and integration across the fields, sectors and interests. In this sense, achieving sustainability is only possible if all the actors (stakeholders) – private and public, but also individuals – are actively involved (Robinson, 2004).

In this chapter, the term 'sustainability' is deemed to mean, similar to Robinson (2004), actively working towards improvements in efficiency and in social and environmental impact. It is contended that this can only be successfully achieved by the change and redefinition of human (and organisational) behaviour and attitudes, and the involvement of all parties. It is also acknowledged that the sustainability spectrum is very broad and that practical applications can vary significantly; therefore, further discussion follows covering the distinction between, and giving examples of, weak and strong sustainability in the construction and the real estate market.

25.2 Dimensions of sustainability

The translation and application of sustainability principles has not been an easy task. On the one hand, sustainability goals may be defined at a specific point in time, hence making the aims achievable. However, on the other hand, over the long term, the perception of sustainability evolves and is adapted to a new status, and therefore achieving sustainability goals could be seen as a continuous process of transformation and evolution (Bagheri and Hjorth, 2007; Berardi, 2013; Emmelin and Cherp, 2016). The second challenge appears when perceiving sustainability as a three-dimensional concept with emphasis on social, economic and environmental dimensions, or emphasis on the dual dimension of the relationship between humanity and nature (Robinson, 2004).

Over many years of debate, it has become clear and accepted by most people that sustainability must be defined by taking three aspects into consideration: economic, environmental and social. Trying to identify how these dimensions should interlock, how to establish value and how these defining aspects should be ranked by importance has not been straightforward (Högberg, 2014). The concept of sustainability can be visualised as three interlocking circles, presenting weaknesses when only two of the three dimension are satisfied (Adams, 2006). When the goals that consider only environmental and economic aspects are satisfied, solutions can be considered 'viable' but not sustainable; similarly, if only social and environmental dimensions are considered, a 'bearable' future can be achieved, whereas engaging only social and economic dimensions gives 'equitable' results (Thomsen and Van der Flier, 2009). Only constructs that build on a balance between economic, social and environmental aspects are considered to have achieved a sustainable outcome (Adams, 2006; Thomsen and Van der Flier, 2009).

The three-dimensional (economic, social and environmental) nature of sustainability has been considered fundamental for sustainable development, and separation of these domains could lead to inadequate outcomes (Kohler, 1999). Furthermore, Adams (2006) argued that economic, environmental and social aspects are not equivalent and should not be considered as adjustable, for if a trade-off is made or imbalance exists, sustainability may shift from strong to weak.

25.2.1 Economic sustainability

Economists have approached defining sustainability in relation to the distribution of equity and its relation to efficiency, ethics and the recognition of existing and future generations' rights (Asheim, Buchholz and Tungodden, 2001; Padilla, 2002; Stavins, Wagner and Wagner, 2003). This discussion is usually held at the macro level and rarely raises the question of how to solve problems of conflicting objectives; therefore, an alternative approach that allows the valuation of costs and benefits has become necessary.

It has been proposed that the approach known as life cycle costing (LCC) may fulfil this need. LCC methodology has been used in the construction industry with the aim of including environmental conditions, for example, for quantification of disposal costs (Abraham and Dickinson, 1998), total energy costs (Sterner, 2002) or energy-efficient measures (Liu, Rohdin and Moshfegh, 2016). However, the application of the environmental context in LCC has been challenged due to perceived limitations of the methodology. LCC is the assessment of investment efficiency; however, significant limitations to this approach include the timeframe and selection of the most appropriate period, and the difficulty of predicting, for example, a future maintenance strategy or pricing (Costanza and Patten, 1995).

Since each stakeholder perceives the time value of money and the cost of capital differently, and has different consumption preferences, LCC calculations made for the same case may have different outcomes due to different discount rate assumptions (Gluch and Baumann, 2004). Additionally, in the case of the built environment, there are uncertainties related to technology and the technical advances applied, such that due to technological development and innovation, the selection of components and materials can become obsolete (Gundes, 2016). This relates to the definition of the lifetime period. There are four different lifetimes that could be considered in the built environment: economic, technical, physical and utility (Gluch and Baumann, 2004). Since the durations of these different lifetimes are not the same, questions arise with regards to the accuracy and interpretation of LCC results.

The economic calculation may have a different outcome depending on system limitations – for example, whether it is performed for a building, a residential or commercial development, or even a suburb or a community. The values, and thus results, may differ depending on the stakeholders' perspectives as well. Estimation of the costs, risks and viable income differ between private and public actors, planners and developers, municipalities and individuals. Some point out the oversimplification of these calculations and problems related to how monetary values are assigned, which may contribute to undervaluing environmental risks and therefore producing inaccurate results (Gulch and Baummann, 2004).

Lind, Annadotter, Björk, Högberg and Af Klintberg (2016) discuss another aspect of economic sustainability: whether a certain renovation strategy leads to higher rents and makes more people dependent on welfare benefits, possibly an unintended consequence.

25.2.2 Environmental sustainability

Environmental, or ecological, sustainability is probably intuitively well understood, and can be defined generally as development that aims at preserving and not compromising natural resources (Anand and Sen, 2000; Högberg, 2014); however, accurate measurement of the impact of human actions on and consequences for the environment has been challenging.

In order to be able to measure the impact of human action, indicators of environmental performance need to be established. There are many different units with which we can measure the effect of actions. One of the commonly used indicators in the built environment is energy. The total energy associated with a building may be divided into energy that is directly connected with the building's operation (energy consumption), energy needed for the building's construction, rehabilitation (or refurbishment or renovation or adaptation) and demolition, and embodied energy, which is the sum of all the energy needed to manufacture and transport goods (that is, all material and technical installations) (Sartori and Hestnes, 2007). The question of how embodied energy and operating energy influence the total energy used in a building's life cycle is the subject of much discussion in the literature. Results differ depending on the building type, the year of construction, the climate zone and finally the energy measures used to analyse a building's performance (Zalejska-Jonsson, 2013). The energy indicator is also dependent on the definition of system boundaries, which may be restricted to an apartment or building, or extended to a development or even a city.

Energy used in buildings can be expressed as end-use energy or primary energy. Primary energy measures energy at the natural source level and indicates the energy needed to obtain the end-use energy, including extraction, transformation and distribution losses (Sartori and Hestnes, 2007), focusing on energy resources and the process in the supply system. Hence, two different buildings may indicate the same end-use energy performance, but differ significantly in performance measured in primary energy due to different energy sources (Gustavsson and Joelsson, 2010).

Energy consumption is relatively easy to monitor and energy performance data is available for many types of real estate. Additionally, energy consumption bills are a part of people's lives and therefore easily understood by the general public (Högberg, 2014). Energy performance can be easily presented in monetary units, and incorporated into an LCC calculation.

25.2.3 Social sustainability

The literature shows that social sustainability is particularly difficult to define. The problems with conceptualising social sustainability have been associated with a lack of clear differentiation between analytical, normative and political aspects (Littig and Griessler, 2005).

The concept of sustainability may constitute development, maintenance or a bridge between the two (Vallance, Perkins and Dixon, 2011; Åhman, 2013). The definition of social sustainability can be rooted in meeting people's basic needs, equity, education and the creation of social capital and justice; but it can also be seen as the maintenance of socio-cultural practices and quality of life and the preservation of the natural landscape (Vallance et al., 2011; Åhman, 2013). Littig and Griessler (2005) suggested that, apart from these indicators, social sustainability should be assessed on aspects of social coherence and suggested measurement of, for example, integration into social networks, involvement in activities as volunteers, as well as measures for solidarity and a tolerant attitude. Social sustainability might also be seen as a bridge between the people and the environment (Vallance et al., 2011). This means that social sustainability could be seen not as an objective in itself, but as an instrument to achieve environmental objectives (Littig and Griessler, 2005).

An important dimension of the theoretical discussions is whether social sustainability is seen from an individual or a general social perspective (Spangenberg and Omann, 2006). A number of academics have based the definition of social sustainability on the institution of work and engagement from both the individual and the community. They define social sustainability as 'a life-enhancing condition' that takes place within communities (McKenzie, 2004: 12), indicating that the critical elements are communities and connection between people (Barron and Gauntlett, 2002), and that work mediates relationships in society and satisfies the goals that the 'set of human needs are shaped in a way that nature and its reproductive capabilities are preserved over a long period of time and the normative claims of social justice, human dignity and participation are fulfilled' (Littig and Griessler, 2005: 72).

25.2.4 Stakeholders and their perspectives

'Sustainability' is categorised as a contested concept (Söderbaum, 2011; Cook and Golton, 1994), which means that it can be defined, understood and interpreted differently by different people (Baumgärtner and Quaas, 2010a, 2010b). Baumgärtner and Quaas (2010b) wrote that 'sustainability is a normative notion about the way humans should act towards nature, and how they are responsible towards one another and future generations'. The normative nature of the concept directly points out the first problem, which is that every individual (or organisation) understands sustainability through the values and propositions they believe in or adhere to (Mebratu, 1998). This also implies that an organisation and an individual, who is employed by this organisation, may have different ideas about, and understandings of, sustainability. Consequently, the objectives and strategies to achieve them would be different.

Allow us to consider the situation in the housing market with three types of stakeholder: developer, local government/municipality (including urban planning) and individual. From an economic perspective, the developer will aim to sell a dwelling at the highest price, while the individual's goal would be to acquire a dwelling at the lowest price. Both perspectives are rational and reasonable, and both aim at an economically sustainable solution as perceived by the company and the individual, respectively. On the other hand, the ideal situation from the government's position is that everybody in society is able to afford housing.

A further example is, if for the sake of this discussion, it is simply assumed that a building's energy performance is the measure of environmental sustainability, and therefore both the developer's and the individual's aims should be allied. However, as Harrison and Freeman (1999) pointed out, the timing and the context affect stakeholders' perspectives. The developer is interested in fulfilling building regulations, and possibly even environmental certifications, at the planning and design phase, but may have little interest in following up building performance if the ownership and responsibility has been transferred at the point of sale to the individual (owner/occupant). This is commonly known as the 'split incentive', whereby one party provides something from which another gains; here, the developer provides the energy efficiency measures and the user benefits from lower operational energy consumption, lower greenhouse gas emissions and, importantly, lower energy bills or costs. The individual or user therefore is mainly interested in the actual energy performance, since this is reflected in operating costs. The government's perspective should encompass the whole building cycle and therefore combines both objectives, aiming at low energy requirement and fulfilling this ambition in the operational phase. In this sense, the government's interest is minimising the energy performance gap.

The social aspect in the housing market touches upon the fulfilment of expectations and needs for shelter and safety. Meeting people's basic needs in the form of a safe home is a requirement that is not always fulfilled. The developer and the individual represent the power of supply and demand in this market, which is often overshadowed by economic expectations. Demand drives supply, but delivery may just relate to the most attractive or profitable segment of the market. The definition of social sustainability depends on the perspective of choice and whose rights are to be protected: the individual's, the organisation's or society's. In a neo-liberal 'free' market world, this often fails the less economically powerful members of society and widens the gap between the rich and the poor (Standing, 2016).

25.2.5 Weak and strong sustainability

Pearce, Markandya and Barbier (1989) were among the first to make the distinction between weak and strong sustainability. Beckerman's (1994) article on the concept of sustainability, and his characterisation of strong and weak sustainability, caused a heated discussion (Beckerman, 1994; Beckerman, 1995; Daly, Jacobs and Skolimowski, 1995; Common, 1996; Ang and Van Passel, 2012). Beckerman (1994: 194) argued that the strong sustainability concept is irrational and extreme as 'a requirement to preserve intact the environment as we find it today' and that the debate that preserving everything at all costs or choosing what to preserve and what not to preserve is morally questionable. Beckerman (1994, 1995) contended that weak sustainability adds little to traditional perceptions of welfare maximisation. It has been pointed out that weak sustainability is flawed and is a direct application of economic concepts (Beckerman, 1994; Gutés, 1996; Gowdy and O'Hara, 1997). The notion of weak sustainability allows people to describe their choices as sustainable, when according to economists' jargon they might be just 'optimal' choices (Beckerman, 1994).

There might be various reasons why a certain form of sustainability has been adopted. One idea is that the perception of sustainability depends on whether an ecocentric or anthropocentric world view is adopted (Wilkinson, 2013). Often, people will adopt positions that express ecocentric or anthropocentric world views without consciously being aware of the distinction (Wilkinson, 2013). Wilkinson identified five distinctive groups representing anthropocentric or ecocentric views, from cornucopian and accommodating environmentalism, to moderate, deep and transpersonal ecology. This conceptual understanding gives an explanation of the spectrum of the concept of sustainability. Wilkinson (2013) argued that, given the political, environmental, social and economic thinking and policies adopted, the ecocentric perspective is more likely to deliver strong sustainability. A pilot study was conducted to test this theory and the sustainability approach of five leading Australian property organisations was analysed. The findings suggested that the Australian sector is adopting a more anthropocentric perspective and is therefore prone to deliver weak sustainability (Wilkinson, 2013).

In Figure 25.1 and Table 25.1, the attributes and actions of weak and strong sustainability are synthesised.

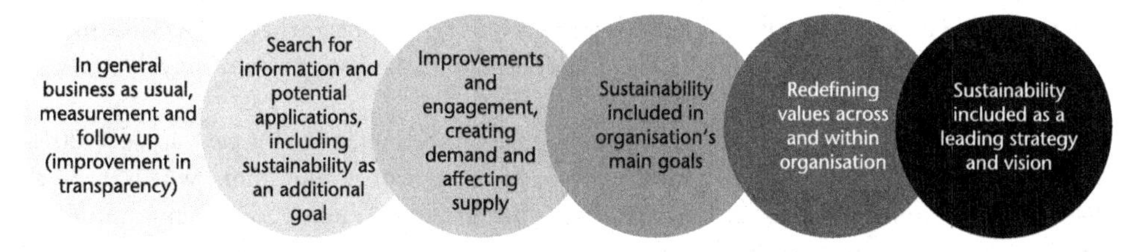

Figure 25.1 Weak and strong sustainability

Source: Adapted from Dixon et al. (2008); Wilkinson (2013); Masalskyte, Andelin, Sarasoja and Ventovuori (2014).

Table 25.1 Examples of weak and strong sustainability adoption by an organisation

	Economic	*Environmental*	*Social*
Weak	Resource usage documentation (water, waste, energy management)	Resource usage documentation (water, waste, energy management)	Communication with stakeholders
	Building certification	Building certification	Communication with employees
	Resource usage monitoring, (water, waste, energy management)	Resource usage monitoring (water, waste, energy management)	Occupant/user information
	Involving sustainability specialists	Involving sustainability specialists	Energy consumption visualisation
	Follow-up and improvement (water, waste, energy management)	Prioritising local production	Engaging workers/ occupants/users in efficient resource consumption
	Life cycle cost	Material certifications option	Green lease
	Green lease	Green suppliers chain	Training and education
	Innovative solutions for resource usage	Strategic sustainable cooperation with suppliers	Cultural change
	Sustainability aims included in practice goals	Life cycle analysis	Local community engagement
	Minimum resource usage and onsite production	Minimum resource usage and onsite production	Optimal resource usage and onsite production
strong	Sustainability decisions made at the strategic level	Engagement and coordination with suppliers and occupants	(Permanent) behavioural change

Sources: Adapted from Dixon et al. (2008); Wilkinson (2013); Masalskyte et al. (2014)

25.3 Applying sustainability objectives in Sweden

In this section, a case study from Sweden is used to reflect on the challenges of applying sustainability aspirations to development projects and renovation projects for residential buildings. Experiences and lessons learned from this case study are used to discuss conflicts between the environmental, social and economic dimensions of sustainability, between stakeholders' perspectives and the various nuances of sustainability described above.

Sweden is a small country located in the northern hemisphere; it stretches between latitudes 58°N and 69°N, and as such the climate in Sweden varies with warm summers and cold winters. The average temperature in the winter months is approximately –4°C and in the summer months +15°C (SMHI, 2016). The Swedish population was approximately 9.8 million in 2016.

Swedish gross domestic product (GDP) was approximately US$46,709 per capita in 2015 (OECD, 2015a), higher than that of the 28 countries in the European Union, which stood at US$38,650 per capita (OECD, 2015a). Sweden is often highly ranked for its environmental performance; for example, greenhouse gas emissions were a low 3.9 tonnes per capita in 2012, whereas the emissions of the European Union's 28 countries was 6.5 tonnes per capita (OECD, 2015b).

Environmental issues have been on the national agenda in Sweden for nearly 50 years. The Swedish Environmental Protection Agency was established in 1967 and the Environmental Protection Act, introduced in Sweden in 1969, is considered to be one of the most comprehensive pieces of environmental legislation (Granberg and Elander, 2007). The oil crisis of 1970 significantly affected policies and environmental goals in Sweden, directing the focus onto alternative energy sources and energy reduction (Granberg and Elander, 2007).

The Swedish Environmental Protection Agency has been responsible for creating good living conditions environmentally, socially and economically, and is therefore considered a crucial actor in shaping climate and sustainable development policies (Granberg and Elander, 2007). At the end of the 1990s, Sweden adopted a set of national environmental objectives, and the Swedish Environmental Protection Agency was set up to be the leading agency for promoting environmental actions (Emmelin and Cherp, 2016). In the 1990s, the action plan for sustainable development – Local Agenda 21, which followed the 1992 United Nations Conference on Environment and Development in Rio de Janeiro – was introduced in Sweden (Lundqvist, 2004). By the end of the century, the majority of Swedish municipalities had adopted Local Agenda 21 in their strategies (Lundqvist, 2004).

One of a set of 16 national environmental quality objectives adopted in Sweden is A Good Built Environment (Environmental Objectives Portal, 2016). The objectives are to ensure, for example, a coherent and sustainable connection between urban and rural areas, infrastructure design adapted to people's needs, efficient and environment-friendly resource management (energy, land, water and other natural resources) and the creation of a healthy built environment that supports people's needs and offers a varied range of housing, workplaces, services and culture (Environmental Objectives Portal, 2016). The Swedish government established milestone targets for each of the objectives. Progress towards the milestones is monitored and evaluated, and guided by a set of indicators such as energy consumption, good indoor environment quality, reduction of hazardous substances in buildings and recycling.

Climate and energy conservation have been among the main objectives of the Swedish agenda. The approach to stimulating energy conservation in buildings, adopted by the Swedish government, has been more market driven than regulation driven (Schade, Wallström, Olofsson and Lagerqvist, 2013). Apart from minimum standards established in the building regulations and energy performance certificates, policy instruments implemented in Sweden have had the nature of a voluntary agreement (Schade et al., 2013).

The most established and applied building environmental rating schemes in the Swedish property market are international: BREEAM and LEED and also the national certification, Environmental Building (Miljöbyggnad). The Swedish certification scheme Environmental Building (Miljöbyggnad) has gained popularity in recent years (SGBC, 2016). The scheme has been developed and adjusted to Swedish norms and standards, enabling the relatively easy application of the certification requirements in the building construction process. Since 2009, over 730 buildings have been certified by the Swedish Green Building Council (SGBC, 2016), of which half consist of newly constructed buildings, including multi-family buildings, offices, schools and retail buildings.

25.4 The case study of Hammarby Sjöstad

A sustainable urban development was selected to illustrate the challenges that stakeholders (private, public and individual) face while attempting to harmonise their objectives within a sustainability framework. The researcher drew on the experiences and lessons learned from the case to discuss the distinction between the application of strong and weak sustainability.

Hammarby Sjöstad is a sustainable urban development in Stockholm whose construction was initiated in the 1990s. Other known examples are: Ecolonia in Alphen aan den Rijn, the Netherlands; Exodus in Delft, the Netherlands; and Rieselfeld and Vauban in Freiburg, Germany (Bayulken and Huisingh, 2015).

Hammarby Sjöstad is an old industrial and harbour area redeveloped into a modern urban housing development. To date, 9,000 apartments have been built, as well as kindergartens, schools, a culture centre, a library and over 250,000 square metres of commercial area. Over 20,000 occupants now live in Hammarby Sjöstad. The development is located in southeastern Stockholm, relatively close to the Stockholm CBD (Stockholm City, 2017).

Planning works for Hammarby Sjöstad began in the early 1990s, and a comprehensive local area development plan was presented in 1991. In the mid-1990s, Stockholm City Council accepted an environmental programme for Hammarby Sjöstad. By 1996, the Quality Programme for the Design of Hammarby Sjöstad was completed and a regulatory detail plan for the district was initiated (Iverot and Brandt, 2011).

The environmental programme had very high ambitions. The plan aimed to achieve 'twice as good' environmental performance, to introduce new technological solutions and to implement a holistic view of planning. The aspiration for Hammarby Sjöstad was to be 'at the forefront … striving for sustainable development in densely populated urban areas' (City of Stockholm, 1996, cited in Iverot and Brandt, 2011: 1044). The environmental programme focused on six directions: land use, soil decontamination, technical supply (energy, waste and water sewage), transport, construction materials and noise (Svane, Wangel, Engberg and Palm, 2011). One of the biggest ambitions of the environmental programme was to reduce the energy requirements in buildings to 60kWh/m^2, which at that time was half of what was expected from new buildings.

The findings show that Hammarby Sjöstad has been successful in utilising 80 per cent of extractable energy from waste and waste water (Iverot and Brandt, 2011). The successful outcome here could be related to the fact that the goals were included in the early planning stages and to the formation and application of a system-based Hammarby Sjöstad model (Iverot and Brandt, 2011). The Hammarby Sjöstad model was formed in cooperation between the municipality and various infrastructure companies in Stockholm: the city energy company managing the thermal plant, the wastewater plant and the city's waste recycling company (Iverot, Vernay, Mulder and

Brandt, 2013). This shows what can be achieved when stakeholders decide to cooperate and collaborate.

The environmental plan for Hammarby Sjöstad aimed at a significant reduction in building energy consumption. This environmental objective proved difficult to meet, and this was mainly due to a conflict of goals between the different stakeholders. On the one hand, the municipality aimed at very low energy consumption in the district. This could have been achieved by utilising new technology and technological innovations in the construction. However, implementation of those advances was not a primary or priority goal in the plan for Hammarby Sjöstad; hence, it received little attention during the construction process and was not adopted as widely as it could have been (Iverot et al., 2013).

25.4.1 Environmental and economic drivers

In order to reach ambitious operational energy goals, buildings should be designed and built adopting a Passive House concept to a large degree, constructing buildings with an airtight, very well insulated envelope and minimising thermal bridges (Krope and Goricanec, 2009). Additionally, in order to maintain a good indoor environment, solar gains and heat losses through windows should be taken into consideration (Krope and Goricanec, 2009; Schroeder, 2009).

The greatest energy losses in a building are through windows, especially those facing north. In the case of the Hammarby Sjöstad housing development, the position of the windows had a significant effect on building energy performance. The Hammarby Lake is located north of the Hammarby Sjöstad development. The view over the lake and the harbour adds to the high attractiveness of the district (Svane, 2008). In order to take advantage of the highly appreciated view, developers installed large windows. Consequently, the north-facing facades of many buildings constructed in the district are mainly glass, contributing significantly to energy losses. This meant that energy performance in the buildings became more difficult to achieve. In order to meet the 60 kWh/m^2 target, designers would have had to implement new technologies and install highly energy-efficient windows (Svane, 2008).

The technological solution needed to achieve very low energy requirements in the buildings – for example, highly energy-efficient triple-glazed windows – was expensive at that time. Additionally, securing a very tight and a very well insulated building envelope requires a high level of accuracy in the construction process, which may have had an effect on labour costs. The material cost and the labour cost for the construction of such energy-efficient buildings could add up to a significant cost increase, even 10–20 per cent higher than for conventional construction (Zalejska-Jonsson, Lind and Hintze, 2012).

The lake view had an economic value for both the developer and the buyer/occupant. Previous studies showed that energy and environmental factors are considered by customers when making their apartment purchase; however, the relevance of these factors is generally not as high as building location or apartment price (Zalejska-Jonsson, 2014). Customers generally acknowledge the importance of environmental and energy performance; however, how these features translate in the housing market and into quality of lifestyle was not as clear.

This example illustrates the effect of the ranking of priorities on the goal and on the achieved sustainability. Even though companies and the municipality aimed at including sustainability in the process and even creating a new demand and affecting supply (Figure 25.1), the promotion of ideas did not match the application, and the possibility of attitude change to deliver strong sustainability was lost. The extra cost for producing very low-energy buildings could not be justified economically and this argument overwrote environmental ambition. The trade-off was made and consequently the sustainability goal was weakened.

25.4.2 Environmental opportunities and technological advances

The research showed that success in achieving environmental targets was related to the Hammarby Sjöstad model (Iverot and Brandt, 2011). The overarching aim of the model was to address energy, waste sewage and water in the district. To a great extent, the model utilisesd the existing infrastructure, but it also included suggestions for new technology. Unfortunately, the model has not been applied to its full potential. The assessment studies indicated that the Hammarby Sjöstad model focused mainly on the efficiency of existing technology and did not sufficiently promote new innovative solutions such as solar cells, solar panels, biogas stoves and green roofs (Iverot and Brandt, 2011). This strategy resulted in missed opportunities to influence and stimulate the innovation market and to further reduce energy dependence (Iverot et al., 2013).

The strong sustainability goals were weakened in the process. The application of new technology, such as solar cells and solar panels, could have positively impacted building energy consumption, social education and community engagement, but were a strain economically. Consequently, the very ambitious goals were not attained, nor was strong sustainability (Figure 25.1).

25.4.3 Environmental targets and social priorities

One of the main goals for Hammarby Sjöstad was to reduce car traffic and to promote and increase the use of public transport, cycling and walking (Iverot and Brandt, 2011). In line with this target, the number of parking spaces was limited to 0.25 per apartment (Mahzouni, 2015). However, the residents, despite supporting the environmental goals, thought that the number of parking spaces should be increased. Their views were considered in the planning and the number of parking places was increased to 0.7 per apartment (Svane, 2008). This example reflects the collision between social and environmental objectives: the clash between the expectations of a comfortable lifestyle, resulting in increasing consumption, and the need for behavioural adjustment.

Hammarby Sjöstad succeeded in accomplishing the reduction of car traffic as 80 per cent of all commuters there are using public transport, cycling or walking (Iverot and Brandt, 2011). This is a great success and certainly can be seen as an achievement of the strong sustainability ambition. The negotiation process shows how difficult it is to redefine needs and limit the convenience.

25.5 Conclusions

There is a risk that the application of sustainability might become just an explanation of old habits and choices and fade away if not strongly synergised with stakeholders' values. The role of a strong third party might be decisive. Housing security, a functioning competitive market, diversity and equality within society are some of the priorities for the government, both national and local. Government, both local and national, has the responsibility and realistic opportunity to influence and redirect stakeholders' approach towards sustainability. This can be achieved by changing regulations and expectations, by education and by redefining the needs of organisations, both private and public, and also those of society, as a group and as individuals. For example, expected building performance can be strengthened with building regulations, responsibility for follow-up can be regulated and car traffic can be reduced by creating alternatives in the infrastructure. Social and environmental expectations can be made explicit in the planning of urban areas.

The vagueness of the definition of sustainability and the commonness of its use poses a danger that the term could lose its meaning and become an ambivalent, empty word (Mitcham, 1995). It follows that if we accept weak sustainability as a step towards achieving sustainability goals, we are deceiving ourselves. This is because the level of sustainability we actually need will not have been reached.

The case of Hammarby Sjöstad shows the difficulty in allying sustainability goals and their hierarchical structure. The balance between economic objectives and environmental aspirations is difficult to achieve in today's market circumstances. It is unreasonable to assume that one of the actors (be they the individual, the developer or the municipality) would, of their own free will, risk their market position and profit in order to fulfil environmental or social objectives. It also indicates that national and local government must take the responsibility of leadership in defining common sustainability goals, in rewarding active participation and in utilising technological advances to their fullest potential.

References

Abraham, D. M. and Dickinson, R. J., 1998. Disposal costs for environmentally regulated facilities: LCC approach. *Journal of Construction Engineering and Management*, 124(2), 146–154.

Adams, W. M., 2006. The future of sustainability: re-thinking environment and development in the twenty-first century. In *Report of the IUCN Renowned Thinkers Meeting*. Available online: http://cmsdata.iucn.org/downloads/iucn-future-of-sustainability.pdf [2017–11–29].

Åhman, H., 2013. Social sustainability: society at the intersection of development and maintenance. *Local Environment*, 18(10), 1153–1166.

Anand, S. and Sen, A., 2000. Human development and economic sustainability. *World Development*, 28(12), 2029–2049.

Ang, F. and van Passel, S., 2012. Beyond the environmentalist's paradox and the debate on weak versus strong sustainability. *BioScience*, 62(3), 251–259.

Asheim, G. B., Buchholz, W. and Tungodden, B., 2001. Justifying sustainability. *Journal of Environmental Economics and Management*, 41(3), 252–268.

Bagheri, A. and Hjorth, P., 2007. Planning for sustainable development: a paradigm shift towards a process-based approach. *Sustainable Development*, 15(2), 83–96.

Barron, L. and Gauntlet, E., 2002. WACOSS housing and sustainable communities indicators project. *Sustaining Our Communities International Local Agenda 21 Conference*, Adelaide, 3–6. Available online: www.regional.org.au/au/soc/2002/4/barron_gauntlett.htm [2015-08-20].

Baumgärtner, S. and Quaas, M., 2010a. Sustainability economics: general versus specific, and conceptual versus practical. *Ecological Economics*, 69(11), 2056–2059.

Baumgärtner, S. and Quaas, M., 2010b. What is sustainability economics? *Ecological Economics*, 69(3), 445–450.

Bayulken, B. and Huisingh, D., 2015. Are lessons from eco-towns helping planners make more effective progress in transforming cities into sustainable urban systems? A literature review (part 2 of 2). *Journal of Cleaner Production*, 109, 152–165.

Beckerman, W., 1994. 'Sustainable development': is it a useful concept? *Environmental Values*, 3(3), 191–209.

Beckerman, W., 1995. How would you like your 'sustainability', sir? Weak or strong? A reply to my critics. *Environmental Values*, 4(2), 167–179.

Berardi, U., 2013. Clarifying the new interpretations of the concept of sustainable building. *Sustainable Cities and Society*, 8, 72–78.

Brundtland Commission, 1987. *Our Common Future: Report of the World Commission on Environment and Development. UN Documents: Gathering a Body of Global Agreements*. Available online: www.un-documents.net/our-common-future.pdf [2016-05-01].

Common, M. S., 1996. Beckerman and his critics on strong and weak sustainability: confusing concepts and conditions. *Environmental Values*, 5(1), 83–88.

Cook, S. J. and Golton, B., 1994. Sustainable development concepts and practice in the built environment: a UK perspective. *CIB TG 16*. Sustainable Construction, Tampa, Florida, USA, 6–9 November.

Costanza, R. and Patten, B.C., 1995. Defining and predicting sustainability. *Ecological Economics*, 15(3), 193–196.

Daly, H., Jacobs, M. and Skolimowski, H., 1995. Discussion of Beckerman's critique of sustainable development. *Environmental Values*, 4(1), 49–70.

Dixon, T., Colantonio, A., Shiers, D., Reed, R., Wilkinson, S. and Gallimore, P., 2008. A green profession? A global survey of RICS members and their engagement with the sustainability agenda. *Journal of Property Investment & Finance*, 26(6), 460–481.

Emmelin, L. and Cherp, A., 2016. National environmental objectives in Sweden: a critical reflection. *Journal of Cleaner Production*, 123, 194–199.

Environmental Objectives Portal, 2016. Available online: www.miljomal.se/Environmental-Objectives-Portal/Undre-meny/About-the-Environmental-Objectives [2017–07–01].

Glavič, P. and Lukman, R., 2007. Review of sustainability terms and their definitions. *Journal of Cleaner Production*, 15(18), 1875–1885.

Gluch, P. and Baumann, H., 2004. The life cycle costing (LCC) approach: a conceptual discussion of its usefulness for environmental decision-making. *Building and Environment*, 39(5), 571–580.

Gowdy, J. and O'Hara, S., 1997. Weak sustainability and viable technologies. *Ecological Economics*, 22(3), 239–247.

Granberg, M. and Elander, I., 2007. Local governance and climate change: reflections on the Swedish experience. *Local Environment*, 12(5), 537–548.

Gundes, S., 2016. The use of life cycle techniques in the assessment of sustainability. *Procedia-Social and Behavioral Sciences*, 216, 916–922.

Gustavsson, L. and Joelsson, A., 2010. Life cycle primary energy analysis of residential buildings. *Energy and Buildings*, 42(2), 210–220.

Gutés, M. C., 1996. The concept of weak sustainability. *Ecological Economics*, 17(3), 147–156.

Harrison, J. S. and Freeman, R. E., 1999. Stakeholders, social responsibility, and performance: empirical evidence and theoretical perspectives. *Academy of Management Journal*, 42(5), 479–485.

Hector, D. C., Christensen, C. B. and Petrie, J., 2014. Sustainability and sustainable development: philosophical distinctions and practical implications. *Environmental Values*, 23(1), 7–28.

Högberg, L., 2014. *Building Sustainability: Studies on Incentives in Construction and Management of Real Estate*. KTH Royal Institute of Technology, Stockholm.

Iverot, S. P. and Brandt, N., 2011. The development of a sustainable urban district in Hammarby Sjöstad, Stockholm, Sweden? *Environment, Development and Sustainability*, 13(6), 1043–1064.

Iverot, S. P., Vernay, A. L., Mulder, K. F. and Brandt, N., 2013. Implications of systems integration at the urban level: the case of Hammarby Sjöstad, Stockholm. *Journal of Cleaner Production*, 48, 220–231.

Kimmet, P., 2009. Comparing 'socially responsible' and 'sustainable' commercial property investment. *Journal of Property Investment & Finance*, 27(5), 470–480.

Kohler, N., 1999. The relevance of Green Building Challenge: an observer's perspective. *Building Research & Information*, 27(4–5), 309–320.

Krope J. and Goricanec, D., 2009. Energy efficiency and thermal envelope. In Mumovic, D. and Santamouris, M. (eds), *A Handbook of Sustainable Building Design and Engineering: An Integrated Approach to Energy, Health and Operational Performance*. Earthscan: London, 23–33.

Lind, H., Annadotter, K., Björk, F., Högberg, L. and Af Klintberg, T., 2016. Sustainable renovation strategy in the Swedish Million Homes Programme: a case study. *Sustainability*, 8(4), 388.

Littig, B. and Griessler, E., 2005. Social sustainability: a catchword between political pragmatism and social theory. *International Journal of Sustainable Development*, 8(1–2), 65–79.

Liu, L., Rohdin, P. and Moshfegh, B., 2016. LCC assessments and environmental impacts on the energy renovation of a multi-family building from the 1890s. *Energy and Buildings*, 133, 823–833.

Lundqvist, L. J., 2004. 'Greening the people's home': the formative power of sustainable development discourse in Swedish housing. *Urban Studies*, 41(7), 1283–1301.

Mahzouni, A., 2015. The 'policy mix' for sustainable urban transition: the city district of Hammarby Sjöstad in Stockholm. *Environmental Policy and Governance*, 25(4), 288–302.

Masalskyte, R., Andelin, M., Sarasoja, A.-L. and Ventovuori, T., (2014). Modelling sustainability maturity in corporate real estate management. *Journal of Corporate Real Estate*, 16(2), 126–139.

McKenzie, S., 2004. Social sustainability: towards some definitions. Working Paper No. 27, Hawke Research Institute: Magill, South Australia. Available online: http://w3.unisa.edu.au/hawkeinstitute/publications/downloads/wp27.pdf [2016-06-21].

Mebratu, D., 1998. Sustainability and sustainable development: historical and conceptual review. *Environmental Impact Assessment Review*, 18(6), 493–520.

Mitcham, C., 1995. The concept of sustainable development: its origins and ambivalence. *Technology in Society*, 17(3), 311–326.

OECD, 2015a. OECD Data. Available online: https://data.oecd.org/sweden.htm [2017–01–21].

OECD, 2015b. OECD Data. Available online: https://data.oecd.org/air/air-and-ghg-emissions.htm [2017–01–21].
Padilla, E., 2002. Intergenerational equity and sustainability. *Ecological Economics*, 41(1), 69–83.
Pearce, D., Markandya, A. and Barbier, E., 1989. *Blueprint for a Green Economy*. Earthscan: London.
Robinson, J., 2004. Squaring the circle? Some thoughts on the idea of sustainable development. *Ecological Economics*, 48(4), 369–384.
Sartori, I. and Hestnes, A. G., 2007. Energy use in the life cycle of conventional and low-energy buildings: a review article. *Energy and Buildings*, 39(3), 249–257.
Schade, J., Wallström, P., Olofsson, T. and Lagerqvist, O., 2013. A comparative study of the design and construction process of energy efficient buildings in Germany and Sweden. *Energy Policy*, 58, 28–37.
Schroeder M., 2009. Energy efficient refurbishment of dwellings: a policy context. In Mumovic, D. and Santamouris, M. (eds), *A Handbook of Sustainable Building Design and Engineering: An Integrated Approach to Energy, Health and Operational Performance* Earthscan: London, 134–151.
SMHI, 2016. Swedish Meteorological and Hydrological Institute. Available online: www.smhi.se [2017–01–21].
Söderbaum, P., 2011. Sustainability economics as a contested concept. *Ecological Economics*, 70(6), 1019–1020.
Spangenberg, J. H. and Omann, I., 2006. Assessing social sustainability: social sustainability and its multicriteria assessment in a sustainability scenario for Germany. *International Journal of Innovation and Sustainable Development*, 1(4), 318–348.
Standing, G., 2016. *The Precariat: The New Dangerous Class*. Bloomsbury: London.
Stavins, R. N., Wagner, A. F. and Wagner, G., 2003. Interpreting sustainability in economic terms: dynamic efficiency plus intergenerational equity. *Economics Letters*, 79(3), 339–343.
Sterner, E., 2002. *Green Procurement of Buildings: Estimation of Environmental Impact and Life-Cycle Cost*. Diss. Luleå Tekniska Universitet. Available online: www.diva-portal.org/smash/get/diva2:999180/FULLTEXT01.pdf [2017-11-20].
Stockholm City, 2017. Hammarby Sjöstad. Available online: http://bygg.stockholm.se/hammarbysjostad [2017-01-21].
Svane, Ö., 2008. Situations of opportunity: Hammarby Sjöstad and Stockholm City's process of environmental management. *Corporate Social Responsibility and Environmental Management*, 15(2), 76–88.
Svane, Ö., Wangel, J., Engberg, L. A. and Palm, J., 2011. Compromise and learning when negotiating sustainabilities: the brownfield development of Hammarby Sjöstad, Stockholm. *International Journal of Urban Sustainable Development*, 3(2), 141–155.
Swedish Green Building Council (SGBC), 2016. Certification. Available online: www.sgbc.se/var-verksamhet [2016–02–21].
Thomsen, A. and van der Flier, K., 2009. Replacement or renovation of dwellings: the relevance of a more sustainable approach. *Building Research & Information*, 37(5–6), 649–659.
Vallance, S., Perkins, H. C. and Dixon, J. E., 2011. What is social sustainability? A clarification of concepts. *Geoforum*, 42(3), 342–348.
Wilkinson, S. J., 2012. Conceptual understanding of sustainability in Australian construction firms. CIB, Montreal, Canada, June.
Wilkinson S. J., 2013. Conceptual understanding of sustainability in the Australian property sector. *Property Management*, 31(3), 260–272.
Wilkinson, S. J., 2016. Understanding sustainability and the Australian property professions. *Journal of Sustainable Real Estate*, 8(1), 95–119.
Zalejska-Jonsson, A., 2013. *In the Business of Building Green: The Value of Low-Energy Residential Buildings from Customer and Developer Perspectives*. Diss. KTH Royal Institute of Technology. Available online: www.diva-portal.org/smash/record.jsf?pid=diva2%3A655904&dswid=9555#sthash.DyP6oZ4s.dpbs [2017-11-20].
Zalejska-Jonsson, A., 2014. Impact of energy and environmental factors in the decision to purchase or rent an apartment: the case of Sweden. *Journal of Sustainable Real Estate*, 5(1), 66–85.
Zalejska-Jonsson, A., Lind, H. and Hintze, S., 2012. Low-energy versus conventional residential buildings: cost and profit. *Journal of European Real Estate Research*, 5(3), 211–228.

26

Sustainable real estate

Where to next?

Sara Wilkinson, Tim Dixon, Sarah Sayce and Norm Miller

> Climate change is destroying our path to sustainability. Ours is a world of looming challenges and increasingly limited resources. Sustainable development offers the best chance to adjust our course.
>
> *Ban Ki-moon (2012), UN Secretary-General, 2007–2016*

> One of the most difficult things is not to change society – but to change yourself.
>
> *Nelson Mandela (2000)*

26.0 Introduction

The two quotes at the start of this chapter reflect not only the imperative to tackle climate change through sustainable development, but also the need for individuals to pause and reflect and change their behaviour. Although Ban Ki-moon was addressing the issue of climate change directly, Nelson Mandela's words are confronting a rather different political context. Yet Mandela's words also hold true in the context of this book: for if we are to tackle climate change and other environmental, social and governance (ESG) risks, and transition to a sustainable future, then our behaviours, both individual and organisational, need to change too. This also implies that, although policies and regulation (or indeed incentives) can drive change, we also need to take individual and collective responsibility for our actions, and to recognise our 'stewardship' role with regard to the built environment (Young, 2012).

In general, the real estate sector globally faces a period of significant transformation, shaped by a complex set of 'megatrends', such as rapid urbanisation, disruptive new technologies, globalised markets and the growing importance of sustainability (RICS, 2015; BPIE, 2016; Dixon, Green and Connaughton, forthcoming). Certainly, as this book has shown, sustainability in today's international real estate sector is much more relevant to decision making and value management than it was even a decade ago. There is now a growing trend in corporate businesses and the public sector to embrace sustainable real estate and 'walk the talk', as demonstrated by the emergence of investor benchmarking schemes such as the Global Real Estate Sustainability Benchmark (GRESB), which has gained considerable traction in Europe, and the increasing number of organisations seeking voluntary sustainability certification for their buildings. In the commercial property sector, as we have seen, market demand from occupiers has proved to be a key driver for sustainable real estate, but corporate responsibility principles have also proved powerful in helping drive change. Despite this progress, however, much remains to be done: large parts of the real estate industry are still highly complex, fragmented and essentially 'conservative' in outlook. Valuers, in particular, have struggled to find the market evidence on which they rely to 'prove' a green premium or a 'brown' discount. The upside is that there continue to be niches of advantage to benefit first movers in the sector, and we are also seeing the emergence of innovative projects in a number of sectors, which may disrupt existing business and finance models and technologies. The downside is that climate change appears to be happening at a faster rate than adoption of deep and meaningful market change.

Although the standardisation of building sustainability assessment measures continues to reflect the nuances of differing national contexts and cultures, there have also been recent moves to enshrine sustainability principles within the real estate sector that work across international boundaries. The World Economic Forum (WEF, 2016), for example, highlights five key principles to which real estate leaders should commit, and which relate to embedding the best in class sustainability standards: ensuring the commitment to collaborative improvements in environmental sustainability; committing to continuous improvement in the environmental performance of construction and development; tracking the environmental performance of real estate assets continuously; and setting explicit targets for sustainability improvement.

What of the future? Can the real estate sector really make a difference, and help reduce carbon emissions globally? The stakes are high: although the Paris Agreement on Climate Change is in place, it is not universally ratified and there is now substantial uncertainty over the continued role of the US, following the election of President Donald Trump. Moreover, recent research (PwC, 2016) shows that even with the Paris Agreement fully in place, in order to prevent warming in excess of 2°C, the global economy needs to cut its carbon intensity (tCO_2/\$m GDP) by 6.3 per cent a year every year from now until 2100. Simply fulfilling the Paris COP21 Agreement would also require a decarbonisation rate (reduction in carbon intensity) of 3 per cent per year, which is more than double the business-as-usual rate of 1.3 per cent (2000–2014) (PwC, 2016). Responding to these immense challenges will require agility and the ability to deal with 'wild-card' or 'black-swan' events (Dixon et al., 2017).

In the final part of this book, therefore, we draw together key findings from the four parts comprising:

1 governance and policy
2 valuation, investment and finance
3 management
4 redevelopment and adaptation.

We then explore the cross-cutting trends that have emerged through the chapters in this book, and the short-term and medium-longer-term trends that will be important in shaping the sector over the next 10–20 years. Then, in conclusion, we draw together our views and thoughts in a final section.

26.1 Part 1: Governance and policy

The first part of this book covers issues connected with governance – primarily, though not exclusively, governance with a little 'g'; that is, through the eye of the individual corporate organisation and through the role of the leading professional body of the land, the RICS. However, the chapters also recognise that changes in 'little g' governance often only come about within the context of actual or likely changes to 'big G' governance (national or international legal and regulatory frameworks): the two are inextricably linked.

A theme throughout the book is that governments, both national and supranational, can, through enforcement measures, bring about improvements to the built stock, but that such action is primarily directed towards new builds and 'deep' retrofits. However, their preferred route is to support or encourage market transformation through voluntary measures. All four parts of the handbook concern the ways in which 'little g' measures are succeeding in delivering positive change.

In Chapter 2, Miller and Pogue explore the rise of corporate social responsibility (CSR) and businesses' response to the well-rehearsed challenges of carbon, energy waste and water usage. They argue that the large corporations have now moved their thinking away from that expressed by Friedman (1962): that the sole social responsibility of business is to use its resources and engage in activities designed to increase its profits. Instead, companies have come to realise, through a series of external 'pushes', the case for adopting an approach that embeds a responsibility to external stakeholders. But, the chapter implies, shifts towards more socially and environmentally responsible policies have often been triggered by legislation brought about in the wake of disasters – such as the Exxon Valdez and Bhopal catastrophes – or by scientific endeavours such as the work of the Intergovernmental Panel on Climate Change (IPCC), which has heightened awareness that there are increasing downsides, both financial and reputational, to corporate 'bad behaviours'. For Miller and Pogue, the upward pressures from consumers in terms of increasing expectations of behavioural standards and the risk of 'big G' (governmental) regulation have been powerful drivers in leading corporates to appoint champions of sustainability, thus spawning a new set of careers focusing on measuring, monitoring and reporting on social and environmental concerns. Whilst sustainability accounting has not become, in the US at any rate, a legal or 'big G' requirement, in the view of Bosteels and Sweatman (2016) corporate real estate owners and investors should recognise that they have a clear fiduciary duty to understand and actively manage ESG and climate-related risks as a routine component of their business thinking, practices and management processes. So, much has changed, and the authors point to this cultural business shift in which the simple pursuit of single-bottom-line economic performance may work against the long-term best business interests: it is now simply too risky a policy to pursue. In this chapter, therefore, we can witness that corporate governance is beginning to change as a result of market push, though with the backdrop of actual or threatened regulatory requirements.

Whilst Chapter 2 deals with corporate governance, Chapter 3 concerns very specific and small-scale governance: that of the contractual relationships between landlords and tenants of commercial premises. In Chapter 3, Patrick, Bright and Janda argue that the very nature of a traditional lease inhibits consensual behaviours that promote environmental improvements. Indeed, describing such improvements as a 'wicked' problem, they detail the 'split-incentive' conundrum in which the responsibility for environmental improvements may lie with the landlord, whilst the benefit may sit with the tenant in the form of a better, more economic building. Unless, or until, such environmental improvements are reflected in rental deals, the incentive for the landlord may be limited. And for this reason, despite significant advocacy, the take-up of so-called 'green leases', which have gained traction in Australia, has been muted in the UK. However, the authors argue, through the findings of their empirical research, the 'big G' measure of minimum energy standards to be introduced for new, and eventually continued lettings in the UK,[1] may provide a spur for building improvements, which, to date, have been hard to implement, thus providing another example of the way in which a regulatory intervention may stimulate market change. However, at the time of writing, there still exists a lamentable lack of knowledge of the legislation among many real estate advisors.

Van der Heijden, in considering the role of voluntary programmes for low-carbon transformation in the US (Chapter 4), concludes that traditional regulatory interventions such as building codes and planning legislation often fall short in accelerating the transition towards sustainable built environments. To overcome this, he argues that government may seek to use positive incentives and voluntary systems, rather than strict compliance regimes. These, he concludes, may provide more effective solutions than a simple regulatory system. However, the question remains as to how effective such voluntary programmes would be were there not the possibility of enforcement of compliance in the event of limited take-up and no market shift.

Chapter 5 by Wilkinson also considers the role of voluntary actions towards retrofits but within an Australian context. In this chapter, the author details a database tracking over 2,000 offices that were retrofitted over the period from 1998 to 2011; the study investigated the role played by various factors including certification. It found that the proactive stance of the City of Melbourne was important in stimulating private-sector action, but that more could be done to encourage the owners of non-retrofitted stock to follow the example of others. An interesting result was the level of awareness and influence of NABERS. The research pointed to a disappointingly low level of awareness of NABERS, but that this is changing as a system of commercial property building disclosure is being systematically brought to tie in with building sales or leases. This requirement to disclose, which in many ways mirrors the European Energy Performance Certificate (EPC), thus provides another example of how a combination of voluntary and compliance measures are required to stimulate market action.

In the last chapter in the section (Chapter 6), Elder explores another type of governance, that of the professional bodies who set and enforce standards. Starting from the perspective of the neo-classical economic market philosophy, Elder argues that steps taken by professional bodies, such as the International Valuation Standards Committee (IVSC) and the Royal Institution of Chartered Surveyors (RICS), together with cross-professional groups such as the International Ethics Committee and the International Property Measurement Group, have a valuable role in better ensuring transparency and efficiency in markets, which in turn can stimulate better use of scarce resources. Elder further argues that such effective and efficient allocations better ensure that sustainable features of real estate are recognised and priced appropriately into the market. Again, the recognition is that the requirement to comply and cooperation between different regulatory bodies are fundamental to positive change. By harnessing this greater transparency and utilising international standards more effectively, Elder concludes that the desire for the environmental characteristics of buildings to be reflected in market pricing is more easily achieved.

In summary, the five take-home findings from these chapters in terms of where to next are:

1 Governance can take the form of both 'big G' (i.e., governmental requirements) and 'little g' (i.e., actions at the level of the corporation).
2 The rise of CSR has been triggered by external shocks leading to public awareness and government regulation. As CSR reporting has grown, a new breed of professionals charged with enhancing sustainability performance has developed. This has positive impacts on the measurement and monitoring of buildings' sustainability performance.
3 Leases provide a 'little g' context that has proved difficult to integrate with a shift towards more sustainable buildings. Whilst green leases have been introduced in some countries, the so-called split-incentive issue has proved a 'wicked' problem in driving cooperation towards more sustainable stock.
4 Moves towards a more sustainable building stock often require outside triggers such as regulation. Schemes in Europe and Australia are beginning to drive change – but are likely to be more effective where combined with punitive provisions such as minimum standards.

5 Professional bodies, which regulate and educate their members, have the power, through the introduction of professional standards, to enhance market transparency, and in turn this can lead to more efficient, effective and thus sustainable resource use.

26.2 Part 2: Valuation, investment and finance

The second part of the book concentrates on matters connected with finance in the broadest sense of the word: it therefore deals with valuation and agency (Chapters 7, 9 and 13); the case for putting a price on 'green' as it is sometimes called (Chapter 12); establishing the economic benefits of investing in retrofits (Chapter 8); and setting up structures to encourage investment in properties that have sustainability attributes (Chapters 10 and 11).

Starting with the theme of value, the role of the valuer is that of a 'reflector,' rather than a 'maker', of markets, a point brought out strongly by Sayce (Chapter 9). Accordingly, both Warren-Myers (Chapter 7) and Sayce (Chapter 9) agree that the reported values of real estate will only reflect sustainability if and when evidence of market penetration has taken place. In Chapter 7, Warren-Myers, working within an Australian context, makes a strong case that, to date, the evidence base in the form of key financial correlations between sustainability and economic return has not been found sufficient to change the practice of valuers. But is this really due to a lack of evidence, insufficiently robust data, or the valuer's own resistance to change or lack of skills? Within the Australian context, it is concluded that there are indeed challenges to genuinely integrating sustainability characteristics within the valuation process, but that partly this is down to a need for valuer education.

Placing these arguments into a global professional body requirement context and with special reference to practice within Europe, Sayce (Chapter 9) agrees with much of this analysis, although she points out that sustainability education is tested through professional competence, albeit only in a limited way. However, the conclusion is that this is partly a matter of appetite by the client: if there is no client appetite to be provided with valuations that pay explicit regard to sustainability, then it is unlikely that the valuer will do so, and if they do not, there will be nothing to aid others in terms of their analysis. However, this does not necessarily mean that the valuer has disregarded the factors – just that they have accounted for them implicitly. And this is one of the issues that, it is argued, needs to be addressed. Valuation reports can be a 'black box' so it is hard to break down how individual attributes have been addressed; conversely, when the valuer is preparing an investment worth rather than market value, the use of discounted cash flow (DCF) techniques enables clear reporting. However, this still does not address the issue of data inadequacy. Nonetheless, the observed lack of integration of sustainability criteria within valuations, despite an encouragement by the professional body, is concerning when placed alongside the results of academic research linking value and sustainability.

The link between sustainability and value from the perspective of a meta-analysis of academic literature is taken up by Dalton and Fuerst (Chapter 12), who argue that the proliferation and rapid growth of voluntary 'eco-labels' (sustainability rating schemes) in and of themselves provide evidence of a market-led environmental agenda. This could potentially create 'green value' that could be factored into prices and rents established via a willingness to pay. From their analysis of the literature, they find that this relates primarily to energy efficiency rather than a wider interpretation of sustainability. But even among the literature they record a disjointed and, at least in part, inconclusive set of results. Although the general incentives and disincentives of energy efficiency and broader sustainability impacts have been examined from a theoretical perspective, there is still a dearth of empirical evidence as to their real effectiveness. Furthermore, the findings that do exist are heterogeneous, which perhaps adds to the reasons why valuers, who deal at the individual level, have found it difficult to incorporate specific premiums or discounts. Dalton and Fuerst then go on to set up a hypothetical DCF commercial office investment example to try to develop a deeper understanding from the findings of the studies. They find that the tenants' tendency to pay a premium for 'green' may at times outweigh the actual cost savings achieved – something that they argue requires greater research attention.

The conclusions of Dalton and Fuerst connect well with Chapter 8, in which Miller and Kok examine the economic payoffs for building retrofits within the US office context. After making a powerful case that retrofitting buildings is key to achieving long-term global sustainability goals – a theme that is recurrent right through this handbook – they conclude that retrofitting makes good economic sense. The case for this lies in the fact that many retrofitted buildings can work well, especially when they achieve sustainability ratings such as LEED and Energy Star, as is increasingly the case. The case for retrofit is based, however, not just on rental and capital value appreciation, but also on workplace productivity (a theme taken up in Part 3 of this book). In terms of the former, Miller and Kok argue that market evidence *does* exist but that the quantum of this depends in part on the local price of energy, which varies from state to state and according to climatic conditions. Where the climate requires above-average heating/cooling and energy costs are also above average, the case will be most convincing. This goes some way to explaining the heterogeneity of findings revealed by Dalton and Fuerst. In general, they conclude that the economic benefits will be in excess of the costs of the improvements but that it is often still a struggle to integrate this with valuation and finance, thus supporting the views of Warren-Myers and Sayce.

Much of the above analysis focuses on building the benefits of sustainability into the valuation process, thus adding transparency to the market and helping support the case for value differentiation based on sustainability criteria. It also has a commercial focus. However, in Chapter 13, Hurst and Wilkinson view the issue through a different lens. Detailing research involving residential sales in Australia, they examine the role of residential estate agents, who they argue are key players in the value chain who can exercise a key role in both educating clients and helping market transformation. By investigating the placement of text advertising sustainability features, they find that overall it does feature highly in marketing, although some pockets of the Melbourne market contain more advertising content related to sustainability. Despite agents purporting to have knowledge of sustainability factors, and the market saying how important sustainability is, it did not feature heavily within their promotion of property. This would imply that transfer (market) prices achieved are not fully reflective of sustainability criteria and that buyers do not seek sustainability features in housing purchases. There is some way to go in order to change this market.

The last theme addressed in this part of the book is that of finance, specifically real estate investment trusts (REITs). In Chapter 10, Parker provides a comprehensive literature review of published work linking the role of REITs to sustainability, but finds surprisingly little. However in Chapter 11, Sah and Miller concentrate specifically on green REITs which, they argue, are a somewhat arbitrary group as they lack clear definition; however, whilst green REITs are not a global phenomenon, they are found in the US, which has a much longer history of REITs than most other countries. Indeed, REITs have been in existence in the US for some 50 years, but only came into existence in the UK ten years ago. Through their analysis, Sah and Miller conclude that green REITs do offer some investment benefits. In particular, they find that, although they may not always exhibit abnormal returns or higher market capitalisations relative to net asset value, they tend to exhibit less volatility in returns. However, the analysis is rendered problematic by the continuing difficulty in deciding what actually constitutes a green REIT – and whether and how green REITs link through to specific certification schemes. Unless and until there is better understanding of these measures, the financial attraction of 'green' will continue to be problematic.

It is interesting that the handbook does not specifically address the relationship between debt funding and sustainability; maybe it will for its next edition, but for now the case for building in differential borrowing rates or other funding mechanisms geared towards promoting sustainability retrofits is only in its infancy.[2]

The eight take-aways we can derive from these chapters are:

1 Research to show a 'green premium' has been extensive but the results are still inconclusive. There is a strong theoretical case, but it now needs greater in-depth dissection as to the differences between the perspectives of the investor and the tenant.
2 The connection between market value and sustainability is developing but far from fully developed. Though it is being facilitated by the professional bodies, market traction is not fully established.
3 Moves within the professional body to encourage greater knowledge and clarity of data and instructions are to be welcomed, but there is still much to be done to enable valuers to say that in reflecting the market, the value of green is specifically identifiable. Greater specificity in reporting and a shift to more DCF techniques would help.
4 Estate agents are also key stakeholders and have a role to play in assisting market transformation.
5 The value case is not just grounded in capital and rental values but also in occupier considerations such as productivity.
6 It is not possible to have a 'one-size-fits-all' approach to value and investment in sustainability. Climatic, regulatory and energy policy and pricing are three key variables that will impact on the matter.
7 Generally, the number of REITS and investment funds dedicated to investing in green products has been growing in the USA, in contrast to other parts of the world. However, a key to market transformation will be investor demand by specialist property companies.
8 Debt funding will also have a part to play, but as yet this is still largely unresearched or unrealised.

26.3 Part 3: Management

The third part of this handbook comprises six chapters that cover retail and office property in the UK, the US and Australia from the perspectives of corporations, facility managers and occupiers.

In Chapter 14, Heywood argues that occupiers are vital to real estate sustainability. Corporate real estate management (CREM) is the 'missing link' in sustainable real estate. To create supply, occupiers need to create demand, and the uptake of sustainable real estate requires organisations to adopt sustainability in their business models and practices. CREM's role, Heywood argues, is to align, enable or add value and greater CREM professionalisation (which is needed to extract optimum value from CRE) to strengthen arguments for sustainable real estate's enabling or added value. This indicates a need to change demand for sustainable real estate. CREM has a role to play, and survey results show that, with advanced CREM practice, this is occurring; however, there is more to do to achieve a sustainability mindset. The challenge is that occupiers focus on their businesses, not the real estate *per se*. Real estate sustainability can be, or is, ignored. Alternatively, the business connection can be leveraged or deleveraged depending on organisational sustainability agendas. Finally, because of CREM's business connectivity with sustainability, there are opportunities, currently untapped, to connect sustainable real estate to wider sustainable business concerns, such as, in green supply chains.

In Chapter 15, Ferm and Livingstone cover the UK retail sector and its sustainability performance, arguing that there is growing awareness and action. Despite significant contributions internationally to carbon emissions, retailers receive little attention in research and professional literature on sustainable real estate. The chapter reveals the drivers and challenges facing UK retailers. The customer-facing nature of retail, competition, increasing consumer awareness, and media mean large, international retailers are showing commitment and emerging as leading drivers of sustainability. Real estate and occupancy concerns are critical for retailers. Branding often drives initiatives along with desires to be a market leader, with risk management and future proofing as other considerations. However, retailer actions in terms of occupancy are limited to aspects within their control, rather than landlord or head office control, and by initiatives that do not weaken the customer experiences that drive bottom-line sales. Split incentives, such as those discussed earlier in the book by Patrick et al., difficult landlord–tenant relationships, shorter leases and concerns over consumer well-being were shown to be barriers to widespread implementation. Changing consumer preferences, improvements to digital and mobile technologies, and changing legislation are thought to be likely to impact retailing in the future, with an emphasis on the flexibility and resilience of buildings, and a shift from the 'sustainable store' to warehousing, distribution and delivery, and carbon management and reduction across operations. This research is UK centric, and larger international samples are needed to understand retailer motivations, landlord–tenant relationships, leasing issues and whether there is a process in becoming more sustainable – that is, moving from 'weak' to 'stronger' models of sustainability. Numerous questions arise from the research and Ferm and Livingstone posit that the 'environmental performance gap' noted in the Australian office market between top-grade and lower-grade office stock may also occur in the retail sector. The growth of online retailing requires research on sustainability in retail to include operations, such as warehousing and logistics. Sustainable retailing is a journey, not a destination, and research needs to reflect this moving forward.

Chapter 16 on sustainable facilities management, by Paul Appleby, shows that sustainability has grown in importance as a framework. British Institute of Facilities Management (BIFM) annual surveys establish how members engage with sustainability. The 2015 survey reported that 36 per cent of respondents had not introduced any standards for environmental and/or energy performance, but that 42 per cent used the international environmental management standard ISO 14001. Most facilities managers use key performance indicators (KPIs). These commonly include health and safety, energy, waste, carbon footprint and training. The survey showed that, though drivers such as legislation and corporate image increased between 2014 and 2015, there was a negative impact on sustainability effectiveness. This is partly due to inadequate communication from senior management and is reflected in lower sustainability knowledge levels amongst lower management and the general workforce. Facilities management (FM) comprises numerous skills and disciplines; successful coordination and programming is key to success. Though many organisations have in-house FM teams, outsourcing is common, and many specialist FM companies provide multidisciplinary services supported by specialist designers and operatives, along with a supply chain of sub-contractors and suppliers. As buildings become more complex, the skills and knowledge required to manage them sustainably will be at a premium, and the facilities manager role is likely to evolve into one of a coordinator of multiple disciplines, akin to project management in the construction industry. Only if this happens will moves to more effective sustainable management be facilitated.

The interface between the Australian office market and Building Energy Efficiency Certificates (BEECs) is the focus of Warren's chapter (Chapter 17). The programme and statutory requirements for owners to obtain a BEEC has had a marked impact on the sector, mirroring the views of Wilkinson set out in Chapter 5. The number of NABERS-rated buildings prior to the introduction of mandatory disclosure stood at 208 in Australia (Warren, 2010), whereas by 2014, 1,153 buildings were certified (Gabe, 2016). From November 2011 to 2013, 1,081 BEECs were issued for 862 buildings (11.1 million square metres of office space) with an average NABERS rating of 3.03 stars. As more BEECs are issued, improvements occur in energy efficiency resulting from improved management and retrofitting of older buildings. The investment market and tenants opting to occupy highly efficient buildings drive the market. Buildings with 4-star NABERS ratings or above have much lower vacancy rates compared to other similar buildings and return on investment is higher (MSCI/IPD, 2016). The combined effect of the mandatory requirement to obtain a BEEC with cost savings and improved building performance shows that buildings managed in a more sustainable and energy-efficient way attract and retain tenants, and provide owners and investors with a greater return on investment whilst reducing the environmental impact of the buildings. The sustainability criteria of Australian office buildings resulting from the combined mandatory government initiatives and the voluntary schemes, such as Green Star, place the region well. 2015 GRESB data shows that Australia has an overall score 25 per cent above the global average (GRESB, 2015). There is, it seems, an opportunity for Australia to show leadership in respect to sustainability improvements in the commercial office sector.

In Chapter 18, Langston and Al-khawaja analyse workplace ecology, which deals with the relationship between job satisfaction, spatial comfort and productivity. They claim that more case studies, in a range of different settings, are needed for greater insight into the relationships. Work completed to date suggests that results are difficult to generalise, as each workplace has unique attributes that affect performance; moreover, these attributes change over time. They need to be measured regularly and improved continuously. Annual occupant surveys are desirable and analysis could be largely automated. Workplace ecology is an emerging field of study for creating or adapting office environments where people are more likely to be happy, efficient and empowered in their work activities. It is now possible to ensure that continuous process improvement for workplaces can occur using a systematic and rigorous procedure they describe. Healthy workplaces require occupants to be satisfied with their job, comfortable in their space, productive with supplied technology and regularly evaluated. This applies to new and existing office environments.

Robinson and Simons (Chapter 19) report on their research into tenant demand and the creation of a new green office building rating matrix in the US targeting buildings below the level of LEED. Index data were drawn from demand-side data from tenant surveys for individual green office building features and from supply-side data from hedonic analysis of rents. Qualitative input from the industry leaders revealed how the matrix could be useful to them. Various models were analysed and tested on 197 office buildings and a reasonable model showed that around a third of non-LEED buildings score higher on this scoring system than the lowest-scoring LEED building. These results show that many office buildings that offer green amenities are unable to signal the presence of those features through current eco-labels. Robinson and Simons' analysis provides a rationale for developing a new US green building index that can be applied to any office building, not just limited to high-performance buildings.

In summary, the seven take-home findings from these chapters in terms of where to next are:

1 With CREM's business connectivity with sustainability, there are opportunities, currently untapped, to connect sustainable real estate to wider sustainable business concerns, such as in green supply chains.

2 The 'environmental performance gap' in Australia's office market between the top-grade and lower-grade office stock may also occur in the retail sector, which, to date, is under-researched.
3 The growth of online retailing requires research on sustainability in retail to include operations, such as warehousing and logistics.
4 The facilities manager's role is likely to evolve into one of a coordinator of multiple disciplines, akin to project management in the construction industry.
5 There is opportunity for other countries to learn from Australia about how to improve sustainability in the commercial office sector through a successful combination of mandatory and voluntary approaches.
6 Workplace ecology is an emerging field of study for creating or adapting office environments where people are more likely to be happy, efficient and empowered. Annual occupant surveys are desirable and analysis could be largely automated to provide on-going benchmarks.
7 Many US non-LEED-rated office buildings perform better than low-rated LEED stock, and there is a rationale for a new US green building index that can be applied to any office building.

26.4 Part 4: Redevelopment and adaptation

The final part of the handbook focuses on the challenges, emerging trends and innovations in the redevelopment and adaptation of sustainable real estate with contributions from Australia, the UK, the Netherlands and Sweden. The chapters focus on housing, commercial energy retrofits, conversion adaptation, sustainable urban redevelopment, and smart growth and real estate development.

To start the section, Susilawati (Chapter 20) describes how a multi-stakeholder partnership approach can promote a sustainable housing supply chain in Queensland, Australia. Home occupiers were found to be not well informed about operation and maintenance and could use heating and cooling in more cost-effective ways. In part, this is because developers do not provide information about sustainability features before purchase or in user manuals. The information is available and kept by others, so improving buyer awareness of value could drive value chains and promote sustainable housing. A pilot project showed how e-building passports can facilitate access to building information in a useful/understandable format. The e-building passport improves documentation accountability, reduces compliance costs and facilitates audits at the design stage and 'as-built' and 'as-occupied' certifications. Susilawati argues that governments have a role to facilitate building information distribution across multi-stakeholders to bridge the knowledge gap in the sustainable housing supply chain. There is a lack of incentives for investors, developers and homeowners to add expensive sustainable features, as there are very limited financial benefits for landlords or tenants. Finally, it is argued that the key to much of this is to stimulate buyer demand for sustainable housing, so that housing without sustainable features will be discounted.

In Chapter 21, Dixon poses the question: what needs to be done in scaling up commercial property energy retrofitting? The sector is complex, diverse and conservative, and rolling out retrofit at scale in the UK is a challenge. Commercial property investors and developers see retrofitting at the individual building and portfolio level rather than at the city scale. This, combined with the diversity of commercial stock and geographical spread, can lead to discontinuities between stakeholders. Achieving consensus on what is meant by retrofit is vital, and, for large-scale commercial property retrofit to succeed, urgent action in both policy and practice is needed. Four key principles are identified as follows:

1 Financing is crucial to success.
2 Actual energy performance should be transparent.
3 Better integrated leadership at the city level is needed.
4 Consistency in standards is needed at a number of levels.

In each instance, Dixon articulates clearly how these principles can be delivered and thus how an agenda for change can be set for this market. There are clear parallels here between this chapter and that of Wilkinson (Chapter 5).

In Chapter 22, Remøy and De Jong examine the conversion adaptation and sustainable real estate in the Netherlands. This chapter sheds light on the sustainability of adaptive reuse, offering insights about obsolescence as opposed to new developments, presenting results about the extent to which adaptive reuse is conceived as 'sustainable development' and linking the quantification of sustainability to adaptive reuse. The authors note that current Dutch models apply a 50-year lifespan for commercial buildings, for which no rationale could be determined. 50 years is very short, arguably too short, as most buildings survive far longer. They assert that the aim should be for a 200-year lifespan to fulfil the demands of quality (urban, architectural), sustainability (embodied energy) and the circular economy (reduction of waste). During this longer lifespan, maintenance and operation costs, including refurbishment, will surpass initial costs, and this requires a new approach to assessing the costs and benefits of the development and use of real estate.

In Chapter 23, Heurkens illustrates approaches to sustainable urban redevelopment in the Netherlands. Although developing sustainable real estate is gaining traction, sustainable urban places are more difficult to realise. Heurkens finds that no consensus exists in Dutch urban development practice about what sustainable urban development is and how it can be achieved. At the urban area scale, many economic, social and environmental sustainability issues can be addressed, as well as smart, circular, energy-neutral, climate-adaptive objectives. Integrated and organic urban development approaches currently coexist and offer the potential to deliver sustainable urban development. These approaches result in top-down, private-sector-led urban development and bottom-up, private-sector-led incremental piecemeal development strategies. A pair of case studies show examples of these strategies in Rotterdam and Amsterdam and the lessons to be learned. The cases indicate that formal legal public–private arrangements and intensive informal public–private interactions are needed to define sustainable urban development for an area and how it can best be achieved. Neither strategy is preferred or superior: both can work. Overall, attitude change and more experience is needed for sustainable urban redevelopment to become the norm.

Alhamoudi and Lee (Chapter 24) focus on smart growth and real estate development in the Middle East. Rapid population and urban growth have resulted in urban sprawl, increasing car dependency, infrastructure costs, pollution levels, housing costs and the rapid transfer of agricultural land to residential and commercial uses. To deal with the issues, governments are implementing smart growth approaches. Alhamoudi and Lee summarise these approaches in ten principles and show the integration of the principles in Saudi Arabia. They raise social issues such as participation, traffic congestion, poor quality services and facilities, reduced sense of community and belonging, offering of unsuitable housing options, affordability and cultural issues. Barriers are analysed and many relate to planning and real estate development professionals unfamiliar with local social and cultural aspects. Challenges include increasing cooperation between government and the private sector, establishing urban observatories and establishing in Saudi Arabia a Real Estate Advisory Council and Ministry of Housing. Their conclusion is that the government could offer better financial options by refining mortgage laws. The holistic use of the smart growth principles will encourage real estate development to be sustainable in Saudi Arabia and elsewhere.

Chapter 25 looks at sustainability in practice in a new urban development in Sweden. Zalejska-Jonsson cites the risk that sustainability might fade away if not aligned to stakeholder values. Housing security, a functioning competitive market, social diversity and equality are identified as national and local government priorities. The chapter argues that governments have the responsibility, opportunity and potential to influence and redirect stakeholders towards sustainability, and that this can be achieved by changing regulations and expectations, by education and by redefining the needs of organisations, both private and public, and those of society, as a group and as individuals. The vagueness of the definition of sustainability poses a danger that the term loses meaning and becomes ambivalent (Mitcham, 1995). Weak sustainability as a step to achieving sustainability goals is arguably a deception, which will deflect from achieving sustainability objectives. To support this, the case of Hammarby Sjöstad is detailed, as it shows the difficulty in aligning economic objectives with environmental aspirations. It is concluded that no stakeholders would willingly risk their market position and profit to fulfil environmental or social objectives. Therefore, national and local government must take the responsibility of leadership in defining common sustainability goals, in rewarding active participation and in utilising full potential of technological advances.

In summary, the take-home findings from these chapters in terms of where to next are:

1 We need to get Queensland homebuyers demanding sustainable housing so that housing without sustainable features will be discounted.
2 Four principles (financing, transparency, integrated leadership and consistent standards) for scaling up UK commercial retrofit are proposed to set an agenda for change for this market.
3 Longer lifespans need to be considered to ensure greater sustainability for adaptive reuse in the Dutch market.
4 Attitude change and more experience is needed for sustainable urban redevelopment to become the norm in the Netherlands.
5 The holistic use of the smart growth principles will encourage real estate development to be sustainable in Saudi Arabia and elsewhere.
6 National and local government in Sweden must lead in defining common sustainability goals, in rewarding active participation and in utilising full potential of technological advances.

26.5 Where to next?

To varying degrees, either directly or indirectly, the big issues currently facing the delivery, quality and standards of global sustainable real estate are social and economic trends such as greater urbanisation, changing demographics and population growth, shifts in economic power, a growing middle class and increased consumption, growing inequality and instability, increasing scarcity of resources and finally an increase in the need for sustainability in the built environment and real estate (RICS, 2015; Standing, 2016). At the time of writing, these issues are manifest in the pressures on urban settlements globally to accommodate growing populations in the form of housing affordability, homelessness and insecure tenure for those less affluent. Growing instability is being exacerbated by unpredictable political events and subsequent pronouncements. The UK Brexit vote in 2016 was a vote for nationalism and a desire to return in some ways to a bygone past. Other EU countries may follow, with France in the grip of Le Pen nationalism and the Netherlands contemplating a Nexit. In the US, a political shift to the right and a controversial set of policies from an unconventional president have added further uncertainties that are at odds with moves towards the common goal of a better world for all. Such discord and the rise of nationalist politics could potentially lead to disruption and unease across economic markets and trading blocs. These are unpredictable and uncertain times, and there is a danger that an important focus on climate change and sustainability will be swept aside and other matters will take precedence.

On the other hand, technology and innovation have created a new business landscape in many respects. Transglobal business has never been so extensive and, despite political will to the contrary, cannot be reversed. This business landscape offers the potential for investment in real estate and infrastructure and changes in business models and practices, especially in real estate and valuations (Wilkinson, Halvitigala and Antoniades, 2017; RICS, 2015). Automated valuation models (AVMs) and other digital technologies are challenging some of the professional services provided by real estate professionals, and may replace individually considered judgement. Big data has both positive and negative implications. On the positive side, there is the potential to collect and analyse data to make more accurate and effective decisions in respect of real estate and infrastructure (RICS, 2015). This includes decisions about sustainability in real estate. There is also the ability to model scenarios and the impacts of development, infrastructure provision and urban regeneration. On the negative side, and a word of caution, is the danger of becoming over-reliant on technology and a diminishing understanding of the metrics and variables influencing decisions, leading to overly simplistic judgements.

A third trend is the changing role of the professions, which is partly implied in many of the chapters of this handbook. As new knowledge is developed and new skills are demanded, the professions must evolve to meet client – but also social and planetary – needs. Even more, they need to lead and advocate for change, to invest in research and innovation and to show others how the future could be. Real estate professionals have a role to play in ensuring that the best possible standards of sustainability are known, advocated and delivered to existing and future generations. At the same time, and in these uncertain times, the ethical standards embodied by professional bodies are needed more than ever to ensure that we make sound solutions for the good of all, not just the private good. As wealth inequality grows, this may be increasingly difficult to do; after all, the power of money can distort.

The question is: where to next? How do we embrace the positive potential in these emerging trends and minimise the negative aspects? First, we must make real estate and sustainability attractive career options for talented, emotionally and academically intelligent people. Second, we should put ethics at the heart of everything we do, which will ensure social, environmental and economic sustainability will feature highly. Third, we can embrace technology and big data use for the good of all, to deliver a more equitable and sustainable environment. If these steps are followed, we will be in a much stronger position to advocate and espouse the delivery of smart, resilient and overall sustainable cities and urban settlements. However, to do this we need strong leadership, and this means speaking out, sometimes loudly, when unethical, unsustainable proposals are advocated in our field, which we know have an enormous impact on global sustainability. In governance and policy, in valuation, investment and finance, in property and facilities management, and in redevelopment and adaptation, we have the opportunity to create a sustainable built environment. This takes courage and conviction. Time will tell if we collectively possess that vision and courage.

26.6 Conclusions

As this book has shown, the response of the real estate sector to increasing climate change and other sustainability pressures has been to focus even more strongly on environmental, social and governance (ESG) issues, and essentially to try to protect and enhance the value of real estate assets from risks associated with these issues. There has therefore been an increasing focus on 'sustainable real estate' – although quite what this is remains contested (Wilkinson, 2013).

In this international handbook, we have brought together academics and other experts from around the world, and from different disciplinary and professional backgrounds, to show *why* we need to tackle climate change and other environmental and related socio-economic risks in the sector, *how* these risks are impacting on buildings and the sector as a whole and *what* can be done to help the sector move to a sustainable future. In doing this, the book has aimed to be truly international in scope, drawing on research and solutions from a range of countries in the developed and developing world.

What has become abundantly clear is that the retrofitting of existing buildings is a greater challenge than that of new-build design. In many ways, new builds are simpler: they fit within a regulatory regime that can be – and often is – tightened year on year. The issue is how to build up the case for such retrofitting. Buildings are heterogeneous so technical challenges exist; one size most definitely does not fit all and buildings subject to continuing occupation present real barriers to upgrades. Further, in addition to technical and logistical challenges, the case to invest still rests on economic factors for the most part and arguments about green added value or brown discounting abound. Both exist but, as reported in the value chapters, might be hard to evidence. Yet evidence a case we must, unless we are to rely totally on government interventions to bring about change. Early signs are that in some countries this is happening, but it takes a brave government to risk the stability of the market by seeking to drive change too quickly.

Ultimately, the future is our hands. The real estate sector has an immensely important role to play in reducing carbon emissions and meeting national and international targets. The Paris Agreement of 2015 aimed to keep global warming below 2°C, and this comes at a time, as was stated earlier in the chapter, when the sector faces a number of other megatrends linked to urbanisation, new technologies and a battle between globalisation and de-globalisation.

Simply to achieve Europe's energy efficiency targets requires an annual investment of €60–100 billion, yet current investments are less than half of this and five times lower than is necessary to deliver a decarbonised built environment by 2050. Current retrofit or renovation rates of 1 per cent need to be tripled if targets are to be met (BPIE, 2016). We are simply not moving fast enough. Time is not on our side; the international community, with some notable exceptions, recognises this. And so it becomes a case not of why, but of how, and at whose cost, can transformation take place. Our collective analyses would show that both top-down (regulatory) and bottom-up (voluntary) measures can be effective in isolation, but a mixed economy of drivers will speed up the process; governments cannot simply wait for the market to move.

Despite these sobering statistics, recent initiatives do offer hope. For example, a recent action framework on Sustainable Real Estate Investment (Bosteels and Sweatman, 2016), which is supported by five of the world's largest investor networks, the RICS and the Global Alliance for Building and Construction, has been developed and widely promoted. The framework is designed to help investors make sense of existing ESG guidance and to accelerate their integration of ESG and climate risks and opportunities into investment decisions. The guide also sets out a range of measures to improve returns or better protect the future value of real estate investments through application of an ESG and climate change management approach. The guide is geared towards large investors; they have a role to help disseminate the framework through to SME sectors.

We stand at a cusp, or turning point, in our global history when we really do need to change, or face the consequences of, our actions. All of those who work in the real estate sector must come together to tackle the tasks we face to help us transition to a sustainable built environment by 2050. Real estate professionals must recognise their stewardship role, which will be founded on market transformation and a step change in behaviours and mainstreaming sustainability in everything that real estate professionals and their clients do.

We all need to remember Mark Twain's words, written in a different context: 'Don't go around saying the world owes you a living. The world owes you nothing. It was here first.'

Notes

1 Under the provisions of the Energy Act 2011, minimum energy efficiency standards (MEES) will be imposed for new lettings from 1 April 2018, and for continuing commercial leases from 1 April 2023.
2 At the time of writing, a number of research projects examining the link between debt funding and sustainability criteria, notably energy efficiency, are underway, but it is too early to report results.

References

Bosteels, T. and Sweatman, P. (2016) *Sustainable Real Estate Investment: Implementing the Paris Climate Agreement: An Action Framework*, Climate Strategy and Partners (Accessed February 2017: www.unepfi.org/fileadmin/documents/SustainableRealEstateInvestment_summary.pdf).

BPIE (2016) *Scaling Up Deep Energy Renovation* (Accessed February 2017: http://bpie.eu/publication/scaling-up-deep-energy-renovation/).

Dixon, T., Green, S. and Connaughton, J. (eds) (in press, 2017). *Sustainable Futures in the Built Environment to 2050: A Foresight Approach to Construction and Development*. Wiley-Blackwell, Oxford..

Friedman, M. (1962) *Capitalism and Freedom*. University of Chicago Press, Chicago.

Gabe, J. (2016) 'Successful greenhouse gas mitigation in existing Australian office buildings', *Building Research & Information*, 44(2), 160–174.

GRESB (2015) *2015 GRESB Report: Australia/NZ Snapshot*. GRESB, Amsterdam.

Mitcham, C. (1995) 'The concept of sustainable development: Its origins and ambivalence', *Technology in Society*, 17(3), 311–326.

MSCI/IPD (2016) 'Australian Property Investment Seminar', *Proceedings of MSCI/IPD Quarterly Update*, Brisbane, August.

PwC (2016) *Low Carbon Economy Index 2016* (Accessed February 2017: www.pwc.co.uk/services/sustainability-climate-change/insights/low-carbon-economy-index.html).

RICS (2015) *RICS Futures: Our Changing World – Let's Be Ready*. RICS, London. (Accessed February 2017: www.rics.org/uk/knowledge/research/insights/futuresour-changing-world/).

Standing, G. (2016) *The Precariat: The New Dangerous Class*. Bloomsbury, London.

Warren, C. M. J. (2010) 'Measures of environmentally sustainable development and their effect on property asset value: An Australian perspective', *Property Management*, 28(2), 68–79.

Wilkinson, S. J. (2013) 'Conceptual understanding of sustainability in Australian property organisations', *Journal of Property Management*, 31(3), 260–272.

Wilkinson, S. J., Halvitigala, D. and Antoniades, H. (2017) 'The future of the valuation profession: Shaping the strategic direction of the profession for 2030', *PRRES Pacific Rim Real Estate Conference*, Sydney, 15–17 January.

World Economic Forum (WEF)(2016) *Environmental Sustainability Principles for the Real Estate Industry* (Accessed February 2017: www3.weforum.org/docs/GAC16/CRE_Sustainability.pdf).

Young, R. (2012) *Stewardship of the Built Environment: Sustainability, Preservation, and Reuse*. Island Press, Washington DC.

Index

Page numbers in italics refer to figures. Page numbers in bold refer to tables.

Printed in the United States
by Baker & Taylor Publisher Services